ECONOMIC BOTANY
Plants in Our World

ECONOMIC BOTANY
Plants in Our World

Beryl Brintnall Simpson
Molly Conner Ogorzaly

The University of Texas at Austin

McGraw-Hill, Inc.

New York   St. Louis   San Francisco   Auckland   Bogotá
Caracas   Lisbon   London   Madrid   Mexico   Milan
Montreal   New Delhi   Paris   San Juan   Singapore
Sydney   Tokyo   Toronto

ECONOMIC BOTANY: Plants in Our World

Copyright © 1986 by McGraw-Hill, Inc. All rights reserved. Printed in the United States of America. Except as permitted under the United States Copyright Act of 1976, no part of this publication may be reproduced or distributed in any form or by any means, or stored in a data base or retrieval system, without the prior written permission of the publisher.

67890 HDHD 998765432

ISBN 0-07-057443-X

This book was set in Meridien by Monotype Composition Company, Inc.
The editors were Mary Jane Martin and Barry Benjamin;
the designer was Charles A. Carson;
cover photographs were done by Fern H. Logan;
the production supervisor was Phil Galea.
The drawings were done by Molly Conner Ogorzaly.
Arcata Graphics/Halliday Lithograph was printer and binder.

Library of Congress Cataloging-in-Publication Data

Simpson, Beryl Brintnall.
    Economic botany.

    Includes bibliographies and index.
    1. Botany, Economic—United States.   2. Botany,
Economic.   I. Conner Ogorzaly, Molly.   II. Title.
SB108.U5S56   1986   581.6'1   85-15370
ISBN 0-07-057443-X

To our parents

Contents

Preface

Recent events have made us all acutely aware of our dependence on green plants as renewable resources and ever-increasing population sizes have made us think about ways to feed large numbers of people, most of whom live under substandard conditions. At the same time, the natural or organic food movement and the increasing skepticism about the use of chemicals in food production has necessitated that individuals become educated consumers of plant products. As a consequence, interest in studying plants of actual or potential use in feeding, clothing, housing, and warming mankind has risen. In universities, this interest has manifested itself in the increased number of "plants and man," "humanistic botany," or "plants and civilization" courses. Unfortunately, a comprehensive text on economic botany suitable for midlevel college students has not been available. The classic texts such as Hill's *Economic Botany* were written many years ago and lack material relevant to modern crop production and utilization. While there has been a recent trend to include substantial sections dealing with economically important plants in general botany texts, such books must limit their coverage to only a select group of important species because their main emphasis is on the study of botany in general.

In this text we provide detailed coverage of the major uses of plants in America today. Rather than being encyclopedic, our concentration is on species of major economic importance in the United States. To keep the text readable and relevant, we have tried to provide a balanced treatment of the plants discussed by including aspects of history, morphology, chemistry, and modern usage. This blend of information is reflected in the detailed illustrations and figure legends which portray the ancient and modern uses of the plants discussed as well as pertinent aspects of plant biology. Plates from Fuchs' *De Historia Stirpium*, Gerarde's *General History of Plants*, Kerner von Marilaun's *A Natural History of Plants*, and Baillon's *A Natural History of Plants* were redrawn in several instances to add historical interest. Whenever possible, Ogorzaly used live material to produce original drawings to show the aspects, development, or growth of plants discussed. Most of the photographs included are original or previously unpublished. Technical data such as current production figures, nutritional content, and chromosome numbers have been put into tables so that they can be easily consulted without encumbering the text.

The material presented is directed toward students with some scientific background, but two introductory chapters, one on plant morphology and chemistry, and a second on general genetics are included to orient students with little or no biological training or those unfamiliar with plant biology. Chapter 3 provides a historical setting for man and his crops by asking when, where, how, and why people adopted agriculture as a way of life.

The body of the text, Chapters 4 to 18, deals with different important crop species. In the first five of these chapters, we group food plants by the parts of the plants (fruits, leaves and stems, roots) that are harvested for food. Legumes and grains are treated individually because of their importance in human nutrition. In the discussions of the plant organs that provide food, we emphasize their primary ecological functions and the various natural modifications that are particularly desirable from the human point of view. An example is fruit pulp, which is important in nature for seed dispersal by animals. The sweet pulp serves as a food that "pays off" these dispersers. Another is starch stored in tubers or roots that serves as a carbohydrate reservoir for some plant species. Such natural accumulations of carbohydrates were independently discovered by human beings in many parts of the world and incorporated into the repertoire of food plants.

In Chapters 10 to 17, we deal with products primarily extracted from plant parts. Included are substances such as volatile oils, alkaloids, latexes, and fibers. These products are grouped according to their use, spices and perfumes, textile fibers, bioactive compounds, etc. For each group of plant products we explore the natural occurrences, chemistries, and functions of the materials in nature. Our treatment of bioactive substances is particularly extensive. We have devoted separate chapters to medicinally important plant products, psychoactive drugs, caffeine-containing plants, and alcoholic beverages.

Chapter 18 concludes our survey of the current uses of plants by explaining the history and current importance of ornamental plants. For many students, their primary contact with growing plants will be ornamentals, either landscape or house plants.

The last chapter provides a look at the exciting possibilities that lie ahead in our use of plants and a consideration of the abuses of nature that human beings have perpetrated. The discussion of the history of selection and the concomitant loss of genetic variation recalls many of the points brought out in Chapter 2. A brief outline of some of the methods of genetic engineering suggests future alternatives to classical breeding practices.

In bringing all of this material together and encouraging us in our efforts, we are indebted to many people. In particular we acknowledge the assistance and past influence of John Averett, Eduoard de la Barre, David Bates, Jerry Brand, Robert A. Bye, Jr., Ron Carroll, W. A. Cote, Norman Dill, Laurence Dorr, Evelyn Edwards, Dan Evans, Garry Fox, Melvin S. Fuller, Doug Gage, Jonathan Gershenson, Raul Guevara, Carol Hoffman, Richard Howard, Duane Isley, William Jackson, Robert D. Koehn, Rose La Farge, J. V. Logomarsino, Ann Lopez, Andy McDonald, Alta Neff, John Neff, Ralph Obendorf, Tim Plowman, Don Pojmann, P. L. Raines, Bernice Schubert, Richard E. Schultes, Randy Scott, David Seigler, Edward Simpson, Brenda Spangler, Maurice Sweatt, Roy Taylor, Carol Todzia, Billie Turner, Garrison Wilkes, Hugh D. Wilson, and Carroll Wood. Last but not least, Jack who made many suggestions, Jonathan and Meghan who learned not to bother mommy while typing, and Bob who did everything from drafting to dishes. All helped us to finish the book.

Beryl Brintnall Simpson

Molly Conner Ogorzaly

ECONOMIC BOTANY
Plants in Our World

Chapter 1

Features of Flowering Plants and Their Products

*M*an's association with plants predates his human condition. Fossil teeth of early hominids show us that man's ancestors were omnivorous and probably consumed, and used as tools, a wide variety of plants. Recent excavations of cave dwellings occupied over 300,000 years ago have revealed that Peking man, the extinct species closest to modern humans, gathered walnuts, hazelnuts, pine nuts, and rose hips, and roasted Chinese hackberry seeds. Such archaeological findings have shown that the history of man's association with plants does not begin at a certain point when humans "discovered" that they could eat or use plants. Rather, these finds indicate that the human ability to manipulate plants became increasingly sophisticated over time.

While a human appreciation of plants is probably innate, the ability to exploit plants successfully is a result of man's unique capacity to transmit knowledge culturally. As a consequence, generations were able to learn from previous ones and knowledge accrued over time. Plants were tried and discarded or added to the repertoire of those used. Because different kinds of plants were available in different parts of the world, various peoples built up their own inventories of useful plants.

It has been estimated that about 3000 species of plants have been used as food by human beings throughout history and about 200 domesticated as food crops. Tragically, the number of species used has decreased, rather than increased, within historical times. Once people began to use some species preferentially over others, or to sow the seeds of selected individuals, they began to alter the plants they used. Over time, people were able to generate crops that produced greater quantities of nutritious substances or were less bitter than their wild progenitors. As a result, wild sources were abandoned in favor of species that humans were able to modify into particularly productive or pleasing crops. This trend has led to our present situation. Today, the world virtually depends on about 15 species, most of which have been highly modified by humans. Chapter 19 discusses some of the ways in which we are now trying to reverse this dangerous dependency.

Initially humans would have chosen plants that intrinsically appealed to them. In some cases, this appeal reflected natural plant adaptations for fruit dispersal by animals. In others, humans learned that certain plant organs contained profitable quantities of substances that could be used for food. By observing naturally occurring fires, humans learned to exploit plants as sources of fuel. Using natural tree and vine structures as models, humans began to use plants to fashion their own crude dwellings and ropes. In all instances, humans capitalized on natural products that are present in plants for a variety of reasons. Some of these products are used by plants for metabolism and growth (carbohydrates) and others to attract animal pollinators or seed dispersers (colors, perfumes, and sugar-rich pulp) or to repel herbivores (toxic compounds). Still others provide suppleness or strength to the plant body (cellulose).

Human beings have also been able to invent novel uses for many plant products. Seed hairs from cotton, functional in nature as a dispersal mechanism, turn into cloth when gathered and processed by humans. Gums exuded by plants to ward off infection become thickening agents for articles as diverse as chocolate milk and drilling muds. Chemicals that naturally deter plant predators are converted by people into substances that regulate human fertility, and wood that provides towering trees with structural support is reduced to pulp for paper on which to record human knowledge. In order to appreciate both the original and current uses of plants, it is necessary to understand the structure of plants, the ways in which they produce and sequester various compounds, and the nature of the compounds themselves. In this first chapter we therefore provide a general introduction to plant morphology and chemistry.

CHEMISTRY OF PLANT COMPOUNDS

Compounds derived from plants fall into the realm of organic chemistry, originally the chemistry of products made by living organisms. Today, the field of organic chemistry has been expanded to include all compounds that contain carbon (abbreviated as C) whether or not they are produced by plants or animals. This focus on carbon stems from the fact that all life as we know it is based on carbon. Although science fiction writers try endlessly to create life forms based on other elements (e.g., silicon), we have as yet found signs of life only on earth, and here, only carbon life forms exist.

That carbon should be the fundamental element in living systems is not surprising in view of its chemical properties. Carbon atoms exhibit several features that allow them to form an almost limitless array of complex molecules. First, carbon atoms can bond to as many as four other atoms. Second, unlike almost all other elements, carbon atoms readily combine with one another to form chains or rings of various sizes. Moreover, the atoms in these chains or rings are able to combine with still other atoms. Strings of carbon atoms thus provide skeletons, or backbones, on which countless combinations of other elements, primarily hydrogen (H), nitrogen (N), oxygen (O), phosphorus (P), sulfur (S), and chlorine (Cl), can be attached. This feature has made possible the enormous proliferation of carbon compounds now found in life on earth. It is estimated that there are more than 10 times as many carbon compounds as compounds of all of the other elements combined.

Atoms consist of a nucleus with one or more protons, usually with associated neutrons surrounded by shells of rapidly orbiting electrons (Fig. 1-1). Protons have a positive electrical charge, neutrons no charge, and electrons a negative charge. In a resting state, atoms have no net charge because the charges on the protons balance those of the electrons. Combinations between atoms involve interactions of their electrons. The way in which the electrons behave when atoms combine determines the kind of bond between the atoms. If one of the atoms in an interaction loses electrons and the other gains them, the bond is ionic. The other major way in which atoms may combine is by sharing pairs of electrons. Bonds involving shared electron pairs are covalent bonds. All carbon-carbon bonds are covalent. If two carbon atoms share one pair of electrons, the bond is a single bond. If two pair of electrons are shared, the bond is double, and, if three pair are shared, triple (Fig. 1-2). No more than three electron pairs are ever shared by two carbon atoms in nature.

The kinds and numbers of atoms combined in a particular compound can be represented by a chemical formula. Such molecular formulas are routinely used in inorganic chemistry and can be used to describe organic compounds as

FIGURE 1-1
Atomic structures of the most
important elements in living matter.
The small letters *p* and *n* in the
nucleus indicate protons and
neutrons, respectively. The capital
letters under each atom are the
standard abbreviations for the
element. The dots surrounding the
abbreviations indicate the numbers
of electrons in the outermost shell.

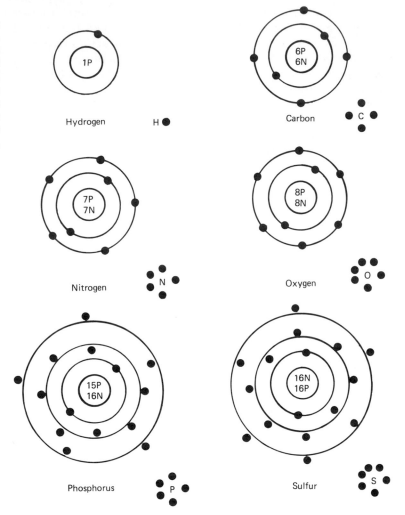

well. As we have seen, however, carbon atoms can form chains or rings of various lengths to which additional rings or chains can be attached. The actual arrangements of the atoms in complex organic molecules are hard to represent with simple molecular formulas. Consequently, organic molecules are often "drawn" rather than simply noted. These drawings, or structural formulas, allow both a complete depiction of the atoms in a molecule and a simplified representation of their three-dimensional structure. Figure 1-3 compares the molecular and structural formulas of various hexose compounds.

Among the thousands of organic compounds that are synthesized, we can define several major classes. These classes are defined by the nature of the portion of the molecule that most actively interacts with other atoms or molecules (the functional group), the numbers and kinds of atoms in the molecule, or both (Fig. 1-4). We mention here only the major kinds of organic compounds that are discussed later in this book. Specific compounds that we procure from plants are treated when we discuss the plants from which they are obtained.

The simplest organic compounds are hydrocarbons, which consist solely of hydrogen and carbon. Hydrocarbons that contain more than three carbon atoms

FIGURE 1-2
Examples of single, double, and triple covalent carbon-carbon bonds.

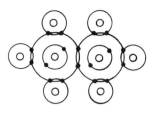

Ethane

can have more than one arrangement (=isomer) of the carbon and hydrogen atoms. In plants, hydrocarbons are often linear chains, although more complex cyclic and branched compounds (such as terpenes) are also common. Figure 1-4A shows the configurations of several common hydrocarbons.

More complex than simple hydrocarbons are organic acids and alcohols. Alcohols are hydrocarbons which have a hydroxyl, or OH, group. This group can be attached to any part of the molecule, but it is usually found on a terminal carbon atom (Fig. 1-4B). An oxygen that is double-bonded to a carbon atom constitutes a carbonyl group. An aldehyde is a molecule with a carbonyl group and a hydrogen atom on a terminal carbon (Fig. 1-4C). Molecules with nonterminal carbonyl groups are ketones (Fig. 4-1D). Organic acids have a terminal carbon that shares electrons with both a carbonyl and an hydroxyl group. This entire unit, COOH, is referred to as a carboxyl group (Fig. 1-4E). Alcohols and acids can combine, with water as the product of their bonding, into esters (Fig. 1-5).

The final major group of compounds of importance here are those with nitrogen as a part of a functional group. The most fundamental of these types of compounds are called amines (Fig. 1-4F). Amines have a nitrogen bonded directly to a carbon atom. In the simplest amines, the nitrogen is further bonded to two hydrogens. Amines are important because they are the starting point from which almost all nitrogen-containing compounds are synthesized. Amino acids are important nitrogenous compounds containing both an amine and a carboxyl group (Fig. 1-6). The amine portion of one amino acid can combine readily with the carboxylic part of another amino acid to form a peptide bond (Fig. 1-6). The amine portion of this second amino acid can then combine with

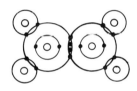

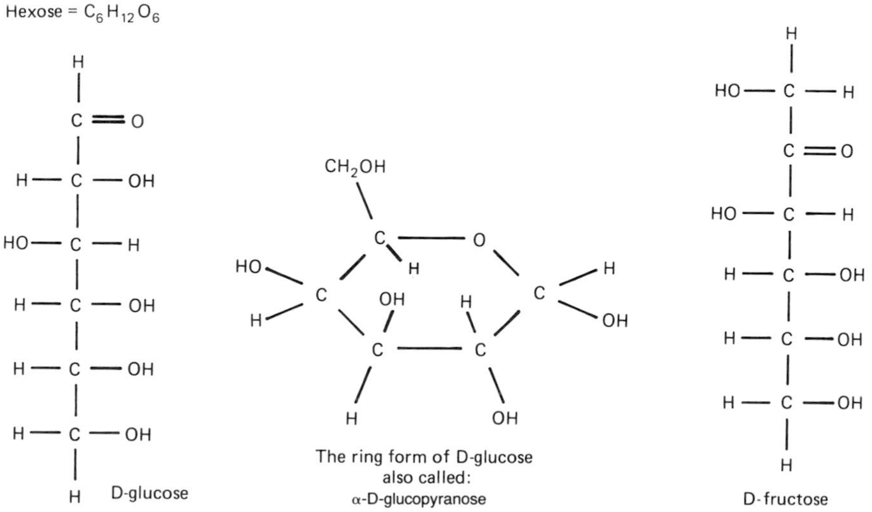

Ethylene

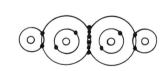

Acetylene

Hexose = $C_6H_{12}O_6$

D-glucose

The ring form of D-glucose
also called:
α-D-glucopyranose

D-fructose

FIGURE 1-3
All of these forms, or isomers, of hexose have the same molecular formula, $C_6H_{12}O_6$, but differ in the ways in which the atoms are attached. By using structural formulas such as these rather than simple molecular formulas, scientists can communicate the relative placements of the various atoms.

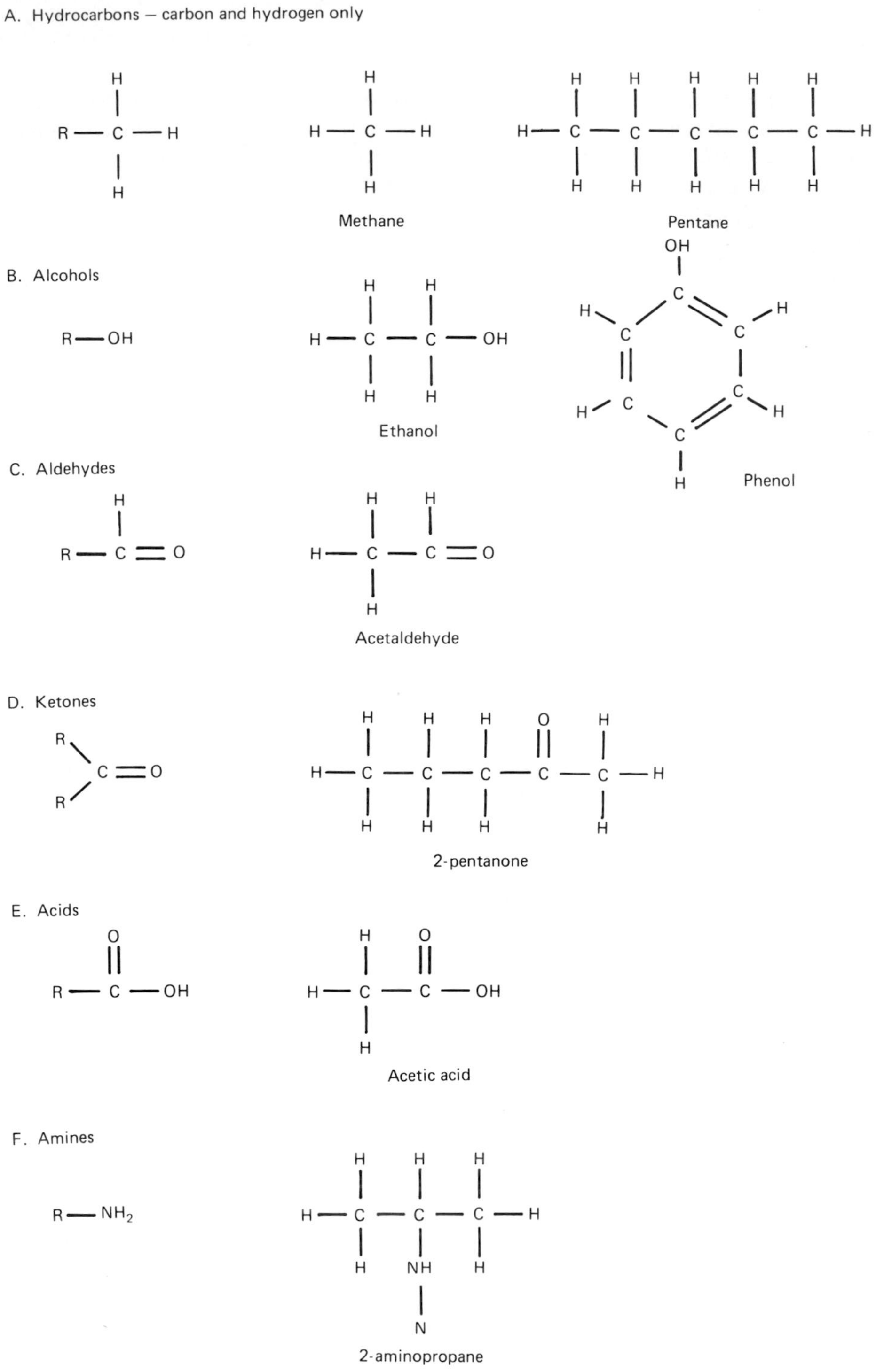

A. Hydrocarbons — carbon and hydrogen only

Methane

Pentane

B. Alcohols

R—OH

Ethanol

Phenol

C. Aldehydes

Acetaldehyde

D. Ketones

2-pentanone

E. Acids

Acetic acid

F. Amines

R—NH$_2$

2-aminopropane

FIGURE 1-4
Major classes of organic compounds. For each of the major groups of compounds, a general formula and a specific example is given. R indicates the "rest of the molecule."

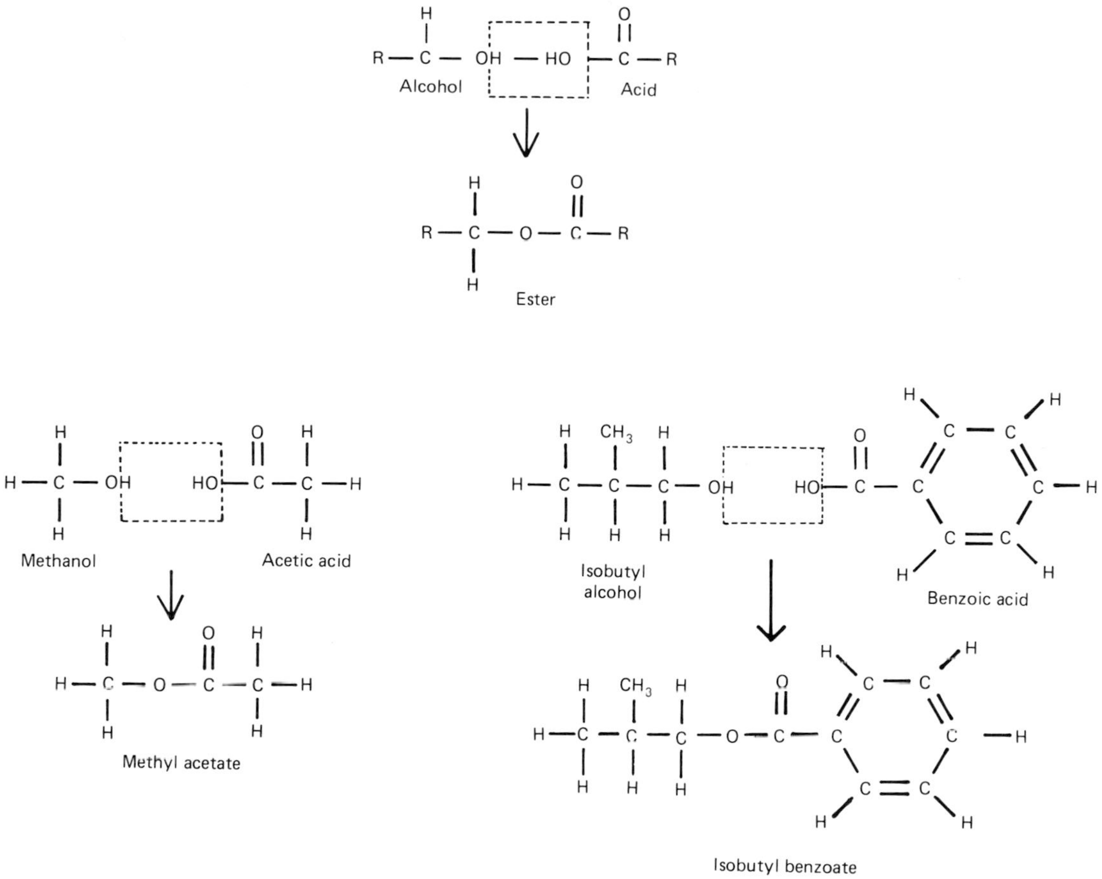

FIGURE 1-5
During the production of an ester from an acid and an alcohol, a molecule of water is shed.

the carboxylic acid of a third amino acid. This process can continue until long strings, or polypeptides, of amino acids are created. Proteins are complete molecules made of amino acids. Twenty different amino acids are commonly found in the proteins of living organisms.

PHOTOSYNTHESIS

Carbon enters the living world via the gas carbon dioxide. During the initial step in the incorporation, or fixation, of carbon, energy from the sun is used to split the hydrogen atoms of water from the oxygen atoms. This reaction, known as the light reaction (Fig. 1-7A), yields chemical energy in the form of adenosine triphosphate (ATP) and hydrogen atoms that enter into glucose synthesis. The energy from this reaction is also used to incorporate carbon dioxide into glucose molecules (Fig. 1-7B). The entire process consisting of the degradation of water with the release of oxygen gas and fixation of carbon into glucose is called

FIGURE 1-6
Amino acids can join together to form peptides. A complete peptide (or polypeptide) is a protein.

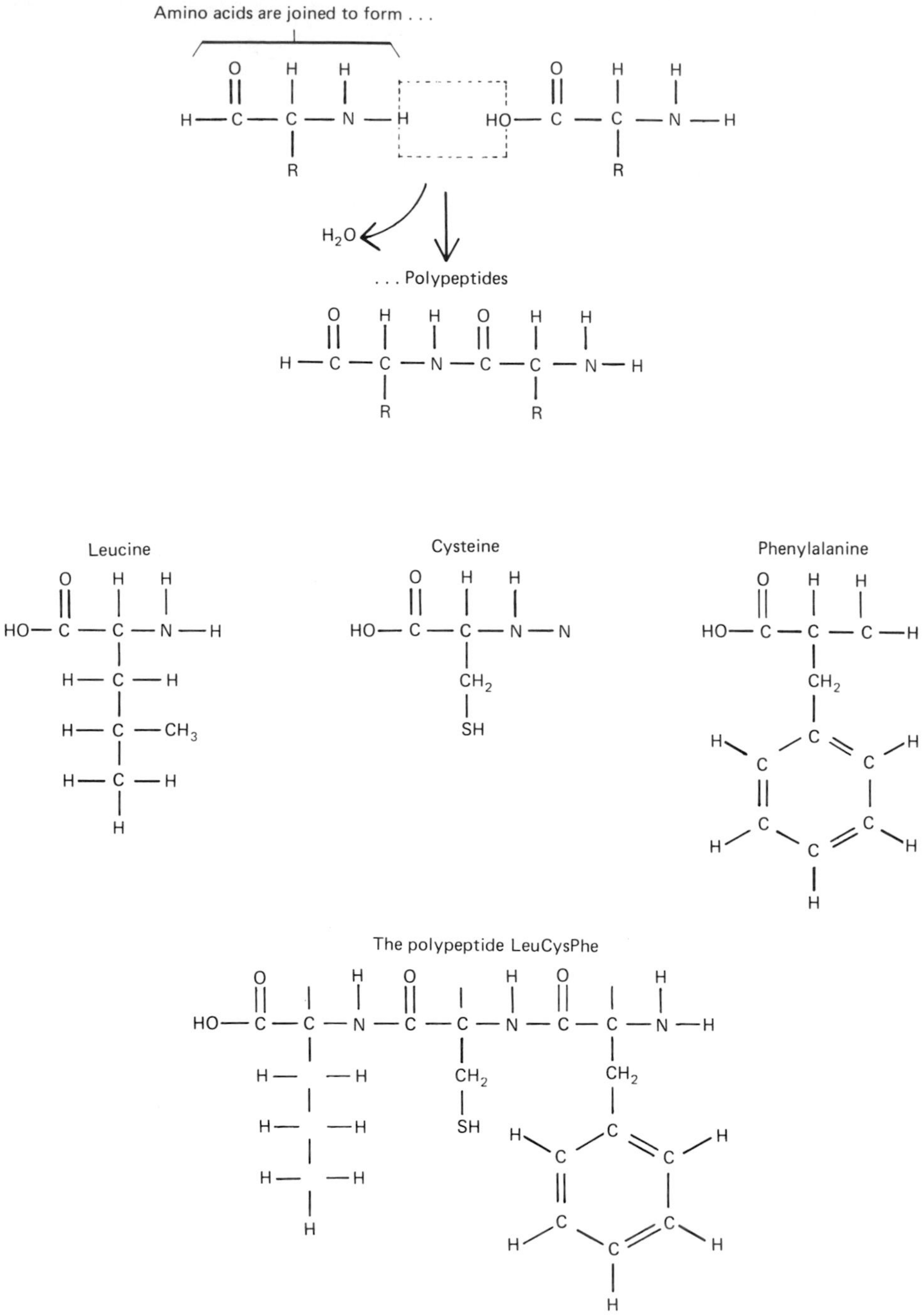

Amino acids are joined to form . . .

. . . Polypeptides

Leucine

Cysteine

Phenylalanine

The polypeptide LeuCysPhe

photosynthesis (Fig. 1-7). Green plants, a few bacteria, blue-green algae, and a few protists are the only organisms that can naturally convert radiant energy into chemical energy. In doing so, they synthesize a molecule, glucose, that yields chemical energy when degraded, or "burned."

The process during which glucose is broken down to provide chemical energy is called respiration. Fortunately, plants photosynthesize more glucose

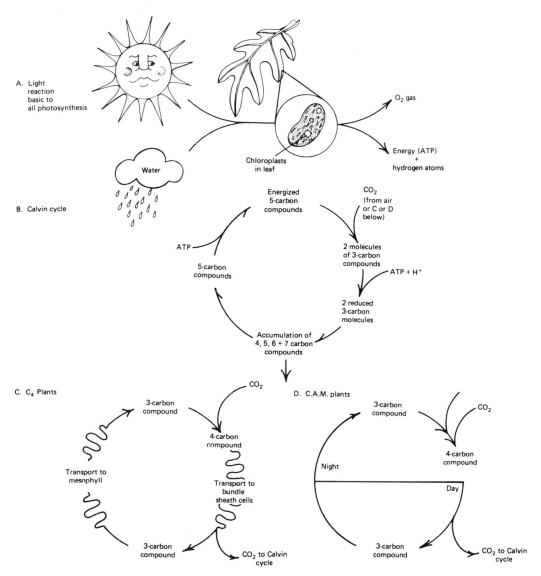

FIGURE 1-7

A schematic diagram illustrating the three ways in which plants carry out photosynthesis. In all cases, there is an initial reaction that splits water using energy from the sun (A). The products of this reaction are energy in the form of ATP, hydrogen ions, and oxygen gas. In C-3 plants which follow the Calvin-Benson pathway, all of the photosynthetic reactions occur in the leaf mesophyll cells (B). In C-4 plants (C), carbon from carbon dioxide is fixed (inserted into an organic molecule) to form a 4-carbon molecule, oxaloacetic acid, in the leaf mesophyll (see Fig. 1-17). A second 4-carbon molecule, malic acid, produced by the reduction of the oxaloacetic acid is then transported to layers of cells around the vascular bundles (bundle sheaths) where the reactions of the Calvin cycle (B) occur. Once hexose (any 6-carbon sugar) is formed, a 3-carbon compound (pyruvic acid) is transported back to the mesophyll. A third photosynthetic system known as crassulacean acid metabolism (CAM) is a modified C-4 pathway. It differs from the C-4 process in that all reactions occur within a single cell but carbon fixation and sugar synthesis are separated in time. Incorporation of carbon occurs at night (D) and the Calvin cycle portion of the process during the day. Since this strategy permits the plants to keep their stomates closed during the day, it is considered to be an adaptation to desertic conditions. CAM photosynthesis is found in many succulents.

than they respire. Animals can therefore crop the excess amounts of photosynthesized material. In addition, photosynthesis is carried out in all of the green aerial parts of a plant. Removal of portions of a plant does not destroy the ability of a plant to carry out its vital processes.

The glucose that is the primary product of photosynthesis is usually found in a cyclic form with an oxygen atom forming one apex of the ring (Fig. 1-3). This molecule, in addition to being the fundamental fuel for all organisms, is a primary building block for two other extremely important products that humans obtain from plants, starch and cellulose. The formation of starches and the plants from which humans obtain them are discussed in Chapters 6 and 8. The chemical nature of cellulose and the major sources of cellulose fibers are the subjects of Chapters 16 and 17.

Ironically, just as plants are the ultimate source of the chemical energy for modern life processes, they are the sources of most fossil fuels (oil, gas, and coal). Unfortunately, we cannot harvest fossilized plant products with the expectation of rapid replacement. Fossilization is a process that takes millions of years. When we talk of renewable resources or sources of energy, we therefore refer to those that can be replenished over very short time spans, tens, rather than millions, of years.

WHAT ARE FLOWERING PLANTS?

To most people, plants are green, insensitive, rooted organisms which normally produce flowers and fruits (Fig. 1-8). While these characteristics are true for the plants we normally encounter, they do not adequately define flowering plants nor convey an idea of the variation that exists among them. The basic differences between plants and animals approach the level of the fundamental structural unit of life, the cell. Green plants have cells that are enclosed in a hard cell wall made primarily of cellulose (Fig. 1-9). Inside this wall is the cell membrane which limits movements of substances in and out of the cell. Animal cells also have cell membranes, but they are not surrounded by a cellulose wall. Inside both plant and animal cells is a fluid cytoplasm in which float organelles, small recognizable bodies that have specific functions, and the nucleus. The nucleus is a relatively large body surrounded by a membrane. Inside the nucleus are the chromosomes, structures made up of several kinds of molecules including deoxyribonucleic acid (DNA), the storehouse of the genetic information for the running of the cell. From a human point of view, the most important organelles in green plants are chloroplasts. These organelles are green because they contain chlorophylls, pigments that are of primary importance in the absorption of light energy from the sun. It is within the chloroplasts that the process of photosynthesis occurs.

There are many kinds of green plants ranging from most algae to mosses, ferns, gymnosperms, and angiosperms. Angiosperms, or flowering plants, are those we most commonly encounter and the group of primary importance to human beings. Consequently, most of this book is concerned with flowering plants of major economic importance.

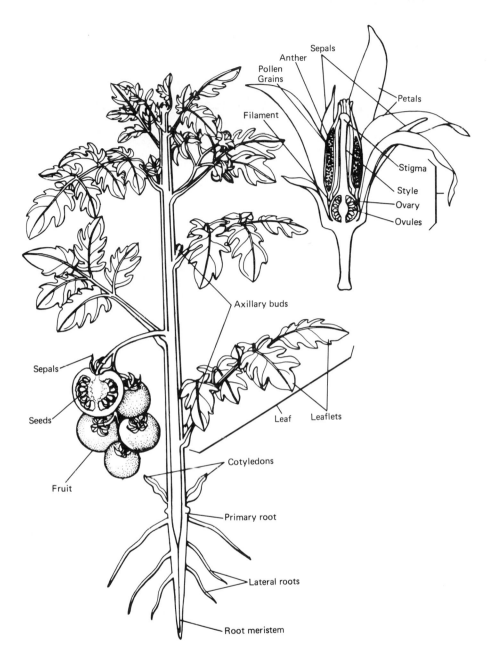

FIGURE 1-8
A tomato plant shows the basic characteristics of a flowering plant: the underground root, a shoot system consisting of stems and leaves, and the reproductive structures borne on the shoots. Axillary buds arise in the axils of leaves and produce lateral branches. The terminal bud at the apex of the plant is responsible for upward growth.

As indicated by the name, flowering plants, unlike other kinds of green plants, produce flowers. However, as described below, it is the structure of the flower and the processes involved in the formation of seeds that set this group apart from all others. Consequently, we include a detailed discussion of flowers and the process of reproduction in this chapter. Before we reach these discussions, we briefly describe the other major plant organs, roots, stems, and leaves. For all of these organs, we emphasize their primary functions in the life of most

FIGURE 1-9
A diagram of a basic plant cell.

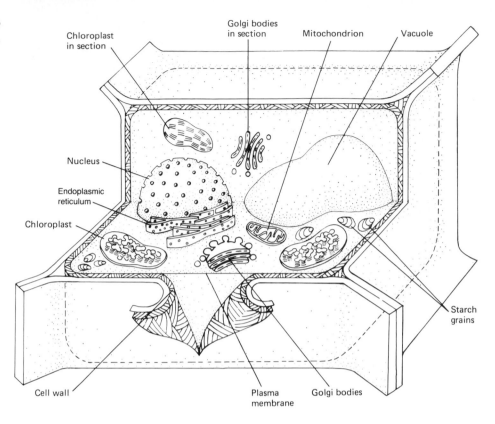

plants and their various modifications for secondary functions. In some cases, natural modifications have altered organs in ways that particularly suit them for human use. Human beings have often intensified such modifications so as to serve their purposes even better. However, humans have historically only been able to capitalize on variations that already existed in nature.

Roots

The first organ of most new plants to emerge from a germinating seed is the root (Fig. 1-10). The roots of most plants form a network of tissue which performs the vital function of absorbing both water and nutrients from the soil. Nutrients can be simple minerals (iron, phosphorus, calcium, etc.) or more complex organic substances such as vitamins, hormones, or amino acids which have leached out of decaying matter. In addition to absorption, the roots of angiosperms normally anchor plants in the soil so that the process of absorption is not interrupted.

The root system of most species consists of a dominant central taproot (Fig. 1-11) from which radiate lateral, or secondary, roots. The secondary roots can also produce lateral roots. These in turn can also branch as the root system expands. Individual roots also grow vertically at their tips. Water from the soil

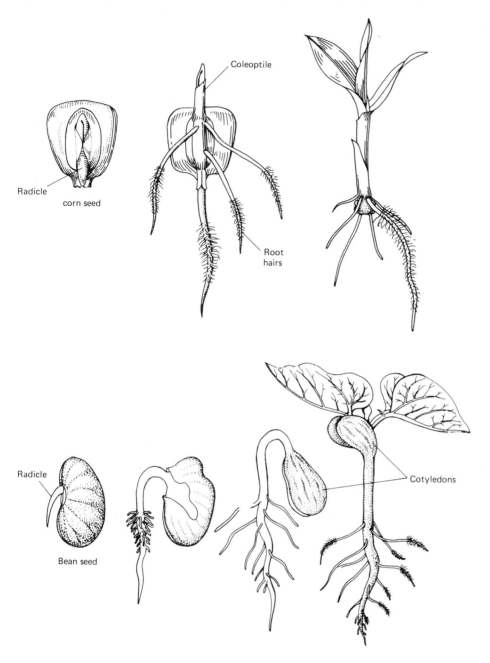

Coleoptile

Radicle

corn seed

Root hairs

Radicle

Bean seed

Cotyledons

FIGURE 1-10
The importance of the root is emphasized during germination. The radicle, or primary root of the embryo, is the first plant part to emerge from the seed. Shown here are germinating corn and bean seeds.

is generally absorbed only by root hairs that are borne along a short region behind the growing tip of a root (Fig. 1-12). Water and nutrients move from the root hairs into the central portion of the root (Chapter 8) and eventually up the stem through a system of conductive tissue known as the xylem.

Not all plants have roots like the typical kind just described. Variations on this theme are numerous and reflect the kind of plant, the environment in which it grows, and its life history. For example, if we pull up a bunch of grass and examine its root mass, it becomes obvious that there is no taproot. The roots form a shallow, very branched, fibrous system (Fig. 1-11). If we look at the roots of many species of tall trees growing in wet, tropical environments, we find that the roots begin to spread out from the tree trunk several feet above the ground and form a system of buttresses. These buttress roots provide stability for the shallowly rooted trees. Another kind of supportive roots known as prop roots (Fig. 1-13) emerge from the stems of a few species of annual monocotyledons such as corn.

In some situations, roots have been modified into structures that do not absorb water from the soil but from other plants. Organs modified for parasitic absorption of water and nutrients are called haustoria. Still another modification of the typical root is common in biennial species, or species that germinate,

FIGURE 1-11
The roots of most lawn grasses form a fibrous mass, whereas a dandelion has a well-developed taproot where starches and sugars are stored to help the plant over winter and produce new leaves in the spring.

grow, reproduce, and die over a 2-year period. Biennials spend the first year producing a leafy plant. As winter approaches, the plant transports sugars to the taproot, which becomes swollen with stored products. This root modification, correlated in nature with the life history or habit of the plant, provides human beings with an easily harvested, abundant supply of food. Most of our temperate ''root'' crops (Chapter 8) have been selected from naturally occurring storage taproots of biennial plants. Some perennials, plants that live and flower over several years, also produce fleshy storage roots. In most instances, such plants are native to arid regions where dry periods cause the dieback of photosynthesizing tissues.

The Shoot System

The shoot system of a plant consists of the stems and leaves. Since these two organs are usually readily identifiable, we discuss each separately.

Stems   Plant stems provide a framework, or body, which bears branches and/or leaves in such a way that the green portions are maximally exposed to the sun. Stems display flowers to pollinators and fruits to dispersal agents. In addition to their primary function of support, stems serve to house the conductive system (Fig. 1-14). As pointed out under the discussion of roots, xylem tissues conduct water and dissolved nutrients from the roots to other parts of the plant. A second conductive system, the phloem, transports products synthesized in the leaves and stems throughout the plant. Stems are generally more or less cylindrical and covered by a protective covering. In herbs, plants that have stems which contain little or no wood, the covering is a relatively thin epidermis.

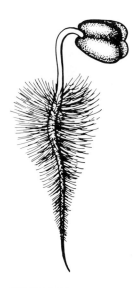

FIGURE 1-12
A radish seedling root covered with root hairs.

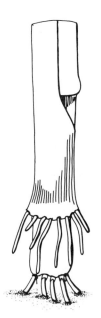

FIGURE 1-13
The aerial roots which develop along the stems of ivy are considered adventitious as are the prop roots which develop on corn.

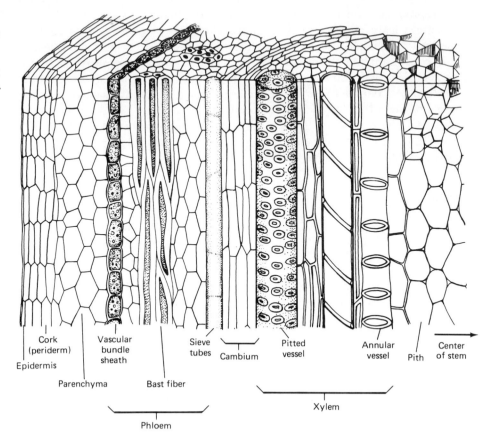

FIGURE 1-14
Cross section of a dicot stem shows the outer epidermal cells, the location of fibers which strengthen the stem, and the cambial cells that produce the xylem and phloem cells.

Woody plants, perennials with permanent stems that increase in girth by adding layers of wood, have a more complex protective bark (Chapter 17). In herbaceous plants, the xylem and phloem occur in bundles with the xylem on the inside and the phloem on the outside. In woody plants, the phloem is usually a continuous ring under the bark with the xylem forming most of the tissue in the central part of the stem. Secondary phloem and xylem cells are produced by a layer of cambial cells that lies between them.

Despite the basic functions common to all stems, many different or accessory functions have evolved (Fig. 1-15). In many desert regions we can find leafless species because the elimination of leaves greatly decreases the surface area of a plant and, consequently, its water loss. However, the lack of leaves presents something of a problem since photosynthesis (see below) normally occurs in leaves. Leafless species have gotten around this problem by shifting photosynthesis to the stems. As a result, stems of such species are green. In cacti, the most diverse group of leafless plants in the New World, the stems are also swollen and used to store water.

A different stem modification occurs in vines such as grapes (Fig. 1-15A). Stems of these plants have become leafless tendrils that aid in holding the weak, clambering plants fast to almost any support encountered. A contrasting modification of stems is that of rigid, pointed stem tips that function as spines, or the conversion of a shoot into a ring of tiny spines (Fig. 1-15C). Modifications

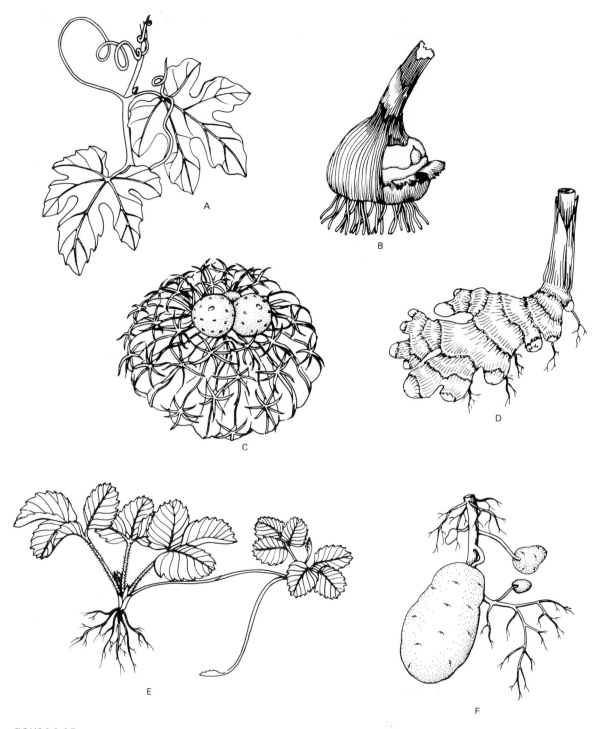

FIGURE 1-15
Natural selection has led to numerous modifications of the normal stem form into the tendril of a grape (A), the corm of a crocus (B) the whorl of glochids at the base of cactus spines (C), the stolons, or "runners" on a strawberry plant (E), the rhizome of ginger (D), and the tuber of a potato (F.).

of stems into such protective structures occur most frequently in leafless desert species and their relatives.

Another kind of stem modification, and an important one from the human point of view, is exhibited by underground stems which are used for food storage. Many plants can reproduce asexually by means of runners (Fig. 1-15E) that can root at some distance from the parent plant. Eventually, the original connection disintegrates and the daughter plant becomes an independent individual. Asexual reproduction can also occur by means of underground stems. These structures often look more like roots than stems, but, unlike true roots, they are capable of producing lateral shoots. Many underground stems are fleshy because they contain stored carbohydrates that provide energy for shoot development once they are separated from the parent plant. Some underground stems are classified as tubers (Fig. 1-15F) if they are short, stout, and nonbranching. Relatively thin, horizontal, underground stems are called rhizomes (Fig. 1-15D). Vertically compressed underground stems are corms (Fig. 1-15B). Human beings take advantage of the carbohydrates sequestered in underground stems by harvesting them when storage is complete but before they begin to initiate shoots. Many common starchy vegetables, such as potatoes or true yams, are underground stems which functioned in the natural ecology of the ancestors of our crops as asexual propagules.

Leaves Leaves come in a variety of shapes and sizes, but most are flat, green, and borne along stems (Fig. 1-16). By exposing flat surfaces to the sun, a plant maximizes the surface area exposed to sunlight and the gases involved in photosynthesis and respiration. The movement of gases into and out of leaves occurs through small openings called stomates. Stomates are usually found on both surfaces of a leaf, allowing gases to move across the protective layers of the upper and lower epidermis and the cuticle (Fig. 1-17). Sandwiched between the epidermal layers is the mesophyll composed of elongate palisade cells and relatively unspecialized parenchyma cells. Photosynthesis occurs primarily in the cells of the palisade layer which contain large numbers of chloroplasts. Bundles of xylem and phloem branch through the parenchyma, bringing water and minerals to the photosynthesizing cells and carrying away synthesized products.

While most leaves serve only as sites of photosynthesis, additional or alternative leaf functions occur in many kinds of plants (Fig. 1-16). In the common poinsettia, the bright red structures resembling flower petals are leaves that are pigmented so as to attract pollinating insects to the inconspicuous flowers they surround (see Fig. 18-38). Where browsing by animals is a threat, leaves bear spines (Fig. 1-16D) or have reduced blades and become rigid. Leaves can also be succulent and store water, particularly in dry environments. Like stems, leaves can be modified into tendrils that help hold a climbing plant to a substrate (Fig. 1-18A).

In some species such as lilies, which have leaves that emerge directly from the ground, leaf bases have been modified for carbohydrate storage. As winter, or a dry period, approaches, photosynthesized material is transported to the leaf bases. When frosts or aridity kill back the exposed portions of the leaves, the fleshy bases remain protected underground. The swollen collections of leaf bases are commonly called bulbs and function to (Fig. 1-18B), provide energy

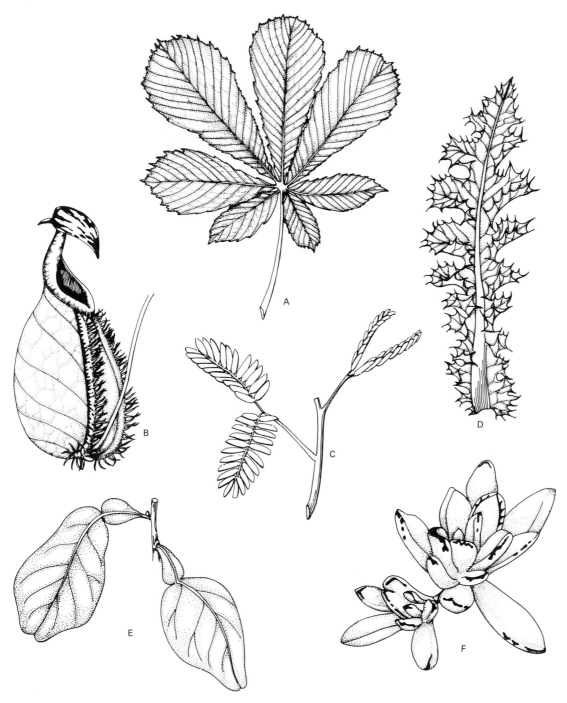

FIGURE 1-16
Leaves exhibit a wide range of variation related to the environmental conditions in which they occur. (A) A palmately compound leaf of a horse chestnut; (B) a *Nepenthes* leaf modified for capturing insects; (C) the pinnately compound leaves of the sensitive plant close in response to a touch; (D) thistle leaves have stiff, spiny tips that deter predation; (E) citrus leaves have flattened petioles and contain packets of volatile oils; (F) the fleshy leaves of many succulents store water in arid climates.

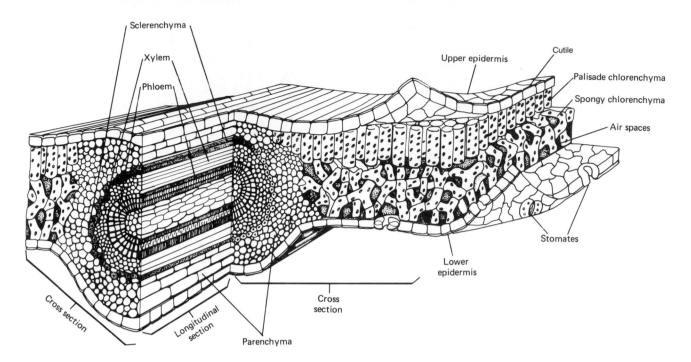

FIGURE 1-17

This leaf section shows how various cells are organized in order to promote photosynthesis. Gas exchange is regulated by the opening and closing of the stomates in the epidermal layer. The elongated cells of the palisade layer contain the chloroplasts where carbon dioxide enters the photosynthetic process and leads to the production of sugar (see Fig. 1-7). Vascular traces throughout the leaf deliver water and minerals and carry away synthesized sugars.

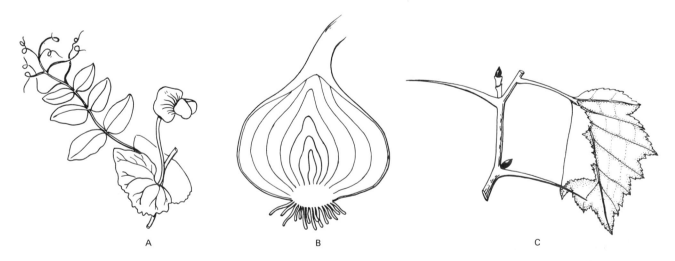

FIGURE 1-18

In peas, leaves have been modified into tendrils (A). The bulbs of onions (B) consist of overlapping, fleshy leaf bases that store nutrients and persist underground during the winter. The clue that the hawthorn spine (C) is a modified leaf lies in the fact that there is an axillary bud at its base.

for sprouting the next season. Like storage taproots, rhizomes, tubers, and corms, bulbs are commonly used by people because they constitute natural, concentrated sources of food.

REPRODUCTIVE STRUCTURES AND SEXUAL REPRODUCTION

Flowers

The most aesthetically appealing part of flowering plants for many people is the flower itself. This appeal is caused by a combination of form, color, and odor (Fig. 1-19). These same features function in nature to attract animals—insects, birds, or even bats (Fig. 1-20) and rodents. Animal attraction is an integral part of the reproductive system of many flowering plants because animals are often used to transport pollen from one flower to another. Unlike most animals, plants generally produce both "male" and "female" reproductive structures on the same plant and most frequently in the same flower. A typical flower with both sexes is considered to be perfect, although some plants produce unisexual (imperfect) flowers. Male flowers lack a pistil (or functional female sex cells) and female flowers lack stamens, or have sterile pollen. A plant that produces separate male and female flowers on the same plant is monoccious ("one house," Fig. 1-21). Plants that bear only male or female flowers are dioecious, a term indicating that there are two kinds ("two houses") of plants present in the species. Flowers normally have an outer ring (calyx) of green sepals and an inner ring (corolla) of colored petals (Fig. 1-22). In the center of the flower is a variable number of stamens and a pistil. The stamens have a stalk or filament supporting the anthers where the male gametes are produced. Pollen grains constitute the fully formed male gametophytes of flowering plants. The pistil commonly has a swollen base (ovary) which contains the ovules where the female gametes are produced. Each ovule is almost completely enclosed by one or two layers of cells, or integuments (Fig. 1-23). The region not covered by the integuments is the micropyle. It is through this small opening that the male gametes enter the ovule. Every pistil also has a stigmatic region where pollen is received. In most plants, the stigma is a relatively small surface capping an elongate style that rises from the top of the ovary.

Reproduction in Angiosperms

We said earlier that a plant cell contains a nucleus in which there are a number of chromosomes. The number of chromosomes per nucleus varies from one kind of organism to another, but it is constant within a kind of organism. This number is called the *diploid number* (written $2n$ = whatever the number is). All normal cells of an organism except the sex cells (and a few other special cases) have this number of chromosomes. The diploid numbers of several economically important plants discussed later in this book are given in Table 1-1. Note that the diploid number is written "$2n$" to indicate that there are two of, or a pair of, something called "n." The letter n represents the haploid number, or the number of chromosomes in a sex cell. While this form of notation might at first seem cumbersome, it is quite useful since it emphasizes that every

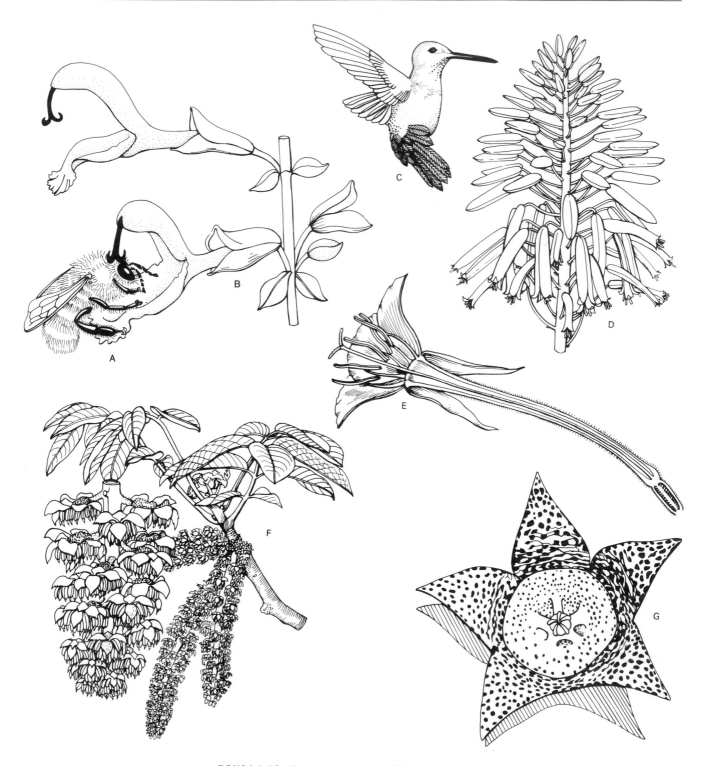

FIGURE 1-19 *(Continued on page 23)*
Bees (A) pollinate about a third of our crop plants. This *Salvia* (B) is typical of bumblebee flowers in that it has a landing platform. When the bee visits a flower to obtain nectar, the anthers touch her head and back to deposit pollen. When she flies to

FIGURE 1-20
Even mammals can serve as pollinators. This agave is pollinated primarily by night-flying bats. (Photo courtesy of USDA.)

chromosome, and hence every gene, is represented twice in every nonsex cell of an organism. The notation also provides the most simple way of indicating the changes in numbers of chromosomes that occur during sexual reproduction.

When a diploid cell reproduces sexually, it first produces sex cells, or gametes (Fig. 1-24), each of which has half the number of chromosomes of a diploid cell. Each sex cell therefore has a haploid number of chromosomes. The production of haploid gametes from diploid cells involves a kind of cell division in which the nucleus undergoes meiosis. As shown in Fig. 1-24, each chromosome in the diploid cell about to undergo meiosis duplicates its DNA strand, finds its partner (or homologous) chromosome, and lies alongside of it during the first part of the division process. The pair of aligned chromosomes

---

another flower, pollen from the previously visited flower is transferred to the new stigma. Hummingbirds (C) are the most important bird pollinators in the New World. Different kinds of birds serve as pollinators on other continents. Aloe (D), a native of South Africa, is pollinated by sunbirds in its natural habitat but by hummingbirds in warm areas of the Americas where it has been introduced as an ornamental. The long corolla tubes and copious nectar found in aloes are characteristic of bird-pollinated flowers. The evening primrose (E) is hawkmoth pollinated and exhibits several characteristics associated with pollination by nocturnal moths. These include opening at night, a sweet scent, a white corolla, a long corolla tube, and copious nectar. Wind-pollinated flowers such as those of the walnut (F) are usually small, green, odorless, and nectarless. In order to facilitate pollen dispersal, the male flowers are borne in long catkins with the stamens well exposed. The female flowers have exposed, feathery stigmas that easily collect wind-borne pollen. *Stapelia* (G) flowers lie close to the ground and are mottled maroon and cream colored. This floral color combined with a fetid odor attracts carrion flies that mistakenly visit the flowers in search of rotting meat. While visiting a flower, a fly will pick up pollen and possibly also deposit some from another flower on the stigma.

FIGURE 1-21
Wax melon flowers, like those of all members of the squash family, are unisexual, but the plants are monoecious.

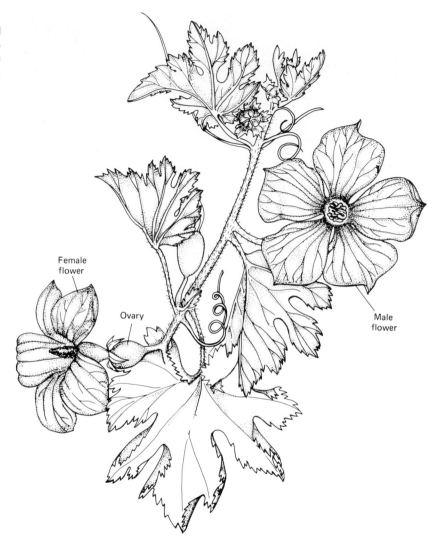

then move to the center of the cell and, finally, are pulled apart, one partner of each pair going to opposite ends of the cell. Two nuclei reform, each with one of every pair of chromosomes. Each nuclei then undergoes a second division in which the duplicated DNA strands of each chromosome disassociate. The final products of meiois are four haploid nuclei. Generally, cell walls form immediately, separating the nuclei into four cells.

While meiosis is the starting point in the process of both male and female sex cell formation, final gamete formation follows different pathways in male and female plant organs.

Formation of male gametes is a relatively straightforward process. As the anthers develop, specific cells inside the mass of dividing tissue begin to undergo meiosis (Fig. 1-23). Each of the haploid cells produced by meiosis will produce a male gametophyte. After the meiotic division, a specialized pollen wall begins to form around each cell. While the wall is forming, the cells enlarge to their

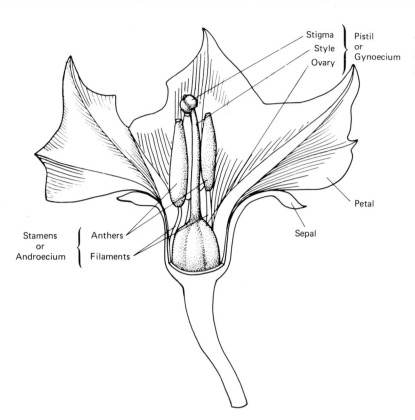

Stigma
Style
Ovary

Pistil
or
Gynoecium

FIGURE 1-22
This diagram of an eggplant flower shows the major parts of a typical perfect flower.

Petal

Sepal

Stamens
or
Androecium

Anthers

Filaments

final size. Once the cell reaches its final size, it divides mitotically (Figs. 1-23, 1-24). Cell membranes, but not cell walls, are produced between the daughter cells within the pollen grain. One of the daughter cells becomes the vegetative or tube cell, the other the generative cell. Some pollen is shed with only these two cells. Others are shed with three cells because the generative cell divides once more to form two gametes.

The process that produces an embryo sac (female gametophyte) containing the egg and its attendant apparatus is more intricate and variable than that leading to pollen production. Within each rudimentary ovule, one cell becomes destined to initiate female gamete production. This cell undergoes a meiotic division, but all of the resulting haploid cells rarely participate in embryo sac formation. The most common type of embryo sac formation appears to proceed with the disintegration of three of the original products of meiosis. The remaining cell then enlarges by absorbing and crushing surrounding cells. The nucleus of this cell then divides mitotically three times (in the most common case) to form eight nuclei. The nuclei roughly position themselves so that three are located opposite the micropyle, three at the micropyle, and two in the center of the grossly enlarged cell. The central nucleus near the micropyle is the egg. The nuclei flanking it are the synergids. Opposite the egg are the three antipodals. In the center, the two nuclei form the polar nucleus. Other types of embryo sac formation occur. In some of these, only two, or none, of the original nuclei resulting from meiosis degenerate (Fig. 1-23). Different kinds of embryo sacs also differ in the final number of polar nuclei.

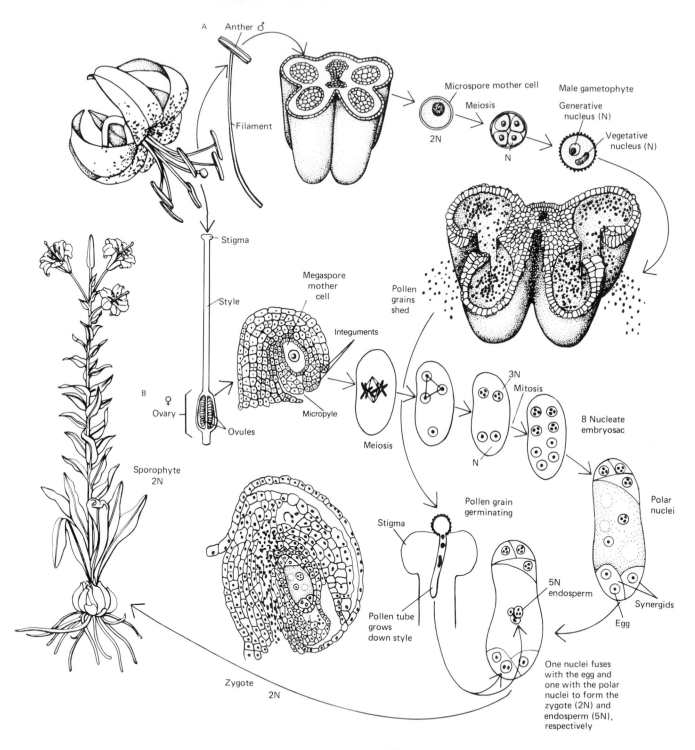

FIGURE 1-23 *(Continued on page 27)*
As exemplified by this lily flower, meiosis in the anthers (A) and ovaries (B) produces haploid sex cells or gametes. Fertilization involves the fusion of a haploid nucleus from the pollen grain with the haploid nucleus of the egg in the ovule and an additional

TABLE 1-1

Diploid Chromosome Numbers of Some Major Crop Species

COMMON NAME	SCIENTIFIC NAME	FAMILY	DIPLOID NUMBER
Apple	*Malus pumila*	Rosaceae	$2n = 34$
Barley	*Hordeum vulgare*	Poaceae	$2n = 14$
Bean, common	*Phaseolus vulgaris*	Fabaceae	$2n = 22$
Beet	*Beta vulgaris*	Chenopodiaceae	$2n = 18$
Cabbage	*Brassica oleracea*	Brassicaceae	$2n = 18$
Cassava	*Manihot esculenta*	Euphorbiaceae	$2n = 36$
Cocoa	*Theobroma cacao*	Sterculiaceae	$2n = 20$
Coconut	*Cocos nucifera*	Arecaceae	$2n = 32$
Corn	*Zea mays*	Poaceae	$2n = 20$
Grape	*Vitis vinifera*	Vitaceae	$2n = 38$
Hemp	*Cannabis sativa*	Cannabaceae	$2n = 20$
Peanut	*Arachis hypogaea*	Fabaceae	$2n = 20$
Rice	*Oryza sativa*	Poaceae	$2n = 24$
Soybean	*Glycine max*	Fabaceae	$2n = 40$
Tomato	*Lycopersicon esculentum*	Solanaceae	$2n = 24$

Source: Adapted from N. W. Simmonds (ed.). 1976. *Evolution of Crop Plants.* New York, Longman.

The second step in sexual reproduction is the fusion of male and female gametes to form a diploid cell. The initial diploid cell resulting from this fusion is the zygote. The cell divisions of the zygote that produce the embryo as well as cell divisions of an adult organism proceed by mitosis. Figure 1-25 outlines this process diagrammatically.

A pattern of haploid gamete formation and gamete fusion into a diploid zygote is common to all sexually reproducing organisms. Angiosperms have a second process that occurs during sexual recombination. As we saw above, the pollen grain contains two or three cells when it is shed. If only two are present when the pollen is dispersed, the third is produced later by a meiotic division. The embryo sac, or female gametophyte, likewise contains more than one cell. When pollen lands on a compatible stigma, it begins to grow down the style toward an embryo sac. It is during the period of pollen tube growth that the second male gamete is produced if it was not present earlier. The vegetative cell in the tube directs the growth of the pollen tube, but it does not fuse with any of the nuclei in the female gametophyte. Once a pollen tube penetrates an embryo sac through the micropyle, the tip ruptures, releasing the two gametes. One of these fuses with the haploid egg cell to form the diploid zygote. The other fuses with the polar nuclei to form a polyploid (usually triploid, but see

---

fusion of the second male gamete and two polar nuclei in the ovule. This process, called double fertilization, produces a zygote from the first fertilization and a food tissue known as endosperm from the second. The zygote is diploid and the endosperm is polyploid. In this example using a lily, the embryo sac is formed from all four of the original products of meiosis. Three of the original haploid nuclei fuse into a triploid nucleus. Both this triploid nucleus and the haploid nucleus undergo two successive mitoses yielding eight nuclei in all. The egg nucleus and its synergids are haploid. The antipodals are triploid. The polar body is formed by a triploid and a haploid nucleus. After double fertilization, the endosperm tissue is pentaploid since it will have been formed from the fusion of five nuclei.

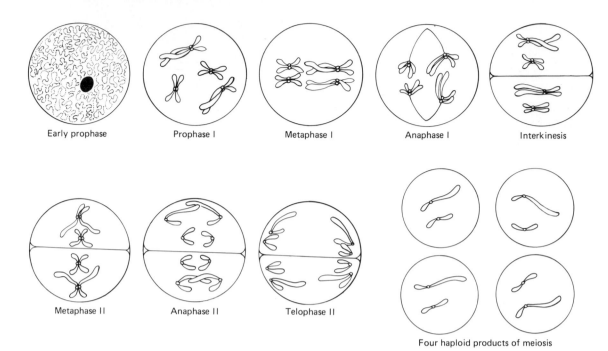

Early prophase Prophase I Metaphase I Anaphase I Interkinesis

Metaphase II Anaphase II Telophase II

Four haploid products of meiosis

FIGURE 1-24
The process of meiosis, shown diagramatically, produces haploid gametes by a reduction division. The new cells produced have the diploid chromosome number. Once the zygote is formed, all growth and replacement of dead cells occur by means of mitotic cell divisions.

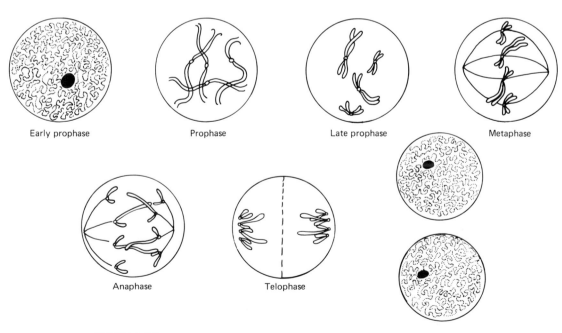

Early prophase Prophase Late prophase Metaphase

Anaphase Telophase

FIGURE 1-25
In mitosis, there is no reduction division.

Fig. 1-23) tissue that also begins to divide mitotically. This tissue, like the zygote, is formed from maternal and paternal components. However, the mass of tissue formed by this series of divisions is not an organism, but rather the food for one. The tissue becomes the endosperm, which is absorbed by the embryo either before it completes development or at the time of germination. Since the process of sexual reproduction in angiosperms involves two separate fusions of haploid nuclei, it has been dubbed *double fertilization.* Double fertilization is unique to the angiosperms.

Before fertilization can occur, pollen must reach a receptive stigma. Since most plants cannot move, they generally depend on external agents to transport the pollen. Such agents include wind, water, and animals. The flower morphology, color, and odor often reflect the kind of agent that effects pollination (Fig. 1-19). The wide array of colors and odors of flowers have been evolved to serve as attractants for animal visitors, but although the pollinators (chiefly insects) use these signals to locate flowers, they actually visit flowers because they derive some reward from them. Without such rewards, there would be no incentive for animals to continue to visit flowers. Floral rewards are usually foodstuffs: nectar, pollen, oils, or even flower parts. In a few special cases, chemicals used in insect mating behavior or in nest construction are collected from the flowers. A few species of plants even deceive insects by mimicking a food source or a female insect of the same species. For example, some flowers emit an odor resembling that of rotting meat, causing female carrion flies to land on them, crawl around (and become dusted with pollen), and even lay eggs. The flies then go off in search of more "rotting meat," which often turns out to be another flower. In the process, pollen gets moved from one flower to another.

While flowers that smell like rotting meat may be repellent to humans, many of the odors that attract birds and bees are often pleasant to humans as well. Compounds that produce floral scents have therefore played an important role in the perfume industry.

Flowers that utilize wind or water to move pollen from one flower to another are almost always green, odorless, and devoid of any reward (Fig. 1-19F). Production of attractant structures or chemicals is unnecessary and selection has favored other investments of the plant's energy.

Fruits

Every sexually reproducing flowering plant bears some type of fruit, all of which serve the same basic purpose, to protect the seeds and disperse them to potential areas where they can germinate. The structure of a fruit often gives a clue as to the mechanism of dispersal. Fruits dispersed by the wind are dry, light in weight, and frequently have wings or tufts of hairlike appendages that help them float in the air (Fig. 1-26A, B, D). Water-dispersed fruits are buoyant and encased in protective waterproof coverings. Animal-dispersed fruits show variable characteristics depending upon the way in which they are transported. In relying on animals for seed dispersal, the plant must use the animal but still prevent the seeds from being destroyed. Seeds can be carried externally by mammals (including humans) or birds. Barbs, hooks (Fig. 1-26F, G), or sticky outer coatings promote adherence of the fruits to skin, fur, or feathers. Eventually, the fruits fall off, possibly in a place where the seeds they contain can germinate.

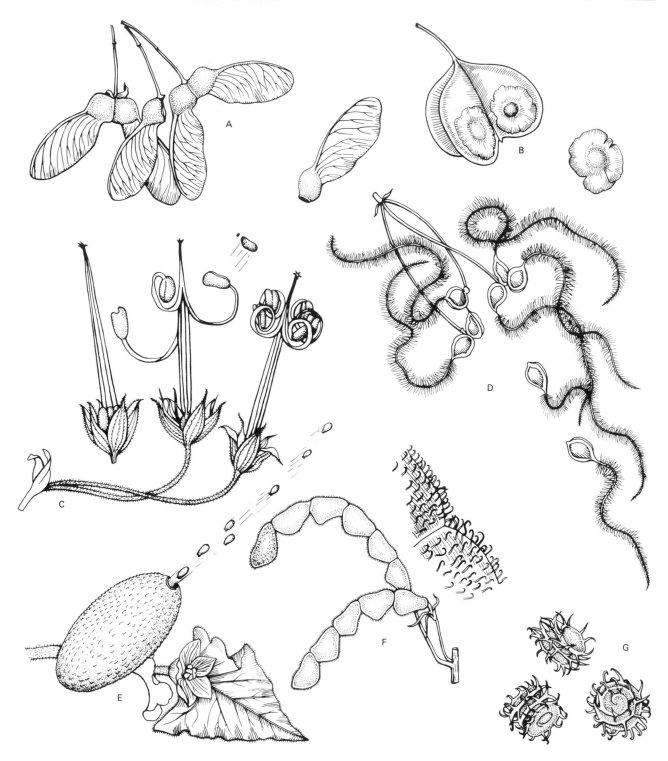

FIGURE 1-26 (Continued on page 31)
Wind-dispersed fruits like the samaras (A) of maple trees, the winged seeds that are
released from *Dioscorea* capsules (B), or the fruits of *Clematis* (D), all have morphological
adaptations that foster movement in the air. The seeds of many geraniums (C) are

Plants also actively attract dispersing animals by producing edible fruits. In order to use edible fruits effectively, however, only the pulp of the fruit and not the seed must be digested. Protection against seed destruction by fruit-eating animals is achieved in several ways. One way is to produce tiny seeds with a seed coat impervious to digestive juices. Animals usually consume the entire fruits of species with such seeds, but the seeds pass through the digestive tract unharmed and are deposited with the feces. An alternative method of protecting the seed is to produce medium-sized, or very large, seeds that are too tough to be crushed by the teeth or beak of fruit-feeding animals. Animals often carry off such fruits, but they eat only the fleshy outer portions and discard the seeds.

Fleshy, animal-dispersed fruits are a natural food source for humans, although within the array of potentially edible fruits there are many that are too bitter or unpleasant for the human palate. All of the items sold in grocery stores as ''fruits,'' and many of our ''vegetables,'' are selected from naturally occurring animal-dispersed, fleshy fruits (Fig. 1-27).

Seeds

Once fertilization has successfully been completed, the embryo and the tissues surrounding it mature into a seed (Fig. 1-28). A seed contains an embryo, variable amounts of stored food, and a protective coat derived from maternal tissues. An angiosperm embryo consists of one or two seed leaves, a region that will develop into the shoot system, and a structure that will grow into the root. One major group of angiosperms, the dicotyledons, have two seed leaves (cotyledons). In some species, the cotyledons absorb all of the stored food by the time the seed is shed (Fig. 1-28B). The two halves of a peanut are good examples of cotyledons replete with stored food. The tiny precursors of the first true leaves can be seen in the notch of one of the two halves. In monocotyledons, the other major group of angiosperms, there is only one seed leaf. In grasses, an especially important group of monocotyledons for humans, food for the developing embryo is stored as a copious mass of starch (Fig. 1-28A). In other monocotyledons such as the coconut, the food supply consists primarily of fats and oils.

Seeds have several features that make them good sources of food for humans. First, they tend to be highly caloric because they contain large amounts of fats or starches. Second, seeds are often produced in large quantities, particularly in annual plants. Third, seeds can be harvested readily. In perennials, the harvesting of seeds does not disturb either the roots or the photosynthetic apparatus of the plant which produced them.

We have outlined in this chapter the general nature of the major plant organs, the unique reproductive system of flowering plants, and the basic chemistry of plant-derived compounds. In the rest of the book we elaborate on

---

catapulted as the follicles dry and dehisce while the squirting cucumber (E) gets its common name from its unusual method of shooting its seeds. The seeds of *Hedysarum canadense* (F) and alfalfa (G) are equipped with barbs or hooks that help them adhere to passing animals that serve as agents of dispersal.

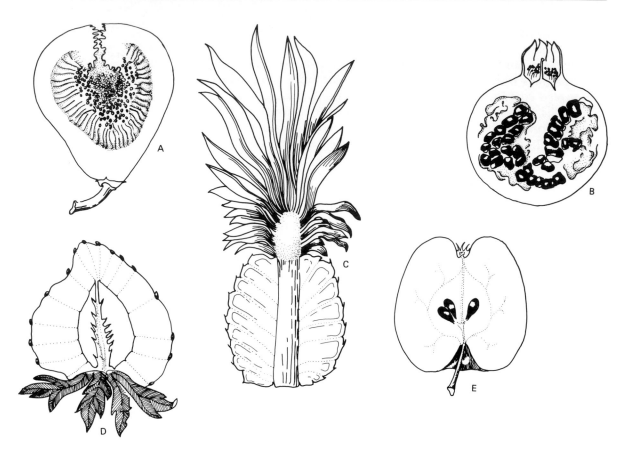

FIGURE 1-27
Animal-dispersed fleshy fruits are preadapted to serve as human foods. Some common fleshy fruit crops are (A) fig, (B) pomegranate, (C) pineapple, (D) strawberry, and (E) apple.

all of these aspects. We do this by treating first products derived from entire plant organs. These include fruits, seeds, stems, and roots, used primarily for food. The second part of the book treats plant products that are extracted from parts of various organs, often different organs of a diverse array of species. In these chapters we concentrate on the chemistry of the products, how we procure them, and the ways in which we process them to meet our needs.

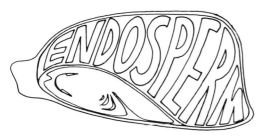

Corn seed section

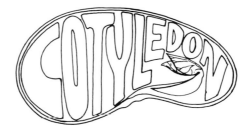

Bean seed section

FIGURE 1-28
Seeds must contain reserves of stored food that provide the energy for germination and seedling establishment. In a common bean, the food reserves are in the cotyledons, or seed leaves. In a grain of corn, the endosperm tissue provides nutritive material for the embryo.

ADDITIONAL READING

Bold, H. C., C. J. Alexopoulos, and T. Delevoryas. 1980. *Morphology of Plants and Fungi,* 4th ed. New York, Harper & Row. A very readable book that describes the basic features of different groups of plants and fungi.

Laetsch, W. M. 1979. *Plants. Basic Concepts in Botany.* Boston, Little, Brown. A basic introduction to the workings of plants and the ecosystems in which they occur.

Proctor, M. and P. Yeo. 1972. *The Pollination of Flowers.* New York, Taplinger. The most complete book on the diversity of flowers and the mechanisms of pollination.

Raven, P. H., R. F. Evert, and H. Curtis. 1981. *Biology of Plants,* 3d ed. New York, Worth. A thorough treatment of the physiology and anatomy of plants as well as of the major groups of plants.

GENERAL BOOKS ON ECONOMIC BOTANY

Baker, H. G. 1978. *Plants and Civilization.* 3d ed. Belmont, CA, Wadsworth.

Janick, J., R. W. Schery, F. W. Woods, and V. W. Rutlan. 1981. *Plant Science. An Introduction to World Crops,* 3d ed. San Francisco, Freeman.

Klein, R. M. 1979. *The Green World.* An Introduction to Plants and People. New York, Harper and Row.

Langenheim, J. H., and K. V. Thimann. 1982. *Botany. Plant Biology and Its Relation to Human Affairs.* New York, Wiley. A general botany book with a strong emphasis on economically important plants.

Schery, R. W. 1972. *Plants for Man,* 2d ed. Englewood Cliffs, New Jersey, Prentice-Hall. A classic text with discussions of almost all plants used today.

Tippo, O. and W. L. Stern. 1977. *Humanistic Botany.* New York, W. W. Norton.

Chapter 2

The Nature of Variation in Plants

Since the time man first began to tend useful plants, he has been selecting for the qualities he desired. This selection has occurred as a result of both unconscious and purposeful efforts. In either case, it initially took place without any real understanding of why various methods were effective. Now we understand why and how selection works, and we have been able to manipulate it to our advantage.

Traditional human genetic manipulation is called *artificial selection* to differentiate it from natural selection, the driving force of evolution in natural populations. For either kind of selection to work there must be inherited variation present in populations on which selection can act. Within this spectrum of variability, natural agents, or humans, carry out selection by differentially allowing certain individuals to reproduce. Throughout this book, we frequently make references to changes brought about by humans in cultivated plants, including the development of new varieties with high yields, increased nutritional value, and/or pest resistance. In order to understand how humans can bring about these changes, it is necessary to be familiar with the causes of biological variation and the way in which variation is maintained.

While we can not include here all of the aspects of the modern theories of evolution and genetics, we can review concepts particularly pertinent to economic botany. These include characteristics of chromosomes, Mendelian inheritance, and the relative contributions of polyploidy and the reproductive system to variation in populations. These features figure into the evolutionary history of all organisms, but some aspects of them are especially important for plants. In particular, the ability of many plants to inbreed, reproduce asexually, and to give rise to individuals with multiple numbers of chromosomes has led to important developments during the course of plant domestication. After a summary of the basic processes of genetics and reproduction, we therefore discuss the breeding systems of plants and the modifications of chromosome structure and number in some detail. We then examine the naming of plants and briefly relate it to evolutionary history.

GENES AND ALLELES

As mentioned in Chapter 1, the nucleus of a plant cell contains chromosomes which house most of its genes, or genetic material. Chromosomes are complex structures made of chains of deoxyribonucleic acid (DNA) and proteins (histones). A gene is usually considered to be a portion of the DNA chain that determines the chemical synthesis of a protein. We often speak, however, of a gene for a certain character rather than the gene for a particular protein because what we see are the results of that protein's action, not the protein itself. Table 2-1 lists several characters of cultivated crops thought to be controlled by one gene. It should be understood that during the time between the initial gene activation and the appearance of the final character, many chemical reactions occur, and often several other gene products play a role. Nevertheless, for ease of discussion, we ignore all of the intermediate steps and treat genes as though they led directly to the characters that we can observe.

We also pointed out in Chapter 1 that the way in which we indicate the diploid chromosome number (i.e., $2n$) emphasizes that every body cell of an

TABLE 2-1

Characters in Cultivated Crops Determined by Single Genes

COMMON NAME	SCIENTIFIC NAME	CHARACTERS
Barley	*Hordeum vulgare*	2- or 6-rowed inflorescences
Bean, common	*Phaseolus vulgaris*	Determinate or indeterminate growth
Beet	*Beta vulgaris*	Annual or biennial habit
Castor bean	*Ricinus communis*	Fruits spiny or spineless
Clover	*Melilotus alba*	Annual or biennial habit
Corn	*Zea mays*	Determinate or indeterminate growth
Cucumber	*Cucumis sativus*	Fruits spiny or spineless
Datura	*Datura stramonium*	Fruits spiny or spineless
Henbane	*Hyoscyamus niger*	Annual or biennial habit
Lettuce	*Lactuca sativa*	Seed heads shattering or not
Pea	*Pisum sativum*	Terminal leaflets leafy or modified into tendrils
Peach	*Prunus persica*	Glands on the petiole bases round or kidney-shaped
Pepper, red	*Capsicum annuum*	Fruit pendant or erect, calyx enclosing the fruit or not, fruit elongate or spherical
Radish	*Raphanus sativus*	Root round or elongate
Rice	*Oryza sativa*	Tall or short stature
Safflower	*Carthamus tinctorius*	Duration of rosette long or short, leaf margin entire or lobed, seed heads shattering or not, achenes with or without pappus
Sesame	*Sesamum indicum*	Fruit locules 2 or multiple, branched habit or nonbranching
Spinach	*Spinacia oleracea*	Bractlets around fruit smooth or horned
Squash	*Cucurbita pepo*	Fruit rind smooth or warty, fruit rind hard or soft, fruit disk-shaped or round
Tomato	*Lycopersicon esculentum*	Number of flower parts 5 or more, locule number 2 or multiple, fruit spherical or pear-shaped

Data in part from Gottlieb, L. 1984. Genetics and morphological evolution in plants. *American Naturalist* 123:685, 686, 691.

organism has chromosomes which occur in homologous (matched) pairs. Since a chromosome contains a particular array of genes, each array of genes, and thus each gene, must be represented twice in a normal cell. This double representation of each gene has important implications. It means that every gene, such as the gene for flower color (assuming that only one gene determines flower color), occurs twice in every cell. However, the fact that the same gene occurs twice does not mean that the same form of the gene occurs twice. Each gene may actually have several different forms, although only two can occur together in a diploid cell. Each form of a gene is called an *allele* of that gene. In the case of flower color, there may be one flower color gene, but it may have a yellow, blue, red, and white allele. The actual flower color will be determined by what pair of alleles occurs in the diploid genotype of a particular individual. If the two alleles of a gene are the same for an individual, we say that the individual is *homozygous* for that gene. If the two alleles differ, the individual is *heterozygous* with respect to that gene. A homozygous individual will obviously exhibit the properties of the gene that is represented twice. If the alleles differ, the expression of the gene depends upon the interaction of the products of the two alleles.

In terms of the interactions between heterozygous alleles, one of the alleles can produce a product that completely obscures or suppresses the effects of the other allele. In such a case, the allele which produces the effect is called *dominant*, and the unapparent allele, *recessive* (Fig. 2-1). An allele is only dominant or recessive in respect to a particular second allele. It may have different interactions with other alleles. Two alleles are considered to be *codominant* when neither is dominant over the other and the final product results from equal actions of both alleles (Fig. 2-2). Within the extremes of dominance and codominance, there is a range of expression of the two alleles (Table 2-2) which determine the phenotype, or physical properties, of an organism. We differentiate between the genotype (genetic makeup) and the phenotype because the phenotype of an organism is attributable to not only genetic, but also environmental, factors. Moreover, it is the expression of the genotype, modified by external influences, upon which selection acts.

We have said that each individual has only two alleles of a gene, but that in a population, there may be many different alleles of that gene. How do new alleles arise in the first place? As far as we know, the only way in which an entirely new allele is produced is by mutation, or an alteration of the chemistry or reading of the DNA strand along which that gene lies. This alteration must be of sufficient magnitude to lead to the production of a different protein. Sometimes, mutations create alleles with such strange end products that the cells containing them cannot function properly and die (or do not reproduce). In some cases, however, a new allele may not have any drastic effect, or may even produce an effect superior (in a given situation) to the effect of the existing alleles. If such a mutation occurs in cells which are involved in reproduction, they can be passed on to the offspring of the individual which has the mutation and subsequently spread within a population. The mutation rates of several genes have been calculated and show that there is a great deal of variation in the frequency with which mutations arise in different genes (Table 2-3).

We have taken advantage of the fact that out of the many millions of mutations that occur, a few produce changes in the organisms possessing them that are "superior" (from the human point of view) to the previously existing types. People can then perpetuate the new types. In order to induce mutations, with the hope that some will produce desirable changes, humans can also subject reproductive tissues to mutagens such as x-rays or chemicals that affect the chromosomes. They can then screen the resultant tissues, or progeny of the exposed parents, for possible "better" crop plants.

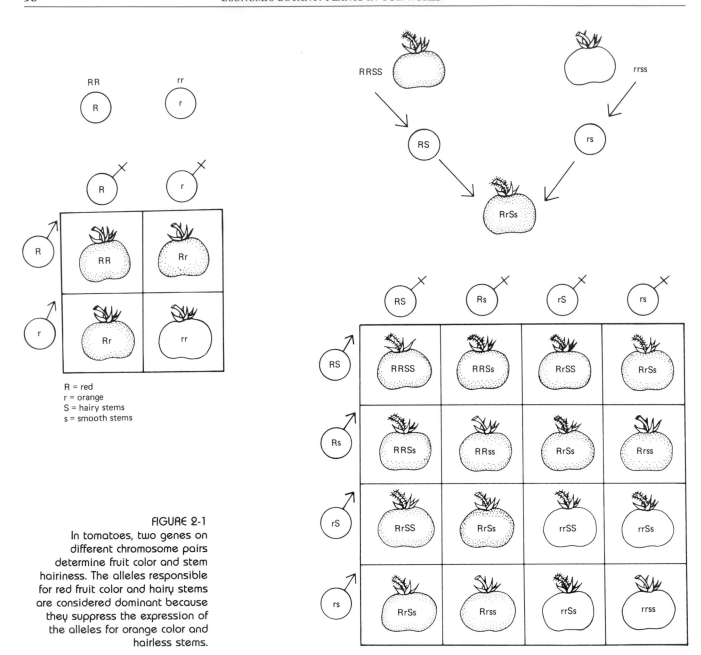

FIGURE 2-1
In tomatoes, two genes on different chromosome pairs determine fruit color and stem hairiness. The alleles responsible for red fruit color and hairy stems are considered dominant because they suppress the expression of the alleles for orange color and hairless stems.

R = red
r = orange
S = hairy stems
s = smooth stems

THE CONSEQUENCES OF SEXUAL RECOMBINATION

We have noted, assuming no immigration, that the only source of new alleles in a population is by mutation. But the occurrence of a mutation will not necessarily lead to its spread to other individuals in the population. For a new allele to produce a range of effects, it must interact with other alleles of its gene and within various combinations of other genes. The only way in which alleles

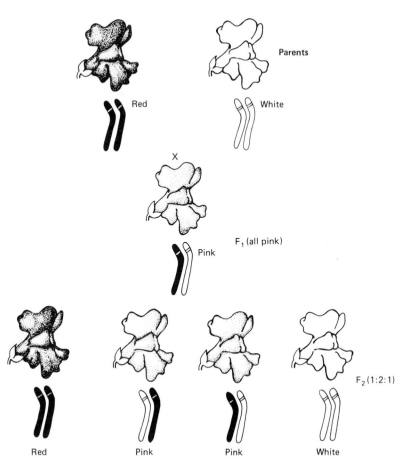

FIGURE 2-2
Flower color in the snapdragon is controlled by one gene. In this case, two alleles of the gene contribute equally to the final color we observe. Consequently, the alleles are considered to be codominant.

TABLE 2-2

Types of Allele Interactions

KIND OF INTERACTION	RESULT
One allele active, the other inactive	Dominance, incomplete dominance
Both alleles equally active and producing the same enzymes	Homozygosity
Both alleles active and producing enzymes, but these are unequal in their effects	Incomplete dominance, intermediacy
Alleles unequally active and producing enzymes with additive effects	Codominance leading toward dominance or heterosis
Alleles unequally active and producing enzymes with antagonistic effects	Incomplete dominance, intermediacy
Alleles equally active with additive effects	Codominance leading to heterosis
Alleles equally active but without additive effects	Codominance leading to dominance or intermediacy

Source: Adapted from V. Grant. 1964. *The Architecture of the Germplasm.* New York, Wiley.

TABLE 2-3

Mutation Rates in Two Species of Cultivated Plants

SPECIES	TRAIT (GENE)*	MUTATION RATE PER CELL OR GAMETE	AVERAGE NUMBER OF CELLS OR GAMETES IN WHICH 1 MUTATION WILL OCCUR
Corn	Shrunken seeds	1×10^{-6}	1 million
	Endosperm maturation (Sh locus)	12×10^{-6}	833,333
	Purple seeds	2×10^{-5}	100,000
	Plant color (R locus)	5×10^{-4}	2,000
Petunia	Self-incompatibility	8×10^{-5}	12,500

*All of these traits have been attributed to a single gene.

Source: Adapted from T. Dobzhansky, F. J. Ayala, G. L. Stebbins, and J. W. Valentine. 1977. *Evolution.* San Francisco, Freeman; and from J. L. Brewbaker. 1964. *Agricultural Genetics.* Englewood Cliffs, New Jersey, Prentice-Hall.

will move into new combinations is by periodic reshuffling of chromosomes or their parts. This shuffling occurs every time meiosis and sexual recombination is completed.

During meiosis (see Fig. 1-24), one partner of each chromosome pair goes to one of the nuclei produced. Of importance to the production of biological variation is the fact that chance determines which partner of every pair ends up in a given daughter nuclei. Figure 2-1 shows the various combinations of chromosomes (genes) possible in the gametes of two individual tomato plants and the diverse products that can arise when zygotes are produced from recombinations of these sex cells. Since there are thousands of genes per organism, and usually numerous alleles of any one gene spread around individuals in the population, sexual recombination tends to produce almost infinite combinations of the various alleles found in a population.

If we are interested in a particular character and the expressions of it within a population, we first assess the frequencies of the phenotypes in the population. If the character has discrete states such as red flowers, blue flowers, etc., each individual will fall into a discrete category of flower color. We can then count the number in each category. Most characters, however, such as height, weight, leaf length, or even some flower colors, show a continuum of values without any marked separation into classes. Such characters are generally the result of the actions of several genes. Consequently, when we measure (or score) individuals for these characters, we usually find that they exhibit a normal distribution with the majority of individuals having values clustered around the mean value, or average type.

INBREEDING AND ASEXUAL REPRODUCTION

Inbreeding

Despite the fact that the majority of angiosperm species have perfect flowers, the most common breeding system appears to involve the fusion of gametes from different individuals. Such a system is called *outcrossing* because the zygotes

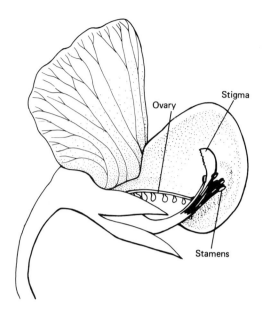

FIGURE 2-3
Although most beans are naturally outcrossing, peas have been selected for self-pollination. The lack of an incompatibility reaction, the close proximity of the anthers and stigma, and the synchronous shedding of pollen and receptivity of the stigma, contribute to successful self-fertilization.

are formed by gametes from outside in combination with gametes formed on the plant. Yet, the fact that most flowers have both male and female gametophytes raises the possibility of zygote formation by gametes produced solely within one flower (Fig. 2-3). In fact, this type of zygote formation does occur and is one kind of inbreeding or self-fertilization. The term inbreeding is also used in agriculture not only for the fusion of gametes within a flower (or within an individual), but also for the fusion of gametes from different, but genetically very similar, plants.

Outcrossing between perfect flowers can be obligatory if the pollen of a given plant is incapable of germinating on its own stigmas, growing down the styles or forming viable zygotes with eggs of its own flowers. Flowers that are unable successfully to self-fertilize are called *self-incompatible*. If fertilization can occur within a plant, flowers (or plants) are considered to be self-compatible. Outcrossing can be fostered in both self-incompatible and self-compatible flowers if pollen release and stigma receptivity are not synchronous, or if the anthers and the stigmas are spatially separated (Fig. 2-4). In nature, most plants appear normally to outcross, although many are self-compatible. Some groups, such as annual herbs and many weeds, characteristically self-pollinate and produce seeds that are largely derived from the fusion of gametes produced within a single flower. Because such plants are mostly independent of external agents for successful pollination and fertilization, seed production tends to be high. In terms of the genetic consequences of self-fertilization, there is a reduction in variability in the offspring of an inbred individual because all of the gametes involved in recombination come from the same individual. Figure 2-5 diagrammatically shows how two populations would differ after three generations if there were complete outcrossing in one case and complete inbreeding (selfing) in the other.

From the human point of view, the fact that flowers of the same plant (or genetically very similar individuals) can be used to produce viable (able to germinate) seeds, is often a desirable character because it permits the rapid

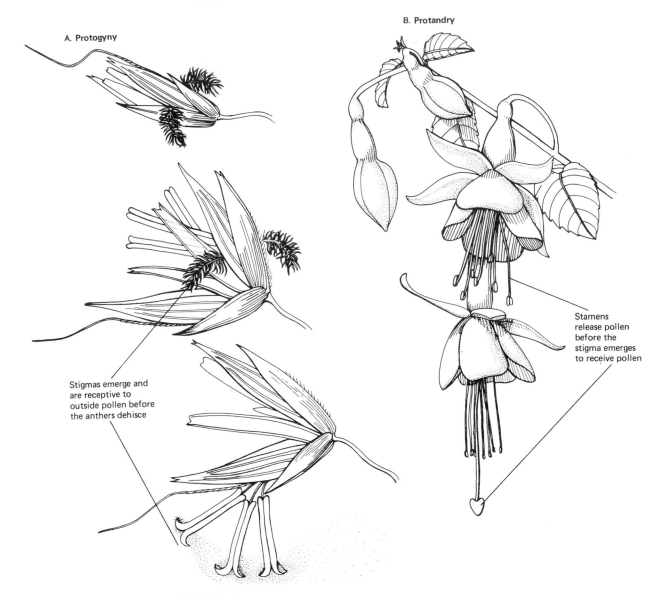

A. Protogyny

B. Protandry

Stigmas emerge and
are receptive to
outside pollen before
the anthers dehisce

Stamens
release pollen
before the
stigma emerges
to receive pollen

FIGURE 2-4
Cross-pollination is fostered in oats by protogyny (A). The stigmas appear and are
receptive before the anthers of the same flowers emerge and release their pollen. In
Fuchsia, outcrossing is fostered by protandry (B). The stamens in a given flower mature
and release their pollen before the stigma of that flower is ready to accept pollen.

production of essentially homozygous individuals (Fig. 2-5). In agricultural
practice, crosses between offspring and one of the parents (or parental types)
are known as back crosses. Crossing between sibs is known as sib or *sib-sib
crossing*. Both of these types of crosses, like self-fertilization, lead to the production
of inbred, homozygous lines. Agriculturalists favor inbred lines for crops because
they produce uniform stands of selected genotypes. Inbred lines are also used
as parental stocks to produce hybrid seed for a particular crop. The purpose of

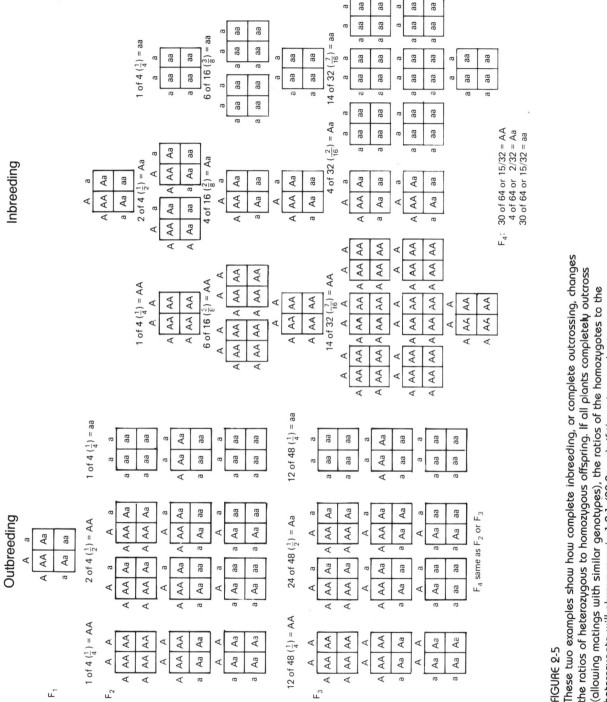

FIGURE 2-5

These two examples show how complete inbreeding, or complete outcrossing, changes the ratios of heterozygous to homozygous offspring. If all plants completely outcross (allowing matings with similar genotypes), the ratios of the homozygotes to the heterozygote will always remain 1:2:1 (AA:Aa:aa), if there is no selection, mutation, or immigration of foreign genes. If all plants obligately inbreed, the proportion of heterozygotes will decrease. Initially, the two homozygotes together constitute only 50 percent of the offspring, but after four generations of complete inbreeding, 94 percent are homozygous. The proportion of homozygotes will continue to rise slowly.

FIGURE 2-6
Even though it takes more effort to produce seed by crossing flowers of plants with known, but different, genotypes than by allowing natural pollination to occur, the resulting hybrid plants can exhibit some qualities superior to those of their parents. In this 1880 advertisement, the hybrid vigor of the Mikado tomato was promised to make it superior to the old varieties. (Courtesy of the Smithsonian Institution.)

producing hybrid crops is related to the fact that plants resulting from the crossing of two inbred lines are often larger and produce larger seed crops than either parent, a phenomenon known as "hybrid vigor," or heterosis (Fig. 2-6). However, if crops are grown from hybrid seed, the seed of the crop cannot be used for further planting because sexual recombination will have produced undesirable combinations in the next generation. New hybrid seed, sold by companies that produce seed solely for planting agricultural crops, is purchased each year. Figure 2-7 shows how inbred lines are used to produce hybrid corn.

The ability of some plants to self-fertilize is of particular significance to humans for another reason. In many cases, a crop is grown for its seed or fruit. Especially in large fields planted with a single crop, it is often difficult to achieve successful pollination of all of the flowers if they are completely dependent on insects or wind for pollen transfer. Humans have, therefore, often selected for mutations that facilitate self-pollination. Figure 2-8 diagrams the postulated changes in the breeding system of the tomato brought about by artificial selection from the natural, outcrossing system to the self-pollinating system now found in all commonly cultivated varieties. Other crops such as wheat, rye, and grapes are all predominantly self-pollinating. If selection for self-fertilization is not

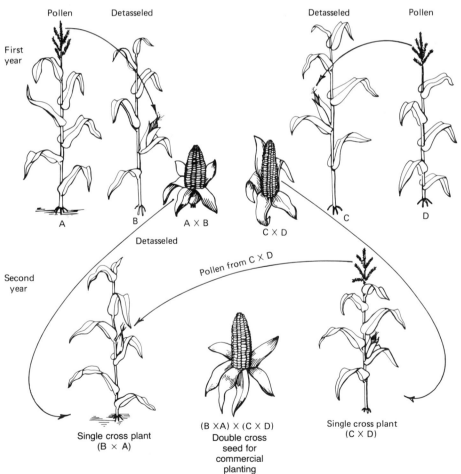

FIGURE 2-7
The production of hybrid corn from inbred, homozygous lines. (After a USDA illustration.)

Wild types ──► Modern domesticates

FIGURE 2-8
Outbreeding was fostered in the wild tomato by the spatial separation of the anthers and the stigma. Pollen shaken from the pores of the anthers did not reach the longer style of the same flower. Humans have selected for a shorter and shorter style until, now, the stigma is below the level of the anthers and pollen shaken from the anthers will naturally fall on the stigma. In this way, we are assured a high level of fruit set in this self-compatible species.

possible, effective pollinators must often be artificially provided during flowering periods. Commercial bee keepers often rent hives to farmers with orchards, cotton fields, or seed alfalfa fields during flowering times in order to ensure a high level of pollination.

Asexual Reproduction

Asexual or vegetative reproduction is another method of propagation which has often been promoted by human beings. When plants reproduce asexually, the "daughter" plants are genetically identical to the "parent." All the resultant plants thus formed can be considered to form a single clone. Vegetative propagation can be accomplished by using pieces of leaves, stems, rhizomes, or tubers which will regenerate new individuals. This form of reproduction, even more than inbreeding, assures crop uniformity. In addition, starting plants with pieces of mature tissue is sometimes more efficient than planting seeds. Young seedlings are more vulnerable to fungal attack than are shoots from rhizomes or tubers. Plants also tend to mature faster if grown from asexual parts than if grown from seed because the time delay involved in seedling establishment is circumvented.

Grafting is an artificial method of asexual reproduction which assures the perpetuation of the genotype of the plant from which the graft was taken and which produces a "plant" that mimics a sapling several years old. In the process of grafting, a branch or bud of a desirable tree or shrub is taken and joined to a rootstock or stem of another individual (Fig. 2-9). Care is taken to assure that the cambium of the scion (branch or bud) and the stock (rootstock or stem) are aligned so that the actively dividing tissues come into contact and grow together. The cut tissues are protected until the graft takes hold. If the graft is successful, the crop-bearing part of the plant will express the genotype of the plant from which the scion was taken. A variety of stocks can be used. Selection of a stock is determined by its cold tolerance, disease resistance, etc., in a particular environment.

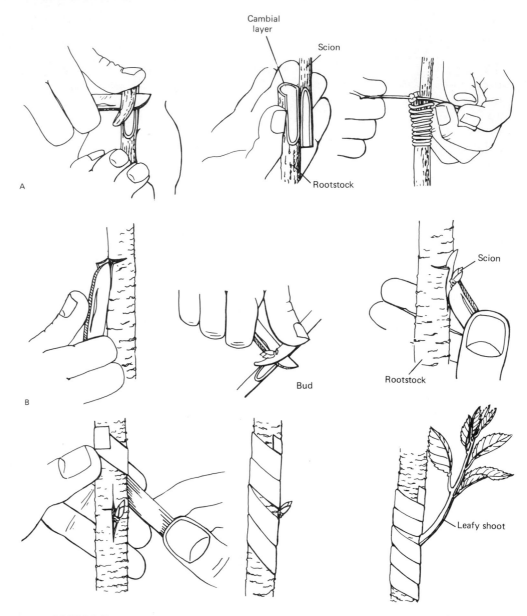

FIGURE 2-9
The essence of grafting is the positioning of the cambial layers of the scion (bud or stem) and the stock (rootstock) on top of one another so that they will grow together. Once the graft has taken, the rootstock will provide the nutrients to a shoot system that resembles the plant from which the scion was taken. (A) Stem graft; (B) bud graft.

OTHER SOURCES OF VARIATION IN PLANTS

We have seen that mutation and sexual recombination are, respectively, the initial source of, and the primary mechanism for maintaining, genetic variation in plant populations. We have seen that humans have been able to use their knowledge of how characters are inherited and how recombination works to produce homozygous lines and make specific hybrid crosses.

Another way in which variation is naturally produced in plant population is by structural changes in the chromosomes of the germ cells which are subsequently passed on to the offspring. Chromosome changes also occur in somatic cells during mitosis, but unless vegetative reproduction occurs, either naturally or via humans, such changes will exist only in the individual in which they originally occurred. In germ cells, on the other hand, the possibility exists for the inheritance of features associated with a variety of changes that can occur in the chromosomes. Such changes can involve alterations in the numbers of chromosomes, or they can involve gross alterations of individual chromosomes (Fig. 2-10).

During the process of DNA replication, pieces of the DNA molecule can be lost (deletion), or replicated twice (duplication). Also, a given segment of the DNA molecule can be replicated in the wrong order (inversion), or a piece of chromosome may be shifted from one place along the chain to another (translocation). Finally, a long arm of an asymmetrical chromosome can replace the short arm of its partner chromosome resulting in two "new" chromosomes, one with two long arms and one with two very short arms. Often the small chromosome is lost when the meiotic chromosomes divide. Some of the resulting gametes will then have one fewer than the normal haploid number of chromosomes.

Another interaction between partner chromosomes is crossing over. Crossing over occurs while the chromosome partners are next to one another in the first stages of meiosis. During this period, portions of the DNA chains of the partner chromosomes can be exchanged (Fig. 2-10). Such interchanges place the alleles

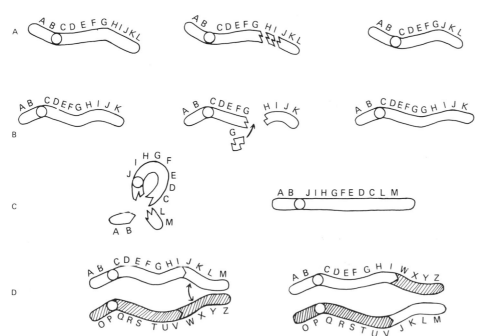

FIGURE 2-10

Chromosomes can change in a number of ways which involve alterations of entire segments of a chromosome. Some of these are (A) deletions, (B) duplications, (C) pericentric inversions, and (D) crossing over. Exchanges of pieces of chromosomes between partner chromosomes during meiosis can also place alleles in new genetic backgrounds. (Adapted from D. Briggs and S. M. Walters. 1969. *Plant Variation and Evolution*. New York, McGraw-Hill.)

of the exchanged segments in different backgrounds from those of the original chromosomes. All of these changes in chromosome sequences have led some workers to proclaim that no two gametes are ever truly alike.

POLYPLOIDY

Other large-scale chromosomal changes that contribute to variability also occur during meiosis. Chromosomes can fail to separate at the first meiotic division, leading to a gamete with the diploid, instead of the haploid, chromosome number. If these gamete cells fuse with gamete cells that have the normal haploid number, zygotes having three times the haploid number of chromosomes will be formed. If gamete cells with a double number of chromosomes fuse with other unreduced gametes, the zygote will have four times the number of chromosomes of the normal haploid cells. Cells which contain multiples of the haploid number (n) greater than two ($2n$) are called *polyploids*. The kind of polyploidy reflects the number of times the haploid number is repeated. A cell with three times the haploid (base) number is a triploid ($3n$), a cell with four times the haploid number is a tetraploid ($4n$), six times represents a hexaploid ($6n$), etc. Table 2-4 lists some cultivated crops that are believed to be of polyploid origin, their chromosome numbers, and presumed ploidy level.

Polyploidy is relatively common in plants and represents a way in which hybrid individuals with unlike chromosomes can become reproductively suc-

TABLE 2-4

Some Major Crop Species of Presumed Polyploid Origin

COMMON NAME	SCIENTIFIC NAME	FAMILY	APPARENT BASE NUMBER	PRESENT DIPLOID NUMBER AND PLOIDY LEVEL
Coffee	Coffea arabica	Rubiaceae	$x = 11$	$2n = 44$ tetraploid
Cotton	Gossypium hirsutum	Malvaceae	$x = 13$	$2n = 52$ tetraploid
Potato	Solanum tuberosum	Solanaceae	$x = 12$	$2n = 48$ tetraploid
Strawberry	Fragaria ananassa	Rosaceae	$x = 7$	$2n = 56$ octaploid
Sugar cane	Saccharum officinarum	Poaceae	$x = 10$	$2n = 80$ octaploid
Tobacco	Nicotiana tabacum	Solanaceae	$x = 12$	$2n = 48$ tetraploid
Wheat	Triticum	Poaceae	$x = 7$	
Durum	T. turgidum			$2n = 28$ tetraploid
Club	T. aestivum			$2n = 42$ hexaploid

Source: Adapted from N. W. Simmonds. 1976. *Evolution of Crop Plants.* New York, Longman.

cessful. Normally, if a plant is formed from two gametes with dissimilar chromosomes, it will have difficulty in producing its own gametes because there will not be a true partner for each chromosome at meiosis. In other words, while there may be an even number of chromosomes, a given chromosome will not recognize, and pair with, another chromosome. If, however, each gamete contains the diploid number of chromosomes and two such gametes fuse to form a zygote, the zygote will have two diploid sets of chromosomes. As a result, every chromosome will have at least one partner, and regular meiosis can occur in the organism with the 4*n* (or tetraploid) number of chromosomes. Polyploids can also be formed by a doubling of the hybrid chromosome complement after the zygote is formed. Fig. 2-11 shows how this process works. A polyploid which has dissimilar diploid sets of chromosomes is called an *allopolyploid.* If all of the diploid sets of chromosomes in the polyploid are the same, it is called an *autopolyploid.*

Human beings have made use, both inadvertently and consciously, of the fact that plants form polyploids relatively easily and that polyploidy is a way in which hybrid types can be made reproductively fertile. Polyploid plants are almost always larger and in many cases have larger fruits and/or seed crops than their diploid progenitors. Thus polyploidy often confers characteristics in plants that humans find desirable, and we have perpetuated them without realizing, until the advent of modern cytology, their polyploid nature. Now that we know how polyploidy acts, and how to induce it, we can experimentally produce polyploids of many crop species. In subsequent chapters, we point out important crop plants that are polyploids. In many cases, the original increase in chromosome number occurred as a chance accident thousands of years ago, but produced a type of plant that humans perpetuated over others. In other cases, modern breeding programs have incorporated the artificial production of polyploids into their methods for the development of new varieties.

GEOGRAPHIC VARIATION IN PLANTS

Despite the fact that humans often want uniformity in crop plants, genetic variability (and, hence, nonuniformity) is the rule in nature. As we have stressed, the entire process of sexual recombination is a mechanism for the continual reshuffling of genes in order to provide a variable array of individuals. Why is variability needed in natural populations? All evidence points to the fact that, in the long run, organisms in nature which produce variable offspring will ultimately leave more survivors than those which produce uniform offspring. The greater success of a variable array of offspring is apparently related to the fact that the environment is not uniform, either in space or time. Since there is no way for a parent to "know" in what kind of a germination site a seed will land, or what climatic and biological factors it will face during its lifetime, it cannot produce a single guaranteed successful genotype. By producing an array of types, the possibility of having at least a few surviving offspring is maximized. In contrast, under cultivation, people can grow enormous numbers of a single genotype (a monoculture) because they can provide a more or less uniform environment and enough care to allow the crops to survive.

FIGURE 2-11
When two gametes from two genetically dissimilar individuals (A and B) fuse to form a zygote (C), the adult plant which results may not be able successfully to form gametes because during meiosis, the chromosomes cannot pair properly and thus cannot undergo a reduction division. If a doubling of the chromosome number occurs in the zygote (D) a tetraploid hybrid will result. Meiosis can occur in the adult which results from this mating because each chromosome will have a perfect partner with which to pair.

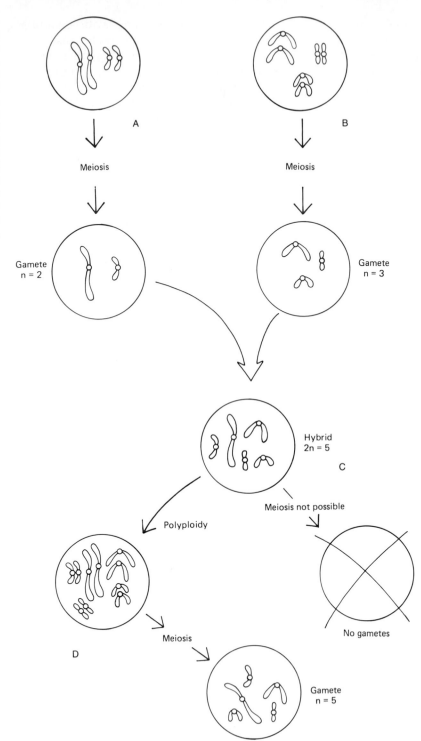

In natural situations, the frequency of some genotypes increases relative to others over time because the plants possessing them have been the most successful at leaving reproductive offspring. In nature, changes in gene frequencies are caused by either biotic or abiotic factors in the environment. Since both factors change over geography, we find that populations of the same organism have different mixtures of genotypes in different parts of the distributional range. Such variation in population composition over space is called *geographic variation.* Eventually, however, there is a limit to the variation in physiological tolerances or competitive ability of a particular kind of organism. For this reason, every kind of organism has a natural distributional limit.

The range of environments in which humans could grow particular crops would normally be restricted by these biological constraints. Often, however, humans have been able to extend the range of a kind of organism either by producing an artificial environment (for example, that in a greenhouse) or by growing a crop as an annual during a favorable season. A common example is the tomato, a native of tropical regions and a crop that cannot survive freezing temperatures. People can grow tomatoes in the northern United States because they are placed in gardens after the danger of frost is past. The tomatoes are harvested before winter again sets in. New plantings must be made every year because both seeds and adult plants are killed by very cold conditions.

People can also extend the range of a crop plant by taking it from the area of the world in which it is native to another with a similar type of climate. Under natural circumstances, the geographic range of these plants was restricted by not only environmental factors but also by physical barriers such as oceans or mountain ranges. People can transport a plant in which they are interested over such natural boundaries. Many important crop plants are, in fact, grown most extensively in areas other than those of their native home.

THE NATURE OF PLANT SPECIES

So far, we have referred primarily to "kinds of organisms" and talked about reproduction among members within such kinds. What we have essentially been discussing are species. People seem to have an intuitive feel for the grouping of organisms into kinds that scientists call species (e.g., crab grass, Bermuda grass, giraffes, horses, donkeys), but scientists disagree among themselves about the definition of "a species." Most zoologists define a species as a set of individuals or populations that are capable of interbreeding among themselves in nature, but which are reproductively isolated from other such individuals or populations. This definition presents problems for plant taxonomists for several reasons, among them, the common occurrence of polyploidy. The production of allopolyploids creates individuals usually unable to produce fertile offspring when crossed with the diploid plants that gave rise to them. Since the "parents" and their offspring are reproductively isolated, they might be considered different species.

A second problem with the use of reproductive barriers as the criterion for defining a plant species is the ease with which plants hybridize, making the phrase "reproductively isolated" difficult to apply. Consequently, botanists tend to consider as a species a group of populations derived from a single ancestor and which can be distinguished morphologically from other groups of populations. Somewhat inconsistently, botanists have often also tended to give names to cultivated species if they have diverged morphologically from their wild progenitors. There has recently been a trend toward placing both a cultigen and its wild ancestor in a single species.

THE NAMING OF PLANTS

Species are named according to an international system of rules called the *International Code of Botanical Nomenclature,* which specifies how correct names are to be determined for various groups of plants. The code does not specify the biological basis for what should be called a species, genus, or family. It merely gives instructions for how to apply a particular name once a decision about the rank of a group of organisms has been determined and a proper description provided. The most commonly used rank is the species. Every species is given a Latin name consisting of two parts. This system, known as the binomial system of nomenclature, was developed by Carl Linnaeus (Fig. 2-12) in Sweden and first used consistently by him in 1753. The first part of the binomial is the genus name and the second, the species name. Both parts of the name are needed to identify a particular species, and both names are always written in italics. The genus name, but usually not the species name, begins with a capital letter. Every species of plant (and animals under the provisions of a different code) has only one correct name. Despite language, colloquial, or common names, the scientific name is the same all over the world. Thus, scientists are able precisely to refer to a particular species even if they do not speak the same language. Often, the Latin name given to a species describes one of its attributes. Table 2-5 gives some Latin descriptive adjectives frequently used as specific names of cultivated plants.

Since the generic name is part of a species' designation, we might ask what is a genus? Unfortunately, there is no real definition of a genus. It is a category that contains a collection of species that are related to one another and therefore share characters. According to evolutionary theory, the species of a genus are all descended from a common ancestor. Likewise, at a higher level of grouping, genera which are believed to have originated from a common ancestor and which tend to share several, but fewer, characters, are clustered together in a plant family. As in the case of species, the names of families are in Latin. Family names begin with a capital letter and normally end with the letters "-aceae" but are not written in italics. The code allows the use of old names, not ending with "-aceae" for several important plant families. Table 2-6 gives the alternative names allowed for families of economic importance that we discuss in this book. Plant families are often easy to recognize because of some trait that is common to almost all members. For example, the bean family includes many genera with many species, but all members have legumes or "beans" for fruits. The dandelion family also contains hundreds of genera and thousands of species, but all have a similar type of dandelion "flower" or head (Fig. 2-13). We shall often use the family as a convenient way to group species of economic importance.

FIGURE 2-12
Linnaeus after his return from Lapland wearing the traditional Lap costume. [Reproduced from *Medicine and the Artist (Ars Medica)* by permission, Philadelphia Museum of Art, Carl Zigrosser, Dover Publications, Inc. 1970.]

TABLE 2-5

Common Latin Names Used for Species and Their English Meanings

LATIN	ENGLISH
aestivus	summer
alba	white
ambiguus	doubtful
americanus	from America
annuus	annual
arabicus	from Arabia
arboreus	treelike
argenteus	silvery
arvensis	from cultivated fields
asiaticus	from Asia
biennis	biennial
blandus	agreeable
bulbiferus	bulb-bearing
caeruleus	blue
communis	common
edulis	edible
erectus	upright
esculentus	edible
foetidus	bad-smelling
glabellus	smooth
japonicus	from japan
luteus	yellow
maritimus	from (near) the sea
minor	small
odoratus	fragrant
officinalis	official, medicinal
oleraceus	aromatic
perennis	perennial
sativus	cultivated
silvestris	from the woods
tinctorius	useful in dyeing
usitatissimus	very useful
vulgaris	common

Note: Species names that are adjectives must agree in gender with their generic names. Those given here are masculine adjectives.
Source: Abstracted from L. H. Bailey. 1933. *How Plants Get Their Names.* New York, Macmillan.

Sometimes, species have geographic populations which are distinctive enough to merit a name. Under the rules of the *International Code of Botanical Nomenclature,* such populations could be classified as subspecies or varieties. We could even have subspecies with varieties since variety is a smaller unit than subspecies (which, of course, means that we cannot have varieties with subspecies). In practice very distinctive populations set apart geographically from others in the species tend to be called subspecies. Less distinctive populations are often treated as varieties.

The word variety used for crops or horticultural plants has a different meaning. Varieties of cultivated plants are distinctive types selected by humans. Under the code, varieties with Latin names are evolutionary entities differentiated from others by natural selection. The agronomist's variety is a human construct analogous to breeds of dogs or cattle. In Chapter 18, we describe the way in which horticultural varieties are named.

TABLE 2-6

Alternatives Names of Important Angiosperm Families:

TRADITIONAL NAME*	RECOMMENDED NAME
MONOCOTYLEDONS	
Palmae	Arecaceae
Graminae	Poaceae
DICOTYLEDONS	
Leguminosae	Fabaceae
Umbelliferae	Apiaceae
Cruciferae	Brassicaceae
Labiatae	Lamiaceae
Compositae	Asteraceae

*Traditional names frequently came from some characteristic aspect of the members of the family. The name Umbelliferae is derived from the shape of the flowering stalk. Leguminosae comes from the type of fruit, a legume, found in all the species of the family. The recommended name is based on a genus in the family, not on a character.

TABLE 2-7

Categories Used in Classifying Economically Important Higher Plants

CATEGORY	EXAMPLE	COMMON NAME
Kingdom	Phyta	Vascular plants
Division	Coniferophyta	Conifers
Class	Pinidae	
Order	Pinales	
Family	Pinaceae	Pine family
Genus	*Pinus*	Pines
Species	P. palustris	Longleaf pine
Division	Anthophyta	Angiosperms
Class	Liliopsida	Monocotyledons*
Order	Cyperales	
Family	Poaceae	Grass family
Genus	Zea	Corns
Species	Z. mays	Corn
Class	Magnoliposida	Dicotyledons*
Order	Fabales	
Family	Fabaceae	Bean family
Genus	*Phaseolus*	Beans
Species	P. vulgaris	Common bean

*The older conceptions of the monocotyledons and the dicotyledons have been somewhat modified in the past few years. It is now recognized that the two groups are not equal lineages. Evolutionarily, the water lilies, historically placed in the monocotyledons, belong with the dicotyledons. *Source:* Based on H. C. Bold, C. J. Alexopoulos, and T. Delevoryas. 1980. *Morphology of Plants*, 4th ed; and on A. Cronquist. 1981. *An Integrated System of Classification of Flowering Plants*. New York, Columbia University Press.

Table 2-7 gives the various categories most frequently used in classifying plants and provides examples of each. Throughout this book we will use species names and indicate the families to which they belong to help in ordering them into evolutionary groups.

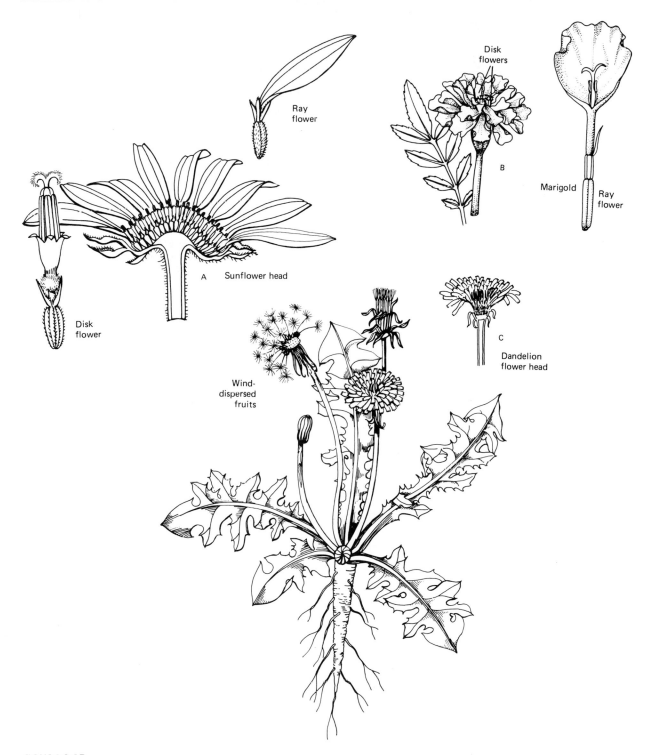

FIGURE 2-13
In the sunflower (A), the head has female ray flowers around the outside and perfect (hermaphroditic) disk flowers in the center. The marigold (B) has few disk flowers and more ray flowers. A dandelion head (C) is composed of perfect flowers, each with a long, strap-shaped corolla reminiscent of the rays in the sunflower.

ADDITIONAL READING

Allard, R. W. 1960. *Principles of Plant Breeding.* New York, Wiley.

Bailey, L.H. 1933. *How Plants Get Their Names.* New York, Macmillan. A somewhat outdated but delightful book that describes the beginnings of the binomial system of nomenclature and how it is used in naming plants.

Briggs, D., and S. M. Walters. 1984. *Plant Variation and Evolution.* (2d ed.) Cambridge, Cambridge University Press. A readable text on the basics of plant genetics and variation.

Grant, V. 1971. *Plant Speciation.* New York, Columbia University Press. An extensive treatment of the processes involved in plant speciation.

Hartmann, H. T., and D. E. Kester. 1968. Plant Propagation, 2d ed. Englewood Cliffs, New Jersey, Prentice-Hall.

Hawkes, J. G. 1983. *The Diversity of Crop Plants.* Cambridge, Massachusetts, Harvard University Press. A well-written, concise essay dealing with the origins of cultivated plants, modern breeding programs, and the attempts to conserve genetic variability.

Jones, S. B., Jr., and A. E. Luchsinger. 1979. *Plant Systematics.* New York, McGraw-Hill. A recent text which describes the methods used by plant taxonomists to determine the best classification of groups of plants and how the plants are to be named. Most of the major vascular plant families are described, and the authors give a synopsis of the terminology used to describe plants.

Chapter 3
Origins of Agriculture

*F*arming is a relatively recent innovation in man's world. Humans (*Homo sapiens sapiens*) have existed for an estimated 100,000 years, but they have practiced widespread cultivation for less than a tenth of that time. The questions we address here, when, where, how, and why people first began to cultivate plants on a permanent basis, have intrigued man for thousands of years. While we cannot hope to do justice to all of the voluminous and scholarly work that has addressed questions about the origins of agriculture, we present in this chapter the most prominent theories of the past and present.

Implicit in all of the questions posed above is the assumption that humans, at some point which we can more or less define, turned from a nonagricultural to an agricultural way of life. To be precise in our use of terms, we should define agriculture in the strict sense, as the caring for, or the cultivation of, plants. As Edgar Anderson and Jack Harlan have pointed out, this caring for plants can be minimal or intensive, ranging from the encouraging of essentially wild individuals to the careful planting and rearing of selected species. In our discussion of the origins of agriculture, we are really talking about the adoption of agriculture as a way of life. Domestication, a word often confused with cultivation, is used to refer to a more intricate process than cultivation which involves the genetic alteration of plants brought about by the activities of humans. In most cases, domestication follows closely on the heels of cultivation since any selection of individuals (for example, the use of certain individuals for seed sources) would lead to genetic changes in the cultivated plants. In some cases, plants have been so altered by domestication that it is no longer possible to link them precisely with a particular wild species.

Once we have discussed ideas about the origin of agriculture in general, we briefly turn our attention to suggestions about places where important crop species were first domesticated.

A TIME FRAME

We can begin with perhaps the easiest question: When did people first consciously begin to care for, rather than simply gather, plants? Or, when did man first become a farmer rather than a hunter-gatherer? This question is "easiest" only because there are methods for dating traces of prehistoric agriculture and most workers are willing to accept the oldest date until an older one is demonstrated.

What are the ways of finding and placing a date on evidences of early agriculture? Naturally, the first step in the process is to find fossil human encampments or resting places. Evidence that agriculture was being practiced might include abundant remains of plants that are known to have been cultivated, or tools used for preparing soil, "cultivating," or harvesting. An analysis showing that the morphology of the fossilized plant material in digs is different from wild forms in nearby areas is considered evidence that humans altered, or domesticated, the material. Plant and animal remains for such analyses are generally found in garbage dumps, strewn around ancient hearths, or as pollen found on pots or pot fragments around the sites. Pollen grains are particularly useful for identifying remains for two reasons. First, they are exceedingly resistant to decomposition, persisting unchanged in deposits for millions of years. Second, many pollen grains have characteristic sculpturing (Fig. 3-1) that allows the botanist to determine the species or genus from which they came. Charred seeds or plant parts can also often be identified by comparing gross or cellular features with those of modern counterparts.

Dating is usually done by what is called the carbon 14 (written ^{14}C) method (Fig. 3-2). This method makes use of the natural occurrence of a carbon isotope that decays at a known rate. All living organisms are primarily made up of carbon atoms. The usual form of carbon (^{12}C) has a molecular weight of 12, but several other atoms reacting chemically like carbon also occur. One of these, ^{14}C, has a molecular weight of 14. This isotope is continuously formed in the atmosphere by the interaction of cosmic rays and nitrogen. It slowly degenerates with a half-life of about 5,700 years back into nitrogen. The process of ^{14}C formation and disintegration has been taking place for so long that an equilibrium between ^{14}C and ^{12}C was reached millions of years ago. When an organism is alive, it is constantly incorporating carbon into its body. In the case of plants, most of the carbon incorporated, or "fixed," is taken into the plant in the form

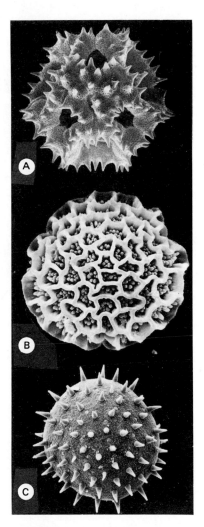

FIGURE 3-1
These scanning electron micrographs show the pollen of (A) dandelion ($\times$ 1830), (B) bougainvillea ($\times$ 2580), and (C) hibiscus ($\times$ 300). Differences in the size and sculpturing of the pollen grain surface allow scientists to identify the plants from which the pollen came. Similarly, pollen recovered from fossil deposits can give an indication of the kinds of plants people used, or which grew in the vicinity of their camps, at the time when the deposit was formed. (Photos courtesy of Joan Nowicke and the Smithsonian Institution.)

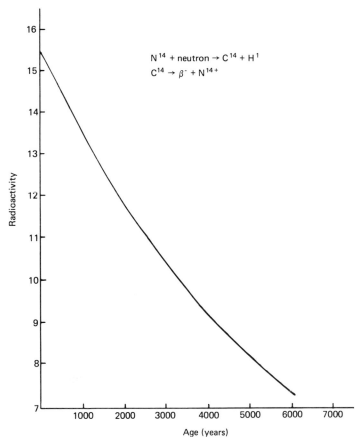

$$N^{14} + neutron \rightarrow C^{14} + H^1$$
$$C^{14} \rightarrow \beta^- + N^{14+}$$

Radioactivity (y-axis)

Age (years) (x-axis)

FIGURE 3-2
The decay curve for carbon 14 clearly shows that the half-life of the isotope is between 5000 and 6000 years (actually 5730 $\pm$ 40). The radioactivity is measured in disintegrations per minute per gram of carbon. (Adapted from W. F. Libby. *Radiocarbon Dating.* 1965 edition.)

of carbon dioxide. Since ^{14}C and ^{12}C occur in a constant ratio in the atmosphere, they are incorporated into plant tissues (with some selectivity) in this ratio. Once a plant dies, or is collected by humans, incorporation ceases. From this time onward, there is only decay of the ^{14}C into nitrogen. By measuring the ratios of ^{12}C and ^{14}C present in a piece of organic fossil material, a formula can be used to calculate the time since the death of the organism. The method can be used even for cooked or burned organic material since neither heating nor charring alters the ^{12}C to ^{14}C ratios. Although the ^{14}C technique sounds simple, accurate dating requires meticulous care, the use of uncontaminated material, and provides reliable dates only for ages between 500 and 50,000 years before present (ybp). However, this time span is perfectly adequate for dates of most archaeological sites.

What have data from such digs told us? The earliest dates firmly ascribed to evidences of agriculture are between 8,000 and 10,500 years ago. Until recently, all dates over 12,000 years old appeared to be from sites that had been inhabited by hunter-gatherers. In 1979, an exciting report appeared which described a barley grain with features of a domesticated kernel. This grain was found in deposits dated to be 18,500 years old, about 8,000 years older than any previous evidence of agriculture. It now appears that the barley kernel was deposited from younger sediments into a mixture of old fossilized material. Consequently, we still have no firm dates for significant agriculture before 10,500 years before present, and we will assume only that by some time between 9,000 and 10,000 ybp, were humans unquestionably practicing agriculture. It is also unquestionable, on the basis of numerous archaeological sites, that by 2,000 years ago, most of the major civilizations of the world practiced agriculture. It was during this 8,000-year period from 10,000 ybp to 8,000 ybp that man's entire way of life and the course of civilization changed. Once people became dependent on their own production of foods, they seldom reversed the process to return to gathering as a means of sustenance.

A CRADLE FOR AGRICULTURE?

The earliest dates indicating the practice of agriculture come from archeological sites located around the eastern edge of the Mediterranean Sea (Syria, Israel, Jordan, etc., and now, possibly Egypt, Fig. 3-3). This region has been variously labeled as the Cradle of Agriculture or the Fertile Crescent. To us today, this looks like an unlikely place to begin cultivating plants. The terrain is hilly and rocky, the soil comparatively poor, and the climate arid.

Although this area has yielded the oldest firm dates for the practice of agriculture, cultivation seems to have begun at about the same time in many other parts of the world. In the New World, fossil pumpkins and gourds suggest some sort of cultivation between 7,000 and 9,000 years ago. In Asia, dates from Spirit Cave (Thailand) imply that cultivation here also began about 9,000 years ago. Cultivation probably began about the same time or only slightly later in northern Africa, but the scanty data available at the present time indicate only that agriculture was not adopted south of the Sahara until much later. Still, with dates from the eastern Mediterranean Sea, southeast Asia, and Central

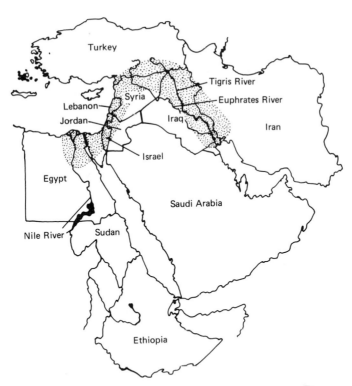

FIGURE 3-3
The Fertile Crescent, a curved area of land extending from the Nile Valley along the Mediterranean to northern Syria and down the Tigris and Euphrates valleys toward the Persian Gulf, is the region in which we have found the earliest records of agriculture.

America so close in time, the question is not really where did agriculture begin, but whether it began in one place and the knowledge of its practice diffused around the world or whether it was independently adopted in several places in the world. While some authors defend a diffusionist view, the majority of botanists, anthropologists, and archaeologists concerned with this subject believe that agriculture was independently adopted in many places around the world.

OF MYTHS AND MAN

How did humans learn to cultivate plants? Both oral and written history are full of speculations of how people first came to know about agriculture. In the Old World, ancient cultures around the Mediterranean favored goddesses as the purveyors of the secrets of agriculture. In ancient Egypt, two of the five children of Geb and Nun (the second couple created by Ra), Isis and Osiris, married one another and the pair took over the ruling of Egypt when Geb abdicated his throne. The couple ruled well and are credited with forbidding humans to eat one another (implying a practice of cannibalism in pre-Egyptian civilizations?) and with teaching people many skills. Osiris taught humans about growing grain and making beer. Isis (Fig. 3-4) developed the practice of embalming, a skill she had to learn, according to legend, when she restored the mutilated body of Osiris.

Perhaps you remember the Greek myth of Demeter's daughter Persephone who was abducted by Pluto, the god of the underworld. Persephone was

FIGURE 3-4
The Egyptian goddess Isis is
shown kneeling with her winged
arms protectively outstretched.
Redrawn from a wall painting in
the Tomb of Seti, ca. 1375 to 1202
B.C., Thebes.

repelled by Pluto and vowed she would not eat until he released her. Demeter roamed the earth searching for Persephone. During these months, she met many mortals, including the family of Metanira who befriended her. Metanira's son Triptolemus was dying, and Demeter, to express her appreciation for the kindness shown to her, tried to make the boy immortal. Metanira, who did not realize that Demeter was a goddess, snatched her son from the ashes where Demeter had placed him as part of his transformation ceremony. While the boy's chance of becoming immortal was thwarted, Demeter nevertheless provided him with the knowledge of plowing and cultivating so that he could teach other men. As she continued her search, Demeter found evidence that her daughter had been drowned, and she cursed the earth so that animals died and plants would not grow. When she finally learned the truth, Demeter pleaded with Zeus to restore her daughter to her. Zeus agreed to do so only if Persephone had eaten nothing during her stay in the underworld. Unfortunately, Persephone had finally given in, accepted a pomegranate offered to her, and eaten the pulp from one seed (Fig. 3-5). As a result, Persephone was forced to marry Pluto, but she had to live with him for only a third of the year, because of the seed she had eaten. This story is part of the Greek agricultural saga that explains the origin of agriculture and provides a rationale for the 4 months of inhospitable weather for growing crops and the 8 month growing season which were linked to the times Persephone was below or above the ground.

In Chinese mythology, Shen Nung, the fictitious second emperor of China (Fig. 3-6) taught humans to use the hoe and the plow. Fire was his symbol, and the Chinese legend also attributes to him the knowledge of how to clear and burn forests.

In Central America, the Aztec god Quetzalcoatl (Fig. 3-7) appears in many myths. In one legend, he disguised himself as a black ant and carried a grain of corn from Tonacatepel to Tamoanchán and gave it to human beings for cultivation. To the south, along the western highlands of the central Andes, the great Inca empire developed its own gods and myths. For the people of this

FIGURE 3-5
Demeter, the Greek goddess of the earth, and her daughter Persephone offer a libation to Triptolemus before he sets out to introduce agriculture to the world. Redrawn from a red-figured vase in the British Museum.

empire, the knowledge of how to sow seed came directly from Father Sun who sent a son and a daughter to teach people to revere him, build houses, and plant seeds.

Needless to say, almost all New World Indian groups as well as various tribes and cultures of Africa developed their various explanations of how people learned about the purposeful cultivation of plants. An interesting exception to the most common central idea of most of these legends, that the knowledge of agriculture was a "gift" from the gods, is found in the book of Genesis (Fig. 3-8). In contrast to most legends, the Judeo-Christian account states that God said to Adam after Eve had eaten the forbidden fruit, "In the sweat of thy face shalt thou eat bread, till thou return to the ground." As punishment, "The Lord God sent him from the Garden of Eden to till the ground from whence he was taken." Obviously, in this case, having to practice agriculture was considered a burden incurred by misbehaving rather than a gift or blessing.

In more recent times, authors have discarded mystical or religious explanations for the origin of agriculture and sometimes replaced them with a "eureka! I've discovered it" theory. In James D. Michener's book *The Source* this idea is perpetuated in the recounting of an innovative woman who hit upon the idea of purposefully planting seeds close to home. This eliminated the need to search for patches of grain. The logical result of this discovery was the subsequent spread of the knowledge to other human populations.

Edgar Anderson, a long-time student of interactions between plants and people, developed a somewhat similar idea in his famous "dump heap hypothesis." He conceived of the knowledge of cultivation as a consequence of the human practice of dumping refuse in a particular spot near the cluster of dwellings (once people began to live in some sort of settlement). Humans might have thrown on it seeds of grains, squashes, and other plants gathered from the wild. Since the wild relatives of many cultivated plants naturally occur in open and often disturbed habitats, the seeds would have readily germinated. Soon, the dump heap would have become a thriving, if untidy, vegetable garden. An enterprising individual would have eventually seen that she or he

FIGURE 3-6
Shen Nung was reputed to be an emperor of China about 3000 B.C. and credited with teaching people how to cultivate plants and about potential uses of different herbs.

FIGURE 3-7
The Aztec god Quetzalcoatl in his manifestation as a plumed serpent, immmortalized in stone at Chichen Itza.

could purposefully create a dump heap and thereby assure a supply of edible plant products close to home. Implicit in this hypothesis is the same idea as that of the eureka school, namely, that something happened which initiated the discovery of how to grow plants.

In contrast to the eureka theories, there is now a considerable body of evidence indicating that nonagricultural people have the knowledge necessary to grow crops. They simply choose not to do so. By extrapolation, many botanists think that before the adoption of agriculture, people also knew how to grow plants (Fig. 3-9). Perhaps, in ancient tribes, curers or shamen even grew a few plants for medicinal uses. We might therefore ask, what tripped the balance? What caused humans seriously to begin the cultivation of crops if they had had the knowledge to do so for some time?

WHY FARM?

Perhaps the most widely debated question about the origin of agriculture, and certainly one of the most interesting, is why humans, at more or less the same time around the world, suddenly adopted cultivation as a way of life. You might say that the answer seems relatively simple. The adoption of agriculture was

FIGURE 3-8
In the biblical story of the origin of agriculture, Adam and Eve were driven from the Garden of Eden as punishment for eating the forbidden apple and from that time onward had to earn their sustenance by practicing agriculture. (Woodcut by Jost Amman; printed by Peter Fabricius in 1587, Frankfort.)

just part of the progress of civilization from primitive cultures to our modern technological society. This kind of philosophy about human determinism was a common notion in the late nineteenth, and even early twentieth, centuries. Primitive humans were conceived of as living a deprived life. The progress of civilization and the improvement of the human lot were viewed as a series of revolutions, each of which led to a better stage in cultural evolution.

However, within the last 20 years, these ideas have completely changed. Recent advances in archaeology, better datings of fossil material, and in-depth studies of modern nonagricultural peoples have indicated that agriculture was not suddenly discovered. Moreover, in contrast to earlier ideas, modern studies have suggested that primitive people are not deprived and that primitive agriculture does not necessarily improve the human lot. Instead, some workers propose that people turned to agriculture only when forced to do so. Obviously, these more recent theories conceive of agricultural practices as neither a technological breakthrough nor as an inevitable part of the human march to higher levels of civilization. In contrast, they view the adoption of agriculture as a logical step for people who know how to grow plants to take when it becomes necessary or the most practical course of action.

If one assumes that humans did turn to agriculture when it best served their ends, the question immediately comes to mind, what ends? There have been two kinds of hypotheses, one emphasizing that agriculture was adopted when it best supplemented a change in life style and the second claiming that some form of stress must have prompted the change in food procurement habits.

One of the leading advocates of the theory that humans began to cultivate crops because it best served their needs was Carl Sauer. Sauer envisioned that people had become sedentary before turning to agriculture and that once settled, collecting wild plant food at greater and greater distances from the settlements became so impractical that they began to raise a permanent supply of plant food nearby. According to Sauer, therefore, the people who developed agriculture were not deprived, but rather were settled and pensive, with enough leisure time to conceive of and implement ways of improving life. In Sauer's thinking, like Darwin's before him, people must have become sedentary before they adopted cultivation.

Many other geographers, botanists, and anthropologists have searched for some event that forced humans to turn to agriculture. One of the most cogently argued of such forces has been climatic change. About 50 years ago, it was proposed that the climate around the eastern Mediterranean Sea became drier following the retreat of the last major Pleistocene glaciation in Europe, forcing people who inhabited the Near East to live crowded together around water sources. Within these oases, the theory states, humans had to practice agriculture in order to obtain enough food for survival. Contrary to this theory, later evidence indicated that the climate around the Mediterranean became wetter, rather than drier, at the end of the Pleistocene, between 10,000 and 11,000 years ago. This increase in humidity would have favored the extension of wild edible grasses into the Near East. An increase in the abundance of wild grains might explain the greater use of them, but it does not explain why humans began to cultivate them.

Another explanation is that humans started out gathering wild grasses,

FIGURE 3-9
The Abyssinian in this frieze (884–859 B.C.) is holding a structure that appears to be a male date inflorescence. Because of this and similar drawings, it has been suggested that humans have understood since very early times that pollination was required to ensure seed set in many crops.

storing any excess that they accumulated. Once they had built up a sizable store of grain, they would have had to stay in one place to use and protect their hoard. Food surpluses could have spurred population increases until population sizes were too large to exist by simply gathering wild grains.

Hypotheses that single out stress resulting from population pressure as the primary factor which caused humans to resort to agriculture have recently received much attention. Data taken to support a view that people turned to agriculture only when hunting and gathering could no longer provide enough food, come from the studies of modern nonagricultural people. The assumption made here is, of course, that these modern hunter-gatherers are similar in many respects to ancient nonagricultural people.

The most famous of these studies have been made by Lee of the !Kung bushmen of Africa (pygmies who live in the Kalahari Desert at the southern end of Africa). These studies indicate that, contrary to prior opinions, groups of hunter-gatherers do not lead lives of hardship or suffer from malnutrition. In fact, the !Kung are selective in the plants that they use and, from a combination of collected plants and a few items of game, receive a very adequate diet. The !Kung recognize about 105 species of edible plants but normally use only the 14 most preferred. In their diet, they usually get an average of 96 grams (g) of protein and 2,355 calories per day. Estimates on modern nutrition lists often indicate that for a 154 lb male, 40 g of protein and about 2,700 calories per day is an adequate diet. In view of the fact that the !Kung are pygmies, their diet would thus seem to be very good. If we use another method of comparison, the amount of work expended to acquire this amount of protein and caloric intake, we again find that the !Kung do not have to work harder than primitive agriculturalists. The !Kung average about 2.5 days per week per person for food acquisition, or between 400 and 1,000 h per person per year. In primitive agricultural societies, 1,000 person-hours per year is the lower limit. Obviously, a switch to agriculture does not provide a life with more leisure time, as had previously been supposed.

When queried about plants, Lee and his associates found that the !Kung were very knowledgeable about many aspects of plant growth and ecology. If asked why they did not grow plants purposefully, the !Kung answer tended to be, why? Why should they indeed. Perhaps you might say the !Kung and such modern nonagricultural people live in especially productive areas, but since many of these groups live in the Kalahari Desert, this seems unlikely. A more pertinent question is, if populations of ancient peoples increased to such levels that they were forced to turn to agriculture, why not the !Kung and similar tribes? As yet, this question has not been satisfactorily answered, although recent evidence indicates that population levels are maintained by extremely frequent nursing which plays a role in preventing ovulation. In fact, the weakest point in the argument of the proponents of population pressure as the stimulus for turning to agriculture is that there is no direct evidence that populations did rise to a critical level at about the time that agriculture was adopted. Circumstantial evidence, such as changes in the composition of the diet, as indicated from the analysis of the contents of ancient refuse heaps, has suggested that there was a tendency toward less "preferred" food—mollusks or snails rather than large vertebrates, at about the time of the emergence of agriculture. Obviously, it will be very difficult ever to know with certainty if populations

10,000 years ago were too dense in certain areas to be able to live by means of a hunting and gathering existence.

As we can see, there is no consensus about why humans became farmers. As Jack Harlan, one of the most respected contemporary researchers working on this question, has pointed out, there is no reason to assume that there is only one reason or that the same reason prevailed in different areas. People and populations under the same conditions may or may not respond in similar ways. Especially in view of the evidence that agriculture arose independently in several parts of the world, it is likely that somewhat different pressures or inducements led people in diverse areas to begin purposefully sowing, harvesting, and resowing plants in relatively permanent settlements.

ORIGINS OF PARTICULAR CROPS

Although there are few definite answers to our questions about the when, where, how, and why of the origin of agriculture, there is better evidence for possible times and sites of the domestication of many specific crops (Fig. 3-10). We seldom think about the fact that the variety of foods that we commonly eat have been drawn from numerous and widely separated parts of the world (Fig. 3-11, 3-12, 3-13), but even typical Italian food such as tomato sauce, chicken cacciatore, and pizza would not have existed before the discovery of America because there were no tomatoes, no red and green peppers, and no zucchini in Italy. Italians obviously had a diet quite different 500 years ago from that they enjoy today. Similarly, what seems more American than apple pie or green peas, both of which were unknown in the New World until Europeans introduced them. It is fascinating to trace the origins of our cultivated plants, some of which have become so modified over the years of domestication that finding their ancestral homes becomes similar to solving a mystery. Furthermore, as we become more aware of the need for maintaining diversity in our cultivated crops (Chapter 19), we realize that such studies are not simply an intriguing pastime, but a necessity.

The first person seriously to undertake a study of plants and their origins was a Swiss, Alphonse de Candolle who, in 1882, published a book titled (in translation), *Origins of Cultivated Plants*. In tracing the original homes of the cultivated species that interested him, de Candolle synthesized information gathered from studies of geography, linguistics, archaeology, and written history, which he combined with his extraordinary knowledge of plants of the world. While his notions about how wild plants were transformed into cultivated crops seem quaint in view of our modern knowledge of genetics and selection, many of his conclusions about places of origin of domesticates have been shown to be correct.

The next comprehensive treatment of crop plant origins came from a Russian about 70 years later. Nikolay Vavilov tried to find centers of origin (gene centers) of cultivated species. As a basis for determining the location of these centers, Vavilov made two assumptions. First, that the areas where wild relatives of cultivated species can now be found are the most likely sites of original domestication. Second, centers should be areas in which one could

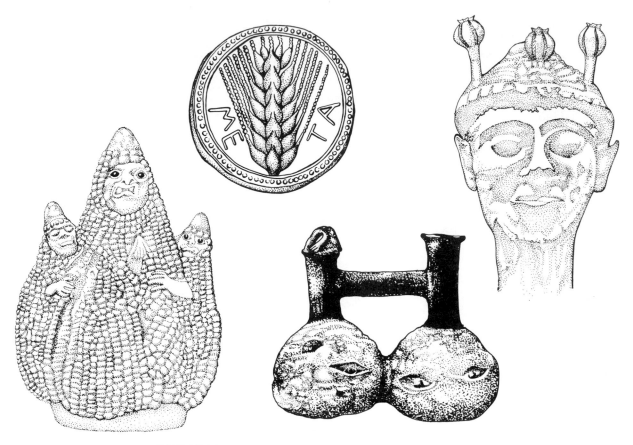

FIGURE 3-10
Archaeological evidence helps in reconstructing the history of particular crop plants.
Barley, an important item of trade in ancient times, as depicted on a Greek coin (above,
left.) Poppy capsules bedecking the head of a stone sculpture from Crete, ca. 1400 B.C.
(above, right). A vase portraying three corn gods left by the Mochica people, who grew
corn under irrigation from the 1st through the 3rd centuries in Peru (left, below) and a
Chimu potato pot redrawn from an artifact in the British Museum (right, below.)

find great amounts of natural variation in crops that are grown. From his
research, Vavilov first delimited six, soon expanded to eight, major centers of
domestication in the world. Later, he and his collaborators increased the list
still further, but the eight centers are the ones which are generally used. These
centers are the Chinese center, the Indochina-Indonesian center, the Mid Eastern
center, the Indian center, the Mediterranean center, the Abyssinian center, the
Mexico–Central American center and the central Andean center. In the original
formulation of six centers, the first two were considered as one, the southeast
Asian center, and the second two as the southwest Asian center. Table 3-1 lists
some of the crops that Vavilov proposed had originated in these centers.

In the light of our modern knowledge of genetics, and data from archaeological findings, many of Vavilov's conclusions seem somewhat naïve. Current
reconstructions of the process of domestication of a crop involve the compilation
of information from cytology, ecology, plant systematics, archaeology, and
ethnology. Careful assessment of the probable areas of origin of many of our

FIGURE 3-11
Plants of Mediterranean or Near East origin. (1) sage, (2) thyme, (3) oregano, (4) parsley, (5) mint, (6) wheat, (7) barley, (8) oats, (9) fig, (10) lentils, (11) pomegranate, (12) olives, (13) leeks, (14) rosemary, (15) lettuce, (16) artichoke.

important crops has now indicated that domestication occurred over a much more diffuse region than Vavilov supposed. Particularly in Africa and South America, there do not seem to be restricted regions in which notable preponderances of crops were first domesticated.

FIGURE 3-12
Plants of New World origin. (1) Jerusalem artichoke, (2) pinto beans, (3) tomatoes, (4) pepper, (5) vanilla, (6) papaya, (7) avocado, (8) wild rice, (9) potatoes, (10) green beans, (11) peanuts, (12) corn (13) jicama, (14) pineapple.

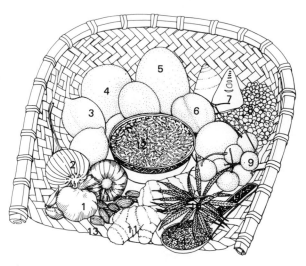

FIGURE 3-13
Plants of Asian origin. (1) garlic, (2) onions, (3) lemons, (4) oranges, (5) mangoes, (6)
peach, (7) bamboo shoots, (8) soybeans, (9) cherries, (10) hemp, (11) ginger, (12) rice,
(13) almonds, (14) tea.

TABLE 3-1

Vavilov's Eight Centers of Origin and Some Crops He Proposed to Have Been Domesticated in Each

CENTER	COMMON NAME	SCIENTIFIC NAME	FAMILY
Chinese	Buckwheat	*Fagopyrum esculentum*	Polygonaceae
	Hemp	*Cannabis sativa*	Cannabaceae
	Mulberry	*Morus alba*	Moraceae
	Orange	*Citrus sinensis*	Rutaceae
	Peach	*Prunus persica*	Rosaceae
	Poppy	*Papaver somniferum*	Papaveraceae
	Soybean	*Glycine max*	Fabaceae
	Tea	*Camellia sinensis*	Theaceae
	Tung	*Aleurites spp.*	Euphorbiaceae
Indian, including Indo-Malayan	Banana	*Musa spp.*	Musaceae
	Breadfruit	*Artocarpus altilis*	Moraceae
	Chick-pea	*Cicer arietinum*	Fabaceae
	Citron	*Citrus medica*	Rutaceae
	Coconut	*Cocos nucifera*	Arecaceae
	Mango	*Mangifera indica*	Anacardiaceae
	Pepper, black	*Piper nigrum*	Piperaceae
	Rice	*Oryza sativa*	Poaceae
	Safflower	*Carthamus tinctorius*	Asteraceae
	Sesame	*Sesamum indicum*	Pedaliaceae
	Sugar cane	*Saccharum officinarum*	Poaceae
	Yam	*Dioscorea alata*	Dioscoreaceae

CENTER	COMMON NAME	SCIENTIFIC NAME	FAMILY
Central Asian (includes NW India)	Apple	*Malus pumila*	Rosaceae
	Carrot	*Daucus carota*	Apiaceae
	Grape	*Vitis vinifera*	Vitaceae
	Onion	*Allium cepa*	Liliaceae
	Pea	*Pisum sativum*	Fabaceae
	Pear	*Pyrus communis*	Rosaceae
	Radish	*Raphanus sativus*	Brassicaceae
	Spinach	*Spinacia oleracea*	Polygonaceae
Near Eastern	Alfalfa	*Medicago sativa*	Fabaceae
	Fig	*Ficus carica*	Moraceae
	Flax	*Linum usitatissimum*	Linaceae
	Hazelnut	*Corylus avellana*	Betulaceae
	Lentil	*Lens culinaris*	Fabaceae
	Melon	*Cucumis melo*	Cucurbitaceae
	Oats	*Avena sativa*	Poaceae
	Quince	*Cydonia oblonga*	Rosaceae
	Rye	*Secale cereale*	Poaceae
	Wheat, einkorn emmer	*Triticum monococcum* *T. turgidum*	Poaceae
Mediterranean	Asparagus	*Asparagus officinalis*	Liliaceae
	Beet	*Beta vulgaris*	Chenopodiaceae
	Cabbage	*Brassica oleracea*	Brassicaceae
	Carob	*Ceratonia siliqua*	Fabaceae
	Lavender	*Lavendula angustifolia*	Lamiaceae
	Leek	*Allium ampeloprasum*	Liliaceae
	Lettuce	*Lactuca sativa*	Asteraceae
	Olive	*Olea europea*	Oleaceae
Abyssinian	Barley	*Hordeum vulgare*	Poaceae
	Castor bean	*Ricinus communis*	Euphorbiaceae
	Coffee	*Coffea arabica*	Rubiaceae
	Millet, African pearl	*Eleusine coracana* *Pennisetum americanum*	Poaceae Poaceae
	Okra	*Abelmoschus esculentus*	Malvaceae
	Sorghum	*Sorghum bicolor*	Poaceae
Mexican-Central American	Avocado	*Persea americana*	Lauraceae
	Bean, common	*Phaseolus vulgaris*	Fabaceae
	Cacao	*Theobroma cacao*	Sterculiaceae
	Corn	*Zea mays*	Poaceae
	Cotton	*Gossypium hirsutum*	Malvaceae
	Potato, sweet	*Ipomoea batatas*	Convolvulaceae
	Pepper, red	*Capsicum spp.*	Solanaceae
	Squash, winter	*Cucurbita moschata*	Cucurbitaceae
Central Andean (including most of South America	Manioc	*Manihot esculentum*	Euphorbiaceae
	Peanut	*Arachis hypogaea*	Fabaceae
	Pineapple	*Ananas spp.*	Bromeliaceae
	Potato, white	*Solanum tuberosum*	Solanaceae
	Pumpkin	*Cucurbita maxima*	Cucurbitaceae
	Rubber	*Hevea brasiliensis*	Euphorbiaceae
	Tobacco	*Nicotiana tabacum*	Solanaceae
	Tomato	*Lycopersicon esculentum*	Solanaceae

Note: While many of Vavilov's determinations about centers of domestication have withstood the test of time, recent evidence has suggested that several of the crops listed in this table arose and/or were domesticated in regions different from those he proposed.
Source: Selected species with names updated from Vavilov, N. I. 1951. The Origin, Variation, Immunity, and Breeding of Cultivated Plants. Chronica Botanica 13. New York, Ronald Press.

Jack Harlan has led in the synthesis of modern data and developed a concept of centers and noncenters. Most simply stated, noncenters are areas so large that designating them as centers is meaningless. A case in point is Africa. Although a variety of crops were first domesticated in Africa, they were brought into cultivation in different parts of the continent. Designating the entire African continent as a center does not convey any information beyond the fact that a plant was probably native to Africa and first used by humans there. Designating Africa as a noncenter is a way of indicating the diffuse nature of the places in which various crops were domesticated. Harlan considered the Near East, north China, and Mesoamerica as centers of agricultural origins, but in his interpretation, these centers are areas in which agriculture independently arose. Each was, in his view, associated with a noncenter over which domestication of individual crops occurred. Nevertheless, while Harlan's views are probably very realistic considering the evidence for the independent origins of agriculture and much recent data about the areas of first cultivation of many important crop plants, workers still refer to Vavilov's centers when they discuss economically important plants. In this text, however, we will try to assess each plant independently rather than trying to force it into a system of centers of domestication.

ADDITIONAL READING

Anderson, E. 1971. *Plants, Man and Life*. Berkeley, University of California Press. A very readable account of Anderson's ideas of human interactions with plants, including the first attempts at agriculture.

Candolle, A. de. 1886. (Reprint of second edition, English translation). *Origin of Cultivated Plants*. New York, Hafner.

Cohen, M. N. 1977. *The Food Crisis in Prehistory*. New Haven, Connecticut, Yale University Press. A very thorough essay which discusses the view that population pressure caused the adoption of an agricultural way of life.

Harlan, J. R. 1975. *Crops and Man*. American Society of Agronomy. Madison, Wisconsin, Crop Science Society of America. A personal view of people's early dealings with plants and changes induced in many of our major crop species.

Heiser, C. B., Jr. 1981. *Seeds to Civilization. The Story of Food*. 3d ed. San Francisco, Freeman.

Heiser, C. B., Jr. 1979. Origins of some cultivated New World Plants. *Annual Review of Ecology and Systematics* 10:309–326.

Pickersgill, B. 1977. Taxonomy and the origin and evolution of cultivated plants in the New World. *Nature* 268:591–595.

Reed, C. A. (ed.). 1977. *Origins of Agriculture*. Paris, Mouton. This impressive collection of articles contains many interesting papers that discuss almost every theory proposed about the origin of agriculture.

Sauer, C. O. 1952. *Agricultural Origins and Dispersals*. New York, The American Geographical Society.

Vavilov, N. I. 1951. *The Origin, Variation, Immunity, and Breeding of Cultivated Plants*. Waltham, Massachusetts, Chronica Botanica 13. New York, Ronald.

Wendorf, F., R. Schield, N. El Hadidi, A. E. Close, M. Kobusiewicz, H. Wieckowska, B. Issawi, and H. Haas. 1979. Use of barley in the Egyptian Late Paleolithic. *Science* 205:1341–1347.

Chapter 4
Fruits and Nuts
of Temperate Regions

*W*hen it comes to fruits, botanical terminology bears little relationship to everyday speech. Nutrition charts group edible plants into categories of fruits, vegetables, grains, and nuts. A typical American grocery shopper would define fruits as sweet plant products generally eaten as dessert or as a first course. The same shopper would probably lump together as vegetables those plant parts eaten as entrees or side dishes and typically seasoned with salt and pepper. Yet, to a botanist, the word vegetable refers to plant matter in general, whereas fruits are specific plant organs. Botanically, green beans and cucumbers are fruits. Tomatoes and eggplants are both fruits classified as berries. Strawberries and blackberries are likewise fruits, but botanists do not consider them berries.

Because all of the foods we commonly call fruits and nuts, and a large proportion of the starches and everyday vegetables we eat, are fruits from a botanical point of view, we discuss fruits in detail in this and the next three chapters. Here, we define fruits, outline their classification, and explore dessert fruits and nuts from temperate regions. In the next chapter we cover dessert fruits and nuts of warm regions. In Chapters 6 and 7 we discuss two particularly important groups of fruits, legumes and grains.

What, then, is a fruit? How are they formed and what kinds are there? A fruit is correctly defined as a mature, or ripened, ovary along with its contents and any adhering accessory structures. A fruit usually includes seeds resulting from the maturation of the ovules following successful fertilization of the egg and the polar nuclei (Chapter 1). However, both seed and fruit maturation can occur without fertilization. Seed production without fertilization is a kind of parthenogenesis known as *agamospermy*. A seed that is produced agamospermously contains genetic and cytoplasmic material derived solely from maternal tissue. While pollen sometimes triggers the development of such seeds, it contributes nothing to them. A fruit that matures without seed formation is called *parthenocarpic*. In some cases, agriculturalists have selected for seedless fruits by propagating parthenocarpic mutants that spontaneously mature fruits without fertilization.

A fruit is classified by the number of ovaries which produced it (one or several), by the position of the ovary (superior or inferior), by whether it is fleshy or dry at maturity, and by the way it releases its seeds (dehiscences). Table 4-1 provides a basic outline of the common fruit types we discuss throughout this book and Table 4-2 is a listing of the plants discussed in this chapter.

Anatomically, an ovary can be simple or compound. In the first case, it has one carpel, in the latter, two or more carpels. The notion of simple or compound and the concept of the carpel comes from an hypothesis called the *foliar theory* of the carpel. According to this theory, the ovary was derived from a leaflike structure on which ovules were borne (Fig. 4-1). During the course of evolution, this leafy structure was supposed to have folded, producing the closed ovary characteristic of the angiosperms. Each simple, folded, leaflike unit is considered to be a single carpel. A simple ovary is derived from one such structure and therefore has one carpel. A compound ovary is one that appears (usually from anatomical study) to have been produced by the fusion of two or more of these leafy structures (Fig. 4-2). The number of carpels in a compound ovary is therefore equal to the number of ancestral leaflike structures fused into the pistil. While this explanation of the carpel's origin is not universally accepted,

TABLE 4-1

Fruit Types

TYPE	EXPLANATION	EXAMPLE
SIMPLE	Fruit from a single ovary in a flower; ovary can be simple or compound.	
Dry fruits	Pericarp dry when mature.	
Indehiscent	Pericarp does not split at maturity.	
Achene	1-seeded, pericarp thin, seed coat separate from ovary wall.	Sunflower, strawberry
Grain (caryopsis)	1-seeded, pericarp fused with ovary wall.	Wheat, rice, corn
Nut	1-seeded fruit with hard or bony pericarp	Walnut, hazlenut
Dehiscent	Pericarp splits open to release seeds.	
Follicle	Podlike fruit that splits along the ventral side.	Lilac, milkweed
Legume	Pods that split along 2 opposite edges.	Peas, beans, soybeans
Silique	Pericarp splits along juncture of 2 carpels.	Mustard
Capsule	Pericarp opens at pores or along slits.	Opium poppy
Schizocarp	Fruit splits into 2 to several achene-like segments with free seeds surrounded by fruit wall.	Dill "seed"
Fleshy fruits	Part or all of pericarp fleshy at maturity.	
Drupe	1- to 2-seeded, mesocarp fleshy, endocarp stony.	Peach, plum
Berry	1- to several-seeded, mesocarp fleshy, endocarp not stony.	Grape, tomato
Hesperidium	A special kind of berry with a leathery rind and oil glands dotting the fruit surface.	Citrus fruits
ACCESSORY FRUITS	Berrylike, but produced from an inferior ovary and thus containing accessory structures.	
"Berries"	Superficially like berries	Coffee, bananas
Pepo	Combined accessory tissues and ovary wall forming a tough rind.	Squashes
AGGREGATE FRUITS	Fruits formed from several separate ovaries within a single flower, each segment like a berry.	Raspberries
MULTIPLE FRUITS	Formed by the fusion or collective shedding of fruits of numerous, independent flowers. Segments similar to berries or drupes.	Figs, pineapples

virtually all plant descriptions use the number of carpels of an ovary as one of the diagnostic characters of a species.

A fruit derived from a single simple or compound ovary is considered to be a *simple fruit.* A simple fruit normally has one or more seeds surrounded by a pericarp, or fruit wall, made up of three layers, the endocarp, mesocarp, and exocarp. The last three terms refer to the inner (endo-), middle (meso-), and

TABLE 4-2

Plants Discussed in Chapter 4

CROP	SPECIES	FAMILY	CHROMOSOME NUMBER
Almond	*Prunus amygdalus*	Rosaceae	$2n = 16$ diploid
Apple	*Malus pumila*	Rosaceae	$2n = 34$ or 51 di- & triploids
Apricots	*Prunus armeniaca*	Rosaceae	$2n = 16$ diploid
Blackberries	*Rubus* spp.	Rosaceae	$2n = 28, 42, 56$ di- to octoploids
Blueberries:	*Vaccinium*	Ericaceae	
High-bush	*V. corymbosum*		$2n = 48$ tetraploid
Low-bush	*V. angustifolium*		$2n = 24$ diploid
Cherries:			
Sour	*Prunus cerasus*	Rosaceae	$2n = 32$ tetraploid
Sweet	*P. avium*		$2n = 16$ diploid
Chestnut:	*Castanea*	Fagaceae	
American	*C. dentata*		$2n = 24$ diploid
European	*C. sativa*		$2n = 24$ diploid
Japanese	*C. crenata*		$2n = 24$ diploid
Cranberries	*Vaccinium oxycoccos*	Ericaceae	$2n = 48$ tetraploid
	V. macrocarpon		$2n = 24$ diploid
Currant:	*Ribes*	Grossulariaceae	
European black	*R. nigrum*		$2n = 16$ diploid
Red	*R. sativum*		$2n = 16$ diploid
Gooseberry	*Ribes uva-crispa*	Grossulariaceae	$2n = 16$ diploid
Grapes:	*Vitis*	Vitaceae	
Wine, table	*V. vinifera*		$2n = 38$ diploid
Fox	*V. lambrusca*		$2n = 38$ diploid
Hazelnut	*Corylus avellana*	Corylaceae	$2n = 28$ diploid
Loganberry	*Rubus vitifolius*	Rosaceae	$2n = 42$ hexaploid
Olive	*Olea europaea*	Oleaceae	$2n = 46$ diploid
Peach	*Prunus persica*	Rosaceae	$2n = 16$ diploid
Pear	*Pyrus communis*	Rosaceae	$2n = 34$ or 51 di- & triploids
Pecan	*Carya illinoinensis*	Juglandaceae	$2n = 32$ diploid
Pistachio	*Pistacia vera*	Anacardiaceae	$2n = 30$ diploid
Plums	*Prunus domestica*	Rosaceae	$2n = 48$ hexaploid
Quince	*Cydonia oblonga*	Rosaceae	$2n = 34$ diploid
Raspberries:	*Rubus*	Rosaceae	
Black	*R. occidentalis*		$2n = 14$ diploid
Red	*R. idaeus*		$2n = 14$ diploid
Strawberry:	*Fragaria*	Rosaceae	
Chilean	*F. chiloensis*		$2n = 56$ octoploid
Virginia	*F. virginiana*		$2n = 56$ octoploid
Walnut	*Juglans*	Juglandaceae	
Black	*J. nigra*		$2n = 32$ diploid
English	*J. regia*		$2n = 32$ diploid

outer (exo-) layers of the ovary wall (Fig. 4-3). The pericarp, with surrounding tissues (if present), can develop in various ways, giving rise to the assortment of fruit types that characterize different species of angiosperms (Table 4-1). For example, the pericarps can remain thin and become dry during fruit maturation. The resultant dry fruits can be shed as entire units, or they can split open to

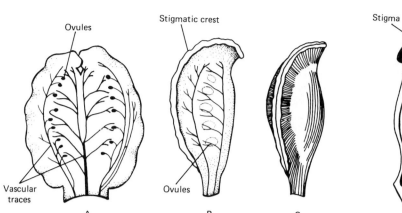

FIGURE 4-1
It is postulated that the carpel was derived from the folding of a leaflike structure on which the ovules were borne (A). The modern carpel of *Drimys winteri* is believed to have resembled this early folded leaflike structure. (B) The vascular traces inside the *Drimys* carpel, (C) an exterior view, (D) cut away view.

release their seeds. In the first case, the fruits are considered to be dry and *indehiscent* (nonsplitting), and in the second case, dry and *dehiscent* (splitting). Ash or elm fruits are indehiscent dry fruits, whereas geranium fruits are dehiscent dry fruits (see Fig. 1-26). Dry fruits, or their seeds, are generally dispersed by wind, water, or gravity. Grains, or caryopses, are a particularly important kind of dry indehiscent fruit in which the seed coat is fused to the ovary wall (see Fig. 6-4). Legumes are dry, dehiscent fruits that split along the two opposite margins of the pod (see Fig. 7-1).

In many cases, the pericarp becomes increasingly fleshy as the fruit matures. The fleshy portion can result from an enlargement of any one (or all) of the pericarp layers or even by a swelling of accessory tissues. If the fleshy portion is derived from the pericarp and the fruit is produced by a single superior ovary within a flower, a *berry* results. A superior ovary is one borne above the insertion of the floral parts. Berries can contain one or several seeds, but the seeds are never enclosed by a hard endocarp. Berrylike fruits with the seed(s) enclosed in a hard, stony endocarp are *drupes*. Citrus fruits are special kinds of berries called *hesperidiums* because they have leathery rinds dotted with packets of oil and specialized fleshy hairs projecting from the inner wall of the endocarp.

If the ovary is positioned below the attachment of the floral parts (sepals, petals, etc.), it is called *inferior*. The walls of inferior ovaries are fused with the bases of the floral parts surrounding them. The fruit wall therefore consists not only of the ovary wall, but also of the extra, or accessory, perianth parts. Fruits produced by inferior ovaries are accessory fruits, but they are usually called by the names of the simple fruits they resemble. Coffee berries, blueberries, and bananas are accessory fruits, but they are generally simply referred to as berries. One group of accessory fruits, those of the squash family, have been given the special name of *pepos* because of their economic importance and distinctive hard rind. In some cases, the perianth parts are fused at the base to form a floral cup around the ovary. This cup can swell as the fruit matures and provide the

FIGURE 4-2
The fusion of several carpels is presumed to have led to a compound ovary. The pea pod (above) is an example of a simple ovary that could be derived from the folding of a single leaflike structure. The fusion of several such structures (or carpels) presumably led to a compound ovary (below.)

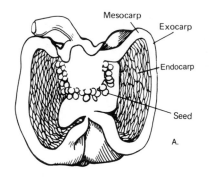

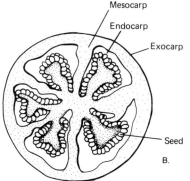

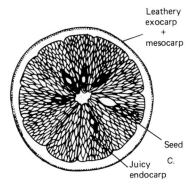

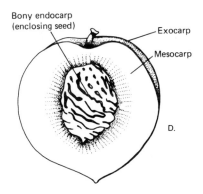

FIGURE 4-3
The exocarp, mesocarp, and endocarp are represented in different ways. (A) Slightly expanded mesocarp gives rise to the fleshy rind of the pepper. The central core is made of the placental tissues on which the seeds are borne. (B) A tomato also has a relatively fleshy mesocarp and an elaboration of the tissues on which the seeds are borne. (C) The fleshy part of an orange is made up of swollen hairs on the inside of the endocarp. (D) A greatly expanded mesocarp produces the sweet flesh of a peach.

majority of the fruit pulp. Fruits of the rose family formed in this way are called *pomes*.

If there are numerous, simple, superior ovaries borne within a single flower, each matures to form part of an *aggregate fruit*. Strawberries, raspberries, and blackberries are therefore aggregate fruits and not true berries.

Finally, a few, mostly tropical, fruits are formed by the tight packing of fleshy fruits from many separate flowers. Fruits such as pineapples and figs constructed in this way are known as *multiple fruits*. In a sense, multiple fruits are false fruits since each is actually composed of many fruits. Each segment of a pineapple, or each tiny achene inside a fig is a complete fruit.

Fleshy fruits have been evolved for the attraction of animals who will eat them, but spit out, drop, or pass the seeds unharmed. Since the succulent fruit tissue is not a source of nutrition for the embryo or for the later development of the seedling, a plant puts as few resources as possible into fruit pulp. Since sugars are the immediate products of photosynthesis, they are less "expensive" for a plant than more complex compounds that require further synthesis (e.g., oils) or that contain limiting nutrients such as nitrogen (e.g., proteins or alkaloids). As a result, fleshy fruit parts tend to consist primarily of flavored sugar solutions held in a cellulose matrix. With the exception of a few fruits, such as the avocado and the olive, fleshy fruits are watery, low in calories, and good sources of water-soluble vitamins.

In contrast to the pulp of fleshy fruits, seeds contain nutrients that are used by the embryo for emergence and establishment. Seeds therefore possess appreciable quantities of high-energy material such as fats, oils, and starches. In nature, any animal that destructively consumes seeds is a predator that reduces a plant's reproductive potential. Consequently, many plants have seeds that are protected by hard shells, seed coats, or toxic compounds. Humans have learned to overcome the various devices for seed protection, thereby taking advantage of the nutrition packed into seeds. Tables 4-3 and 4-4 highlight the nutritive differences between fruit pulps and seeds by comparing the caloric values of fleshy fruit crops with nuts, an important group of foods derived from seeds.

TABLE 4-3

Nutritional Composition of Some Temperate Fruits Based on 100 g Edible Portion

FRUIT	WATER, %	CALORIES	PROTEIN, g	FAT, g	CARBOHYDRATE, g	VITAMINS A, units	VITAMINS C, mg
Apple	84	58	0.2	0.6	15	90	4
Apricot	85	51	1.0	0.2	13	2,700	10
Blackberry	85	58	1.2	0.9	13	200	21
Blueberry	83	62	0.7	0.5	15	100	14
Cherry, red	84	58	1.2	0.3	14	1,000	10
Cranberry	88	46	0.4	0.7	11	40	11
Grape, table	82	69	1.3	1.0	16	100	4
Olive, green	78	116	1.4	12.7	1	300	—
Peach	89	38	0.6	0.1	10	1,330	7
Pear	83	61	0.7	0.4	15	20	4
Plum	81	66	0.5	Trace	18	300	—
Quince	84	57	0.4	0.1	15	40	15
Raspberry, red	84	57	1.2	0.5	14	130	25
Strawberry	90	37	0.7	0.5	8	60	59

Source: Data from *Composition of Foods.* USDA handbook no. 8. 1975 printing. Vitamin A is given in international units.

TABLE 4-4

Nutritional Composition of Some Temperate Nuts Based on 100 g Edible Portion

NUT	WATER, %	CALORIES	PROTEIN, g	FAT, g	CARBOHYDRATE, g	VITAMINS A, units	VITAMINS C, mg
Almond	5	598	19	54	20	0	Trace
Chestnut	53	194	3	2	42	—	—
Hazelnut	6	634	13	62	17	—	Trace
Pecan	3	687	9	71	15	130	2
Pistachio	5	594	13	60	21	230	0
Walnut	4	651	15	64	16	30	2

Source: Data from *Composition of Foods.* USDA handbook no. 8. 1975 printing. Vitamin A is given in international units.

APPLES AND THEIR RELATIVES

Apples are the most important temperate fruit tree crop in the world. Apples and their relatives, pears, peaches, plums, cherries, strawberries, and raspberries, all come from the same family, the Rosaceae, or rose family. These fruits differ from one another, however, in their structures. Because of these differences, apples and pears are placed in a separate subfamily from plums and peaches. Strawberries and raspberries are put in a third group.

The Rosaceae is divided by most botanists into four subfamilies, the Spiraeoideae, Pyroideae, Prunoideae, and Rosoideae. The last three are all important sources of edible fruits (Fig. 4-4). The Pyroideae (also called the Pomoideae) produces pome fruits. The Prunoideae contains species which

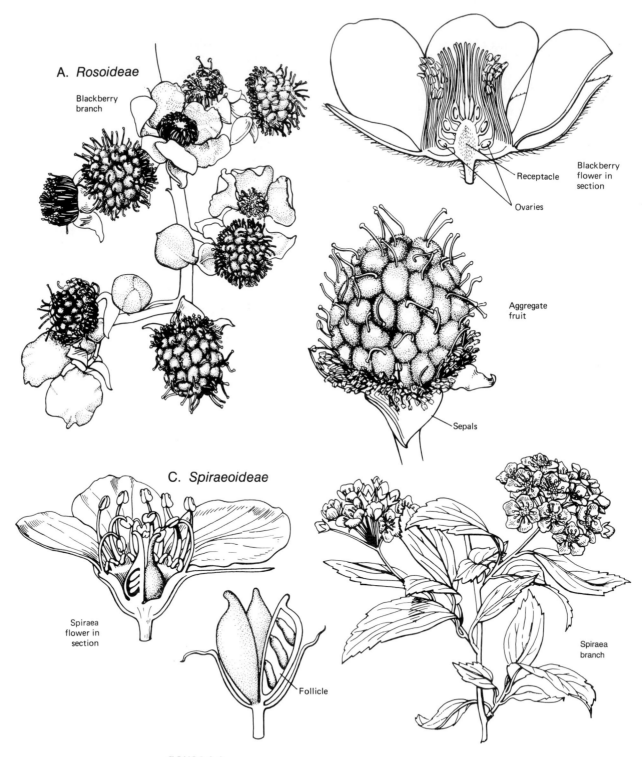

A. *Rosoideae*

Blackberry branch

Receptacle

Ovaries

Blackberry flower in section

Aggregate fruit

Sepals

C. *Spiraeoideae*

Spiraea flower in section

Follicle

Spiraea branch

FIGURE 4-4
The Rosaceae is divided into several subfamilies on the basis of flower and fruit characters. These include (A) the Rosoideae, which has fruits formed primarily from

B. *Pyroideae*

Pear flower
in section

Remnants
of
sepals
+
petals

Floral
cup

Vascular
tissue

Pome

Flowering
pear branch

D. *Prunoideae*

Peach
flower
in
section

Drupe

Peach branch

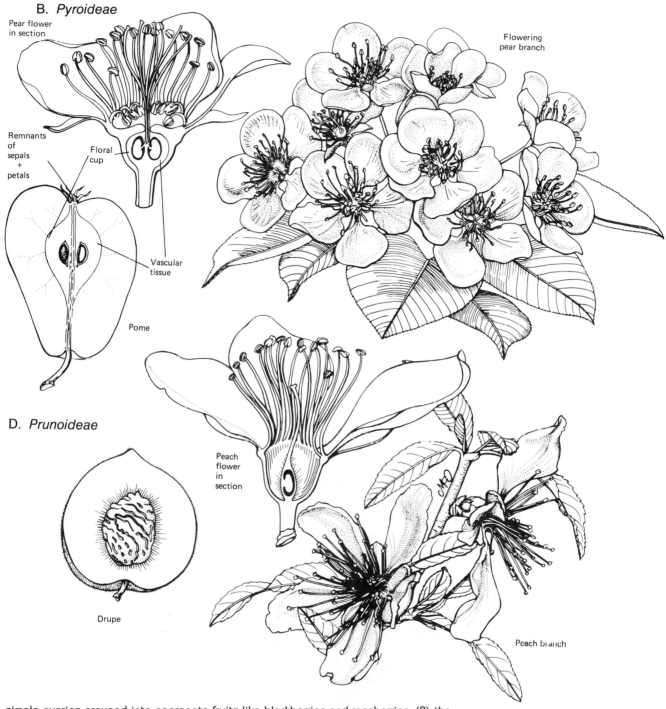

simple ovaries grouped into aggregate fruits like blackberries and raspberries; (B) the
Pyroideae, with fruits like the apple and pear that develop in part from a floral cup
surrounding the compound ovary; (C) the Spiraeoideae, with capsulate or follicular fruits
like those of the ornamental *Spiraea;* and (D) the Prunoideae, with fruits formed from
simple, superior ovaries developing into drupes like cherries, peaches, and plums.
(*Spiraea* and Blackberry cross section, after Baillon, the rest was drawn from nature).

produce drupes as fruits, and plants in the Rosoideae produce aggregate fruits. Figure 4-4 illustrates the structural differences in the fruits of these three subfamilies.

Pyroideae

The Pyroideae contains two important genera of fruit plants, *Malus*, apple, and *Pyrus*, pear, as well as the less well known quince, *Cydonia*. Apples (Fig. 4-5) account for over 50 percent of the world's deciduous fruit tree production (Table 4-5). Cultivated apples are usually lumped under the species *Malus pumila*. It is believed that the genus originated in western Asia but spread to North America long before any human arrived on the continent. Nevertheless, despite our expression "as American as apple pie," our cultivated, edible apples are all natives of the Old World. Apples have been eaten in Eurasia since prehistoric times, and selection must have been practiced very early since the Romans are known to have had at least 22 apple varieties. By 1670, enough varieties had been selected to allow the Grand Duke of Tuscany to give a banquet during which he served 56 varieties of apples. Today, more than 6,500 horticultural varieties have been registered. Unfortunately, in modern markets we seem to see fewer varieties than our grandfathers. Your parents or grandparents can probably list many kinds of apples—Cortlands, McIntosh, Granny Smiths, Jonathans, Winesaps, Gravensteins, and Pippins (among others)—that they used to see regularly. Now, a few varieties, especially the Red and Yellow Delicious, developed in the United States, dominate in world trade, although two or three others can commonly be found.

Apples normally have an ovary with five carpels, but one or more often contain no seeds. The fleshy part of the apple that we eat is the floral cup which expands as the fruit develops following fertilization (see Fig. 4-4B). The wall of the ovary is the thin, papery layer that surrounds the core.

FIGURE 4-5
An apple harvest in Japan.
(Courtesy of the Japan National
Tourist Association.)

TABLE 4-5

World Production of Temperate Fruits and Nuts

| CROP | TOP FIVE COUNTRIES | | TOP FIVE CONTINENTS | | WORLD, 1,000 t |
	COUNTRY	METRIC TONS (t)	CONTINENT	METRIC TONS (t)	
Almonds	United States	178	Europe	410	888
	Italy	175	Asia	237	
	Spain	160	North & Central America	178	
	China	90*	Africa	53	
	Iran	52	Oceania	1	
Apples	U.S.S.R.	7,200*	Europe	13,968	36,799
	United States	3,718	Asia	8,679	
	China	2,740*	North & Central America	4,497	
	Italy	1,990	South America	1,386	
	France	1,950	Oceania	543	
Apricots	Italy	180	Europe	776	1,746
	Turkey	171*	Asia	456	
	U.S.S.R.	170*	Africa	169	
	Spain	161	North & Central America	98	
	Greece	180	Oceania	40	
Chestnuts	China	230*	Asia	396	567
	Italy	70	Europe	153	
	Turkey	60*	South America	12	
	Japan	59*	—		
	Korea, Republic of	40*	—		
Currants	Poland	164	Europe	485	556
	Germany, Federal Republic of	129	Oceania	2	
	U.S.S.R.	70*	—		
	United Kingdom	35	—		
	Austria	31	—		
Grapes	Italy	12,255	Europe	36,832	65,167
	France	8,550	Asia	7,434	
	U.S.S.R.	7,200*	North & Central America	5,354	
	Spain	5,046	South America	5,339	
	United States	4,791	Africa	2,216	
Hazelnuts	United Arab Emirates	370*	Asia	3777	551
	Italy	118	Europe	161	
	Spain	31	North & Central America	7	
	Greece	8	Africa	2	
	Iran	7*	—		
Peaches and nectarines	Italy	1,700	Europe	3,655	7,406
	United States	981	Asia	1,249	
	Spain	492	North & Central America	1,206	
	Greece	480	South America	577	
	France	474	Africa	240	

TABLE 4-5 (continued)

World Production of Temperate Fruits and Nuts

CROP	COUNTRY	METRIC TONS (t)	CONTINENT	METRIC TONS (t)	WORLD, 1,000 t
	TOP FIVE COUNTRIES		TOP FIVE CONTINENTS		
Pears	China	2,098*	Europe	4,143	9,523
	Italy	1,235	Asia	3,368	
	United States	702	North & Central America	782	
	Spain	559	South America	296	
	U.S.S.R.	550*	Africa	2354	
Plums	Yugoslavia	1,038*	Europe	3,411	6,187
	U.S.S.R.	950*	Asia	964	
	Romania	665	North & Central America	693	
	United States	606	South America	78	
	China	512*	Africa	67	
Raspberries	U.S.S.R.	100*	Europe	139	267
	Poland	29	North & Central America	26	
	Germany, Federal Republic of	24	Oceania	2	
	United Kingdom	24	—		
	Yugoslavia	23*	—		
Strawberries	United States	405	Europe	923	1,861
	Japan	196*	North & Central America	516	
	Poland	191	Asia	297	
	Italy	150	South America	15	
	U.S.S.R.	100*	Oceania	7	
Walnuts	United States	172	Asia	269	784
	Turkey	121*	Europe	267	
	China	120*	North & Central America	178	
	U.S.S.R.	57*	South America	7	
	Italy	48	Africa	6	

*Indicates an estimated or unofficial figure.
Note: Only crops of major importance in international trade are included. U.S.S.R. is treated as a country only. Its figures are not added into those of either Europe or Asia.
Source: Data from *FAO Yearbook for 1983*, vol. 37. 1984. FAO, Rome.

Most of the apples grown today are triploids. Because the initial production of these triploid hybrids is time-consuming and costly, superior varieties that have been developed are perpetuated by grafting. Grafting also permits the use of a variety of rootstocks which have different adaptations for hardiness or pest resistance and which are therefore better suited to a variety of localities.

The apple has been involved in folklore since ancient times. In Greek mythology, one of Hercules's tasks was to obtain the golden apples of Hesperides which were well guarded because they were supposed to impart immortality. In the book of Genesis, Eve was tempted by a serpent and took a bite of a forbidden fruit which most people have assumed was an apple. The generic name of the apple, *Malus* after the Latin "malus" meaning bad, was chosen because of the presumed role of the fruit in the downfall of man. There is some dispute, however, about the identity of the golden apples of mythology or the fruit in the Garden of Eden because some authorities maintain that apples did not grow in the Near East or around the Mediterranean Sea when these episodes were supposed to have occurred. According to this view, a late climatic change allowed the movement of wild apples into this region. To these workers, apricots are a more likely candidate than apples to be the fruit mentioned in these stories. Apricots are called golden apples today in the Near East and were the most abundant fruit in Palestine (next to figs) in early historical times. Still, most authorities believe apples to be native to the Middle East and see no reason to doubt that they figured prominently in early legends.

There are also more recent tales involving apples. William Tell, a Swiss hero, is said to have been forced to shoot an apple from his son's head because he refused to pay homage to Austrian conquerors. It is rumored that Newton discovered gravity when an apple fell on his head. In America, the legend of Johnny Appleseed, based on the life of John Chapman (1774–1845), recounts the migration of the professional arborist from Massachusetts as he trekked westward planting his beloved fruit trees.

While less popular than apples in the United States, pears are the second largest deciduous tree fruit crop on a worldwide basis (Table 4-5). There are many species of pears which appear to hybridize freely and to have been involved in the production of the most widely cultivated species, *Pyrus communis*. Opinions are divided as to whether the pear had a center of origin in eastern Europe or whether, like the apple, western China was its primary place of origin. Both areas are regions of high diversity of *Pyrus* species. Unlike apples, there are no pears native to the New World. The fruit was introduced to America by either the English or the Spanish. In Eurasia, pears have been cultivated since at least classical Grecian times. Selection in pears has been primarily for different fruit appearances. Propagation is generally by means of grafting. The slightly gritty texture of pears is caused by the presence of stone cells, short, squarish cells with thick cell walls. These hard cells are grouped throughout the flesh of the fruit.

The third economically important fruit of the Pyroideae is the quince (*Cydonia oblonga*, Fig. 4-6). Quinces are sometimes used as rootstocks for pears, and like pears are more popular in Europe than in the United States. Quince fruits tend to have a hard flesh with noticeably more stone cells than pears. Consequently most quinces are made into preserves and conserves.

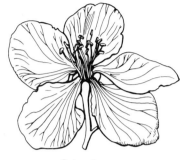

Quince flower

Flower in section

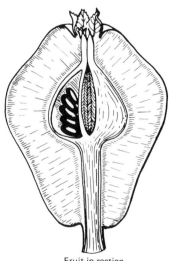

Fruit in section

FIGURE 4-6

Quince trees can reach 15 to 20 feet in height and are among the first plants to bloom in the spring. Their lovely pink and white flowers develop into yellow-skinned pome fruits similar to pears in form and size. Unlike pears, however, quinces contain up to 20 seeds in each carpel and their flesh is too hard and tart to eat raw. Fortunately, the pulp softens and turns pink and delicious when cooked with sugar. (After Baillon)

Prunoideae

Within the Prunoideae, there are several important fruit species, all belonging to the same genus, *Prunus*. These are plums (primarily *P. domestica*), cherries (*P. avium* and *P. cerasus*), peaches (*P. persica*), and apricots (*P. armeniaca*). In all of these fruits, the part of the fruit eaten is the fleshy mesocarp (see Fig. 4-4D). The exocarp is the "skin" of the fruit. The endocarp is modified into the "pit," a hard, stonelike casing serving as a protection for the seed.

The center of origin of all of the various fruits in this group seems to have been central or western China. Plums and cherries have native species in the New World, whereas peaches and apricots did not occur in North or South America before the arrival of Europeans.

Plums, peaches, and apricots have been cultivated for 2,000 years or more. Plums in particular are an important European crop (Table 4-5) with hundreds of developed varieties. In U.S. supermarkets we can find a few varieties such as Italian, purple, damson, Mirabell yellow, and sometimes various Japanese cultivars. Because the New World species tend to be small and to have very watery flesh, most selection has been carried out using the Old World species *P. domestica*. Nevertheless, over 600 varieties derived from native American species have been registered. In plum orchards in the United States today, trees that mature the same kind of plum but at different times of the summer are planted in separate rows. This arrangement allows a prolonged harvest without having to alter any procedures or machinery. Plums are cultivated in the United States in every state except Alaska, but the primary area of production is along the west coast.

Prunes are plums which have been dried without allowing them to ferment. Despite the common modern association of prunes and a laxative effect, their production dates back to times before modern methods of preservation because, once dried, they can be kept for long periods without spoiling.

Like plums, there are New World cherries, but cultivated cherries are essentially all derived from Old World species. Their cultivation in the United States (Fig. 4-7) began with the early settlers, but the most important date for the U.S. cherry industry is 1847, when two brothers named Lewelling settled in Oregon after a 300-mile oxcart trip from Iowa with several hundred cherry seedlings. One of the brothers eventually selected both the bing cherry, the most important cultivar in the United States today, and the Lambert, the second most important U.S. variety.

Cherries have a special meaning for Americans because of their association with young George Washington. The famous story of Washington's chopping down the cherry tree and confessing the deed to his father is, unfortunately, fictitious. Parson Weems created the story for the fifth edition (1806) of his book *The Life and Memorable Actions of George Washington*. Weems seems to have been more concerned with providing a good example of honesty than an accurate account. Today, cherry trees which are cultivated for their flowers rather than their fruits adorn the Tidal Basin of Washington, D.C. These trees, a gift from the Japanese government, signal the advent of spring with a display that draws visitors from all over the country.

Cherries are grouped according to taste into sweet and sour types. Sweet cherries are usually considered to belong to *Prunus avium* and sour types to *P.*

FIGURE 4-7
Cherry trees can be shaken
mechanically during harvest which
causes the fruits to fall onto a
tarpaulin placed around the base
of the tree. (Photo courtesy of the
USDA.)

cerasus. Within each of these groups, cherries are classified on the basis of the color and firmness of their flesh.

Maraschino cherries are made from sweet cherries, usually the Royal Anne variety, which have been bleached, deseeded, and soaked in a sugar solution to which red food coloring and flavoring have been added. The production of maraschino cherries dates back several hundred years to Italy where a cordial, maraschino (named after the marasca cherry most commonly used in its preparation), was produced. The French later developed the sweet type of maraschino that has now replaced the original bitter Italian liqueur.

Peaches are the third most important fruit crop in the United States, lagging behind oranges and apples. They are native to China where they have been cultivated for thousands of years and where they symbolize long life. Spaniards introduced peaches to Mexico in the late 1500s. Since that time, their cultivation in the New World has spread, and many selections have been made that allow their growth in a variety of climates. The United States is a leading peach producer (Table 4-5). Peaches are an important crop in California, Arkansas, Texas, Georgia, and New Jersey.

Peaches fall into two major classes, freestones and clings. Freestones are peaches that have pits which separate easily from the flesh. The mesocarp of cling peaches adheres tenaciously to the pit. Like apples and plums, peaches are generally propagated by grafting, once a superior cultivar has been selected or discovered. Nectarines, an increasingly popular fruit, are peaches without their furry coats. They are all descended from a mutant that was noted by a nurseryman and subsequently propagated by grafting.

Apricots were first brought to Greece from Persia by Alexander the Great and to the New World by Spaniards who established orchards in California in 1770. In ancient times, they figured prominently in folklore. In Persia, apricots were called "seeds of the sun." Chinese legends ascribed prophetic powers to them. Confucius is said to have worked out his philosophy under an apricot tree, and the indiscretions of the Duchess of Malfi in John Webster's English blood tragedy were uncovered when she ate apricots and claimed that they were bitter, a sign that she was pregnant. Apricots have also been associated with healthiness. In Hunza, a small kingdom in the Himalayas, apricots are an important part of the diet. The astounding long life and the robust health of the people in this tiny country have often been attributed to apricots.

Rosoideae

The last important subfamily of the Rosaceae, the Rosoideae, contains the false berries of the fruit trade, strawberries, raspberries, and blackberries. Although all of these fruits contain many separate mature ovaries from a single flower, their fleshy portions have different origins (Fig. 4-8). Most of a strawberry is the swollen, convex receptacle on which the ovaries were borne. Consequently, when a strawberry is picked, it is detached under the receptacle. The fruits are tiny achenes that dot the surface of the swollen portion. Blackberries and raspberries consist of a conglomeration of swollen ovaries. The receptacles of these fruits are not swollen and remain on the plant when they are picked. These berries are consequently hollow. Each bump on a raspberry or blackberry is an individual, fleshy, mature ovary with a seed inside.

Strawberries are one of the most beautiful fruits in the Rosaceae and the most important commercial fruit crop in the Rosoideae. The delicate, yet distinctive flavor of strawberries, their beautiful color, shape, and odor, and their ability to combine with a variety of foods make them an increasingly popular fruit in American markets (Table 4-5). The generic name *Fragaria* is derived from the Latin "fragrans" because of the fruit's delightful fragrance. The Anglo-Saxon name strawberry probably refers to the plant's habit of sending out runners, or "strewing" itself across the ground. Strawberries occur wild in Eurasia and the Americas. The fruits of all species are edible and were eaten by native peoples of both the Old and New Worlds. Romans relished strawberries, and, for the peasants during the Middle Ages, they were one of the few delicacies available. European strawberries tend to be small, however, and few berries are produced per plant. As North and South America were explored, new species of strawberries were found and sent to Europe. One such species was the Virginia strawberry, *F. virginiana*, found on the North American east coast. Other strawberries were found in Canada, and, in 1646, an exceptional species, *F. chiloensis*, was located in Chile. A spontaneous cross between the Virginia and the Chilean species in a European garden about 1750 produced our most commonly cultivated species, *F. ananassa*. Today, strawberry cultivars have been developed that yield more fruit per unit area per time than any other temperate dessert fruit.

FIGURE 4-8
The many separate ovaries within a strawberry flower become tiny, black achenes when they mature. These dot the surface of the red, fleshy part of the fruit produced by the swelling of the receptacle on which the ovaries were borne.

Many people simply lump raspberries and blackberries together as "brambles," but both actually consist of a number of species of the genus *Rubus* that occur natively in North America and Europe. The principal raspberries grown commercially are *R. idaeus* (red), a European species, and *R. occidentalis* (black), a North American taxon. Black raspberries are rarely marketed as fresh fruit, but they are grown in the Pacific Northwest as a source of purple dye. This dye, used to label meats with the familiar USDA ratings of prime, choice, etc., is completely edible since it is nothing more than raspberry juice. A cross between an octoploid California blackberry and a European raspberry first found in the California garden of Joshua Logan gave rise to the loganberry, a popular fruit in the western United States used for pies and jams.

Currants and gooseberries

Both currants and gooseberries belong to the same genus, *Ribes* (Grossulariaceae, Fig. 4-9). Their generic name is derived from the Arabic word *rheum* for

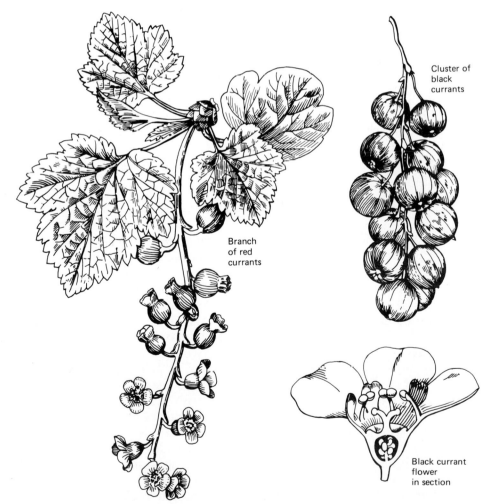

Cluster of
black
currants

Branch
of red
currants

Black currant
flower
in section

FIGURE 4-9
Currant flowers are produced in pendulous racemes. Each flower has five small petals and five stamens, which are alternately attached to a cup formed by the fused petals. The simple ovary and adherent floral cup develop into a round fruit capped with the shriveled remains of the flower parts. (After Baillon)

rhubarb, referring to a plant that produces an acid juice. Because of this pronounced tartness, all are better when cooked and used in jams, jellies, or pies. There are numerous species of both currants and gooseberries that occur natively across northern Eurasia and North America. While the New World members have been eaten locally and used for breeding purposes, the most commonly grown species of currants and gooseberries are derived from European taxa. Currants and gooseberries are accessory berries borne on long-lived shrubs, but currant bushes tend to have a sprawling habit that has made mechanical harvesting difficult. Gooseberry bushes are erect, but often heavily armed with spines.

Currants are small (3 to 6 mm), red or black berries with smooth skins. The most commonly cultivated black currant, the European black currant (*R. nigrum*), has an especially strong, smoky flavor which improves when combined with sugar in jams or liqueurs such as creme de cassis. The red "cherry" currants (*R. sativum*) are generally made into jam or are dried. In dried form, currants resemble small raisins and are often substituted for them in cookies, fruitcakes, and puddings. Because currants serve as a host for the pine blister rust (*Cronartium rubicola*), their cultivation is banned in the United States where pine trees are grown commercially. Breeding has led to increased resistance of currants to the fungus, but currant cultivation in the United States is still much lower than in Europe.

Gooseberries (primarily *Ribes uva-crispa*) have apparently been cultivated commercially for only about 100 years, although they were grown on small farms for at least a century or more before. The berries are green, translucent, and marked with stripes. Because they are exceedingly tart and rather bland, they are used exclusively in pies and jams. Like currants, most of the breeding and cultivation of gooseberries have been in Europe. In the United States they are grown commercially only in Oregon.

BLUEBERRIES AND CRANBERRIES

Although blueberries and cranberries are not important on a worldwide basis, both are popular in the United States. Blueberry pie is almost as American as apple pie, and cranberries are an indispensable part of a Thanksgiving meal. Both of these berries belong to the Ericaceae, or heath family. The family is probably best known for its ornamental shrubs: azaleas, rhododendrons, and mountain laurels. Like these ornamentals, cranberries and blueberries thrive best in acid soil. Blueberry cultivation is limited not only by its soil preferences, but also because the berries are difficult to pick. Two types, high-bush blueberries, *Vaccinium corymbosum,* and low-bush blueberries, *V. angustifolium,* yield commercial fruit crops (Fig. 4-10). Both species are native to North America and have been introduced into Eurasia.

Cranberries are usually placed in the same genus as blueberries, although some botanists separate them as the genus *Oxycoccus.* Two species—*Vaccinium macrocarpon,* of North America, and *V. oxycoccos,* a widespread north temperate species—are cultivated. Cranberry plants are creeping shrubs that are grown in boggy areas which can be periodically flooded. Flooding prevents winter desiccation, protects plants from freezing, and helps to regulate fruiting times.

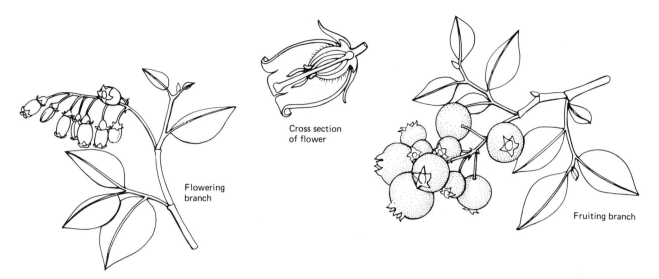

Cross section
of flower

Flowering
branch

Fruiting branch

FIGURE 4-10
Blueberry plants have small, urn-shaped flowers characteristic of many members of the heath family.

Fields are also flooded to facilitate harvesting. The ripe berries arc shaken loose from the plants as they float to the surface of the water and are scooped up by large machinery (Fig. 4-11).

GRAPES

One of the fruits that has most inspired man, yet cqually often led to his downfall, is the grape. Both of these effects are, however, from wine, the form in which grapes are most commonly consumed. We discuss wine and its derivatives in detail in Chapter 15. Undoubtedly, the eating of grapes preceded the discovery of their potential for producing a fermented beverage. The most widely cultivated species of grape is *Vitis vinifera* (Vitaceae), a perennial vine native to the eastern Mediterranean region (Fig. 4-12). Many of the table, or eating, varieties of this species produce poor wine, although some, such as the muscats, produce both an interesting wine and a fine eating grape. There are at least 175 table grape cultivars. Thompson seedless and Perlette are among those developed in the United States. Seedless grapes are particularly popular eating grapes because they are easy to eat. Like most seeded grapes, seedless grapes are propagated by grafting.

FIGURE 4-11
The wet method of harvesting cranberries involves flooding the bogs and beating the bushes to free the berries which float to the surface of the water where they are scooped up mechanically. (Courtesy of Ocean Spray Cranberries, Inc., Plymouth, Massachusetts.)

FIGURE 4-12
Grapes are an important crop on
the lower mountain slopes in
Afghanistan. (Photo courtesy of the
United Nations.)

The most important native New World grape is *V. lambrusca,* the fox grape. Early settlers in the northeastern part of the country selected the Concord and Catawba grapes from this species. Concord grapes are now the most important grape in America for juice, jams, and specialty wines. This grape initially became popular because of its robust flavor. In order to prevent grape juice from fermenting spontaneously, it must be pasteurized or processed by ultrafiltration. Pasteurization was formerly the most common method of preventing fermentation, but it altered the taste of delicate juices such as that of *V. vinifera.* The hardy flavor of Concord grapes was able to survive the effects of pasteurization. Consequently, it became the traditional juice grape. In Europe, where ultrafiltration is standardly employed to process grape juice, regular wine grape varieties are generally used.

Raisins are grapes that have been carefully dried (Fig. 4-13). Drying is sometimes carried out in a sulfur dioxide atmosphere to prevent fungal and bacterial growth. Varieties of grapes that are used for raisin production have been selected for a soft texture, reduced stickiness, and pleasing flavor. Common grape varieties that produce good raisins are the Sultana, Black Corinth, and Muscat of Alexandria. In the United States, almost all raisins are produced in California.

OLIVES

Despite their stereotype as martini decorations, olives (Fig. 4-14) have been an extremely important fruit and source of oil (Chapter 10) for western people for over 5,000 years. Olives, *Olea europaea* (Oleaceae), are native to the region

FIGURE 4-13
Raisins are dried, nonfermented grapes. The first stage in drying consists of spreading the bunches of fruit on sheets of paper between the rows of vines and periodically turning them. (Photo courtesy of California Raisin Advisory Board.)

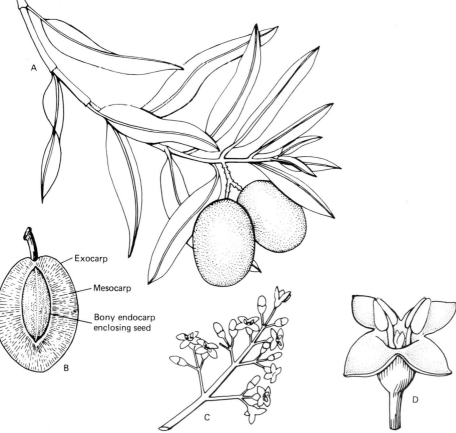

Exocarp

Mesocarp

Bony endocarp enclosing seed

FIGURE 4-14
Olives are drupes with a single seed borne on low, stout trees native to the Mediterranean region. (A) Branch, (B) cross section of a fruit, (C) a cluster of flowers; (D) enlargement of a fruit.

bordering the Mediterranean Sea where they still grow wild. Olive seeds, dated to 3000 B.C. have been found in excavated settlements in various parts of Israel. Fresh olives are extremely bitter because they contain oleuropein, a compound that must be broken down or altered to make olives palatable. In ancient times, olives were processed by drying, salting, or pickling. These procedures removed some, but not all, of the bitterness. About 1900, a process was developed in California (where olives had been cultivated since their introduction in 1769) that produced the first truly tender, nonbitter olives. The process consists basically of treating ripe olives with sodium hydroxide to hydrolyze the bitter oleuropein. The alkali is then flushed from the fruits. If the treated olives are subsequently allowed to oxidize by exposing them to the air, they turn into black olives. Green olives are kept submerged after processing so that they will retain their color. Spanish olives are fermented in brine after hydrolysis. Pitting is the last step in processing before the olives are packed in jars or cans. Despite these newer methods of producing edible olives, only about 1 or 2 percent of the world's crop is eaten as fruit. Most olives are pressed for oil (Chapter 10).

TEMPERATE NUTS

When we think of nuts, we tend to think of walnuts, pecans, almonds, chestnuts, filberts, and peanuts. All of these nuts except peanuts (see Chapter 7) are native to north temperate regions and most to the Old World. Pecans are the only important temperate nut to come from North America. Nuts differ from the fruits we have already discussed because they are basically dry fruits. In addition, the edible portion of a nut is the embryo, which has greatly enlarged cotyledons that have absorbed all of the endosperm during development. Table 4-4 shows the nutritional composition of the major temperate nuts, and Table 4-5 lists their primary areas of production.

Walnuts, chestnuts, filberts, and pecans are all borne on large trees that are important native components of the deciduous forests of Eurasia and North America. All of these trees are monoecious with small, green, wind-pollinated flowers.

Pecans and walnuts belong to the Juglandaceae, or walnut family. Despite its name, the English walnut, *Juglans regia,* is native to southeastern Europe and Asia. The black walnut, *J. nigra* (Figs. 4-15 and 1-19), a North America native, is a relatively unimportant commercial species because it has a strong taste and exceedingly hard shell. Walnuts were blanched, pulverized, and soaked in water in Europe in the eighteenth century to provide a milk substitute that was a staple in many households. While walnut fruits were being used to nourish humans, the trees were used to ward off insects. Walnut trees are notorious for inhibiting the growth of other plants surrounding them. This type of inhibition, known as allelopathy, is caused by chemicals present in the leaves that are leached out by rain and soak into the ground around the trunk. Apparently, this property led farmers to believe that the trees had insecticidal properties because they planted them near barns to keep flies away from the animals. Unfortunately, the effect of walnuts on other plants does not extend

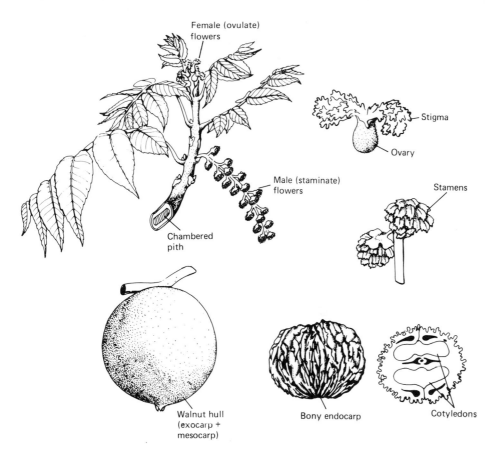

Female (ovulate) flowers

Stigma

Ovary

Male (staminate) flowers

Stamens

Chambered pith

Walnut hull (exocarp + mesocarp)

Bony endocarp

Cotyledons

FIGURE 4-15
Walnuts are produced by large, monoecious trees that can reach over 30 m in height at maturity. Up to 40 male flowers are borne in catkins that dangle in the wind. The female flowers occur singly or in small groups. Following pollination, the ovaries mature into green, drupelike nuts. The stone of the drupelike nuts is the shell of walnuts we purchase in stores.

to insects. In the United States, production of the introduced English walnut is centered in California, but there are also commercial orchards in the northeastern part of the country and adjacent Canada.

Pecans and their relatives hickory nuts and butternuts belong to the same family as walnuts but are placed in the genus *Carya*. The most important commercial species is *Carya illinoinensis* (Fig. 4-16), native to Mexico and the American southwest. Pecans are now extensively grown in the southeastern part of the United States, where they have become a prominent part of the regional cuisine. Commercial breeding programs with pecans led to the development of pecans with thin shells for easy cracking. Selection has also improved disease resistance and uniformity of fruit production. Despite its relatively recent introduction into trade, pecans have over 300 named varieties. Annual production in the United States exceeds 52,000 tons.

In addition to the pecan, North America once produced a significant quantity of chestnuts. Prior to 1900, roasted American chestnuts (*Castanea dentata*, Fagaceae) were common in the northeastern United States. About 1890, the chestnut blight, caused by a fungus of the genus *Endothia*, was introduced into the United States from Asia. The blight virtually destroyed all of the U.S. fruit-bearing trees within a few years. Besides this American chestnut, there are two other commercially important species, the European chestnut (*C. sativa*) and the Japanese species (*C. crenata*). The Japanese chestnut, which is more resistant

FIGURE 4-16
Pecan trees, like walnut trees, are large, monoecious, and deciduous. They have gray, furrowed bark and produce a handsome, spreading crown. Male flowers are produced on branched, pendulous catkins and female flowers in spikes that are clustered into groups of 2 to 10. The ovaries mature into fruits that split at maturity to release the seed encased in the endocarp.

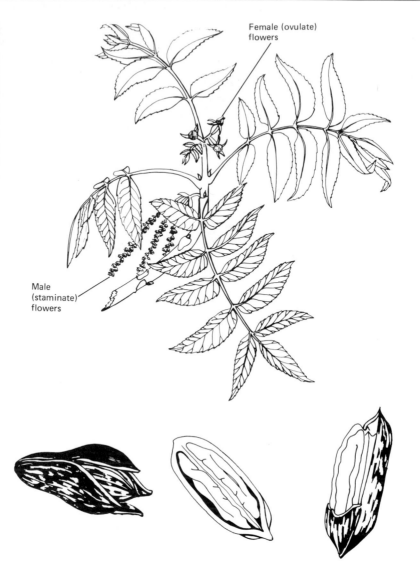

Female (ovulate) flowers

Male (staminate) flowers

to the fungus than either the American or the European species, is now being planted in North America.

Chestnuts are borne three to a cluster (Fig. 4-17). Each nut is produced by a single flower, but three flowers are grouped together in a compact influorescence surrounded by fleshy bracts. When the nuts are mature, the three bracts spread apart, revealing the chestnuts. The nuts are usually picked before they are completely mature, but when the bracts are beginning to separate. Ripening is completed after harvesting. Unlike most nuts, chestnuts are usually eaten without appreciable drying. In Europe they are considered one of the most versatile and delicious nuts and are used in candies and pastries, are pureed as a vegetable, and eaten as a plain or roasted nut.

Hazelnuts, also known as filberts or cob nuts (Fig. 4-18), occur natively across the north temperate zone, but only the European species (*Corylus avellana*, Corylaceae) is extensively cultivated as a crop. There is little agreement as to

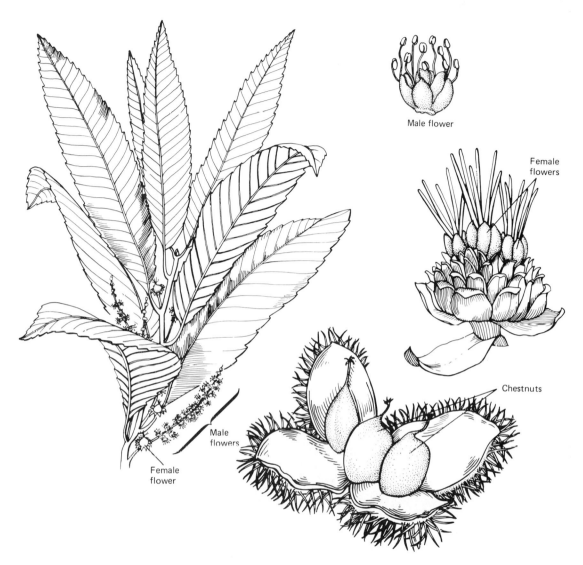

FIGURE 4-17
Chestnuts are borne on tall forest trees with coarsely toothed leaves. Male flowers occur in catkins 10 to 30 cm long. The female flowers, borne on the same plant, occur in groups of three. At maturity, the spiny bracts surrounding each group of female flowers split to release the nuts. (After Baillon)

the origin of the common names of this nut. Various authors ascribe the name filbert to St. Philibert because the nuts ripen in Europe around St. Philibert's day. Others say that the nut was named for King Philibert of France. Still others claim that filbert refers to ''full beard,'' a descriptive term for the bract that accompanies the nut. The origins of the names hazelnut and cob nut are just as obscure. In ancient times, filberts were considered to have special powers that allowed divination. Divining rods were made from hazelnut wood. In Greece, the nuts symbolized fertility, and in other parts of Europe, the filbert tree was believed to be effective against lightning and witches.

FIGURE 4-18
Filbert nuts are enclosed by
distinctive leafy bracts. (After
Baillon)

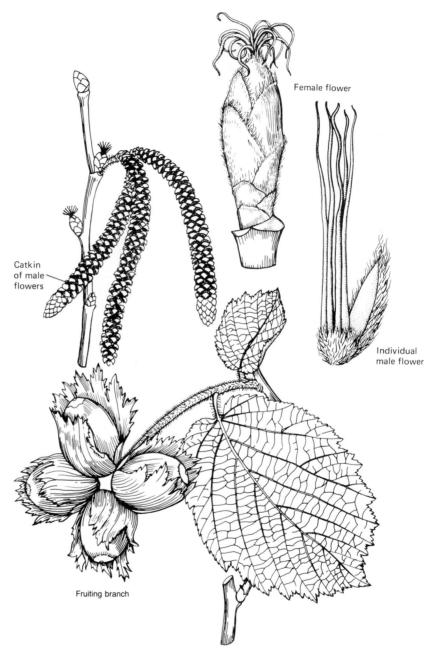

Female flower

Catkin
of male
flowers

Individual
male flower

Fruiting branch

Filberts are much more popular in Europe than in the United States. In Italy and France, the nuts are ground into a fine powder that is used to flavor ice cream and candies. Hazelnut paste is also used much like peanut butter. Almost all of the commercial filbert production (95 percent) in the United States is in Oregon.

Among the temperate nuts, almonds (Fig. 4-19) are unique because they belong to a family (Rosaceae) noted for its showy, insect-pollinated flowers and

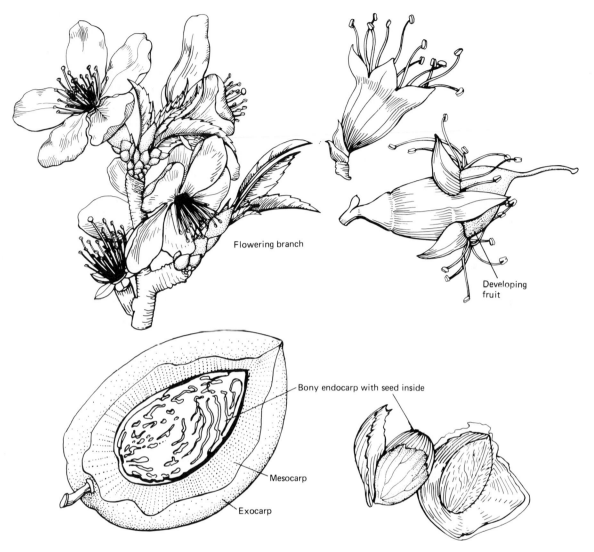

Flowering branch

Developing fruit

Bony endocarp with seed inside

Mesocarp

Exocarp

FIGURE 4-19
The almond has the same kind of lovely flowers as a peach or cherry tree, but the mesocarp of the almond fruit is tough and leathery.

fleshy fruits. Almonds belong to the same genus as plums, apricots, and peaches, but to a distinct species, *Prunus amygdalus*. In contrast to their relatives with fleshy drupes, almonds have a mesocarp that expands little as the fruit matures. Removal of the leathery mesocarp leaves the familiar almond inside the endocarp (shell). Like apricot and peach seeds, almond seeds contain amygdalin, a cyanogenic compound which gives them the characteristic "almond" flavor and odor. The seeds of peaches and apricots are not eaten because they are too bitter for most people's taste and contain dangerously high levels of amygdalin.

Almonds are native to Eurasia where they have been cultivated for thousands of years. In the United States, almonds are cultivated in California where the climate is similar to that of the almond's native region.

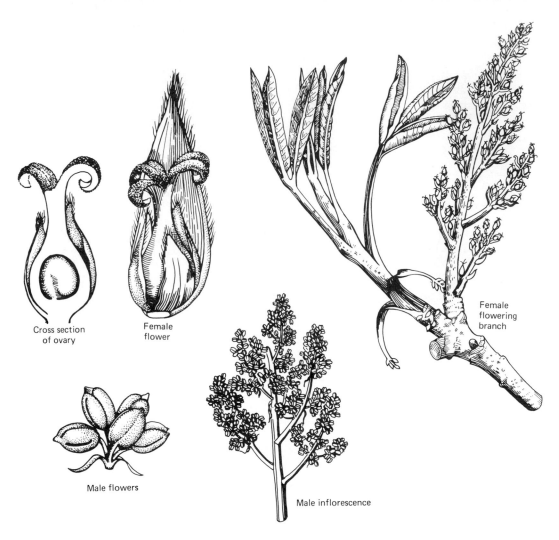

Cross section
of ovary

Female
flower

Female
flowering
branch

Male flowers

Male inflorescence

FIGURE 4-20
Pistachio nuts are produced on deciduous trees that will grow to 10 m in height. Male
and female flowers are borne on axillary racemes of separate trees. The small flowers
lack petals and contain a single ovule each which develops into the familiar pistachio
nut. (After Baillon)

Slightly more exotic than the common dessert nuts is the pistachio. The
distinctive, slightly resinous flavor of the pistachio (*Pistacia vera*, Anacardiaceae,
Fig. 4-20) has endeared it to many who love the yellow-green kernels salted
or incorporated in ice cream, cakes, or nougat candies. Native to the eastern
Mediterranean and central Asia where they have been cultivated for over 3,000
years, pistachios are borne on small dioecious trees. Female trees produce highly
reduced flowers in axillary inflorescences. The fruits are small, ovoid drupes.
What we consider the "nut" is the seed enclosed by the endocarp "shell." The
surrounding pulp has been removed before marketing.

The practice of dyeing pistachio nuts was begun by an American entrepreneur from the east coast who wanted to hide the blemishes on the shells of imperfect nuts. The practice has been perpetuated, especially in the eastern United States where consumers have become accustomed to red pistachios. However, the red dyes traditionally used (red dyes nos. 3 and 40) have been implicated in the production of hyperactivity and cancer, suggesting that the practice of dyeing should now be abandoned.

SELECTED REFERENCES

Hartmann, H. T. 1948. The olive industry of California. *Economic Botany* 2:341–362. An interesting article chronicling olive cultivation in California.

Janick, J., and J. N. Moore (eds.). 1975. *Advances in Fruit Breeding.* West Lafayette, Indiana, Purdue University Press. This collection of articles contains detailed accounts of every major fruit and nut cultivated in the United States.

Simmonds, N. W. 1976 (ed.). *Evolution of Crop Plants.* New York, Longman. Most of the major crops of the world, and many of the minor crops, are included in this concise, authoritative book. Each crop is treated by a leading world specialist.

Chapter 5

Fruits and Nuts of Warm Regions

*P*apayas, mangoes, sour sops, and custard apples bring to mind visions of humid tropical rain forests and thoughts of exotic flavors. Such fruits seem to have little in common with tomatoes, green peppers, squashes, and cucumbers. Yet all of these fruits are native to tropical and warm regions of the world, and none will survive winters that have hard freezes. In agriculture we have circumvented the lack of winter hardiness in the cases of tomatoes, squashes, and peppers by planting them as annual crops that germinate, flower, and produce harvestable fruit during the spring or summer season. These fruits are exceptions, however. Most tropical fruits come from woody perennials which cannot germinate and produce fruits in a few months. Growth of such species is therefore restricted to areas that do not experience freezes. The tropics have many more kinds of edible fruits than temperate regions, yet temperate fruits dominate our markets and constitute an appreciable part of the commercially available fruits in tropical regions as well. One reason for the limited production of tropical fruits is their perishability. As a result, most are gathered from the wild, or cultivated on a limited scale, and consumed locally. A few exceptions include bananas, pineapples, and avocados, all of which have become readily available in the past century when methods of shipping them without spoilage were developed. Luckily, the selection of tropical fruits offered in this country is expanding. We therefore begin this chapter with well-known fruits native to warm regions and then treat the more exotic fruits occasionally found on our grocer's shelves (Table 5-1).

CITRUS FRUITS

Orange juice is an integral part of the "all-American breakfast." Yet, at the turn of the century, oranges were considered luxuries in northern markets and were prescribed as cold or consumption remedies by physicians. Today, oranges are the largest perennial fruit tree crop in the United States (Table 5-2). Oranges and their relatives, limes, lemons, citrons, grapefruits, and tangerines, all belong to the same genus *Citrus* (Rutaceae). All of these citrus species are trees with shiny, evergreen leaves. Most occur naturally in the region extending from north-central China southward to Australia and New Caledonia, but the cultivated citrus are believed to be derived from species native to southeastern Asia. Placement of the domesticated members into biological species is difficult because hybridization and human selection for mutants has obscured natural boundaries.

All citrus species share the same fruit type, a hesperidium, which is basically a berry with a leathery skin formed by the combined exocarp and mesocarp (Fig. 5-1). This skin, like the leaves and other vegetative parts of the plant, is covered with pockets filled with aromatic oils. These oil-filled cavities are characteristic of the citrus family and are responsible for the aroma given off when citrus peels are bruised, grated, or twisted. The endocarp of citrus fruits bears highly modified into fleshy hairs that fill the segments of the ovary and surround the seeds. These hairs, or juice sacs, are the parts of the oranges, grapefruits, and tangerines that we eat. They contain a watery solution of sugars and acids. In some species, the acid concentrations decrease with age and produce fruits that we eat without the addition of sugar.

TABLE 5-1

Names of Plants Discussed in Chapter 5

COMMON NAME	SPECIES	FAMILY	CHROMOSOME NUMBER
Avocado	Persea americana	Lauraceae	$2n = 24$ diploid
Banana	Musa spp.	Musaceae	$2n = 33$ triploid (others as well)
Bitter melon	Momordica charantia	Cucurbitaceae	Unreported
Bitter orange	Citrus aurantinum	Rutaceae	$2n = 18$ diploid
Bottle gourd	Lagenaria siceraria	Cucurbitaceae	$2n = 22$ diploid
Brazil nut	Bertholletia excelsa	Lecythidaceae	$2n = 34$
Breadfruit	Artocarpus altilis	Moraceae	$2n = 56, 84$ di- & triploids
Carambola	Averrhoa carambola	Oxalidaceae	$2n = 22, 24$ diploid
Cashew	Anacardium occidentale	Anacardiaceae	$2n = 42$ diploid
Chayote	Sechium edule	Cucurbitaceae	$2n = 24$ diploid
Citron	Citrus medica	Rutaceae	$2n = 18$ diploid
Coconut	Cocos nucifera	Arecaceae	$2n = 32$ diploid
Cucumber	Cucumis sativus	Cucurbitaceae	$2n = 14$ diploid
Custard apple	Annona squamosa	Annonaceae	$2n = 14$ diploid
Date	Phoenix dactylifera	Arecaceae	$2n = 36$ diploid
Eggplant	Solanum melongena	Solanaceae	$2n = 24$ diploid
Fig	Ficus carica	Moraceae	$2n = 26$ diploid
Grapefruit	Citrus paradisi	Rutaceae	$2n = 18$ diploid
Guava	Psidium guajava	Myrtaceae	$2n = 22$ diploid
Jackfruit	Artocarpus heterophyllus	Moraceae	$2n = 56$ tetraploid
Kiwi fruit	Actinidia chinensis	Actinidiaceae	$2n = $ ca. 160 high polyploid
Lemon	Citrus limon	Rutaceae	$2n = 18$ diploid
Lime	Citrus aurantiifolia	Rutaceae	$2n = 18$ diploid
Litchi	Litchi chinensis	Sapindaceae	$2n = 30$ diploid
Loquat	Eriobotrya japonica	Rosaceae	$2n = 32, 34$
Luffa:	L. acutangula	Cucurbitaceae	$2n = 26$ diploid
	L. cylindrica	Cucurbitaceae	$2n = 26$ diploid
Macadamia nut	Macadamia integrifolia	Proteaceae	$2n = 28$ diploid
Malay apple	Syzygium malaccense	Myrtaceae	$2n = 22$ diploid
Mammey sapota	Pouteria sapota	Sapotaceae	Unreported
Mango	Mangifera indica	Anacardiaceae	$2n = 40$

TABLE 5-1 (continued)

Names of Plants Discussed in Chapter 5

COMMON NAME	SPECIES	FAMILY	CHROMOSOME NUMBER
Melon (cantaloupe, honeydew, etc.)	*Cucumis melo*	Cucurbitaceae	$2n = 24$ diploid
Okra	*Abelmoschus esculentus*	Malvaceae	$2n = 72–132$ polyploid
Orange:			
Sweet	*Citrus sinensis*	Rutaceae	$2n = 18$ diploid
Bitter	*Citrus aurantinum*	Rutaceae	$2n = 18$ diploid
Papaya	*Carica papaya*	Caricaceae	$2n = 18$ diploid
Passion fruit	*Passiflora edulis*	Passifloraceae	$2n = 18$ diploid
Pepper, sweet	*Capsicum annuum*	Solanaceae	$2n = 24$ diploid
Pineapple	*Ananas comosus*	Bromeliaceae	$2n = 50$ diploid
Pomegranate	*Punica granatum*	Punicaceae	$2n = 16, 18$ diploid, euploid
Pummelo	*Citrus maxima*	Rutaceae	$2n = 18$ diploid
Rose apple	*Syzygium jambos*	Myrtaceae	$2n = 28–66$ variable
Sapodilla	*Manilkara zapota*	Sapotaceae	$2n = 26$ diploid
Squash	*Cucurbita maxima*	Cucurbitaceae	$2n = 40$ diploid
	C. moschata	Cucurbitaceae	$2n = 40$ diploid
	C. pepo	Cucurbitaceae	$2n = 40$ diploid
Tangerine	*Citrus reticulata*	Rutaceae	$2n - 18$ diploid
Tomato	*Lycopersicon esculentum*	Solanaceae	$2n = 24$ diploid
Watermelon	*Citrullus lanatus*	Cucurbitaceae	$2n = 22$ diploid
Wax gourd	*Benincasa hispida*	Cucurbitaceae	$2n = 24$ diploid
Zucchini	*Cucurbita pepo*	Cucurbitaceae	$2n = 40$ diploid

The cultivated *Citrus* species are not natives of lowland wet tropical regions, although none can tolerate real freezing temperatures. Most do best where there is good soil moisture, but abundant sun and reasonably dry air. In the past, most citrus fruits were grown from seed, but with the rise in popularity of certain cultivars, many of which are partially or completely sterile, bud grafting has become the most common method of propagation. Of the various species, bitter oranges, pummelos, and citrons are still frequently grown from seed.

The sweet orange (*Citrus sinensis*, Fig. 5-1) is now the most widely grown citrus fruit in the world, but with the wild ancestors gone, we can only speculate about its origin. Some authors believe the orange is derived from an unidentified or extinct Chinese species, but others assert that it resulted from selection of a hybrid between a tangerine and a pummelo. Oranges were considered by some

TABLE 5-2

World Production of Fruits and Nuts of Warm Regions

CROP	TOP 5 COUNTRIES, 1,000 t		TOP 5 CONTINENTS, 1,000 t		WORLD, 1,000 t
	COUNTRY	t	COUNTRY	t	
Avocados	Mexico	448*	North & Central America	983	1,586
	United States	218*	South America	327	
	Dominican Republic	136*	Asia	153	
	Brazil	125*	Africa	117	
	Peru	67*	Oceania	3	
Bananas	Brazil	6,692	Asia	15,708	40,700
	India	4,500*	South America	11,529	
	Philippines	4,200*	North & Central America	7,275	
	Thailand	2,035*	Africa	4,547	
	Ecuador	2,000*	Oceania	1,151	
Chiles and green peppers	China	1,605*	Asia	3,172	7,064
	Nigeria	695*	Europe	2,269	
	Turkey	683*	Africa	1,181	
	Spain	579	North & Central America	810	
	Italy	450*	South America	171	
Coconuts	Indonesia	11,100*	Asia	29,142	34,890
	Philippines	9,200*	Oceania	2,175	
	India	3,900*	Africa	1,554	
	Sri Lanka	2,300	North & Central America	1,443	
	Malaysia	1,200*	South America	576	
Cucumbers and gherkins	China	2,980*	Asia	5,767	11,467
	U.S.S.R.	1,450*	Europe	2,755	
	Japan	1,075*	North & Central America	1,101	
	United States	817*	Africa	336	
	Turkey	558*	South America	46	
Dates	Saudi Arabia	440*	Asia	1,679	2,776
	Egypt	440*	Africa	1,066	
	Iraq	400*	North & Central America	19	
	Iran	302*	Europe	12	
	Pakistan	218*	South America	1	
Eggplant	China	1,732*	Asia	3,941	5,007
	Turkey	729*	Europe	572	
	Japan	652*	Africa	420	
	Italy	330	North & Central America	66	
	Egypt	300*	South America	8	

*Indicates an estimated or unofficial figure.

TABLE 5-2 (continued)

World Production of Fruits and Nuts of Warm Regions

CROP	TOP 5 COUNTRIES, 1,000 t		TOP 5 CONTINENTS, 1,000 t		WORLD, 1,000 t
	COUNTRY	t	COUNTRY	t	
Grapefruit and pummelo	United States	2220	North & Central America	2,559	4,088
	Israel	450	Asia	862	
	China	157*	South America	319	
	Argentina	146	Africa	292	
	Cuba	142*	Oceania	33	
Lemons and limes	United States	921	North & Central America	1,629	5,477
	Italy	770	Europe	1,478	
	Mexico	580*	Asia	1,325	
	Spain	522	South America	758	
	India	500*	Africa	250	
Mangoes	India	8,700*	Asia	10,917	13,954
	Pakistan	683	North & Central America	1,362	
	Mexico	665*	Africa	836	
	Brazil	610*	South America	828	
	Philippines	550*	Oceania	11	
Melons	China	1,825*	Asia	3,822	7,707
	United States	844*	Europe	1,558	
	Spain	668	North & Central America	1,277	
	Iran	536*	Africa	718	
	Egypt	400*	South America	332	
Olives	Italy	3,080	Europe	5,599	7,733
	Spain	1,297	Africa	1,080	
	Greece	1,052	Asia	815	
	Tunisia	600	South America	138	
	Turkey	450*	North & Central America	98	
Oranges	Brazil	9,515	South America	11,924	38,171
	United States	8,631	North & Central America	11,082	
	Italy	1,945	Asia	6,645	
	Spain	1,895	Europe	4,493	
	Mexico	1,480*	Africa	3,284	
Papayas	Brazil	460*	Asia	799	1,982
	Indonesia	310*	South America	620	
	India	270*	North & Central America	323	
	Mexico	230*	Africa	222	
	Zaire	156*	Oceania	18	
Pineapples	Thailand	1,439*	Asia	4,768	8,665
	Philippines	1,300*	South America	1,306	
	Brazil	841	Africa	1,257	
	India	660*	North & Central America	1,206	
	Mexico	549	Oceania	127	

*Indicates an estimated or unofficial figure.

TABLE 5-2 (continued)

World Production of Fruits and Nuts of Warm Regions

| CROP | TOP 5 COUNTRIES, 1,000 t | | TOP 5 CONTINENTS, 1,000 t | | WORLD, 1,000 t |
	COUNTRY	t	COUNTRY	t	
Squashes	China	990*	Asia	2,310	5,355
	Egypt	460*	Europe	1,023	
	Turkey	374*	Africa	962	
	Argentina	370*	South America	741	
	Romania	360*	North & Central America	237	
Tangerines	Japan	3,167	Asia	4,163	8,047
	Spain	1,117	Europe	1,615	
	Brazil	580*	South America	985	
	United States	546	North & Central America	718	
	Italy	400	Africa	533	
Tomatoes	United States	7,567	Europe	15,239	54,240
	U.S.S.R.	7,250*	Asia	13,250	
	China	4,559*	North & Central America	9,771	
	Italy	4,550	Africa	5,465	
	Turkey	3,700*	South America	2,999	
Watermelons	China	4,600*	Asia	14,843	26,996
	Turkey	4,500*	Europe	2,993	
	U.S.S.R.	4,200*	Africa	2,182	
	Egypt	1,200*	North & Central America	1,727	
	United States	1,184*	South America	1,006	

*Indicates an estimated or unofficial figure.
Note: Only crops of major importance in international trade are included. U.S.S.R. is treated as a country only. Its figures are not added into those of Europe or Asia.
Source: Data from *FAO Production Yearbook for 1983*, vol. 37. 1984. FAO, Rome.

to be the "golden apples" of Greek mythology that the Goddess of Fertility gave to Hera when she married Zeus. Hera planted the seeds in the garden of Hesperides, hence, the designation "hesperidium" for the citrus fruit type. Oranges were transported along caravan routes from the orient to the Persian empire. The Moors brought them to Spain and used them medicinally and in religious services. The Spanish and Portuguese later introduced them into their New World territories.

Orange juice, like lime juice, is effective in preventing scurvy, and the spread of oranges followed the paths of seafaring explorers who wanted to ensure having supplies of the fruit along their routes. Nevertheless, up through the eighteenth and nineteenth centuries, sweet oranges were a delicacy reserved for the affluent. When it was discovered that oranges could be grown in temperate climates if protected from freezes, wealthy individuals began to grow them in glasshouses. The possession of such "orangeries" became a status symbol (Fig. 5-2).

Oranges were first grown in Florida in 1565, but it was not until the United States took possession of the peninsula in 1820 that sweet oranges became an important U.S. commodity. There are three main types of hybrid sweet oranges:

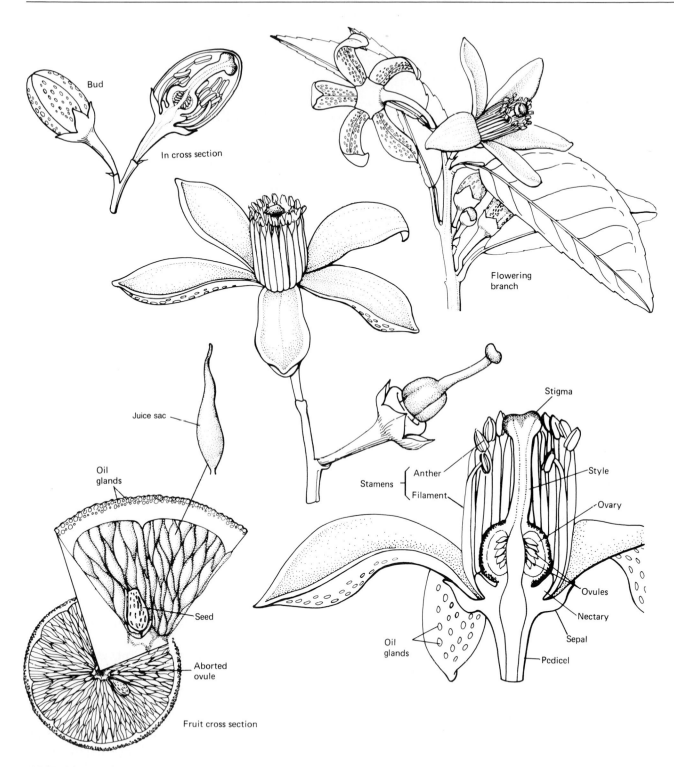

FIGURE 5-1
A branch bud, flower, and fruit of a sweet orange showing development from the initiation of the bud to the production of a mature fruit.

FIGURE 5-2
In the summer, plants taken from
an orangerie are set out in a
formal arrangement in the gardens
of the Versailles near Paris.
Oranges were so prized in northern
Europe that the wealthy built
glasshouses to protect the trees
and ensure a winter supply of the
fruit.

bloods, normals, and navels. Bloods have bands of red in the pulp which makes them unattractive to many Americans, although they are popular in Europe. The most commonly grown normal type is the Valencia orange. Modern Valencia cultivars were introduced in California in 1876 and in Florida in 1877 by an Englishman who had become familiar with the variety in the Azores. Valencia oranges are now the most important variety grown in Florida. They produce a richly flavored juice that sets the standard by which other orange juices are judged. Because we have come to expect orange juice to taste a certain way, commercial orange juice is often blended in order to ensure consistency in flavor and texture. In many cases, the juice is even filtered, freeze-dried, and reconstituted with controlled amounts of citrus peel and pulp.

Large-scale navel orange production is a recent phenomenon, but a type of navel orange was known in Europe at least 300 years ago (Fig. 5-3), and similar seedless oranges were either purposefully propagated or arose spontaneously in most orange-growing regions. Navel oranges are formed because the flowers produce an abnormal, small ovary on top of the regular ovary. When the fruit matures, the second ovary becomes the button, or navel. Correlated with the aberrant production of the navel is sterility. The pollen produced by navel orange flowers is nonfunctional, and the ovules almost never produce viable seeds. New plants must be produced vegetatively, usually by grafting. Navel orange production in Florida began in 1835 and in California 3 years later. Navels are now the leading orange variety grown in California. Unlike the Valencias, navel oranges are primarily eating oranges because of their lack of seeds and the ease with which they can be peeled.

The common English name "orange" is derived from the usual color of the fruit, but fully ripened oranges are not necessarily orange. In areas where temperatures never get cool (e.g., Thailand), oranges remain green to maturity. Cool temperatures thus contribute to the orange color because they promote the release of carotene, an orange pigment, and the breakdown of chlorophyll. If temperatures in an area fluctuate while the oranges are ripening, the fruits can change from orange to green and vice versa. Merchants, realizing the public desire for "orange" oranges, even though it makes no difference in their taste, will spray batches of green or blotchy oranges with ethylene to induce a color change, or they may even resort to dyeing the fruits.

Despite the modern preeminence of sweet oranges in the new world, they were not the first citrus grown in the Americas. Three others, the lemon, *Citrus limon*; the bitter orange, *C. aurantium*; and the citron, *C. medica* arrived almost 100 years earlier. Columbus brought seeds of these three species to the new world on his second voyage in 1493 and appears to have established them successfully on the island of Hispaniola. Bitter oranges are now used almost exclusively for the making of preserves, marmalade, and orange liqueurs. The pulp of citrons is rarely eaten, but the peel is often candied and used in confections such as fruitcakes. Although native to southeastern Asia, citrons were known to the ancient Greeks, and Linnaeus used Pliny's name "citron" as the basis of the generic name for all of the "citrus" fruits. Citron seems to have given some of its genes in addition to its name to other citrus fruits. Taxonomic studies indicate that the citron figures in the ancestry of both the lemon and the lime.

Like citrons and bitter oranges, lemons are practically never eaten alone, but their juice is used for flavoring everything from beverages to meat dishes. Of all of the citrus fruits, lemons are most used for ancillary purposes. The oils

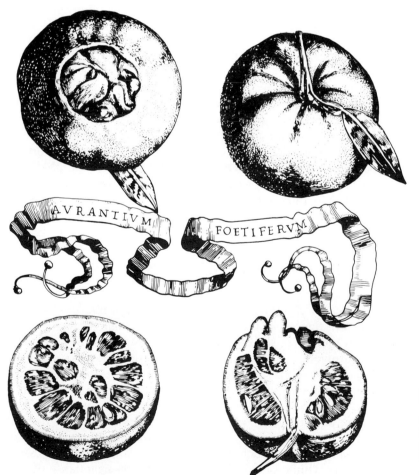

FIGURE 5-3
The original of this drawing of a navel orange is used as evidence that navels were known in Europe at least 300 years ago. (Redrawn from a woodcut by Ferrari in *Hesperides*, published in 1646.)

in lemon rinds are extracted for perfumes, cleaning products, and deodorants. The inner portion of the rind is a valuable source of pectin (see Fig. 11-11).

Morphologically the most distinctive member of the citrus group is *Citrus aurantiifolia,* the lime, originally cultivated in the East Indies. The Arabs were using limes by the tenth century A.D. and introduced them into Europe in the twelfth or thirteenth century. Women in the Renaissance French court carried limes with them for cosmetic purposes, to whiten their skin and redden their lips. Lime cultivation in the New World began in the seventeenth century in the Spanish and Portuguese territories.

When the correlation between the consumption of fresh fruits or fruit juices and the absence of scurvy was realized, British sailors were ordered by the admiralty to drink a daily ration of lime juice. This juice kept especially well on long voyages and appeared to be particularly effective in small doses. We now know that it is the high vitamin C content of limes that prevents scurvy, but without knowing why, the British saved hundreds of lives with their kegs of lime juice. To this day, English sailors are known as "limeys."

Grapefruits (*Citrus paradisi*) are thought to have originated as the result of a cross between a pummelo (*C. maxima*) and a sweet orange (*C. sinensis*) which spontaneously occurred in the West Indies, probably on Barbados. Thus, grapefruits can be considered a New World product and constitute a "species" that is only a few hundred years old. One of the ancestors, the pummelo, is little known in America today, but is relished in its native Thailand. Pummelos were apparently brought to the West Indies by the English or Dutch.

Commercial production of grapefruits began in the United States in 1880. Grapefruits with pink flesh arose as a bud mutation in a Florida plantation and were propagated by asexual grafting. The Texas "ruby reds" (with deep red-pink flesh) were developed in McAllen, Texas, in 1929.

The last important citrus fruit, *Citrus reticulata,* the tangerine, or Mandarin orange, was originally cultivated in China and is still an important fruit in the Far East. Recent numerical and chemical studies indicate that this species is a biological species and not a natural or artificial hybrid. Cultivation in the United States began in 1850, but the fruit has never enjoyed the popularity of oranges or grapefruits.

SQUASHES AND THEIR RELATIVES

In both the Old and the New World, members of the squash family (Cucurbitaceae) have figured in the early histories of man. Corn, beans, and squash, known as the "three sisters," were the mainstays of early agricultural peoples of the southwestern United States (see Fig. 6-29), Mexico, and Peru. For native New World peoples, squashes provide not only edible flesh, but also seeds relatively rich in the sulfur-containing amino acids. The blossoms are collected in some areas and used in soups. Fossilized edible squash remains from Peru have been dated to 5 or 6 thousand years before present. Undoubtedly, wild squashes were collected and eaten long before they were cultivated. In Africa, hunter-gatherer tribes such as the !Kung in the Kalahari Desert gather wild watermelons. Other melons figure in the earliest Egyptian writings.

As implied above, squashes are all native to the New World and most melons are indigenous to Africa and Asia. Both melons and squashes are "squashes" in that they belong to the squash family. All cultivated members of the family are vines. Most are annuals, and the domesticated species discussed here are all monoecious. Since the plants are annuals, they are planted each year from seed. All members of this family have fruits called pepos produced from inferior ovaries. The skin, or rind, of a pepo is therefore composed of tissues of the ovary wall fused with the lower parts of the corolla (Fig. 5-4).

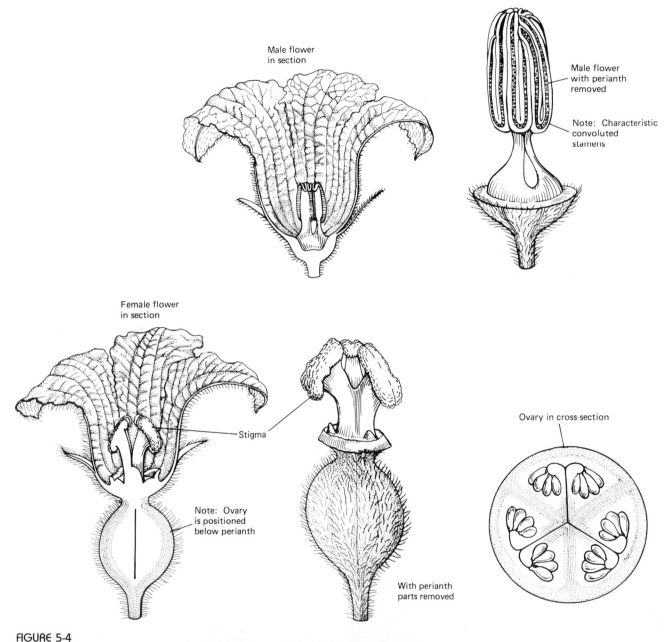

FIGURE 5-4
Squash fruits are formed from inferior ovaries with the rind of the fruit a composite structure of the ovary wall and lower parts of the calyx and corolla. (After Baillon.)

Because of their hard rinds, many squashes and melons store well if not bruised or cut.

The naming of true squashes (members of the genus *Cucurbita*) has been in a state of hopeless confusion for years, in part because common names of the edible squashes bear little relationship to the species from which they come. There appear to be four widely cultivated species, three of which, *Cucurbita pepo, C. moschata,* and *C. maxima,* are commonly encountered in our supermarkets. All of these squashes are known only as domesticated species. Wild cucurbits usually have thin rinds and bitter flesh. *Cucurbita pepo,* initially domesticated in Mexico, is now the most widely cultivated and versatile species of squash in the United States. It provides us with Halloween pumpkins as well as yellow crookneck squashes, and, in immature form, zucchini and summer squash. Fossil remains believed to belong to this species have been uncovered in Mexico and are dated to be between 7,500 and 9,000 years old.

Closely related to *Cucurbita pepo, C. moschata* has turned up in fossil deposits over 7,000 years old. Since the earliest of these fossils are from Mexico, this species also appears to have first been domesticated in Central America and subsequently taken to South America. Many kinds of winter squashes including butternut and winter crookneck come from this species (Fig 5-5). Although *C. pepo* and *C. moschata* are naturally trailing vines, there has been selection for bush types in both species so that they take up less space in fields and are comparatively easy to tend and harvest.

The last of the trio, *C. maxima,* has the least extensive fossil record of the commonly cultivated species. The oldest evidence of its domestication is Peruvian material that dates to only 2000 A.D. Because it is not found in ancient Central American deposits, scientists think its cultivation spread to North America from

FIGURE 5-5
A variety of squashes: Pumpkins, banana, acorn, hubbard, and butternut squashes. (USDA photo by D. W. Watt)

FIGURE 5-6
In the produce section of the supermarket, it can be difficult for a novice to choose a good melon. A melon that will ripen should have no stub of the stem still attached because a melon picked too soon may get soft but will never become sweet. Melons develop all the sugar they will ever contain while still on the vine. Home gardeners can pick their melons when they are "fully ripe," or when they "slip" from the vine without effort. For commercial sales, melons can be picked slightly before they are fully ripe, but they should still be at the "full slip" stage (detaching readily flush with the melon). Several other characters to look for are softness, a sweet aroma, and a sloshing sound when the melon is shaken, which indicates that the seeds are loose within the cavity.
Muskmelons or cantaloupes should have a netting that does not rub off easily. Watermelons are a bit more difficult to judge. The common practice of thumping a watermelon to see if it produces a hollow sound is not necessarily a good criterion since it can lead to the purchase of an overripe melon.
(Photo courtesy of Patti Anderson.)

South America only after European colonization. Nevertheless, *C. maxima* now provides us with hubbard, buttercup, and turban squashes (Fig. 5-5). Several of its fruit forms are also called pumpkins. In fact, since the flesh of winter squash is usually deeper orange in color and less fibrous than that of *C. pepo,* it is often canned and sold as pumpkin pie filling.

While the New World cucurbits provide important staple food crops, the Old World members provide luxuries. Except for the cucumber, all of the commercially exploited Old World members of the family tend to be bland or sweet and, in this country at least, most are eaten as dessert fruits. Watermelons (*Citrullus lanatus,* Fig. 5-6), now grown extensively in California and Texas, are African natives that were appreciated by Europeans in ancient times. By the eleventh century A.D., the Chinese were also growing them. Watermelons contain between 87 and 92 percent water and have a high sugar content relative to that of other melons. In addition, watermelon pulp is acidic, which helps to explain why it is not used as a flavoring for ice cream.

Other melons, varieties of *Cucumis melo,* seem also to be of African origin. References to melons do not appear in Egyptian or Greek writings but do occur in texts produced at the end of the Roman Empire. Selection eventually led to a wide array of different varieties including cantaloupe, and Persian, musk, Cranshaw, and honeydew melons. These sweet, refreshing melons may not seem very similar to cucumbers (*Cucumis sativus,* Fig. 5-7), but they are actually closely related to them. Cucumbers are often eaten with sugar in their native region of southern Asia. The long, narrow fruits might be considered the prototype of the modern thermos since they were often carried along on caravan journeys as a source of water. Egyptian slaves were reportedly provided with rations of leeks and garlic as protection against sunstroke, and cucumbers as a source of water. High water content and a bland flavor were prized qualities in the arid regions of Asia Minor and travelers introduced cucumbers throughout the Mediterranean region. Cucumbers are mentioned in the Bible and are a much more important item in the diet of peoples of western Asia than in the United States where they are limited to use in salads and for pickle making (Fig. 5-8).

Several other members of the Cucurbitaceae are important in Latin America. The white-flowered calabash, or bottle gourd, *Lagenaria siceraria,* has continued to baffle biologists interested in the origins of domesticated plants because it

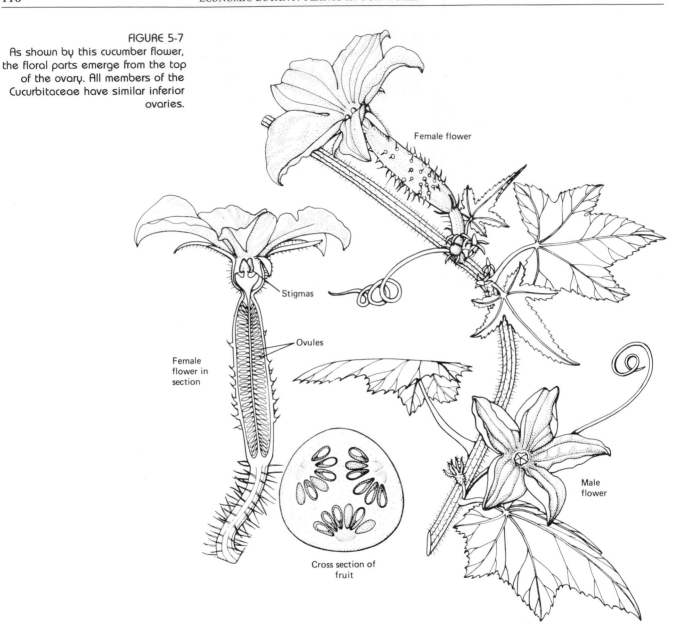

Female flower

Stigmas

Ovules

Female
flower in
section

Cross section of
fruit

Male
flower

appears to have been cultivated in Ecuador and Peru over 7,000 years ago and in Egypt over 3,000 years ago. How the species managed to be transported across either the Pacific or Atlantic Ocean before European contact has been the subject of many debates. Most workers now consider the disjunct distribution to be a natural one, attributable to the rafting of fruits across an ocean. Archaeological collections from Peru show a striking diversity of gourd shapes, several of which had apparently been selected by Indians for specific functions. Gourds of a certain shape were used as *chicha* (corn beer) cups, while others served as water vessels and still others as bowls.

In Mexico, not only various species of *Cucurbita,* but also the chayote (mirliton or chowchow) (*Sechium edule,* Fig. 5-9) was domesticated in pre-

FIGURE 5-8
Two hundred bushels of cucumbers are carried on this conveyor belt each hour during harvest on a Michigan farm. Mechanical harvesters saved the cucumber industry, which was threatened by a shortage of hand labor. Courtesy U.S.D.A.

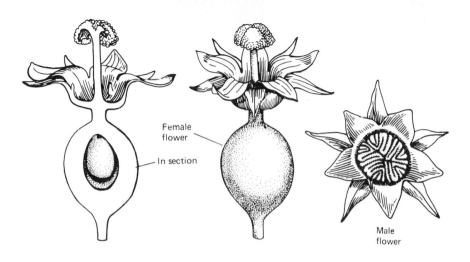

Female flower

In section

Male flower

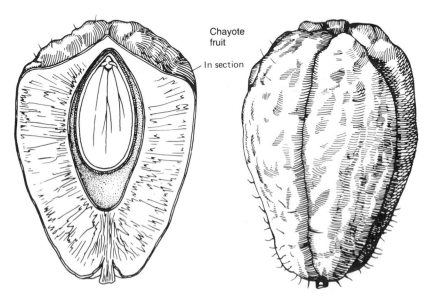

Chayote fruit

In section

FIGURE 5-9
Because each chayote contains a single large seed, entire fruits are planted to establish new vines. (After Baillon.)

FIGURE 5-10
The luffa gourd is eaten when young and tender, but as the fruits mature, they become tough and fibrous. The mature fruits can be picked or allowed to remain on the vine until after the first frost. The soft portions of the fruits are removed from the fibers by retting. The network of rough tissue that remains serves as a body scrubber.

Columbian times. The pear-shaped, green fruits that are occasionally seen in U.S. supermarkets can be boiled, fried, stuffed, or eaten in salads. Unlike most other members of the squash family which have numerous seeds embedded in the endocarp tissues, chayotes have only a single large seed. As the fruit matures, the seed becomes bitter and imparts this flavor to the fruit. Consequently, chayotes are almost always eaten when they are young.

Other cucurbitaceous genera, *Luffa* (primarily *L. acutangula* and *L. cylindrica*), *Momordica,* and *Benincasa* (Fig. 5-10), are commonly used in Asia. Luffa fruits are eaten only when they are very young. Old fruits are allowed to rot partially, and the soft, disintegrating cells are washed away. The remaining network of fibrous tissue yields a luffa "sponge" used for scrubbing. *Momordica,* or bitter melon, is generally used in soups. *Benincasa hispida,* the wax gourd (see Fig. 1-21), is a mild, watery melon eaten in a variety of ways in Malaya.

TOMATOES, PEPPERS, EGGPLANTS

It is hard to think of Italian food without evoking visions of fragrant pots of tomato sauce, yet tomatoes are American and their inclusion in Old World cuisines came only after the discovery of the New World. Initially, acceptance of the fruit and its incorporation into regional cooking was slow because of superstitious beliefs about it. In fact, tomatoes, which are today the most important vegetable grown in the United States (in terms of tonnage), were thought to be poisonous in this country until the 1800s.

While there is some dispute about whether the tomato is native to Peru or Mexico, current opinion favors eastern Mexico as the area of first domestication. The Mayans called the fruit "xtomatl" or "tomatl," corrupted by the Spanish into "tomate." The English substituted the "o" for the "e." Once the colorful fruits were brought back to Europe, the Spanish and the Italians were the first to accept them. Elsewhere in Europe and in the British colonies, acceptance was much slower because of persistent misconceptions. Some believed that tomatoes had aphrodisiac properties. The French called tomatoes *pommes d'amour* (love apples), but this was apparently a misinterpretation of the Italian name *pomo d'oro* (meaning golden apples) or a variant of *pomi dei Moro,* a name that referred to the introduction of the fruit into Europe by the Moors.

Tomatoes were initially thought to be poisonous because many European members of their family (the Solanaceae) have bitter fruits containing toxic and/or hallucinogenic compounds. The German common name "wolf-peach" reflected the belief that the fruits could be used in cabals to evoke werewolves. Linnaeus formalized this early appellation by giving the name *Lycopersicon esculentum,* Latin for "juicy wolf peach," to the species.

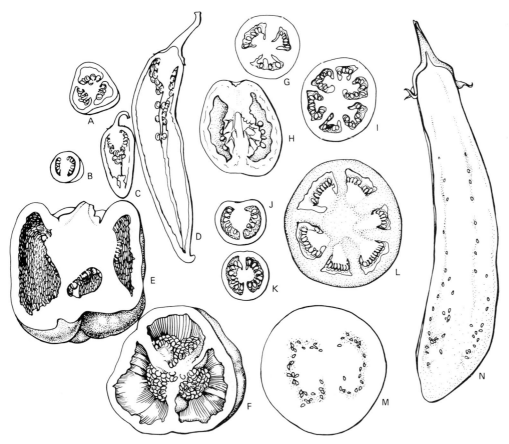

FIGURE 5-11
Cross sections of fruits of the Solanaceae (nightshade family). (A–D) hot peppers; (E–F) sweet bell peppers; (G–H) paste (plum) tomatoes; (I & L) salad tomatoes; (J, K) cherry tomatoes; (M, N) Japanese eggplant. All are berries.

Tomatoes were brought to temperate North America by the British, but they were initially grown only as ornamental plants. Until 1800, the fruits were relegated to a role as a pustule remover. According to an old farm journal, doubts about the fruit's edibility were finally dramatically removed in 1820 when Colonel Robert Gibbon Johnson announced that at noon on September 26, he would eat a bushel of the dreaded fruits. Two thousand people thought him mad and turned out to witness the event. To their astonishment, he survived. In the ensuing 20 years, tomatoes became a relatively popular fruit,

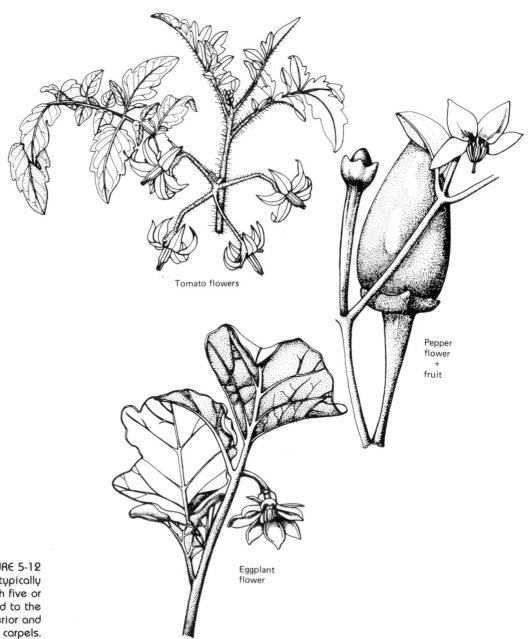

Tomato flowers

Pepper flower + fruit

Eggplant flower

FIGURE 5-12
Flowers of the Solanaceae typically have five fused petals with five or fewer stamens attached to the corolla. The ovary is superior and generally has two carpels.

FIGURE 5-13
Children of migrant workers harvesting bell peppers in a North Carolina field. (Photo by S. Rotner. Courtesy of the United Nations.)

but the spectacular surge in tomato cultivation came after 1920. Most of this increased productivity went into processed goods such as tomato juice, canned tomatoes, tomato paste, and catsup.

Wild tomatoes are outbreeding perennial herbs with small red berries which have two carpels (Fig. 5-11 J, K). Cherry tomatoes are more similar to wild types than the giant-fruited varieties that are more commonly grown today. During the process of domestication, there has been selection for self-pollination (see Fig. 2-8) to ensure high fruit set. Even so, tomato flowers have to be vibrated in order to shake the pollen from the tubular anthers (Fig. 5-12). In open fields, wind currents are sufficient to dislodge the pollen, but in greenhouses, the plants must be artificially shaken. Man has also selected for resistance to bacterial and fungal infections, nematode damage, and for a variety of fruit colors, shapes, and sizes. Unfortunately, tomatoes are a crop in which taste has been sacrificed for durability. Tomatoes grown on a large scale for mechanical harvesting and shipping long distances tend to be tough, dry, and flavorless compared to home-grown fruits.

Like tomatoes, sweet peppers (*Capsicum annuum*, Solanaceae, Figs. 5-11, 5-12) are native to the New World and were probably also first domesticated in Mexico. Archaeological sites at Tehuacan, Mexico, have yielded pepper seeds dated to be almost 8,000 years old. Early peppers all seem to have been pungent types (see Chapter 9). Selection for sweet varieties must have occurred later. Most of the peppers produced in the United States today are sweet, or bell, varieties (Fig. 5-13).

The last commonly encountered edible fruit of the Solanaceae is the eggplant (*Solanum melongena*, Figs. 5-11 and 5-12), known only as a cultivated species. India, or perhaps southern China, is purported to be its native home. Sometime in the fifteenth century, eggplant cultivation spread to Europe and later to the New World. Eggplants have remained a very important dietary item in India but usually serve as an accessory food in other countries. Characteristics of eggplants that have tended to hold down their popularity in America are the browning of the flesh once the fruits are peeled or skinned, and a tendency toward bitterness. While the name eggplant may seem inappropriate for the large, ovoid, black-purple fruits that are now marketed, varieties common a few hundred years ago had small fruits that more closely resembled true eggs.

COCONUTS

There is a South Seas proverb, "He who plants a coconut tree, plants food and drink, vessels and clothing, a habitation for himself, and a heritage for his children." Because of the versatility reflected in this adage, coconuts have earned the designation as the greatest provider in the tropics. Coconuts (*Cocos nucifera*, Arecaceae) can yield oil, fiber, food, and drink. Since we discuss oil production in Chapter 10 and fiber in Chapter 16, we mention here only the history of the coconut and its use as a primary food. Discussing its origin is not, however, an easy task. Controversy has raged for years about the coconut's native home, because the fruits were present on the Pacific coasts of South America, southeast Asia, and Polynesia before Europeans reached the New World. Various geographers have suggested that ancient voyagers crossed the Pacific before 1492, but evidence currently supports the hypothesis that coconuts are native to the Indo-Pacific region and that they dispersed across oceans passively by means of currents. Coconuts are now planted wherever the climate supports their growth.

Coconut palms are monocotyledons with a trunk composed of sheathing leaf bases. Plants begin to flower and fruit several years after germination and establishment. The species is monoecious with numerous male and a few female flowers on each inflorescence. The fruit is formed from a flower with three carpels (each represented by an "eye" of the coconut), only one of which develops. The mature fruit contains one seed, the largest known. The embryo itself is small and located near the stem end. Initially, the endosperm is a liquid containing free nuclei. This is the liquid drunk from green coconuts in many tropical countries. As the endosperm matures, cell walls form around the nuclei, and the endosperm solidifies into an oil-rich layer of coconut "meat" inside the seed coat. It is the solidified endosperm that we eat in pieces or grated in cookies, cakes, and pies. Coconut "milk" is made by soaking grated coconut meat in water and squeezing out the liquid.

If mature coconuts are not utilized or harmed, the embryo can germinate within the coconut since there is no dormancy period. The germinating seedling eventually extends the tip of the cotyledon through one of the eyes. The base of the embryo swells into an absorbing organ that eventually fills the entire cavity of the coconut as it digests the endosperm (Fig. 5-14). The swollen organ, called a coconut apple, can also be eaten. In addition to providing coconut meat, or copra, for eating or oil production (Chapter 10), the stalk can be fed to animals. The inflorescence stalk can also be tapped and the exuded sap used as a basis for fermentation. Occasionally, the growing tip of a coconut palm is removed and eaten as a vegetable, but since this results in the death of a valuable tree, it is rarely done. Hearts of palm are therefore usually obtained from other palm species.

FIGURE 5-14
Coconuts are harvested from trees which can reach up to 80 feet in height. Once the embryo has germinated, it produces an absorbing organ that enlarges as it consumes the rich endosperm. This organ, known as the "coconut apple," is sometimes eaten. (Germinating coconut modified after W. H. Brown. 1935. *The Plant Kingdom.* Boston, Ginn.)

DATES

The fruits of the date palm (*Phoenix dactylifera*, Arecaceae) were an important food in biblical times. Fruits of wild relatives of the modern commercial date were presumably gathered for thousands of years by tribes wandering across the arid regions surrounding the Mediterranean, but by about 4500 B.C., dates had been domesticated. The Muslims believed that the date palm was made from the dust left after God made Adam. Certainly, dates would have seemed like a divine gift to desert travelers because they indicated the presence of water, which meant food and relief from the parching heat. Dates are also highly nutritious, containing 75 percent carbohydrate and 2 percent protein. It is thus not surprising that the date palm was considered the "tree of life" to Bedouin people.

Date palms are dioecious, but as early as 2300 B.C., agriculturalists had learned to hang a male inflorescence in a female tree to enhance pollination. Modern machines now blow pollen across the female flowers when the stigmas are receptive.

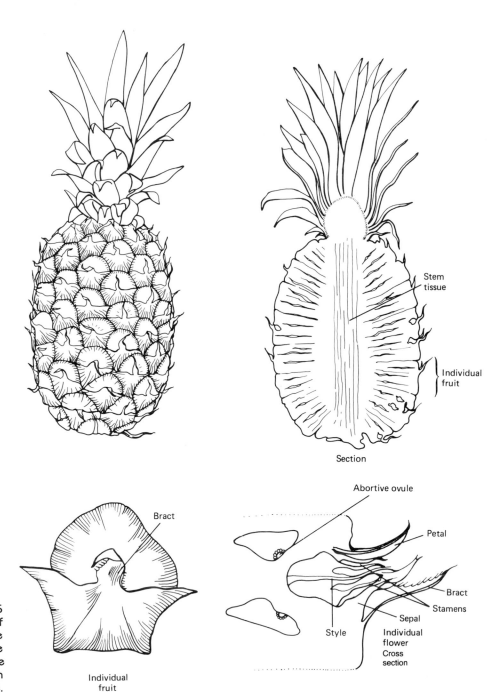

Stem tissue

Individual fruit

Section

Bract

Individual fruit

Abortive ovule

Petal

Bract

Stamens

Sepal

Style

Individual flower Cross section

FIGURE 5-15
A pineapple fruit is composed of 100 to 200 ovaries of separate flowers fused together and to the flowering stem. The remnants of the floral parts produce the prickles on the knobbly surface.

PINEAPPLES

Pineapples (*Ananas comosus*, Bromeliaceae) are indigenous to the New World and were widely cultivated by native people by the middle of the fifteenth century. Columbus described pineapple fruits and noted during his second voyage that they resembled pine cones, a similarity that led to the common English name. American Indians considered the pineapple a symbol of hospitality, and they used the sweet juice for making an alcoholic beverage, a poultice, and as a component of arrow poison concoctions.

Pineapples are multiple fruits formed by the fused ovaries of 100 to 200 independent flowers borne on a stout stalk arising from a tuft of stiff, swordlike leaves (Fig. 5-15). Wild pineapples are pollinated by hummingbirds, but modern cultivars are parthenocarpic, although seeds can be obtained by carefully pollinating the flowers. Since most cultivated varieties are seedless, commercial plantings are made by inserting either the leafy portion that is borne above the fruiting stalk or the severed axillary branches into the soil.

Pineapples were spread around the world in the 1500s by Portuguese, Dutch, and Spanish traders. They were introduced into Hawaii in the early nineteenth century, but it was only after a young yankee entrepreneur, J. D. Dole, encouraged the natives to grow the plants that Hawaii began to rise to its present position as the world's largest producer of the fruits (Fig. 5-16). Pineapples are easily canned (Fig. 5-17) or pressed for juice, but they contain a proteolytic enzyme, bromelain, which prevents their being used in gelatin desserts. Because of this protein-degrading enzyme, pineapples have been used as a meat tenderizer. Fibers can be extracted from the leaves and made into a strong, flexible cloth used for clothing in Brazil, Malaya, and the Philippines.

FIGURE 5-16
A pineapple's whorl of sword-shaped leaves, seen in these plants being harvested, is typical of bromeliads. (FAO photo by Banoun/Caracciolo.)

FIGURE 5-17
A production line in a Taiwanese pineapple canning factory. The stem portion of the fruit, which tends to be tough and tasteless, is removed before the fruit is canned. (Courtesy of the United Nations.)

BANANAS

Bananas are so readily available in the United States today and so important a food in some Latin American countries that it is easy to forget their recency in the New World. Bananas and their relatives in the genus *Musa* (Musaceae) are all native to eastern Asia and Australia. Various species of the genus have been used for fiber (see Chapter 16), food, and ornament in these areas since prehistoric times. It was even proposed by Sauer that agriculture first arose in southeast Asia with the banana among the first plants cultivated. While few people would now support Sauer's idea, there is no doubt about the importance of wild bananas in the lives of preagricultural and early agrarian peoples in southeast Asia.

Domestication and eventual commercial production of the modern edible banana involved hybridization, polyploidy, and the development of seedlessness. This complicated history has obscured the ancestry of the common banana so thoroughly that Simmonds has suggested that it should not be given a specific name but should merely be referred to as *Musa* followed by cultivar name. Other authors, however, still prefer to use Linnaeus's species name, *Musa paradisiaca* for sweet and starchy bananas.

From their native southwestern Pacific home, bananas spread to India by 600 B.C. They were probably introduced independently into Africa and eastward across the Pacific. In 1522, bananas were taken from the western coast of Africa to the Canary Islands as food for slaves being sent across the Atlantic. The first clones to occur in the New World were planted in the West Indies in the seventeenth century. Included in these early clones was at least one type of sweet banana, the "Silk Fig." Other early cultivars were starchy and were generally best when cooked before eating. While bananas are used for beer and are steamed, boiled, dried, or roasted in their native region of Asia, they are usually simply eaten fresh or fried in the New World. The common eating bananas found throughout the world today belong primarily to the "Gros Michel" variety, a type derived within the last 200 years and introduced into Dominica in the early nineteenth century.

The story of the banana's rise to a prominent agriculture position is intimately linked to the history of the United Fruit Company, which was formed in 1899 by the merger of the Boston Fruit Company with Keith Interests. Both of these earlier companies had been transporting bananas to the United States from Central America for 10 years before the merger. Yet, since bananas do not keep well, and neither company had a good distribution system, their operations remained small. In 1900, the newly formed company developed a successful system of transport, and the United Fruit Company began to expand into Panama (1903), Guatemala (1907), El Salvador, and Costa Rica (in the 1920s). Nevertheless, it was 1929 before the first shipments of bananas were sent directly to the United States. In the 1930s and 1940s the company extended its operations into Colombia and Ecuador.

Even with a good shipping system, cargos of bananas were often ruined because carbon dioxide and ethylene produced by bruised and ripe bananas accelerated ripening and caused rotting of entire shipments. Experiments eventually showed that if bananas are kept at 10°C (56°F) and 90 percent humidity, ripening is delayed. Conversely, ethylene can be profitably used at a later time to induce ripening of green bananas. Once the physiology of banana ripening was understood, shippers were able to prevent large-scale spoilage and even ship green bananas that they could ripen at will when they reached their destination. As a result, banana production boomed and the United Fruit Company continued to expand.

For several decades, United Fruit essentially owned the land and controlled the workers on huge acreages of several Central and South American countries, causing them to be referred to as the "banana republics." By World War II, in the face of public pressure and political reorganization within these countries, the influence of the United Fruit Company began to wane. Still, the importance of banana cultivation in many tropical American countries is reflected by the fact that bananas now constitute the major source of carbohydrate for millions of their people.

Bananas are planted using any of numerous vegetative parts, corms, pieces of corms, or suckers. Adult plants are essentially giant herbs, athough they are called trees and their cylindrical stalks commonly considered trunks (Fig. 5-18). Lateral stems form at the base of the main stem producing suckers, most of which are removed in plantations. Plants start to flower 2 to 6 months after initial shoot elongation. The flowers are borne on long, pendant, monoecious inflorescences with the female flowers at the base of the stalks and male flowers at the apex. Female flowers are produced in clusters, each protected by a large bract. As a cluster of 12 to 20 female flowers becomes mature, the subtending bract peels back. Each cluster of female flowers produces a "bunch," or "hand," of bananas when it matures. In some regions, immature bananas are covered with plastic bags to prevent damage from soil, sun, dust, birds, or insects.

Bananas are often cut by hand by two workers, one of whom cuts the inflorescence stalk, while the other stands ready to receive the heavy mass of fruit on his back. In this way, fruit damage is minimized. Care must be taken to avoid bruising or wounding the bananas, because such injuries cause the fruits to release ethylene. For shipment, only perfect bananas are used. They are picked green, washed, dipped in pesticide, and packed.

After fruiting, the main banana stalk is usually cut at the base. Banana plants can be replanted each year, or a sucker of each of the previous year's

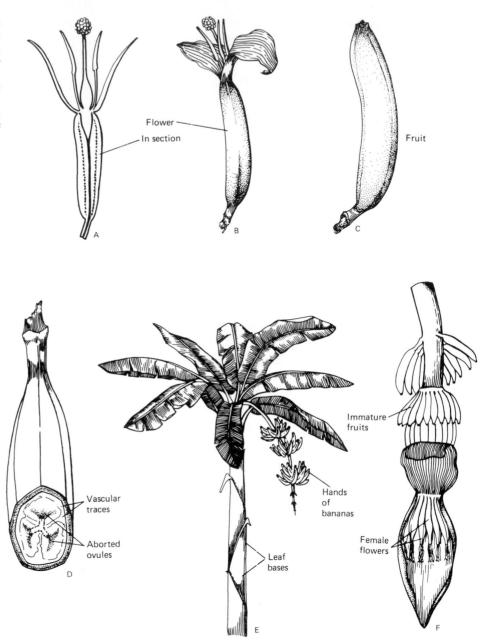

plants can be allowed to develop. If the initial planting was in rows, a sucker from the same side of each of the plants can be left to develop. The following year, the rows simply shift a few centimeters. This practice of allowing natural vegetative reproduction to produce subsequent plants can continue over a 5- to 25-year period. Bananas can yield impressive amounts of carbohydrate per unit area, with yields similar to those of potatoes (between 14 to 47 tons per acre).

FIGS

The same family, Moraceae, that provides us with mulberries contains figs, an important subtropical fruit. The fig genus *Ficus* contains over 800 species, many of which are cultivated, but only one of which, *Ficus carica*, produces a commercially important fruit. This fig, a native to the Mediterranean region, is mentioned in Egyptian documents (2700 B.C., Fig. 5-19), and is frequently referred to in the Bible. Adam and Eve were supposed to have sewed together fig leaves to cover themselves once they had learned to be ashamed of their naked bodies. To the Greeks, an edible fig was a "sykon," a designation that led to the modern term of syconium for the fruit, which is actually an inflorescence enclosed inside receptacular tissue.

Wild figs are disseminated by birds or bats who eat the fruits and later defecate the tiny seeds. People generally use cuttings for propagation of cultivated varieties. A few figs are either self-pollinating or parthenocarpic. The popular Kadota and Mission figs, for example, can produce fruits without pollinators, but many wild figs, and most of the edible varieties, have a very elaborate pollination system involving tiny wasps that mate inside the fruits (Fig. 5-20). Newly emerged and mated females carrying pollen from male flowers of their natal figs, fly to immature figs and enter. While laying eggs in modified female flowers of the immature figs, females pollinate normal, fertile female flowers. Smyrna figs, historically the most popular edible fig, contain only female flowers in their syconia. In order to produce fruits, the flowers must be pollinated with pollen from male flowers in syconia on different trees. These pollen-donor trees are called "caprifigs." In the Mediterranean region, Smyrna and caprifigs naturally grow together. When Smyrna figs were brought to the New World, the pollination system was not fully appreciated and the inedible caprifigs were left behind. Naturally, the Smyrna trees failed to produce fruits. Eventually caprifig trees and the wasps were imported and successful American fig cultivation began.

FIGURE 5-19
Figs and monkeys. (Redrawn from an Egyptian frieze of the twelfth century B.C.)

FIGURE 5-20
The pollination of most figs is an intricate process involving species-specific wasps that mate and lay their eggs in the ovaries of short-styled female flowers within the inclosed inflorescences, or syconia. When the young wasps emerge, they escape from the syconia where they were born, but they carry some pollen from this syconium to the female flowers in the syconium where they will mate and lay their eggs. (After von Marilaun)

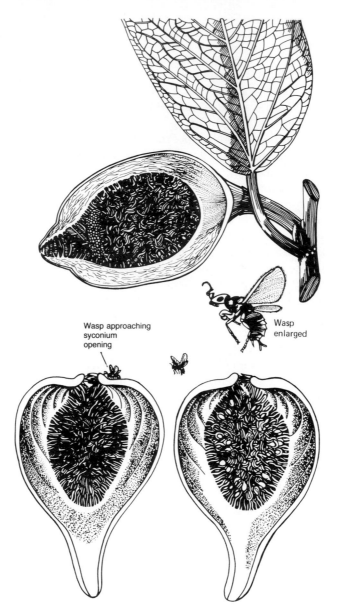

Wasp approaching syconium opening

Wasp enlarged

BREADFRUIT AND JACKFRUIT

Like bananas, breadfruits constitute an important carbohydrate source for people in many tropical areas. A native of Polynesia, breadfruit (*Artocarpus altilis*, Moraceae), has been cultivated since ancient times. Breadfruits are tall (to 20 m) trees that have compound, unisexual inflorescences. Female inflorescences develop into a round, multiple fruit about 10 to 30 cm in diameter (Fig. 5-21). The moist, edible pulp is formed by abortive flowers. The few seeds that are produced are not normally eaten but can be used in times of food shortages.

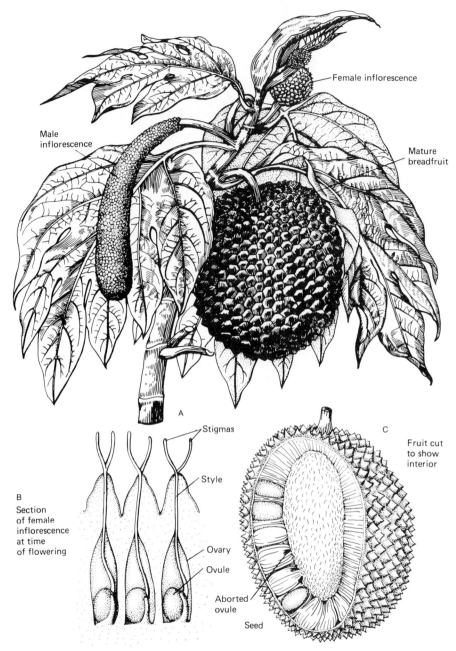

FIGURE 5-21
The breadfruit is a multiple fruit, 10 to 30 cm in diameter. Cooked, the flesh resembles potatoes in flavor and texture. (A) Flower and flowering branch (after Baillon); (B) cross section of a portion of a female flower; (C) cross-section of part of a fruit.

An average fruit usually contains about 20 percent carbohydrates, and when cooked is similar in taste and consistency to a white potato.

Early European explorers were amazed when they first encountered breadfruits, describing the plants as producing masses of "bread" (hence the English name). In an attempt to capitalize on this bounteous (and cheap) natural source of carbohydrates, Britain sent Captain Bligh (Fig. 5-22) and *H.M.S. Bounty* to Tahiti to collect breadfruits for introduction into the New World colonies. In 1773, Captain Bligh headed out from England with his ship

FIGURE 5-22
H.M.S. Bounty, the ill-fated ship on
which Captain Bligh attempted to
bring breadfruit trees to the West
Indies. (Photo courtesy of the
National Maritime Museum,
London.)

equipped to be a floating greenhouse. It took Bligh and his men 6 months to collect seedling breadfruits and prepare them for the long journey to the West Indies. During their stay in Tahiti where they collected the plants, the crew began to savor the easy island life and compliant women. Soon after setting sail for the West Indies, they mutinied and set Bligh and 18 loyal seamen adrift in a longboat with a small cache of rations (Fig. 5-22). Bligh survived the mutiny, crossing 3,618 nautical miles of the Pacific Ocean to Timor. Undaunted, he soon returned to Tahiti in another ship, *H.M.S. Providence*, and during his second voyage successfully established the trees in Jamaica. Today, they grow throughout the humid tropics.

Jackfruit (*Artocarpus heterophyllus*, Moraceae) is similar in general morphology to the breadfruit, but its fruits contain a higher percentage (up to 38 percent) of carbohydrates than its relative. Jackfruits are commonly eaten as a dessert fruit. Large-scale cultivation has tended to remain restricted to its native range of India and Sri Lanka.

AVOCADOS

Avocados or alligator pears (*Persea americana*, Lauraceae, Fig. 5-23) have a controversial history and seem to defy our often repeated theories about animal-dispersed fleshy fruits. Fleshy fruits are generally low in calories and consist primarily of water and sugars, or in the cases of bananas and breadfruits, starch as well. Avocados, in contrast, have a mesocarp that is extremely rich in oil. Up to 30 percent of the pulp of cultivated varieties (on a dry weight basis) can be oil. As a result, avocados have the highest energy containing fruit pulp (from 2,000 to 2,800 calories per kilogram) known. Moreover, the seed is not protected in any way from the sharp teeth or digestive juices of an animal that might feed on the fruit. It has been suggested that extinct large animals were the original dispersers of avocados, but the postulation of a large animal does not explain how a seed without a hard endocarp is protected in its journey from

FIGURE 5-23
In American markets, we can find smooth-skinned West Indian avocados grown in Florida (*left*) and smaller, warty California avocados (*right*), which are a cross between Mexican and Guatemalan varieties.

the mouth through the digestive system of such an animal. The natural dispersal of avocados thus remains a mystery.

Avocados are known only as a cultivated species and yet they appear in some of the oldest archaeological deposits in southern Mexico (7000 B.C.). It is possible that avocados were independently domesticated three times in the Americas, giving rise to what are now known as the West Indian, Guatemalan, and the Mexican (or Drymifolia) varieties. The Mexican variety has a small (about 250 gm, or 8 oz) fruit with a thin, smooth skin, a seed loose within the cavity, and a 30 percent oil content. Guatemalan fruits weigh between 500 and 1000 g (1 to 2 lb) and have thick, warty skins. The West Indian types are about the same size with thick but smooth skins and a mesocarp with only 8 to 10 percent oil.

Avocados were first planted as a commercial venture in the United States in Florida in 1893, but acceptance was slow. It is often said that the recent popularity of the avocado is a direct result of a calculated publicity campaign. Having failed to combat historical, adverse feelings about the fruit, growers (on the advice of an advertising agent) began loudly to deny that the fruits had any aphrodisiac properties. In the 1920s, few other claims would have been able to guarantee an immediate demand. The campaign was a success, and the popularity of avocados has been increasing ever since. The relationship between avocados and sex had, in fact, been claimed by native American Indians. The word avocado comes from the Aztec word "ahuacacuahatl" meaning testicle tree, which refers either to its stimulating properties or the appearance of the fruits that commonly hang in pairs.

California and Florida now supply the country with an abundance of avocados during good fruiting years. In California, most fruits are hybrids between the Mexican and Guatemalan varieties. Florida produces predominantly the large, smooth-skinned West Indian types (Fig. 5-23).

MANGOES

Of all of the truly tropical fruits, mangoes (*Mangifera indica*, Anacardiaceae) are probably the most popular on a worldwide basis. They are particularly appreciated in their native region of southeast Asia, where not only the pulp but also

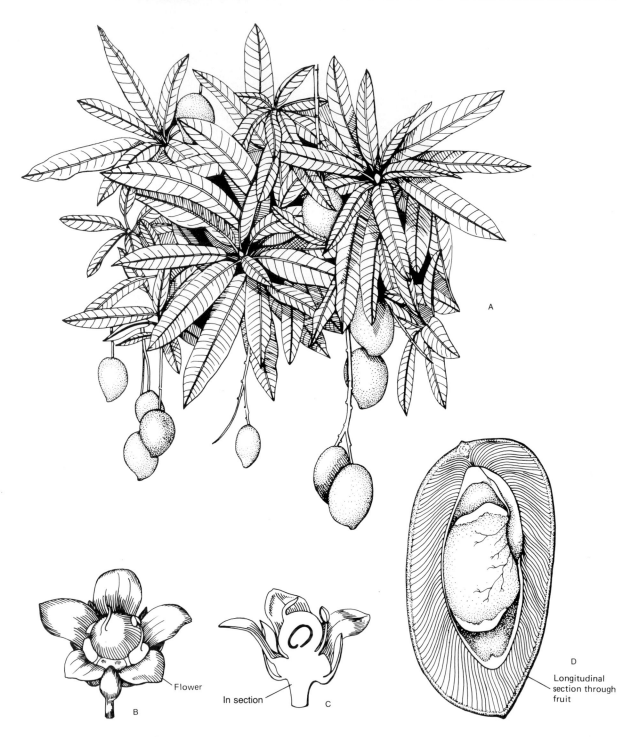

FIGURE 5-24
(A) Mangoes dangle from trees on long stalks. (B) A single mango flower shown (C) in cross section. (D) The fruit in cross section shows the fleshy mesocarp and the hard endocarp.

the seeds have been used for food. In times of emergency, the cotyledons of the seed can be ground into a flourlike substitute and the leaves of the trees can be used for cattle food.

Mangoes are naturally large trees, reaching 34 m in height. Where they are planted and irrigated in relatively dry areas, the trees are much smaller and the plantations tend to resemble orchards. The flowers of mangoes are small and borne in long clusters (Fig. 5-24). Usually only one flower per cluster develops a fruit. In full fruit, a mango tree looks as if someone has tied fruits on the branches because the fruits hang on long, bare stalks. The usually stringy pulp of a mango is the mesocarp that surrounds a stony endocarp enclosing the large seed.

Mangoes belong to the same family as poison ivy and, like many members of the family, produce a latex that is irritating to some individuals. Fortunately, latex production is restricted to the leathery skin, so peeling the fruit can generally eliminate the source of irritation.

Mangoes were introduced to Brazil by the Portuguese in the early 1700s and brought from South America to the West Indies in 1742. They now constitute an important seasonal food source for the poor populaces of Jamaica and Haiti. Until recently, few were exported, but they are now commonly found in many U.S. supermarkets.

OKRA

Although the native region of okra (*Abelmoschus esculentus,* Malvaceae) is still uncertain, it is thought to be either southwest or south-central Asia. The spread of the domesticated species must have been very early because okra was used in ancient Egypt and spread from there to the Far East and Europe. Okra first reached the New World in the 1600s, and in the 1700s it was adopted as a crop by French immigrants in Louisiana. They referred to the vegetable by its African name, "gumbo," and used it with such frequency that it became an integral part of Cajun cooking.

Okra is a member of the Malvaceae, or mallow family with hibiscus-like flowers that produce capsules resembling elongated cotton bolls (Fig. 5-25). Many people find the mucilaginous coating on okra seeds objectionable but the gum acts as a thickening agent in soups, stews, and gumbos. Okra has maintained a place in American agriculture because it can tolerate even the hottest southern months and still produce a substantial fruit crop at the end of the long growing season.

POMEGRANATES

Pomegranates (*Punica granatum,* Punicaceae) are subtropical fruits native to the Near East and Africa (Fig. 5-26). Their importance in the ancient world was as much symbolic and artistic as nutritional. The Hebrews embellished their buildings with pomegranate motifs, and the globular fruits with their numerous

FIGURE 5-25
The appearance of okra flowers betrays the species' kinship with *Hibiscus* and cotton. Because of its shape, the Greeks named the fruit "okra," meaning lady's fingers.

juicy red seeds (see Fig. 1-27) eventually became a symbol of fertility and abundance. This symbolism was carried to Asia where fruits were offered to wedding guests who were expected to throw them onto the floor of the honeymoon suite so forcefully that they shattered, scattering their seeds. This practice was believed to ensure fertility and a large number of offspring for the newlyweds. The broadcasting of pomegranate seeds has another, more sobering, connotation. The French word for the fruit, "grenade," is applied to a hand-thrown weapon used extensively since the seventeenth century that scatters not seeds, but deadly metal fragments.

The Moors took pomegranates with them when they conquered Spain in 800 A.D. They used the seeds for food and the rinds of the fruit as a source of

FIGURE 5-26
A flower and bud of the pomegranate.

tannin for leather. The Spanish introduced the fruits into the New World many centuries later. Pomegranates are now grown from the southern United States to northern Chile and Argentina. In many areas, the trees are planted primarily as ornamentals because their flowers as well as their fruits are attractive. In addition to being eaten fresh, the seeds can be expressed to yield a juice. If sweetened with sugar, the juice becomes grenadine syrup.

PAPAYAS

Papayas (*Carica papaya*, Caricaceae) are smooth-textured, musky-flavored fruits native to Central America but little known in the United States until recently. In parts of Central and South America, they find their way into practically every meal during the fruiting season. The demand in the United States for papayas has not traditionally been for the fresh fruit but for papain, an enzyme extracted from the latex exuded when the skin of the fruit is scored. Papain is very effective in breaking down proteins and thus forms the basis of commercial meat tenderizers. It is also used in chewing gum and cosmetics. Medicinally, papain is effective for digestive ailments.

Papaya plants are short-lived, soft-wooded trees that have a characteristic umbrella-shape growth form. The flowers are borne along the stems in leaf axils. By the time the fruits are mature, the leaves have been shed, leaving clusters of the ovoid fruits emerging directly from the stem. Inside each fruit is a mass of slimy (each 0.5 cm in diameter) black seeds (Fig. 5-27). For latex tapping, green fruits are used, beginning when they are 10 cm in diameter. The skin is scored diagonally in such a way that only the outer part of the exocarp is cut. The latex which oozes out is collected the next day. A single fruit can be tapped several times. The fruit can still be eaten after having been subjected to tapping.

A SAMPLE OF MORE EXOTIC TROPICAL FRUITS

While many of the tropical fruits we mention here can be tasted fresh only where they are grown, they are of interest since some do occasionally appear in specialty food stores or even supermarkets. Within their native ranges, some are important foods for local peoples during the fruiting season.

The sops, species of the genus *Annona* (Annonaceae), are American natives. Several, such as the sugar apple (*A. squamosa*), the custard apple or sweet sop (*A. reticulata*), and the cherimoya (*A. cherimola*), are eaten as fruits or used for juice throughout the Neotropics. The flowers of *Annona* species have numerous pistils. Consequently, annonas produce compound fruits that have a characteristic segmented appearance on the outside. The custard apple is the most widely consumed of the annonas because selection has led to consistently high quality in the fruits and to seedlessness. The species also tolerates a wider range of climates than its relatives. It has been introduced to Florida and is grown there on a very limited scale. To some, however, the cherimoya is the finest of all of

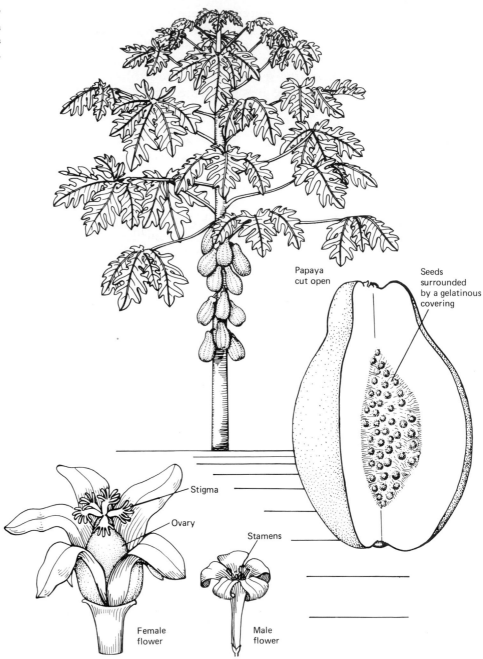

FIGURE 5-27
Because papaya flowers are
cauliflorous, the fruits are borne
directly on the plant's stem.

the sops, but its growth is restricted to montane tropical regions between elevations of about 1,400 to 2,000 m.

Loquats (*Eriobotrya japonica,* Rosaceae) are among the few tropical rosaceous fruits. Native to eastern and southern Asia, loquats have since been carried to the Mediterranean area and to the southern United States where they are planted primarily as ornamentals. The large shrubs are valued for their coarse-textured foliage, clusters of white flowers, and the orange, ovoid fruits that mature in the spring if winters are mild. The sweet flavor of loquats is best appreciated when the fruits are fully ripe, but slightly immature fruits are used in the preparation of jams, jellies, and liqueurs.

The carambola (*Averrhoa carambola*, Oxalidaceae) is a tart tropical fruit. The tartness is caused by the presence of calcium oxalate crystals in the flesh, which dissolve in the saliva when the fruit is eaten, forming oxalic acid. The fruits are yellow, 3 to 6 cm long, and star-shaped, reflecting the five carpels characteristic of the oxalis family (Fig. 5-28). Native to Indonesia, star fruits are now planted throughout the tropics.

Most Americans encounter litchis as canned fruits accompanying fortune cookies at the end of a Chinese meal. The so-called nuts, (*Litchi chinensis*, Sapindaceae), are natives of China. They are cultivated primarily in southern China and, to a limited extent, in subtropical areas of Florida. Litchis are produced on medium-sized trees (10 to 12 m tall). The edible portion of the warty, mauve-colored, apricot-sized fruits is the aril that covers the seed.

A recent introduction into U.S. supermarkets is the kiwi (*Actinidia chinensis*, Actinidiaceae, Fig. 5-29). Despite the fact that most Americans associate the fruit with New Zealand, it is a native of Asia. It has been grown commercially in New Zealand and marketed as the Chinese gooseberry for 30 years. However, fruit sales remained poor until New Zealand held a national contest to rename the fruit. Under the name of kiwi, supplied by the contest winner, the fruit has steadily increased in popularity. Orchards of the dioecious vines now cover large acreages in New Zealand and a limited area of California. The pulp of the ovoid, fuzzy green fruits is delicate in flavor and blends well with other foods. In addition, slices of the fruit with their translucent, pale green flesh surrounding a narrow ring of tiny black seeds are striking and produce attractive desserts.

Because of advertising for a canned fruit drink that includes their juice, passion fruits have become familiar to most Americans. To many people, they seem to represent the essence of tropical fruits. It was the flowers, however, and not the fruits that originally attracted Europeans to passion flowers (Fig. 5-30). The intricate and showy flower parts were considered by religious immigrants to represent the instruments of Christ's passion. Plants were consequently taken to Europe where they were grown as curiosities in green-

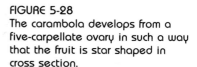

FIGURE 5-28
The carambola develops from a five-carpellate ovary in such a way that the fruit is star shaped in cross section.

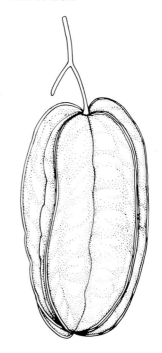

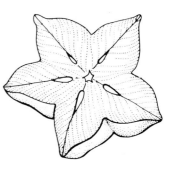

FIGURE 5-29
Kiwi fruits have been hailed as the "fruit of the eighties" by some. Unknown to most Americans prior to 1980, kiwis have now found a place in most supermarkets around the country.

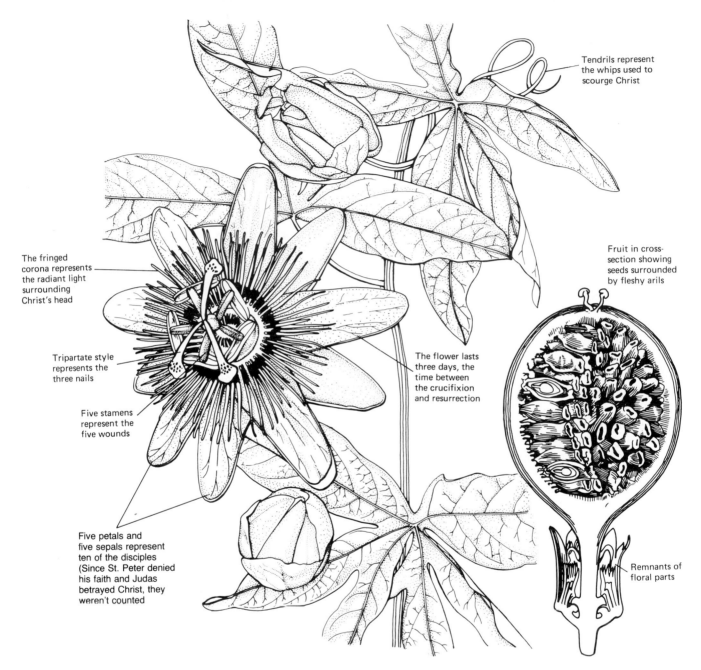

Tendrils represent the whips used to scourge Christ

Fruit in cross-section showing seeds surrounded by fleshy arils

The fringed corona represents the radiant light surrounding Christ's head

Tripartate style represents the three nails

Five stamens represent the five wounds

The flower lasts three days, the time between the crucifixion and resurrection

Five petals and five sepals represent ten of the disciples (Since St. Peter denied his faith and Judas betrayed Christ, they weren't counted

Remnants of floral parts

FIGURE 5-30
The exotic floral morphology of the passion fruit led to its association with the crucifixion of Christ. (Cross section of fruit, after Baillon. Vine and flower, from nature.)

houses. Although the fruits had been eaten by native Americans long before European contact, systematic cultivation of passion flowers as a fruit crop began only a few hundred years ago. Of the 55 or so edible species, *Passiflora edulis* (Passifloraceae) is now grown on the largest scale. This species, a native of Brazil, was introduced to Hawaii in 1810. Since the species grows best in tropical montane habitats at about 2,000-m elevation, Hawaii provides excellent growing conditions. In northwestern South America, *P. mollissima* is commonly grown.

The parts of passion fruits that are eaten or pressed for juice are the arils surrounding the seeds. Each fruit has a tough pericarp surrounding numerous seeds, all of which are embedded in fleshy, aromatic, red-yellow arils. Since the plants are vines, plantations of passion fruits resemble vineyards. The perennial plants are evenly spaced and the branches trained along wires that are strung a meter or so above the ground.

Malay apples (*Syzygium malaccense*, Myrtaceae) and rose apples (*S. jambos* are popular West Indian fruits. The species, introduced from southeastern Asia, have a tart, fresh, crisp flavor. The fruits are pear-shaped and are red or purple on the outside.

Another myrtaceous fruit is the guava (*Psidium guajava*, Fig. 5-31), which has spread from its native home in South America to the Old World tropics. The pear-shaped fruits of the guava are yellow when ripe and contain more vitamin C than most citrus fruits. Because guavas have a distinctly pungent taste when eaten raw, they are most commonly stewed or processed into jams, jellies, or pastes.

Sapote is a name given to an assemblage of species, the most common of which belong to the Sapotaceae. Two of these, the mammey sapote (*Pouteria sapota*) and the green sapote (*P. viride*) are both native American species that produce fruits about 5 to 20 cm (3 to 6 in) long with one to four seeds. The edible portion is the firm, reddish mesocarp surrounding the seed clusters. These two species are related to the more common sapodilla (*Manilkara zapota*), the fruit of the tree from which we obtain latex for chewing gum (see Chapter 11).

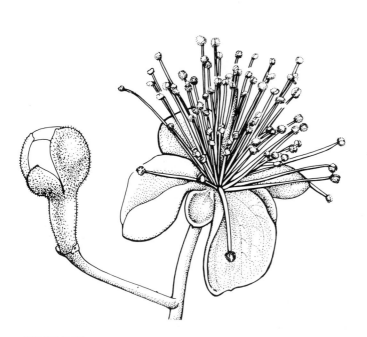
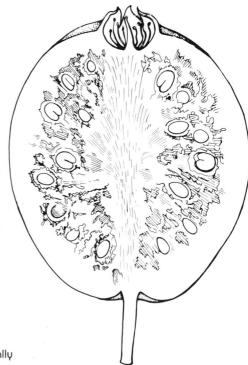

FIGURE 5-31
Because they contain many small seeds and tend to be bitter, guavas are usually processed into jams and jellies. Flower from nature, section after Baillon.

TROPICAL NUTS

Although it seems that most of our common nuts are produced by trees of temperate regions (see Chapter 4), several important nuts come from warm subtropical or tropical areas. One of the most expensive, and to many the most delicious, of tropical nuts is the macadamia (*Macadamia integrifolia*). Macadamias are members of the Proteaceae, a family known primarily for its exotic inflorescences (see Fig. 18-30). Although native to Australia, macadamias are now produced primarily in Hawaii, where they have been cultivated as ornamentals since the 1800s. Hawaii provides well-drained soils and abundant rainfall that are ideal for macadamia growth.

Commercial macadamia fruit production was initially hindered by the fact that macadamia trees do not produce a discrete, abundant fruit crop at one time of the year. In addition, the shells of the seeds are extremely difficult to crack. The taxonomy of macadamia was also confused for many years, and one of the bitter-seeded relatives (*M. ternifolia*) of the sweet macadamia was occasionally grown by mistake. Careful study of the biology and morphology of the macadamias has now shown that the edible macadamias differ from the bitter types by several characters. The problem of freeing macadamia nuts successfully from their shells was eased somewhat when it was discovered that drying the fruits before cracking facilitates the operation. Nevertheless, a hefty 300 lb/in^2 must still be exerted in order to crack the shells. While macadamia nut production is still limited by climatic conditions, asynchronous fruiting patterns, and tenacious shells, Hawaii now enjoys a 6-month (July through January) season during which orchards yield an abundant harvest of the sweet, buttery nuts.

Cashews, another popular nut in the United States, are native to forests of South America. The cashew (*Anacardium occidentale*, Anacardiaceae) is related to poison ivy and mangoes and also contains toxic compounds in various parts

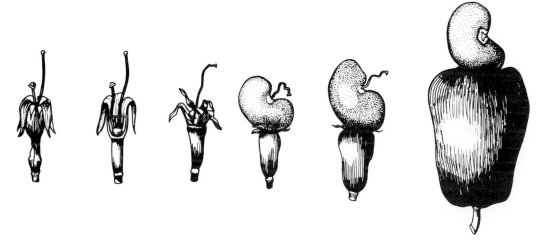

FIGURE 5-32
The development of a cashew. One of the stamens surrounding the ovary is characteristically larger than the others. The cashew "nut" develops from the fertilized ovule inside the ovary and the "cashew apple" from the stem and receptacle. (Modified after W. H. Brown. 1935. *The Plant Kingdom.* Boston, Ginn.)

of the plant. The latex in the seed coat, for example, is irritating, but the seed itself is not. Cashews are produced on curious fruits. The "nut" is the embryo borne inside a hard seed coat. This structure sits atop a large, fleshy, inverted pear-shaped organ (Fig. 5-32). Presumably, the fleshy portions attract animals that disperse the seeds. In Brazil, the fleshy portion of the fruit, called the apple, is occasionally crushed and fermented into a wine called kaju.

Brazil nuts (*Bertholletia excelsa* and other species of the Lecythidaceae) are likewise native to the South American tropics and are also borne in a peculiar fashion. The flowers have a striking appearance, and the fruits are large woody spheres weighing 1 to 2.5 kg (2 to 5 lb) each (Fig. 5-33). Within the massive fruits are 12 to 24 wedge-shaped seeds arranged in concentric rings. Each seed is encased in a stony seedcoat. The tall, rain-forest trees can each produce over 300 pods. Needless to say, standing under a Brazil nut tree when the pods fall can be dangerous, and pickers are often injured during the harvest season. Almost all of the world's supply of the nuts is collected from wild trees. Despite their long association with the trees, native Brazilians rarely eat Brazil nuts. In view of the fact that the nuts contain up to 66 percent fat, it is strange that such a nutritious food source has not been more extensively exploited by local peoples.

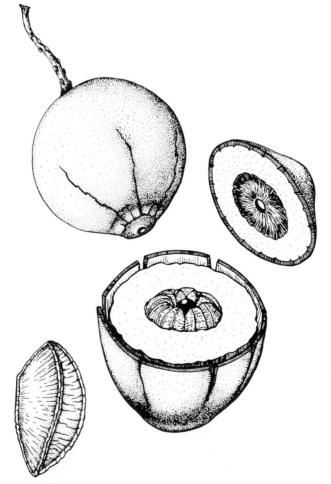

FIGURE 5-33
Brazil nuts are the seeds of species of large forest trees. Between 12 and 20 seeds develop within the 2 to 4-lb fruits. The natives in many regions of Brazil call the fruits "monkey pods" because the inquisitive primates apparently get their hands stuck inside the hard exocarp when they reach inside fruits that have been gnawed open by rodents.

ADDITIONAL READING

Choudhury, B. 1976. Eggplant, in N. W. Simmonds (ed.), *Evolution of Crop Plants.* New York, Longman, pp. 278–279.

Hodgson, R. W. 1950. The avocado—A gift from the Middle Americas. *Economic Botany* 4:253–293.

Moore, O. K. 1948. The coconut palm—Mankind's greatest provider in the tropics. *Economic Botany* 2:119–144.

Pickersgill, B. 1969. The archaeological record of chili peppers (*Capsicum* spp.) and the sequence of plant domestication in Peru. *American Antiquity* 34:54–61.

Purseglove, J. W. 1974. *Tropical Crops,* Dicotyledons. London, Longman. This scholarly work includes a detailed description of most of the tropical fruits, nuts, and vegetables included here.

Reuther, W., W. J. Webber, and L. D. Batchelor (eds.). 1967. *The Citrus Industry,* vol. 1. Davis, California, University of California Division of Agricultural Science. An authoritative work on all aspects of citrus cultivation.

Rick, C. M. 1978. The tomato. *Scientific American* 239(2):77–87. A beautifully illustrated summary of tomato biology and history.

Robinson, T. R. 1952. Grapefruit and pummelo. *Economic Botany* 6:228–245.

Sauer, C. O. 1969. *Agricultural Origins and Dispersals.* 2nd ed. Cambridge, M.I.T. Press. A classic work on the origins of agriculture.

Simmonds, N. W. 1966. *Bananas,* 2d ed. London, Longman. An extremely detailed monograph on the evolutionary history, biology, and agronomy of the species of *Musa* used by human beings.

Whitaker, T. W., and G. W. Bohn. 1950. The taxonomy, genetics, production and uses of the cultivated species of *Cucurbita. Economic Botany* 4:52–81.

Chapter 6

Grains and Forage Grasses

Man may not live by bread alone, but he does devote more than 70 percent of the world's farmlands to the growing of grains (see Fig. 19-8). Harvests from these cereals provide at least half of the world's calories today. Not only humans, but their domesticated animals as well are fed primarily by grains and forage grasses. In this chapter we examine these remarkable plants and discuss why grasses have become the most important group of food plants in the world today.

Grains are considered to be among the first cultivated crops. In the opinion of some anthropologists, their use was a prerequisite for civilization. Wheats and barley sustained early Near Eastern cultures, rice the Far Eastern populations, and corn the pre-Columbian New World civilizations. Contrary to the popular view that Rome fell because of decadence and debauchery, its decline can be attributed, in part, to the failure of its North African grain supplies. Grain failures have devastated many other cultures in the past and threaten modern economies as well.

By definition, grains are the fruits of members of the grass family, Poaceae (Table 6-1). They are indehiscent, one-seeded fruits in which the seed coat is fused with the ovary wall. Grain amaranths and buckwheat, often grouped with the grains, are fruits of species belonging to the Amaranthaceae and Polygonaceae, respectively. Consequently, they are not true grains. The commonly cultivated grasses that yield grains are known as cereals, in deference to the Greek goddess Ceres (Fig. 6-1). The grass family contains over 9,000 species

FIGURE 6-1
Because of the importance of grains in international trade, an art deco interpretation of Ceres, the goddess of grain, was placed on the top of the Chicago Board of Trade building. (Courtesy of the Chicago Board of Trade.)

TABLE 6–1

Grains [Poaceae (Graminae)] Discussed in Chapter 6*

COMMON NAME	SPECIES	CHROMOSOME NUMBER PLOIDY LEVEL
Barley	*Hordeum vulgare*	$2n = 14$ diploid
Corn	*Zea mays*	$2n = 20$ diploid
Goat grass	*Aegilops speltoides*	$2n = 14$ diploid
	A. longissima	$2n = 14$ diploid
Millet:		
Pearl	*Pennisetum americanum*	$2n = 14$ diploid
Finger	*Eleusine coracana*	$2n = 36$ tetraploid (cultivated)
Oats	*Avena sativa*	$2n = 42$ hexaploid
Rice:		
African	*Oryza glaberrima*	$2n = 24$ diploid
Common	*O. sativa*	$2n = 24$ diploid
Wild	*Zizania aquatica*	$2n = 30$ diploid
Texas wild	*Z. texana*	$2n = 30$ diploid
Rye	*Secale cereale*	$2n = 14$ diploid
Sorghum	*Sorghum bicolor*	$2n = 20$ diploid
Teosinte	*Zea mays* subsp. *mexicana*	$2n = 20$ diploid
Tripsacum spp.	*Tripsacum* spp.	$2n = 36$ diploid
Triticale	*Triticosecale* spp.	$2n = 42, 56$ hexaploid and octoploid
Wheat:		
Einkorn	*Triticum monococcum*	$2n = 14$ diploid
Emmer	*T. turgidum*	$2n = 28$ tetraploid
Bread	*T. aestivum*	$2n = 42$ hexaploid

* Names mentioned in Tables 6–4 and 6–5 are not included.

TABLE 6–2

World Production of Grains

CROP	TOP 5 COUNTRIES, 1,000 t		TOP 5 CONTINENTS, 1,000 t		WORLD, 1,000 t
	COUNTRY	t	CONTINENT	t	
Barley	U.S.S.R.	54,000*	Europe	64,623	167,176
	United States	11,300	North & Central America	22,450	
	Canada	10,616	Asia	16,596	
	United Kingdom	10,094	Oceania	5,260	
	Germany, Federal Republic of	8,944	Africa	3,592	
Maize	United States	106,781	North & Central America	129,299	344,103
	China	64,135*	Asia	90,350	
	Brazil	18,756	Europe	56,503	
	U.S.S.R.	14,000*	South America	31,289	
	Mexico	13,928*	Africa	22,383	
Millets (combined)	India	10,500*	Asia	18,709	29,563
	China	7,004*	Africa	8,415	
	Nigeria	2,300*	South America	179	
	U.S.S.R.	2,200*	Europe	31	
	Niger	1,325*	Oceania	30	
Oats	U.S.S.R.	16,000*	Europe	12,479	43,101
	United States	6,928	North & Central America	9,771	
	Canada	2,773	Oceania	2,418	
	Poland	2,377	Asia	1,349	
	Australia	2,360	South America	838	
Rice	China	172,184*	Asia	417,135	449,827
	India	90,000*	South America	12,396	
	Indonesia	34,300*	Africa	8,551	
	Bangladesh	21,700	North & Central America	6,983	
	Thailand	18,535*	Europe	1,709	
Rye	U.S.S.R.	13,500*	Europe	15,070	32,194
	Poland	8,781	Asia	1,909	
	German Democratic Republic	2,064*	North & Central America	1,546	
	Germany, Federal Republic of	2,064*	South America	152	
	China	1,400*	Oceania	11	
Sorghum	United States	12,270	Asia	23,164	62,483
	India	12,000*	North & Central America	19,167	
	China	10,014*	South America	9,460	
	Argentina	8,250*	Africa	8,986	
	Mexico	6,367	Oceania	991	
Wheat	U.S.S.R.	82,000	Asia	170,931	498,182
	China	81,392*	Europe	102,037	
	United States	66,010	North & Central America	96,662	
	India	42,502	Oceania	22,061	
	Canada	26,914	South America	15,517	

* Indicates an estimated or unofficial figure. *Note:* Only crops of major importance in international trade are included. Russia is treated as a country only. Its figures are not included in those of either Europe or Asia.
Source: Data from *FAO Yearbook for 1983*, vol. 37. 1984. FAO, Rome.

distributed worldwide, but only an estimated 35 have been cultivated as cereals. Of these, less than a dozen are important today (Table 6-2). Of the 35 species cultivated at some time, only five were domesticated in the New World, and of the top half-dozen major cereals in the world today, corn is the only New World native.

While the natural characteristics of some wild grasses encouraged their initial domestication by humans, thousands of years of artificial selection have exaggerated the differences between cultivated and wild forms. Yet, within grains as a group, we can trace several directions of human selection depending, in part, on the characteristics of the original wild species. In order to appreciate why people around the world chose grasses as their primary food crops and how, under the action of human selection, grains rose to their modern preeminence as mainstays of the world's populations, it is necessary to understand the biology of grasses.

THE GRASS PLANT

Grasses are ubiquitous. In nature, the matted roots of wild grasses hold down soils from the Arctic tundra to tropical savannas, while planted grasses carpet our manicured suburban lawns. Yet, perhaps because of their green, inconspicuous nature, we never seem to notice grasses, even though they are constantly underfoot. Once examined closely, however, this group of monocotyledons proves to be elegant in design and different in many respects from the "generalized" plant we outlined in Chapter 1.

Starting at the roots, we find that grasses have fibrous roots systems (see Fig. 1-11) and lack dominant taproots. Grasses also have a different pattern of growth from most other kinds of plants (Fig. 6-2). Many species branch only at the base, producing more or less equally sized vertical stems called *tillers.* Still other grasses have one or a few major stems that produce sequential branches as they grow. Individual stems or tillers are made of fused, clasping leaf bases surrounding a thin growing axis. Although we tend to call only the flattened, divergent part of a grass blade "a leaf," the leaf actually extends for a considerable distance below this point. The regions along the stem where leaves differentiate from it are called *nodes.* Somewhere above the nodes, the leaves bend outward. At this point, there is often a rim of hairs or a membranous structure called the *ligule.*

Vegetative reproduction is common in perennial grasses. Many species have rhizomes that grow outward as the grass clump ages. Still others produce horizontal branches (*stolons*) that grow along the ground, often rooting at the nodes as they go. A stoloniferous vegetative growth permits the rapid spread of grasses that are inserted into lawns in the form of starter plugs.

Grass floral structures are even more highly specialized than their stems and leaves. Most grass flowers are perfect, although those of corn and wild rice are notable exceptions. Grass flowers are borne in compound inflorescences composed of series of *spikelets* (Fig. 6-3). Each spikelet is a flowering branch with one to several florets. A *floret* is an individual grass flower which typically has a superior ovary topped by two feathery stigmas, three stamens (sometimes

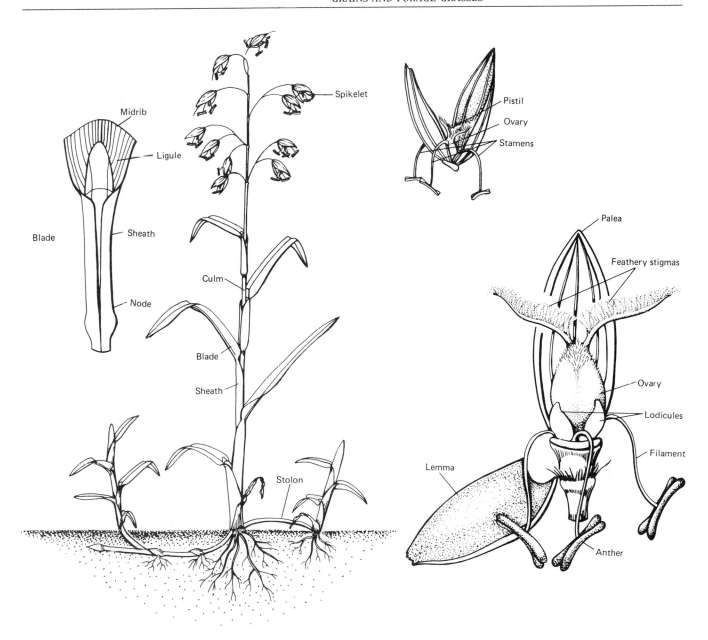

FIGURE 6-2
A generalized grass plant.

six), and three scalelike remnants of the petals. Each floret is enclosed in two protective bracts. The inner bract is the *palea,* the outer is the *lemma.* The nerves of the lemma can extend beyond the body of the lemma, forming a long structure called an *awn* that aids in dispersal. At the base of each spikelet (or cluster of florets) are additional (sterile) bracts called *glumes* (see Fig. 6-15). Once fertilized, the single ovary matures into a fruit in which the seed coat and

FIGURE 6-3
Grasses usually have compound
inflorescences with spikelets
arranged in a panicle, raceme, or
spike (*left* to *right*).

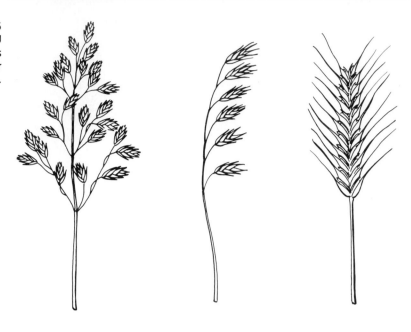

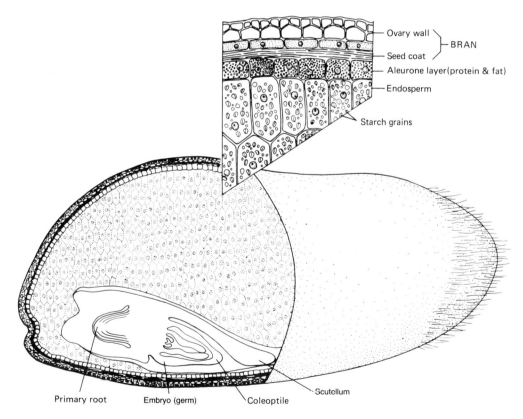

FIGURE 6-4
Cross section of a generalized grain. Note that the ovary wall and the seed coat are
distinguished here even though they are fused and not discernible to the naked eye.

TABLE 6-3

Nutritional Composition of Grains Based on 100 g Edible Portion

GRAIN	WATER, %	CALORIES	PROTEIN, g	FAT, g	TOTAL g
Barley, pearled	11.1	349	8.2	1.0	78.8
Corn, field	13.8	348	8.9	3.9	72.2
Oats, rolled	8.3	390	14.2	7.4	68.2
Rice, brown	12.0	360	7.5	1.9	77.4
Sorghum	11.0	332	11.0	3.3	73.0
Wheat, whole grain:					
Hard red winter	12.5	330	12.3	1.8	71.7
Durum	13.0	332	12.7	2.5	70.1

Source: Data from *Composition of Foods*. Agricultural Handbook No. 8. USDA, Washington, D.C. Revised 1963.

the ovary wall (pericarp) are fused into a complex structure known, in lay terms, as the *bran*. The different components of the bran can be distinguished only microscopically (Fig. 6-4). Inside the bran is a layer of cells known as the *aleurone layer*. The cells of this layer are typically rich in protein and fats. The aleurone layer plays a key role in the germination of grain seeds because it secretes the enzymes that break down the endosperm starch into sugars absorbed by the expanding embryo.

In contrast to legumes (Chapter 7) which store their food in the cotyledons, grasses store energy for germination and seedling establishment in the endosperm (see Fig. 1-29). This storage product is predominantly starch, although it can also contain protein and traces of fat. The embryo itself contains high levels of proteins, fats, and vitamins. Thus, when eaten whole, grains contain carbohydrates, proteins, fats, and some vitamins (Table 6-3). When grains are polished, the bran, the aleurone layer, and the germ are removed, leaving the starchy endosperm and any protein it might contain.

Many grasses are self-compatible annuals that invest a large proportion of their energy in fruits. A high fruit set and a highly caloric endosperm make grains particularly attractive as crops. Humans are interested in the highest possible yields per unit area of cultivation, and grains are among those species with the maximum known yields. Modern yields greatly exceed those of the wild ancestors of the cultivated grains because people have made use of the natural characteristics of grass species in the selection for greater yields.

DIRECTIONS OF SELECTION IN GRAINS

There have been two major ways in which humans have modified the habit of cereals to increase the total amount of grain that can be easily collected at one time. First, in grains that ancestrally had a tillering habit, we have selected for synchrony of tiller formation. Since each tiller is terminated by an inflorescence (Fig. 6-5), tillers that begin growth at the same time tend to mature seeds at the same time. Thus, by replacing sporadic tiller formation with simultaneous

FIGURE 6-5
In tillering grains which have been
selected for simultaneous tiller
formation, all of the inflorescences
tend to mature at the same time.

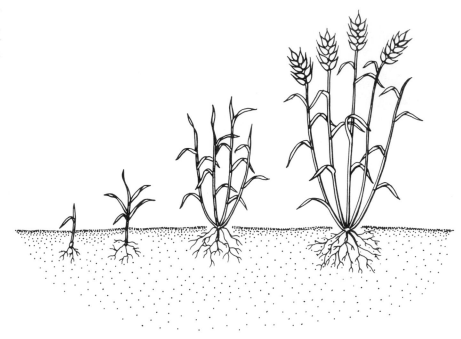

tiller initiation, we have been able to ensure that a single harvesting of grains such as wheat, rye, and barley would collect almost all of the seeds produced. In grains that have branched stems, the lateral branches as well as the main stem usually bear one or more inflorescences. Since the plant grows upward, lateral branches are produced sequentially from the base to the tip. Thus it is normally impossible for all of the lateral shoots and the main shoot to mature fruits at the same time. In cereal crops that ancestrally had branched primary stems, there has been a selective trend for the elimination of branching and the production of single-stalked individuals with high fruit production per stalk. Modern cultivars of corn, sorghum, and pearl millet all have a large, single, unbranched stem, whereas their ancestors or early cultivars were branched.

There has also been selection in recent times for short plant stature. A major problem of many grains, particularly when they are grown in areas much wetter than their native habitats, is *lodging*. Lodging is the matting together of stalks which have been bent over by wind or rain. The matted stalks can not disentangle and return to an upright position. Once lodging has occurred, the stalks generally rot. The native areas of the Old World cereals are relatively dry, at least during the growing season. When humans tried to grow these grains in wet areas of the world, lodging became a serious problem. Intensive selection programs for increased stalk strength and for dwarf varieties of wheat, rice, and even sorghum, have allowed a dramatic increase in the cultivation of these cereals, primarily in the tropics.

Under the influences of human selection, the inflorescences of the cultivated cereals have undergone even more drastic changes than the vegetative portions. One of the initial changes in the cultivated cereals was the retention of fruits on the inflorescence once they matured. All flowering plants face the problem of dispersing their seeds to areas suitable for germination. Many grasses disperse their seeds by having inflorescences that become brittle at maturity so that they shatter when jostled by wind, pressure from passing animals, or neighboring plants. From the human point of view, shattering is an undesirable character because the seeds fly in all directions when harvesting is attempted. Any mutation, or series of mutations, that dampened or prevented shattering would have been selected because seeds of the nonshattering type would have preferentially been collected and subsequently used as seed sources for the next year's planting. Since nonshattering is a genetically controlled character, over time it would become the predominant type used in cultivation. Whether or not the trait was intentionally selected for by humans, all of the major cereal crops now have nonshattering inflorescences.

Another major change in the morphology of grains during their domestication is related to the ease of fruit separation from the inflorescences once the stalks have been cut. Grains can separate from the stalks below the attachment of the bracts, or the fruits can fall free from the enclosing bracts (chaff). Primitive and wild grains come free from the stalk still enclosed in their bracts. Free threshing grains are those that fall free of the bracts. Removing the fruits from the bracts is known as *threshing* (Fig. 6-6). Threshing was formerly done by hand by beating the stalks. Hand threshing is an arduous job, and any mutation that lessened the task would have been seized upon by humans. The appearance of "free-threshing" types in the archaeological record documents early selection for mutations that facilitated the separation of the fruits from the chaff. Free-threshing types solved the dilemma of needing a grain that would not shatter before harvest, but which separated easily from the enclosing bracts once collected.

Light fragments of broken chaff that are mixed with the threshed grain are eliminated by *winnowing*, an operation that involves throwing the grain-chaff mixture into the air (Fig. 6-7). The light-weight chaff fragments are blown away by the wind, while the heavier grain falls back down again. In industrialized countries, modern combines thresh and winnow many grains as they are harvested in the field, but threshing and winnowing by hand are still commonly practiced in many parts of the world.

Separation of any remaining pieces of the fruit wall from the seeds is even

FIGURE 6-6
Women threshing millet in a village in central Niger, Africa. (WFP/FAO photo by F. Mattioli.)

FIGURE 6-7
This Egyptian wall frieze from about 1450 B.C. shows workers carrying the grain in baskets (*lower left*) to the threshing floor, where it is spread out (*lower right*) for cattle to trample (*upper right*) in order to separate the seeds from the husks. The grain is cleaned by subsequent winnowing (*upper left*) in shallow trays.

FIGURE 6-8
A woman grinding grain in Ethiopia, Africa. (Photo by Ray Witlin. Courtesy of the United Nations.)

more difficult than the separation of the fruits from the inflorescence and chaff because of the nature of the coat of the grain. In ancient times, the fruit wall was removed by rubbing the grains (wheat, rice, or occasionally barley) between two stones with enough pressure to abrade off the unwanted layers without crushing the seeds (Fig. 6-8). Today, these grains are hulled, dehusked, or "pearled" to various degrees by machines. The complete removal of the joint fruit and seed coat (which generally also pulls away the embryo) is very common for wheat and rice and leads to white flour and polished (white) rice. The removal of the germ serves a purpose in some cases because degermed grains keep longer than those with germs. The oils in the germ tend to become rancid over time, particularly if the grain is crushed. Groats are grains, often broken, from which only a minimum amount of the fruit wall has been removed.

Additional modifications of grass inflorescences include selection for increased tiller production, restoration of fertility to sterile florets, increase in individual grain size, and changes in the kinds of storage products in the endosperm. When we discuss each of the major grain crops, we point out the particularly important kinds of artificial selection to which each has been subjected.

MAJOR GRAIN CROPS

Barley

On the basis of archaeological evidence, barley (*Hordeum vulgare*) appears to have been the first cereal domesticated. Archaeological digs in the Fertile Crescent (primarily in Syria and Iraq) have uncovered charred barley kernels about 10,000 years old. Wild forms of barley are all two-row (Fig. 6-9). In the Fertile Crescent, the undisputedly oldest cultivated forms are also two-row. This

FIGURE 6-9
Two- and six-row barley.

type of barley has only one fertile floret in each spikelet of three on the flowering stalk. Since each node bears two spikelets, the net effect of suppression of two of the three flowers in each spikelet is the production of an infructescence with two rows of grain. If all of the florets are fertile, the final infructescence bears six rows of grain (Fig. 6-9). By 6000 B.C., six-row forms appear in the archaeological record.

Initially, barley appears to have been ground and made into pastes. However, since raw starch, even mixed with water, is relatively unappetizing, the grain was probably first toasted by heating it on stones. The pastes might also have been boiled into porridges over coals in stone-lined pits. By Egyptian times, barley had become such an integral part of the culture that it had its own hieroglyphic symbol. Likewise, the ancient Greeks depended primarily on barley and greatly expanded its uses. In addition to pastes, they produced variously flavored, baked barley breads. They also learned to soak the grains in water before drying and grinding in order to make them more digestible. This process was the forerunner of malting. The ground meal, once remixed with water was highly susceptible to fermentation. Baking and beer making thus became part and parcel of the same operation. Barley flour was also commonly mixed with ground flax seed and baked on griddles. Common items in the diet of ancient Egyptians and Greeks were therefore barley pastes, gruels, and breads complemented by figs, olives, and goat's milk and cheese.

Barley reigned as the king of the cereals until about the second century B.C., when it was supplanted by tetraploid wheats. More than half of the barley grown in the United States today is used as feed for livestock. About a quarter is used in the brewing of beer and whiskey (Chapter 15). Modern selection has been directed toward synchronous tiller production, short stems, and tolerance to cold and saline soils. For grain used in malting, there has been selection for uniformity of the endosperm composition.

Wheat

Wheat is now considered to be the "staff of life." Its initial cultivation was probably synchronous with, or slightly later than, that of barley, since the wild wheats that were taken into cultivation occurred in the same regions as wild barley. Wheat was originally less popular than barley, and the grain on the staff of Demeter was barley, not wheat. But by the time the Book of Genesis was written, wheat had become the dominant cereal of the Mediterranean area and was the grain from which Adam and Eve were forced to make their own bread.

Wild and early domesticates of wheat were diploid ($2n = 14$) and have been classified as *Triticum monococcum*. At some point during the early cultivation of this species, a natural mutation occurred that suppressed shattering. This mutant was quickly adopted and soon became the major cultivated type (Fig. 6-10). Today these wheats, known as einkorns, are still cultivated in a few

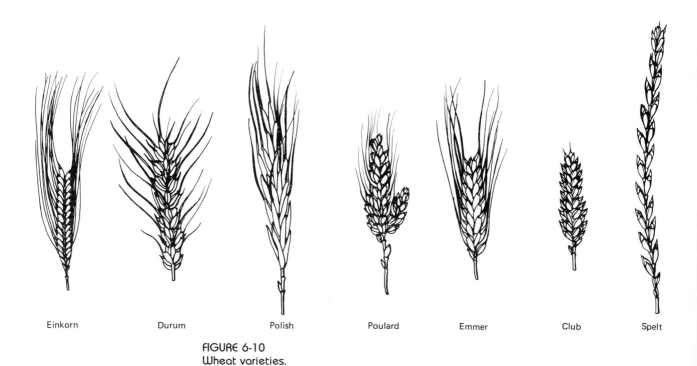

| Einkorn | Durum | Polish | Poulard | Emmer | Club | Spelt |

FIGURE 6-10
Wheat varieties.

FIGURE 6-11
The differences in loaf sizes in this picture show the effect of gluten in producing a light loaf of bread. In the top row are breads baked with flour containing normal amounts of fat. The breads in the bottom row are made with defatted flours. In both rows, the loaf on the far left is made from an average flour containing 13% gluten (protein). Next to the end loaf on the left are gluten free loaves and continuing to the right, loaves made with gluten-free flour to which 10%, 13% (as in the left loaf), and 16% protein have been readded respectively. The differences in loaf heights (a measure of lightness) is obviously due to the amount of protein and not to the possession of fat or starch. (Photo courtesy of Y. Pomeranz and K. F. Finney.)

remote areas of Yugoslavia and Turkey. By the eighth century B.C., an einkorn wheat and another species [thought by some to be a wild goat grass (*Aegilops speltoides*), but by others to be *Aegilops longissima*], hybridized. The determination of the species which hybridized with this einkorn wheat is difficult because the chromosomes contributed by this species appear to have undergone many changes since the initial cross. The first change which occurred was a doubling of the chromosome complement, which produced tetraploid emmer wheat (*T. turgidum*). Unlike einkorn wheat, one variety of emmer wheat (*T. turgidum* var. *dicoccum*) underwent a mutation which caused the bases of the glumes to collapse at maturity. This change made the separation of the fruit from the chaff relatively easy and thus gave rise to the free-threshing type of wheat known as durum wheat (*T. turgidum* var. *durum*). Perhaps even more important than the free-threshing nature of these grains was the change that polyploidy made in their ability to produce raised bread.

Wheat contains two proteins, gliadin and glutenin, along with starch, that combine to form a tenacious complex known as *gluten*. Because of these proteins, flour from bread wheat becomes elastic when it is mixed with water and kneaded. If yeast is added to the mixture, the yeast cells carry out their normal process of fermentation using the small amount of natural sugars in the flour. However, when the yeast cells release carbon dioxide as a product of fermentation, the gas becomes trapped in the spongy mass and the dough rises (Fig. 6-11). Baking "sets" the dough, by drying the starch and forcing the gas out of the pockets in the dough. Cooking also kills the yeast cells.

Prior to the evolution of free-threshing wheats, wheat was commonly heated to facilitate threshing. As a side effect of this process, the proteins coagulated and became inelastic and unable to trap carbon dioxide. Flour made from grain treated in this way was useless for making raised bread. Wheat naturally has more gluten than other grains and is the only one capable of forming a light dough if the proteins remain in their natural state. However, until free-threshing types occurred, the advantages of gluten were obscured. The formation of the tetraploid emmer wheats simultaneously produced a grain with an especially large amount of gluten and the free-threshing condition. These features combined to produce a wheat that was easily separated from the bracts and capable of producing light, raised bread. Durum wheats are known today as hard, or macaroni, wheats and are primarily grown in areas of low rainfall such as the Mediterranean region, central Russia, and the northern Great Plains. These wheats are now used primarily for pasta and noodle making because another event in the history of wheat produced a grain superior even to durum wheat for the making of bread.

Shortly after the appearance of tetraploid wheats, a hexaploid species arose that had three sets of chromosomes, two from a tetraploid emmer wheat and another set from a diploid species generally assumed to have been *Triticum tauschii*. These free-threshing hexaploids, known botanically as *T. aestivum*, have an especially hard, proteinaceous endosperm (Fig. 6-11) that surpasses other wheats in bread making. There are now over 20,000 cultivars of bread wheat (Fig. 6-10), including the hard red and the club wheats.

The common names of wheat varieties are confusing since, in some cases, the common name of one variety is the same as the cultivar name for another. In other cases, the common names refer strictly to the time of sowing, not necessarily to a specific variety. For example, spring and winter wheats are so named because the former are sown in the spring and the latter in the fall (Fig. 6-12). Winter wheats germinate in the fall and survive the winter as seedlings. When spring arrives, the seedlings rapidly send up numerous tillers. Where winters are too severe for seedling survival, spring wheat is planted as early as possible for a fall harvest. The harvesting sequence of the various wheats across North America is such that combines can start harvesting wheat in May in Texas and other southern wheat-growing areas, and then follow the maturation of the crop northward until fall.

Wheat has not always been a major crop in North America. The Spanish brought the grain to the New World in 1520, and English colonists also tried to grow it in their early settlements. Almost all of these early agricultural attempts failed. Mennonite immigrants in westen Pennsylvania were one of the few groups to be successful with wheat in the eastern United States. As settlers moved westward across North America, they carried seeds with them, but prior to the 1870s, wheat growing in the midwest was plagued with problems similar to those experienced in the east. Periods of drought, strong winds, and various pests took heavy tolls on the wheat crops. In the 1870s immigrants began to pour out of central Russia hoping to dodge military service or to find religious freedom. Like many other groups of immigrants, they settled in regions similar to the ones they had left behind. In this case, they chose Kansas, North Dakota, Montana, Oklahoma, and parts of Texas. These areas had climatic conditions ideal for wheat farming: temperatures averaging above freezing in the winter and an annual rainfall of 32 to 87 cm (13 to 35 in). Most of the wheat brought by the immigrants from Russia was "Turkey" wheat, although it is debated

FIGURE 6-12
A wheat combine harvesting hard red wheat in Kansas. (Photo courtesy of the USDA.)

FIGURE 6-13
At harvest time, the amount of wheat is often so great that there is insufficient storage space and some must be stored on the ground. (Photo courtesy of the USDA.)

whether the variety originally came from Turkey or the Crimea. Because of its superior qualities, this particular variety made wheat farming feasible in the United States. It subsequently became the ancestor of the successful American varieties.

Two later developments contributed to the success of wheat growing in the United States. One helped to alleviate pest problems, and the other led to better qualities of flour, the principal product obtained from wheat. Although selection improved many aspects of cultivated wheat, it failed until recent times to produce disease-resistant varieties. A major group of wheat pests are the rusts, fungi that infest the fruiting stalks and ruin the crop. The U.S.D.A. and the Minnesota Agricultural Experiment Station began looking for ways to combat black stem rust (caused by a fungus, *Puccinia graminis*) in 1907. Chemical pesticides proved unfeasible, and destruction of the other necessary host of the fungus, barberry (*Berberis vulgaris,* Berberidaceae), did not significantly reduce infestation. Breeders finally learned that resistance could be conferred by a change in a single gene. Enormous breeding programs were begun and still continue. Agricultural agencies now collect and maintain stocks of mutants of this gene that can be used to breed resistance into current varieties.

Just as resistance to rust is conferred by a single gene, the ability of the rust to overcome the resistance appears to be as simply controlled. Consequently, mutations that permit the rust to infect previously immune wheat varieties continually arise, necessitating the incorporation of a different form of the resistance gene into the crop. The understanding of the genetics of rust immunity and the ability to interject alleles for resistance quickly into seed stocks has allowed wheat to become one of the major crops of the United States (Fig. 6-13).

The second important event encouraging wheat production was the development of the modern industrial flour mill. Steel rollers replaced traditional mill stones and successive layers of silk bolting allowed the rapid sifting of a fine white flour. The purpose of a flour mill is to produce finely ground wheat. To produce white flour, the starchy endosperm must be separated from the bran and the germ. In modern mills, whole grains are fed between grooved steel cylinders that move at varying speeds. The first rollers break and tear open the grains. The "chop" produced by this operation is sieved into three components: (1) first-break flour, (2) coarse nodules of starch, and (3) large pieces of the grain wall with some endosperm still attached. The coarse nodules are called *semolina* and are a "harder" starch than the first-break flour. Semolina flour is preferred for fine pasta. The large pieces of grain are fed through as many as four or five sets of rollers with semolinas sifted out after each break. The semolinas and the first-break flour are then finely ground between smooth rollers and sieved to remove any lingering, flattened, pieces of the bran. With the improved industrial milling of flour, white flour, traditionally associated with the rich classes, became readily available to all. The germ and the bran are now often sold as animal feed.

In addition to its widespread appeal, white flour has a longer shelf life than brown (whole wheat) flour because it lacks the embryo oils that tend to become rancid over time. Graham flour, which contains the germ and the endosperm but not the bran, also becomes rancid quickly. Nevertheless, white flour is much less nutritious than whole wheat flour. It is lower in protein and fat, and lacks the fiber of the bran and the vitamins of whole flour. Most white flour sold in the United States is "enriched" with a mixture of vitamin B1 (thiamin), niacin, and iron.

With the modern increasing awareness of good nutrition, whole wheat or partially refined flour has experienced an increase in popularity. Wheat "berries" (soaked, whole grains), sprouted wheat, cracked wheat, and wheat germ are all now frequently added to specialty breads.

Rye

In contrast to wheat and barley, which were both deliberately brought into cultivation, rye (*Secale cereale*) appears to have begun its association with man as a weed in wheat and barley fields. Instead of fighting the weed, people simply adopted it as a cultigen. Consequently, it is commonly referred to as a secondary crop. As might be expected from this historical account, rye, like barley and wheat, is native to southwest Asia. Initial changes in rye paralleled those of the primary crops with which it grew. By about 3000 B.C., it appears to have been fully domesticated and its cultivation spread across Europe during the next 1,000 years.

Rye is a hardy grain capable of germinating at 1°C above freezing and maturing when temperatures are as low as 12°C (55°F). Its roots can reach 2 m (over 6.5 feet) in length, allowing it to grow in even relatively dry habitats. Rye has been called "poor man's wheat," because it has historically been grown in northern Europe in areas where wheat would not grow well. Northern European black breads are a reflection of the widespread cultivation of rye in these regions.

Unlike wheat, the protein of rye is useless for making light bread. Consequently, even black bread usually contains wheat as well as rye flour. In the United States "rye" breads are made with half wheat and half rye flour. Rye flour contains comparatively large amounts of lysine, but rye is still cultivated less than most of the major cereals (except triticale) discussed in this chapter. In addition to its use as a bread flour, rye is consumed as forage and feed, planted for erosion control, and fermented in the production of rye whiskey and Dutch gin.

Triticale

Wheat and rye share not only a host-weed relationship, but also a common ancestor. This kinship is reflected by the fact that they can be crossed (albeit with some difficulty) to produce fertile offspring. While this potential for hybridization has been known for over 100 years, it is only recently that plant breeders have taken advantage of the fact and produced a fertile hybrid crop. The result, *Triticosecale,* or triticale, has been hailed as the first truly man-made cereal (Fig. 6-14). Both hexaploid triticale, resulting from the doubling of the chromosome complement of hybrids between tetraploid wheat and diploid rye, and octoploids, resulting from the doubling of hybrids from hexaploid wheat and rye, are now being grown.

You might ask why such a cross was attempted or why triticale has become a commercial crop. Like most crosses that have been found useful to humans, triticale combines desirable attributes of its parents and exceeds both of them in some characters. Triticale produces higher yields than either parent in marginal grain-growing regions, is nutritionally superior to wheat, and has the hardiness of rye. Although early reports claimed that triticale was extremely high in protein, they were based on poor data.

Oats

Like rye, oats are generally considered a secondary crop. In Ethiopia today, they are simply gathered and resown with the wheat. In most other areas, they are now cultivated alone. Of all of the cereals native to the eastern Mediterranean, oats appear to have been the last domesticated, perhaps as late as 1000 B.C.

FIGURE 6-14
Wheat and rye are crossbred to produce triticale. This manufactured hybrid combines the high yields of wheat with the hardiness of rye. The grains of triticale are larger than those of either of its parents. *Left* wheat, *center* rye, *right* triticale.

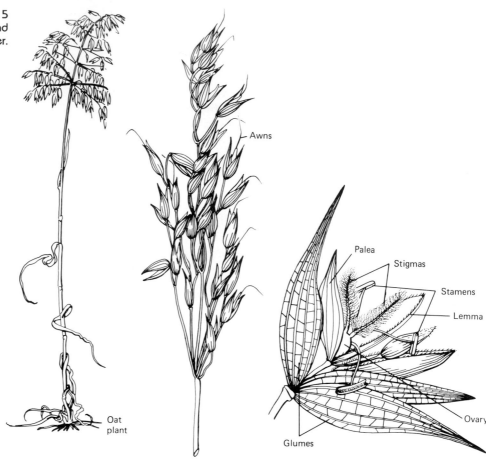

FIGURE 6-15
An oat plant, inflorescence, and flower.

Domestication also probably occurred first in Europe rather than in the Near East. Early cultivars were diploid or tetraploid, but by the first century A.D., the hexaploid, *Avena sativa*, became established as the major cultigen (Fig. 6-15).

Until recent times, oats were a major cereal of temperate areas. Part of this former prominence was due to the use of oats as feed, particularly for horses. Horses were introduced into Europe as draft animals about 200 B.C., but initially did poorly because the native grasses did not provide good forage and, in the wet southern regions, horses suffered from foot problems. Only after the eighth century A.D. did several developments lead to adoption of the horse as the primary draft animal (replacing oxen). First, horseshoes, a protection against foot injuries and diseases began to be used. Second, a type of harness that did not choke horses when they were hitched to a plow or wagon was introduced from China. Peasants preferred horses to oxen because they were much faster than the lumbering ungulates. The widespread use of horses necessitated a source of good feed, and oats served this purpose (Fig. 6-16). The Romans took up cultivation of oats for their animals but referred to the Germans as "oat-eating barbarians," reflecting the Roman opinion that the grain was unsuitable for human consumption. Oats are high in protein and fat, and provide a straw

FIGURE 6-16
A common sight in former times was a horse with a feedbag filled with oats.

that makes good roughage. Oat cultivation thus became an integral part of the former European agricultural system.

In the United States as elsewhere, oats were extremely important until mechanized equipment obviated the need for farm draft animals after the 1930s and 1940s. Vestiges of the former widespread cultivation and use of oats are evident in the American love of oatmeal and in the popular song of the 1940s:

Mares eat oats
Does eat oats
Little lambs eat ivy
A kid'll eat ivy too
Wouldn't you?

Rice

Rice (*Oryza sativa*) can quite properly be called the world's most important crop. Wheat out ranks it in terms of the number of tons produced per year, but more people in the world eat rice than wheat. It is estimated that 1.6 billion people depend on rice with over 200 million tons of rice consumed each year. The cereal has been an integral part of eastern cultures for thousands of years. It is considered sacred and a symbol of fertility in the Orient, a feeling expressed by the throwing of rice at weddings. Many Asians do not consider that they have completed a meal unless rice was among the items eaten.

The fact that *Oryza sativa* was originally domesticated in Asia is unequivocal, but the exact time and place within the continent are uncertain. China, India, and Indochina have all been suggested as possible places where rice cultivation first began. The greatest diversity of rice varieties is in the region of northeastern India and across southeastern Asia.

While rice must have been an important food item in Asia several thousand years before the birth of Christ, it was unknown in ancient Egypt and not mentioned in the Bible. Alexander the Great's conquests led to the first mention of the grain in Europe about 320 B.C. During the next 300 years before Christ, rice cultivation spread across the Middle East and into Egypt. By the fifteenth century, Spain and Italy were both growing rice. Shortly thereafter, the Portuguese introduced rice into Brazil and West Africa. In the sixteenth century, the English started importing rice from the island of Madagascar off the east coast of Africa. One hundred years later (1695) one of these British ships, sailing home loaded with rice, was blown off course and landed in Charlestown, South Carolina. This unplanned visit led to the first plantings of rice in the United States, but commercially profitable operations began only much later, in 1912 in California. Australia, now a large rice producer, began commercial cultivation as late as 1925. In the United States today, rice constitutes a major crop in some areas of California, Arkansas, Texas, Louisiana, and Mississippi.

There are now several major kinds of rice and numerous methods of rice cultivation. Rice is often divided into three subspecies, each of which produces a slightly different cooked product. One of the subspecies, commonly called the *indica* type, is labeled in this country as long-grain rice. When cooked, the grains are relatively dry and separate easily from one another when "fluffed." The *sativa* or *japonica* type is called short-grain rice in the United States. When it is steamed or boiled, it becomes quite soft and often slightly gluey. The third type, *javanica,* is grown only on a very limited scale in the equatorial part of Indochina.

Rice needs large quantities of water in order to grow well, but it does not need to grow in standing water. Rice grown without standing water is known as *upland rice,* but upland rice will produce a harvestable crop only where there are 1.5 to 2 m of rain per year. Brazil is now the world's largest producer of upland rice. Huge acreages of humid tropical rainforest in Brazil have been cleared and planted with rice. Unfortunately, there are usually not enough nutrients in the soil to support rice cultivation for more than a year or two. After the rice potential has been depleted, fields are converted to pastures often seeded with African tropical grasses.

FIGURE 6-17
Rice paddies in Japan. (Courtesy of the Japan National Tourist Organization.)

Most rice is *wet rice,* or rice grown in standing water. Rice is able to grow partially submerged because its specialized stem anatomy allows oxygen to reach the roots. We tend to call flooded fields of rice, "paddies," (Fig. 6-17), but the word "paddy" is also applied to unhusked rice grains. Wet rice cultivation can permit almost continuous cultivation (assuming that temperatures are favorable). The standing water rots the plant material remaining after harvest and harbors algae that provide green fertilizer. Flooding the fields also prevents the growth of terrestrial weeds. In eastern countries, rice cultivation often involves extensive hand labor (Fig. 6-18). The seed is sown in nursery beds and the seedlings individually transplanted into the fields for maturation (Fig. 6-19). In the United States, seeding is often done by planes that broadcast the grains directly across the fields. Special harvesters have also been designed that mechanically harvest the crop when ripe.

A third method of cultivation makes use of natural flood water. In floodplain areas, seeds are sown on the plain at times of low water. As the water rises, the rice grows apace with it. In some cases, the plants can reach 4 to 5 m in height before they finish growing. Harvesting is often done in such areas by boat.

With the advent of modern milling procedures, the world switched from brown to polished rice. The loss of the bran and germ was a subtle dietary change that had profound consequences. In the nineteenth century, a number of Japanese navy men experienced a loss of muscle tone in their arms and legs as a result of nerve inflammation. The disease, later diagnosed as beriberi, soon broke out among thousands of people in Asia and Africa who subsisted on a rice diet. A Dutch doctor, Christian Eijkmann, working in Java was among those who concluded that the disease resulted from a dietary deficiency aggravated by eating foods high in carbohydrates. It was later shown that a lack of vitamin B1 was the cause of the illness. The brans of grains are naturally high in vitamin B1, and brown rice is particularly rich, containing 0.40 mg per 100 g of grain. Polished rice contains only 0.04 mg per 100 g of grain. Several ways have been devised to get around the vitamin deficiency. In Japan, fish oils compensate to some degree. It was also learned that parboiling unmilled

FIGURE 6-19
A single rice flower at anthesis. The lemma and palea have just opened, and six anthers can be seen near their tops. The feathery stigma extends near the bottom of the opening between the two glumes. (Photo courtesy of the USDA.)

FIGURE 6-20
Freshly harvested rice is examined by a World Food Programme (WFP) officer in Bolivia. (WFP photo by P. Johnson.)

rice causes the outer layers of the endosperm to absorb some of the vitamin before polishing. Rice can also be enriched by spraying it with synthetically produced vitamin B1 after polishing. Puffing rice (Fig. 6-21) was an American invention that uses enriched, partially milled rice.

Rice is native to warm regions, but for many years, production in the wet tropics was unsuccessful because of lodging and/or because soils in such regions are poor and require large amounts of nitrogenous fertilizers. The development by the International Rice Institute in the Philippines of dwarf varieties such as IR8, which do not lodge, expanded cultivation in tropical wet areas (Fig. 6-20). Newer varieties, such as IR36 developed by the Institute, combine the qualities of IR8 with increased pest resistance and a wide tolerance of poor soil conditions.

The genus *Oryza* has several other species, most of which occur in Africa. One of these, *O. glaberrima,* was domesticated in western Africa where it is still grown, although its cultivation is being replaced by that of *O. sativa.*

Wild Rice

Wild rice is not a member of the genus *Oryza,* but rather a New World species, *Zizania aquatica.* The species, like its relative Texas wild rice (Fig. 6-22), was used as a source of grain by New World Indians, although it was never really

FIGURE 6-21
"The cereal that's shot from guns!" Quaker Oats promoted their new puffed cereal at the 1904 World's Fair by loading several cannon barrels from the Spanish-American War with rice. After being rotated in gas-fired ovens, the sealed barrels were unplugged. With the sudden release of pressure, the grains exploded and shot out of the cannon with a loud bang. In modern cereal factories, automatically loaded, self-firing, multiple-barrel devices are used to puff grains.

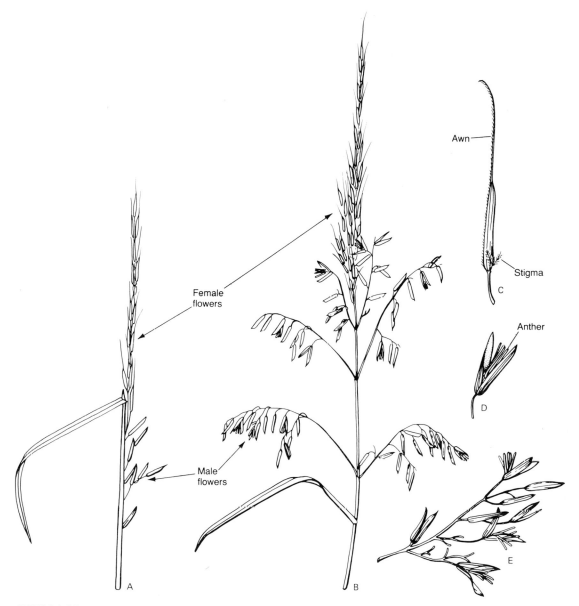

FIGURE 6-22
Wild rice is one of the few monoecious grain crops. It bears female flowers on the top of the flowering stalk and male flowers below. (A) The female flowers emerge earlier than the male. (B) and (E) Male flowers releasing their pollen. (C) A female flower, and (D) a male flower. Shown is *Zizania texana*, Texas wild rice. This species, also used in former times by native Indians, is on the endangered species list because its populations have been so reduced by destruction of its habitats by humans.

domesticated by them. The inflorescences of wild rice shatter, and the grain has traditionally been collected by beating mature inflorescences while they were held over canoes. Native peoples roasted the grain, poured it into deer-skin-lined pits, and trampled it to remove the husks.

Wild rice cultivation in the United States began only in 1959 with the diking and flooding of areas of poor farmland in the northern part of the country. Plant breeders were finally successful in selecting a nonshattering strain. Nonshattering varieties raised yields from 100 to 700 pounds per acre because they reduced losses incurred during harvesting of the older shattering types. Combines equipped with oversized wheels now harvest the grain. Over 17,000 acres are currently under wild rice cultivation, but the demand for the gourmet treat still exceeds the supply, and prices remain very high.

Sorghum

Sorghum (*Sorghum bicolor*) is a grain native to Africa, but beyond this simple statement there is little agreement among botanists about its history. Some workers have favored a single point of initial domestication within Africa and a diffusion of its cultivation outward. Others, such as Jack Harlan, postulate as many as three independent episodes of domestication. The date of this event or events is also speculative, with dates ranging from 2200 to 4000 B.C. In any case, since it was brought into cultivation, sorghum has been subjected to intensive divergent selection leading to numerous, distinctive types (Fig. 6-23). There has been much discussion and little agreement about the best method for grouping these types. A recent simple botanical solution suggests that the cultivated types should be grouped into five main biological races. Numerous intermediates between these races occur because of crossing practices. Nevertheless, cultivated sorghums are usually grouped by farmers into four main types based primarily on the use of the sorghum. These groups are (1) grain sorghums, (2) sweet sorghum or sorgo (used for animal feed), (3) Sudan grass (a different but related species), and (4) broomcorn or broom millet. The hybrid sorgos grown in the southern part of the United States are crosses between members of two of the biological races (kafir and durra).

Traditionally, sorghum has been a grain of hot regions with too little rainfall (500 to 1,500 mm per year) for other grains. The waxy leaves of *Sorghum bicolor* roll up to retard evaporation and thus aid in coping with water stress. Sorghum is today the major grain consumed by many Africans and Indians. Dried grains

FIGURE 6-23
A Food and Agricultural Organization (FAO) agronomist holds a variety of sorghum (*left*) which yields about six times more grain than the local variety (*right*). (Photo by F. Mattioli. Courtesy of the United Nations.)

FIGURE 6-24
An Ethiopian farmer harvesting a
millet known as tef. (FAO photo.)

are often ground and used to make flat breads. There are even pop sorghums similar to our popcorn. The grain is also used to make a beer. If sorghum is cut while in the milk stage (the same stage at which we pick sweet corn), it can be used for silage. In the United States, almost all of the sorghum grown is used for feed, although in some localities special syrup varieties are cultivated.

For syrup, or sorghum "molasses," the sorghum is harvested before flowering and the stalks crushed. The expressed juice is not purified, so the final, thick, sweet syrup resulting from the evaporation of most of the water retains a characteristic flavor.

Broomcorn is a type of sorghum with an open inflorescence. The tillers are cut and stripped of grain and the stiff inflorescence "branches" used for making whisk brooms.

There has recently been an increased interest in sorghum because of its drought tolerance and potential use on land that is marginal for other grains. It is also a potentially useful crop for the making of alcohol.

Millets

The name millet does not refer to a single grain but to several grasses (Fig. 6-24). Numerous forage grasses and even sorghum are often called millet (Fig. 6-25). A list of some of the most widely grown species known as millets is given in Table 6-4. Two of the species in this list, *Eleusine coracana* and *Pennisetum americanum*, are important compared to other millets as human food. The first of these, known most commonly as finger millet, was domesticated in northeastern tropical Africa. It is now a staple in the eastern and central parts of the continent. One of its important characteristics is its ability to withstand exceedingly long periods of storage, up to 10 years or more. Weevils avoid the grain and the grain itself does not deteriorate. In Africa, it is generally ground into flour and made into large, flat breads (Fig. 6-26), or eaten as porridge. Once ground, however, the flour has to be consumed quickly because the germ is not removed and the flour rapidly becomes rancid. Finger millet can also be malted and used to make beer.

The second species that constitutes an important human food is *Pennisetum americanum*, also known as pearl, bullrush, spiked, or cattail millet (Fig. 6-27).

FIGURE 6-25
Panicum miliaceum, common broomcorn or hog's millet, as it appeared in Fuch's *De Historia Stirpivum*, 1542.

FIGURE 6-26
An Ethiopian woman making injera, a pancakelike bread made from teff flour. (United Nations photo by Ray Witlin.)

FIGURE 6-27
The fruiting heads of pearl millet. (FAO photo by P. N. Sharma.)

This species of grain is more drought-resistant than any of the other cereals. Moreover, it will give economically profitable yields on exhausted and nutrient-poor soil. Consequently, pearl millet has become the food upon which millions of Africans living on the edges of the Sahara now depend. Like finger millet, pearl millet stores well, but it is susceptible to weevil infestation and bird predation. It can also be ground into flour, eaten like rice, or malted.

TABLE 6-4

Common Millets of the World*

SCIENTIFIC NAME	COMMON NAME	USES
Panicum:		
P. miliaceum	Common millet, proso, broom-corn millet, hog millet, Hershey millet	Food, feed
P. ramosum	Brown top millet	Feed
P. virgatum	Switch grass	Pasture grass
P. obtusum	Vine mesquite	Pasture grass
P. texanum	Texas millet	Pasture grass
P. maximum	Guinea grass	Pasture grass
P. purpurascens	Pará grass	Pasture grass
Setaria italica	Foxtail millet	Food
Echinochloa crus-galli	Japanese barnyard millet	Food
E. colona	Shama millet	Food
Eleusine coracana	Finger millet, ragi, African millet	Food, primarily Africa, southern Asia
Pennisetum americanum	Pearl millet, cattail millet, bajra, bajri	Food
Paspalum	Koda millet	Forage grass
Eragrostis tef	Teff	Food

* Some sorghums are also called millets.

Corn

In contrast to the Old World, the New World provided only one major domesticated cereal, maize (*Zea mays*, Fig. 6-28). Maize formed the basis of all of the major new world civilizations, the Mayan, Aztec, and Inca, although *Amaranthus* species (Amaranthaceae) were very important in some regions in

FIGURE 6-28
A corn plant (A) showing the staminate tassels (B) with an individual staminate flower, (C) a young ear (D) at the silk stage, and a magnification of an individual silk onto which pollen grains have fallen (E).

FIGURE 6-29
Sand paintings, an art form developed by the Indians of the American southwest, often portrayed the four plants sacred to them, clockwise from the upper left: corn, squash, beans, tobacco.

pre-Columbian times. The decline in the use of amaranths can be attributed in part to the fact that the early Spanish forbid their use because the Mayans used them in religious services as well as for food. Of the important modern cereal grains, corn is the most efficient in converting water and carbon dioxide into foodstuffs. Moreover, it can be cultivated in both tropical and temperate areas.

As children we learned that the Indians showed the Pilgrims how to grow corn and beans using fish as fertilizer. The planting of beans and corn together (usually without the fish) was a common practice throughout North and South America by the time Europeans reached the New World (Fig. 6-29). Presumably, the practice developed without any real knowledge of the mutual benefits that the system provides. An Indian legend says the association came about when man-corn was looking for a wife. Squash asked to be considered, but she was rejected because she had the habit of wandering haphazardly over the ground. Bean, on the other hand, clasped the corn stalk so dearly that corn was assured of her fidelity, and consequently chose her for his bride. Whatever the origin of the biculture, the association proved to be highly successful in terms of both agronomy and nutrition. All of the legumes consumed as pulses have associations with nitrogen-fixing bacteria (Chapter 7) which provide an excess of nitrogenous compounds that the corn can use. The two kinds of food also complement one another in terms of human nutritional needs (see Fig. 7-6).

Because of its importance, there has been great interest in the origin and subsequent evolutionary history of corn. It has been agreed for some time that corn was initially domesticated in the highlands of Mexico and that its cultivation spread north and southward. Charred cobs over 7,000 years old have been found in Mexico.

The 7,000-year-old archaeological corn cobs excavated from dry, pre-Columbian caves in Tehuacan, Mexico, are small and similar in appearance to a kind of corn called Argentine pop corn that is still cultivated in some remote areas of Mexico (Fig. 6-30*A*). These corns have relatively large, soft glumes around the small kernels. Less than 2,000 years later, corn found in archaeological digs is very similar to modern corn. By the time of Columbus, corn was being grown from Canada to Chile and an estimated 300 major corn races had already been selected by native peoples.

Still, there has been a century-long debate about the ancestry of modern corn and its relationship to one or both of two possible parents, *Tripsacum* and teosinte, both giant New World wild grasses. One group of researchers, led by Paul Mangelsdorf and Edgar Anderson, produced a tripartite theory which postulated that *Zea mays* was a direct descendant of a wild South American form of corn that is now extinct. Teosinte, they suggested, was a "species" that resulted from hybridization between *Tripsacum* and corn once corn reached Central America. Modern corn itself resulted from subsequent introgression with *Tripsacum*. Other workers, including Harshberger, George Beadle, Hugh Iltis, and Walter Galinat, have amassed evidence that indicates corn must have been derived from teosinte (Fig. 6-31). They initially placed teosinte in the same

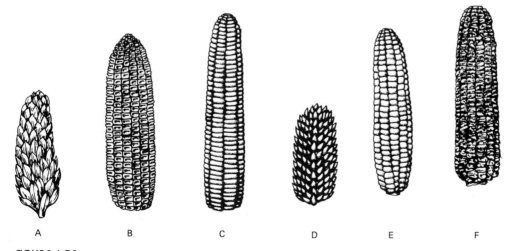

A   B   C   D   E   F

FIGURE 6-30

The six main corn varieties are (A) pod corn, with entire, husklike glumes surrounding each kernel; (B) dent corn, with a small dent in the back of each dried kernel that is caused by the differential drying of the soft and hard endosperm in the grain; (C) flint corn, with long kernels composed primarily of hard endosperm and widely grown in former times in northern regions of North America; (D) popcorn, which contains mostly dense endosperm inside a tough coat; (E) flour corn, which contains only soft endosperm and formerly was widely grown in South America; and (F) sweet corn, the variety eaten as a vegetable in the United States because part of the sugars in the endosperm are not converted to starch until some time after picking.

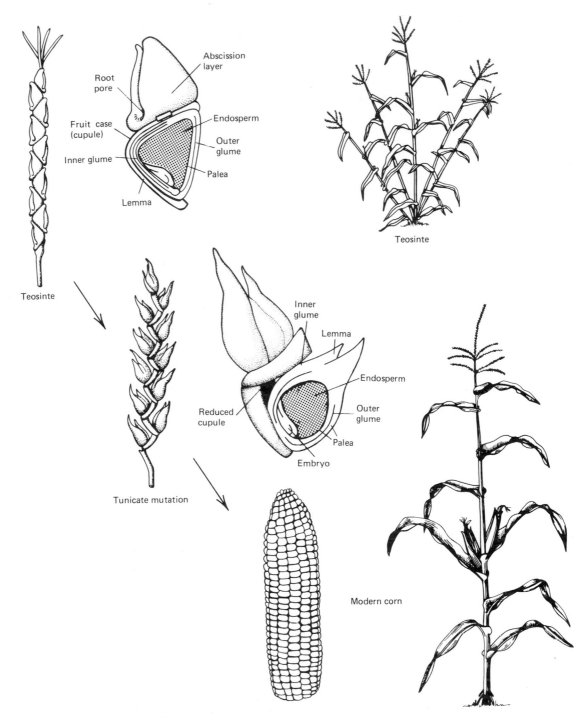

FIGURE 6-31
Teosinte is considered by most botanists to be the ancestor of corn. One reason for believing that corn is derived from teosinte is that a single mutation will convert the hard case around a teosinte kernel into the tunicate form of corn. This type of corn is considered "primitive." Modern corn may differ by perhaps as few as five mutations from teosinte.

genus as corn, calling it *Zea mexicana,* and later in the same species as *Zea mays* subsp. *mexicana.* Many American and Mexican scientists have emphasized how easily the two taxa cross. Analysis of the progeny resulting from the crosses suggested to some that corn and teosinte differ by as few as five genes and that the conversion of a teosinte ear into a corn ear could have happened relatively easily by converting the hard teosinte bracts into soft structures.

Most recently, Hugh Iltis has proposed that corn did not originate from the female teosinte ear, but that a sudden mutation caused the feminization of a male inflorescence on a lateral branch. Iltis believes the appearance of true maize in the fossil record is more consistent with a catastrophic sexual trans-mutation hypothesis (Fig. 6-32) than with a hypothesis that proposes a gradual selection of the corn cob from a female ear of teosinte. Iltis's theory helps explain the commonly seen abnormal corn tassels in which the central spike of male flowers forms an ear and the equally common occurrence of tassels in abnormal ears of corn.

Although there are now many thousands of corn cultivars, many of them hybrids, there are six main historical types of corn: pod, dent, flint, pop, flour, and sweet corn (Fig. 6-30). In North America north of Mexico, two of these types, southern dents and northern flints (eight-row corn) were most commonly grown at the time of European settlement.

Corn was taken back to Europe on the first voyage of Columbus but did not gain much attention. Even today, corn is less popular in Europe than in other parts of the world, partly because most of Europe is too far north for corn to grow well. The European prejudice against corn was so strong that the Irish are said to have refused corn when dying of hunger following the decimation of the potato crop in the 1840s (Chapter 8). Europeans insisted on grinding dried corn into cornmeal, which has limited uses compared with wheat flour. While this was the same method of preparation practiced by the early European settlers in the New World (and used to make the classic cornbread, johnny cakes, etc.), native Americans had more interesting and nutritionally superior ways to use the grain. The Aztecs ate sweet maize porridge or boiled the corn with a little charcoal (forming lime) to separate the grain coats from the endosperm. The endosperm was then mashed and the resultant dough formed around pieces of meat or a mixture of beans, peppers, and meat. The mass was wrapped with leaves and steamed to produce tamales. Ground, dried cornmeal was used to thicken the Aztec's chocolate drink (Chapter 14).

Corn does not have the same nutritional value as other whole grains. It is deficient in the amino acids tryptophan and lysine, and it is relatively low in total protein. Since corn has no gluten, it can produce only flat breads such as tortillas or crumbly cakes. The nutritional deficiencies manifest themselves wherever corn has become the main, or the sole, part of the diet. Pellagra, a disease characterized by dermatitis, diarrhea, and dementia, was originally thought to be caused by a niacin (a member of the vitamin B complex) deficiency. When pellagra broke out among people in parts of Europe, northern Africa, and the Americas who were subsisting on practically pure corn diets, scientists were puzzled. It was known that niacin deficiencies caused pellagra, but corn does not have particularly low levels of niacin. It was subsequently discovered that tryptophan deficiencies were also a factor in causing pellagra. The diet of many rural Americans of molasses, corn bread, and salt pork made

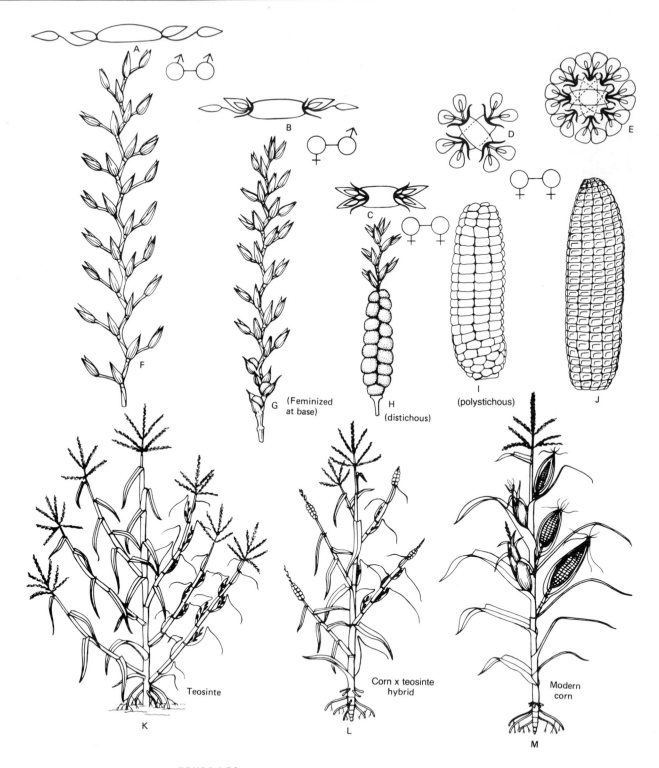

FIGURE 6-32
The catastrophic sexual transmutation theory for the origin of modern corn hypothesizes that the ear of modern corn (J) is not derived from a teosinte cob but from a central spike

FIGURE 6-33
One of the most unusual uses for corn is as an art media as demonstrated by the Corn Palace in Mitchell, South Dakota. Since 1892, the residents of Mitchell have decorated this building with 2,000 to 3,000 bushels of corn along with other grasses. By using variously colored ears of corn which are sawed in half lengthwise and nailed onto tar paper designs, a different motif is created each year. The last week in September is set aside for the Corn Palace Festival. (Photo courtesy of Mitchell Chamber of Commerce.)

pellagra a serious problem in the south until 1914 when yeast was introduced as a food supplement in this diet. Concerted efforts to improve dietary practices in this region led to a 74 percent reduction in pellagra in some southeastern states between 1928 and 1938.

Most of the United States corn crop goes into animals. The entire plant can be chopped and fermented into silage. Corn is also used for a wide variety of other purposes (Fig. 6-33). Corn starch can be hydrolized to make corn syrup, which contains glucose and fructose used in baby formulas and intravenous feeding. Purified, ground corn endosperm yields corn starch used to thicken cooking liquids and even to powder baby bottoms. Maize beer, or chicha, is produced by fermenting hydrolyzed corn starch. Traditionally, corn kernels were chewed before fermenting by Central and South American Indians in order to introduce salivary amylase to break down the starch. Corn is an important adjunct in modern brewing operations (Chapter 15) and corn is a major ingredient in the production of bourbon, and in the production of industrial alcohol.

The cytology and genetics of corn are today probably better understood

of a lateral male teosinte inflorescence (F), which undergoes progressive feminization and suppression of the other male spikes. Only the terminal male spike remains in corn. Shown are (A) a cross section of a male inflorescence (F) of Chalco teosinte (K); (B) a cross section of a male inflorescence (G) of a hybrid between modern corn and Chalco teosinte (L); (C) a cross section of a cob (H) produced by a hybrid between popcorn and teosinte which resembles primitive cobs found in archaeological digs; (D) a cross section of a female polystichous ear (I) of an early corn with eight rows of kernels; (E) a cross section of a modern cob (J) of hybrid corn (M). Each plant at the bottom is drawn so that the left side shows the external features and the right side shows the cross-sectional features of the reproductive structures.

than that of any other plant species. Breeders now produce superior hybrids, often by making crosses using inbred, or highly homozygous, lines (see Fig. 2-7). Because corn is monoecious, hybridization is relatively easy. Plants to be hybridized are planted in alternate rows. The rows that serve as the female parent (those with ears to be used for seed) are detasseled by workers riding through the fields. Self-pollination is thus prevented, and all the seeds produced are of hybrid origin. In the late 1960s it was discovered that a male-sterile gene could be bred into corn lines. Using male-sterile plants, detasseling became unnecessary. At first, the use of male-sterile lines saved money, but it soon became dramatically clear that the cytoplasmic sterility gene carried with it a susceptibility to a new race of the southern corn blight (*Helminthosporium maydis,* Dematiaceae). Once this fungus started infesting fields, it spread rapidly. In 1970, warm, moist weather encouraged fungal growth, leading to devastating crop losses amounting to over $4 billion in the United States. Luckily, the next season was dry, averting a second major outbreak. New immune hybrid crosses without this particular male-sterility gene were redistributed during the next few years. This episode startled plant breeders into realizing the need for genetic reserves of this and other major crop species (Chapter 19).

FORAGE GRASSES

It would be impossible to discuss all of the forage grasses used throughout the world. In many cases, statistics about them are lacking, because forage grasses are usually consumed on the farm and thus do not enter into trade. Nevertheless, it is estimated that in the United States, forage crops (which includes forage legumes as well as forage grasses), cover five times the acreage of all cereal grain crops combined (Fig. 6-34).

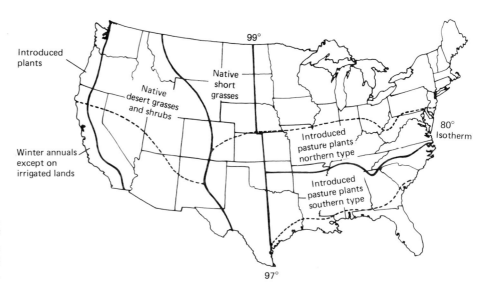

FIGURE 6-34
Comparative distributions of introduced and native forage grasses important in the United States today.

Land that is not well suited for high-intensity agriculture or that is "resting" between periods of cultivation is often allowed to become grassland. In addition to renewing the humus in the soil, grasses help prevent erosion and increase soil drainage.

In order to maximize the food value of forage land and the improvement of the soil, grasses are normally sown with legumes. Forage can be grazed directly, made into hay, or used for silage.

Hay is produced by dehydrating green forage to a moisture content of 15 percent water or less. Hay quality is determined by what plants are incorporated into the hay, the amount of leaf material relative to stem material, the time the forage was harvested, and the amount of weathering and handling to which it has been subjected. The form in which the hay is presented to animals (loose, pelletized, etc.) also has an effect on its suitability for different animals.

Grasses have a high cellulose content and, for most animals, the cellulose has to be digested via bacterial fermentation before the hay can be utilized. In addition to reducing cellulose to simpler compounds that can be digested by vertebrates, the intestinal bacteria produce necessary amino acids and vitamins. Ruminants have large gastrointestinal tracts that contain abundant bacteria. The rumen, or primary area where bacterial fermentation occurs, is at the beginning of the digestive tract of ruminants. Consequently, hay with a high percentage of grass relative to legumes provides adequate feed for ruminants. In non-ruminants, or animals with simple stomachs, fermentation takes place near the end of the digestive tract. Consequently, these animals cannot take full advantage of bacterial fermentation and their feed must contain proportionately higher quantities of proteinaceous foods. Since over 95 percent of the hay produced in this country is for ruminants, most hay has a high proportion of grasses relative to legumes.

Silage is the product of controlled anaerobic fermentation of undried forage, or the stalks of corn, sorghum, or cane. The process of making silage is called *ensiling* and the place where silage is made, a *silo*. Silage can be classified into high moisture (over 70 percent water), wilted (between 60 and 70 percent moisture), or low moisture (40 to 60 percent water), but the basic process involved in the production of all three types is the same. The cut plant material is loaded into a silo. For a while after the plants are cut, aerobic bacteria present on the plant surfaces multiply, using carbohydrates exuded by the cut plants. Both heat and carbon dioxide are produced by the respiration of these bacteria. Once these microorganisms have exhausted the oxygen supply, a second kind of bacteria that can grow without oxygen begin to multiply. These bacteria produce lactic acid and some additional organic acids. The acid levels in the silo eventually reach 8 to 9 percent. This concentration of acid eventually "pickles" the plant matter and keeps it from rotting. If the acid content is high enough and air is excluded, silage can be kept for several years.

Many of the major forage grasses of the United States were introduced from Europe or Africa (Fig. 6-34), although native species are also used. Usually, forage grasses are perennials, but some such as corn, sorghum, millet, and Sudan grass are annuals. Table 6-5 lists many of the major forage grasses of the United States.

TABLE 6-5

Important Forage Grasses of the United States

SCIENTIFIC NAME	COMMON NAME
PERENNIALS	
Agropyron cristatum	Crested wheat grass
A. intermedium	Intermediate wheat grass
Andropogon gerardi	Big bluestem
Bouteloua curtipendula	Sideoats grama
B. gracilis	Blue grama
Bromus inermis	Smooth broomgrass
Cynodon dactylon	Bermuda grass
Dactylis glomerata	Orchard grass
Digitaria decumbens	Pangola grass
Elymus junceus	Russian wild rye
Eragrostis curvula	Weeping love grass
Festuca elatior	Tall fescue
Lolium perenne	Perennial rye grass
Paspalum dilatatum	Dallis grass
P. notatum	Bahia grass
Pennisetum americanum	Pearl millet
Phalaris arundinacea	Reed canary grass
Phleum pratense	Timothy grass
Poa pratensis	Kentucky bluegrass
Schizachyrium scoparium	Little bluestem
ANNUALS	
Lolium multiflorum	Italian rye grass
Sorghum sudanensis	Sudan grass
Sorghum bicolor	Sorghum
Zea mays	Corn

Source: From H. J. Hodgson. 1976. Forage crops. *Scientific American* 234:61–68.

ADDITIONAL READING

Anderson, E., and J. H. Martin. 1949. World production and consumption of millet and sorghum. *Economic Botany* 3:265–288.

Beadle, J. 1980. The ancestry of corn. *Scientific American* 242:112–119. A well-illustrated account of the studies that indicate few genetic differences between corn and teosinte.

Briggs, D. E. 1978. *Barley.* London, Chapman & Hall.

Feldman, M. 1976. Wheats, in N. W. Simmonds (ed.). *Evolution of Crop Plants.* New York, Longman, pp. 120–128.

Feldman, M., and E. R. Sears. 1981. The wild gene resources of wheat. *Scientific American* 244(1):102–112.

Galinat, W. C. 1971. The origin of maize. *Annual Review of Genetics* 5:447–478. A summary of the studies up to this time that showed teosinte was an ancestor of maize.

Harlan, J. 1967. A wild wheat harvest in Turkey. *Archaeology* 20:197–201. Harlan's personal experience with the harvesting of wild grains with primitive tools in the Near East.

Harlan, J. R., and J. M. J. de Wet. 1972. A simplified classification of cultivated sorghum. *Crop Science* 12:172–176. A workable system for dealing with the confused taxonomy of the cultivated sorghums.

Harlan, J. R., J. M. J. de Wet, and E. G. Price. 1973. Comparative evolution of cereals. *Evolution* 27:311–325.

Hulse, J. H., and D. Spurgeon. 1974. Triticale. *Scientific American* 231(2):72–80.

Iltis, H. H. 1983. From teosinte to maize: The catastrophic sexual transmutation. *Science* 222:886–894. An alternative theory to that proposed by Galinat and Beadle about the origin of maize from teosinte.

Kahn, E. J., Jr. 1984. The staffs of life. III. Fiat panis. *New Yorker* December 17 (1984): 57–106.

Karper, R. E., and J. R. Quinby. 1947. Sorghum—Its production, utilization and breeding. *Economic Botany* 1:355–371.

Mangelsdorf, P. C., and R. G. Reeves. 1938. The origin of maize. *Proceedings of the National Academy of Sciences (U.S.A.)*24:303–312. The original expostulation of the tripartite theory for the origin of corn.

Purseglove, J. W. 1975. *Oryza* in *Tropical Crops. Monocotyledons.* New York, pp. 161–198.

Shands, H. L., and A. D. Dickson. 1953. Barley—Botany, production, and economics, harvesting, processing, utilization, and economics. *Economic Botany* 7:3–26.

Wet, J. M. J. de. 1981. Grasses and the culture (sic) history of man. *Annals of the Missouri Botanical Garden* 68:87–104. An interesting summary of human historical association with grasses.

Chapter 7
Legumes

*P*lants of the legume family are second only to those of the grass family in terms of their importance to humans. Every major civilization since the development of agriculture has had a legume as well as a grain as part of its support system. Rice and soybeans, barley and lentils, and corn and beans are classic grain-legume combinations that not only taste good, but make good nutritional sense as well.

Legumes are members of the Fabaceae (Leguminosae), or bean, family (Table 7-1). The family is large and diverse, containing perhaps as many as 12,000 species that botanists usually group into three subfamilies on the basis of their floral characteristics (Fig. 7-1). Despite differences in their flowers, all members of the Fabaceae share the same kind of fruit formed by a single carpel which splits along two opposite longitudinal margins at maturity to release its seeds. The fruit, botanically called a *legume,* is more commonly known as a bean pod (Fig. 7-2), although some members of the family have fruits so modified that they only slightly resemble a common bean.

TABLE 7-1

Names of Legumes Discussed in Chapter 7

COMMON NAME	SCIENTIFIC NAME	CHROMOSOME NUMBER PLOIDY LEVEL
Alfalfa	Medicago sativa	$2n = 32$ autotetraploid
Bean:		
Broad	Vicia faba	$2n = 12, 14$ diploid
Common	Phaseolus vulgaris	$2n = 22$ diploid
Lima	Phaseolus lunatus	$2n = 22$ diploid
Bird's-foot trefoil	Lotus corniculatus	$2n = 12, 24, 26$ diploid tetraploid, aneuploid
Carob	Ceratonia siliqua	$2n = 24$ diploid
Chick-pea	Cicer arietinum	$2n = 16$ diploid
Clover:		
Red	Trifolium pratense	$2n = 14$ diploid
White	Trifolium repens	$2n = 32$ autotetraploid
Cowpea	Vigna unguiculata	$2n = 22$ diploid
Lentil	Lens culinaris	$2n = 14$ diploid
Lespedeza:		
Korean	Lespedeza stipulacea	$2n = 20$ diploid
Sericea	L. cuneata	$2n = 18, 20$ diploid
Striate	L. striata	$2n = 22$ diploid
Pea, green	Pisum sativum	$2n = 14$ diploid
Peanut	Arachis hypogaea	$2n = 20$ diploid, 40 autotetraploid
Pigeon pea	Cajanus cajan	$2n = 22, 44, 66$ diploid, tetraploid, hexaploid
Soybean	Glycine max	$2n = 40$ diploid
Sweet clover:		
White	Melilotus alba	$2n = 16, 24, 32$ polyploid complex
Yellow	M. officinalis	$2n = 16, 32$ diploid, tetraploid
Tamarind	Tamarindus indica	$2n = 24$ diploid

Forages discussed in this chapter are listed in Table 7-4.

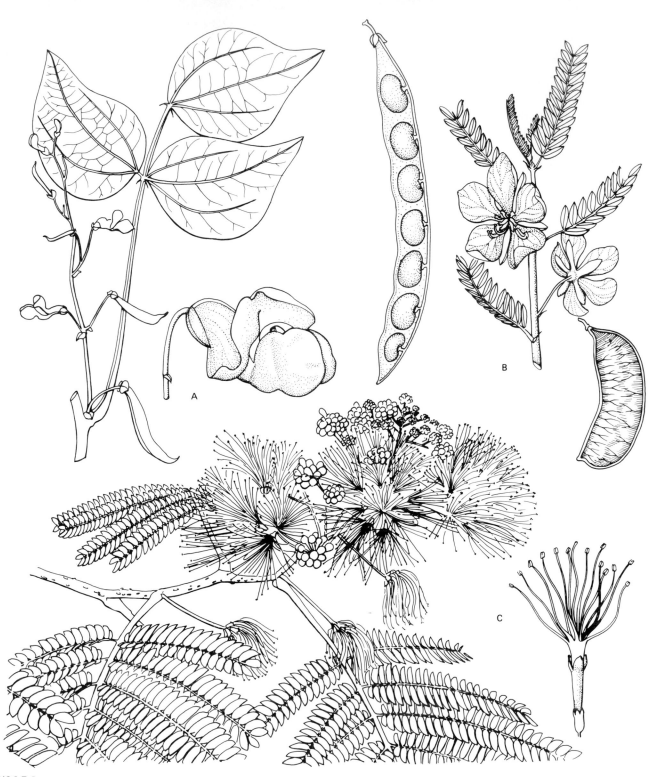

FIGURE 7-1
The subfamilies of the Fabaceae differ in floral characters. (A) The common bean, like most members of the Faboideae, has irregular flowers with the upper petal or standard exterior to the wing petals. The keel, formed by the joining of the two lower petals, houses the style and the variously joined stamens. (B) The Caesalpinoideae, illustrated by *Cassia*, has zygomorphic flowers in which the upper petal is inside of the others and the stamens are usually free. (C) *Mimosa*, similar to other members of the Mimosoideae, has tiny, regular flowers borne in clusters. The stamens are often the showy part of the flowers.

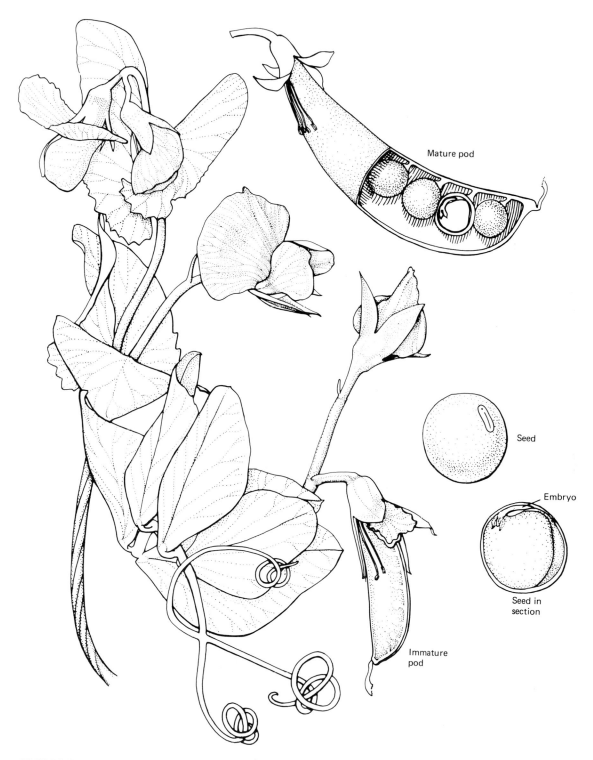

Mature pod

Seed

Embryo

Seed in
section

Immature
pod

FIGURE 7-2
Snow peas are simply a variety of the common pea that are picked while they are
immature. Sweet green peas are also picked before the pod is completely mature, but the
seeds have expanded to their final size.

FIGURE 7-3
Root nodules formed on the roots of a black-eyed pea. (Courtesy of the USDA.)

FIGURE 7-4
The numbers printed on bags of fertilizer refer to the relative proportions of nitrogen, phosphorus, and potassium. Thus, 10-20-10 has 10 percent available nitrogen, 20 percent phosphorous, and 10 percent potassium by weight. The other information on this label is required by law.

Of the three subfamilies, the Faboideae is the most important as a source of food crops. Members of the other subfamilies are occasionally used for food, but more commonly to improve soils, as forage, for their gums (Chapter 11), as sources of dyestuffs (Chapter 16), or as fuel (Chapter 19). Here we deal primarily with the pulses (dried legume seeds used for human food) and forage crops, but we also include tamarind and carob, two other foods obtained from the Caesalpinoideae.

An important characteristic of most species in the Faboideae, and some species of other subfamilies as well, is their propensity to form root associations with bacteria, primarily members of the genus *Rhizobium* (Azobacteriaceae). These bacteria infect the roots of these legumes, causing the production of swollen areas called *nodules* (Fig. 7-3). The bacteria live within these nodules and absorb nutrients from the host plant. It has been shown by growing plants which normally have root nodules in conditions free of *Rhizobium* (but with abundant fertilizer) that the bacteria use resources that the legume host could put into its own growth. Under natural conditions, however, abundant fertilizer is not present. In particular, one component of fertilizer, nitrogen, is often in short supply. Although nitrogen gas is the most abundant substance in the earth's atmosphere, flowering plants cannot use elemental nitrogen. Instead, they must absorb nitrogen through their roots in the form of ammonia, nitrite, or nitrate ions, or, perhaps, as organic compounds from decayed organisms. It is the supply of these nitrogenous ions that is often lacking in the soil. Hence, in cultivated fields, we generally have to add fertilizers high in nitrogenous compounds (Fig. 7-4). Plants with nodulating bacteria have overcome the problem of obtaining sufficient usable nitrogen. The *Rhizobium* bacteria, during their metabolic activities, convert atmospheric nitrogen into ammonia (Fig. 7-5). In nature, therefore, the cost incurred by a plant of supplying its nodulating

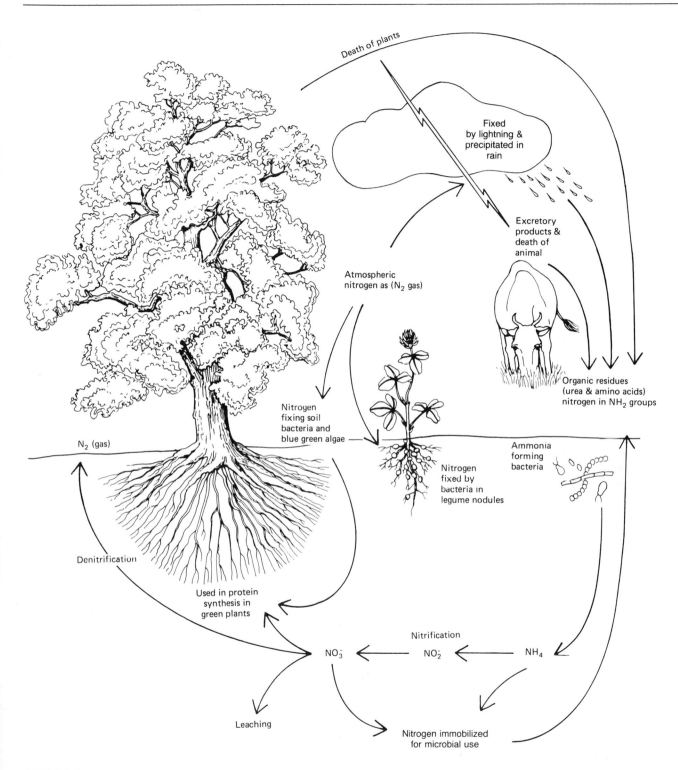

FIGURE 7-5
The nitrogen cycle showing the role of legumes and their associated nitrogen-fixing bacteria.

TABLE 7-2

Composition of Food Legumes Based on 100 g Edible Portion

NAME	PERCENT WATER	FOOD ENERGY, CAL	PROTEIN, g	FAT, g	CARBOHYDRATE	
					TOTAL, g	FIBER, g
Beans:						
Pinto, raw	8.3	349	22.9	1.2	63.7	4.3
White, raw	10.9	340	22.3	1.6	61.3	4.3
Broad beans, dry	11.9	338	25.1	1.7	58.2	6.7
Carob:						
Flour (pulp only)	11.2	180	4.5	1.4	81.0	7.7
Flour & germ	9.8	326	47.3	2.8	34.0	3.8
Chick-peas	10.7	360	20.5	4.8	61.0	5.0
Cowpeas, dry or raw	10.5	343	22.8	1.5	61.7	4.4
Lentils, raw	11.1	340	24.7	1.1	60.1	3.9
Lima beans, raw	10.3	345	20.4	1.6	64.0	4.3
Peanuts, raw, with skins	5.6	564	26.0	47.5	18.6	2.4
Peas	11.7	340	24.1	1.3	60.3	4.9
Pigeon peas	10.8	342	20.4	1.4	63.7	7.0
Tamarind, raw	31.0	239	2.8	0.6	62.0	5.1
Soybeans, raw	10.0	403	34.1	17.7	33.5	4.9

Source: Composition of Foods. USDA Handbook No. 8. 1975 printing. In all cases, mature seeds are considered.

bacteria with nutrients appears to be more than offset by the benefits gained from having a guaranteed source of nitrogen ions. For this reason, farmers sometimes inoculate legume seeds with an appropriate *Rhizobium* species before sowing them into a new field.

The ability of legumes to form associations with bacteria that convert, or "fix," atmospheric nitrogen into a usable form leads to two features of importance to humans. First, since nitrogen is not limiting to them, legumes seem to be more lavish in its incorporation into tissues than other groups of angiosperms. For example, proteins (see Fig. 1-6) all contain nitrogen, and legume seeds and foliage are especially protein-rich compared with those of other plant species. This contrast is shown in Tables 6-3 and 7-2, which give the nitrogen (protein) levels of important grains and legumes. It should be noted that, while legumes are good sources of protein in general, they nevertheless tend to be low in some of the amino acids necessary for humans such as isoleucine, lysine, and the few amino acids that contain sulfur. By eating other kinds of plants containing the amino acids that are deficient in legumes, it is possible to obtain a complete spectrum of amino acids necessary for human nutrition. Figure 7-6 shows the amino acid profiles of several pulses discussed in this chapter compared to beef steak, and Fig. 7-7 provides examples of the use of complementary amino acid sources to provide all the protein components necessary in the human diet.

The second feature of legumes of particular value to humans is the fact

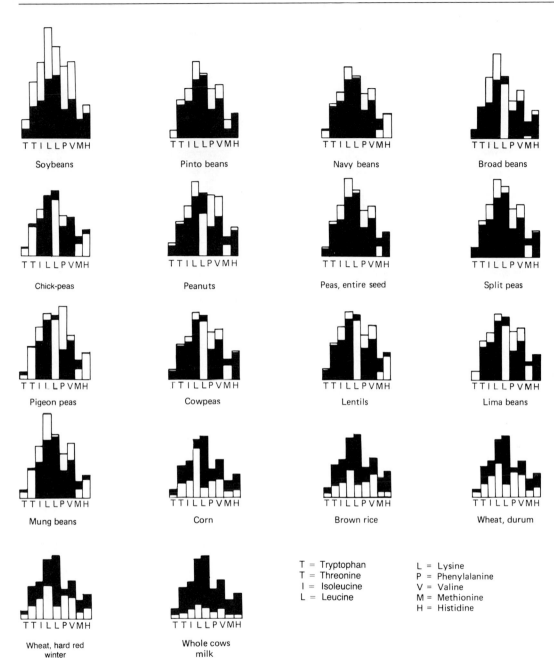

T = Tryptophan
T = Threonine
I = Isoleucine
L = Leucine

L = Lysine
P = Phenylalanine
V = Valine
M = Methionine
H = Histidine

FIGURE 7-6
The amino acid content of various legumes and other foods in 100 g of an edible portion of each item. The black silhouette represents the value for chuck steak, considered to be a complete protein source. Note that the graphs of the legumes superimposed on the graph of steak are comparable in their amino acid profiles in all but methionine. Corn, wheat, and rice also shown are low in most amino acids but contain amounts of methionine comparable to legumes. Consequently, eating these grains together with legumes adds variety to the diet and makes up the deficiencies of methionine. (Data from *Amino Acid Content of Foods*. Home Economics Research Report No. 4., USDA.)

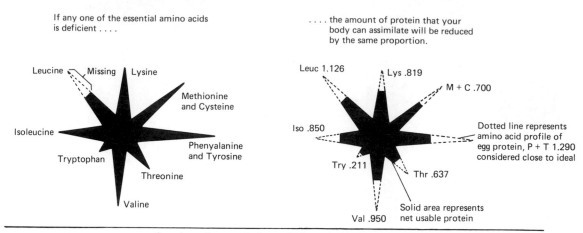

If any one of the essential amino acids is deficient . . . .

. . . . the amount of protein that your body can assimilate will be reduced by the same proportion.

Complementing proteins by combining grains and legumes

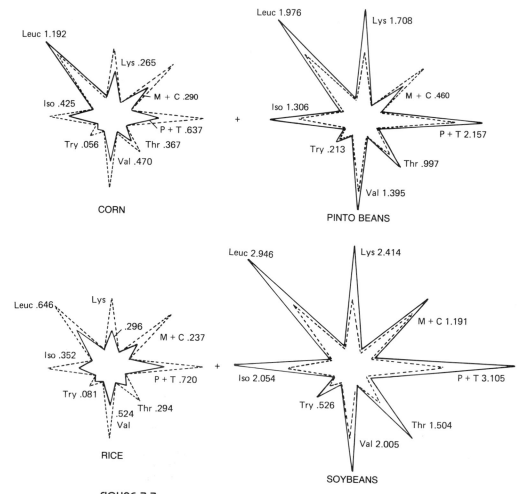

CORN

PINTO BEANS

RICE

SOYBEANS

FIGURE 7-7

The stars at the top of the chart above show how a deficiency in any one of the essential amino acids will reduce the utilization of the others in the same proportion. The stars at the bottom of the chart demonstrate how grains and legumes can

TABLE 7-3

Annual World Production of Pulses

CROP	TOP 5 COUNTRIES, 1,000 t		TOP 5 CONTINENTS, 1,000 t		WORLD, 1,000 t
	COUNTRY	t	CONTINENT	t	
Beans, common	India	2,900*	Asia	5,869	12,938
	China	1,745*	North & Central America	2,543	
	Brazil	1,592	South America	2,152	
	Mexico	1,427	Africa	1,596	
	United States	692	Europe	690	
Beans, broad	China	2,300*	Asia	2,390	4,222
	Ethiopia	600*	Africa	1,149	
	Egypt	295*	Europe	466	
	Italy	175	South America	123	
	France	150	North & Central America	85	
Chick-peas	India	5,092	Asia	6,243	6,827
	Pakistan	491	Africa	278	
	Turkey	250*	North & Central America	200	
	Mexico	200*	Europe	86	
	Burma	197	South America	20	
Lentils	India	550*	Asia	1,189	1,496
	Turkey	450*	North & Central America	107	
	Syria	61	Africa	85	
	Canada	53	Europe	68	
	Bangladesh	48	South America	37	
Peas	U.S.S.R.	5,200*	Asia	2,399	9,362
	China	2,000*	Europe	872	
	France	400	Africa	338	
	India	330	North & Central America	267	
	Ethiopia	150*	Oceania	196	
Peanuts	India	7,300*	Asia	13,410	19,792
	China	4,036*	Africa	4,099	
	United States	1,485	North & Central America	1,625	
	Sudan	900*	South America	598	
	Indonesia	760	Oceania	35	
Soybeans	United States	43,421	North & Central America	45,023	78,566
	Brazil	14,582	South America	19,334	
	China	9,770*	Asia	12,352	
	Argentina	3,750	Europe	757	
	Mexico	721	Africa	391	

*Indicates an estimated or unofficial figure.
Note: Only crops of importance in international trade are included. Russia is treated as a country only. Its figures are not included in those of either Europe or Asia.
Source: Data from *FAO Yearbook for 1983,* vol. 37. 1984. FAO, Rome.

complement each other to provide a relatively complete amino acid profile while allowing a diet high in grams for the provision of calories. (Adapted from *Amino Acid Content of Foods and Biological Data on Proteins.* 1970. FAO, Rome.)

that the legume-bacteria association usually produces an excess of usable nitrogen in the soil. This means, in effect, that growing legumes can serve to provide a food crop and simultaneously to fertilize the soil. Because of this property of legume crops, farmers often rotate (alternate) them with other crops that would normally require the application of nitrogenous fertilizers. After growing nonlegume crops for 1 or 2 years, the supply of nitrogen ions in the soil is depleted and legumes are replanted.

While we could have included many of the legume species used by man, we have chosen first to discuss 10 of the most important pulses and 2 caesalpinoid fruit crops. We then look at several of the major legume forages grown in the world today. In our discussion of the important pulses, we begin with native Old World species and finish with American contributions to the world's supply of these protein-rich plant foods. Table 7-3 shows the production figures for 6 of the 10 pulses we discuss which are of importance in international trade.

PULSES

Lentils

Lentils are among the most ancient plants known to have been cultivated by man. Carbonized lentils found in Neolithic villages in the Middle East have been dated as being between 8 and 9 thousand years old. Analysis of the fossilized seeds indicates a period of domestication had occurred years before the time when the seeds were deposited. The seeds found in the fossil excavations are larger than the seeds of wild lentils which still grow in the area. After initial cultivation of the crop in the Middle East, lentil use began to spread around the Mediterranean. By 2200 B.C., lentils began to appear in Egyptian tombs. Lentils are the first pulse to be mentioned in the Bible. According to the story in Genesis (Genesis 25), Esau, the first-born twin of Rebecca and Isaac, sold his birthright to his brother Jacob for a meal of red lentils.

Both the common name lentil and the scientific name *Lens culinaris* refer to the flattened shape of the seed, which is similar to that of the human eye lens (Fig. 7-8). One to three seeds are produced per pod, a smaller number than produced by most other pulses. However, as shown in Table 7-2, lentils rank fifth among the major legumes in protein content, and they are among the most digestible of all the commonly eaten pulses. Over the many thousands of years during which they have been cultivated, lentils have been selected by humans for a range of colors and shapes. Today, lentils are particularly important in India where black, green, yellow, and red lentils form the basis of most dals (or dhals), purees derived from several pulses, but usually lentils or pigeon peas. Dals are an extremely important component of almost every Indian meal.

Because of their relatively high drought resistance, lentils are grown in semiarid regions scattered throughout the world. In the United States, they are grown primarily in the dry portions of the northwest where they serve as a rotational crop with wheat.

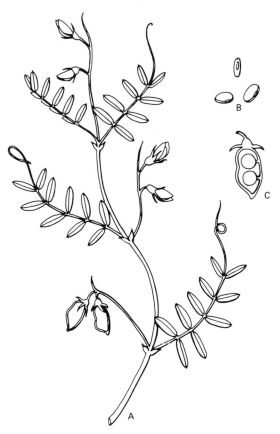

FIGURE 7-8
Lentils. (A) A branch; (B) seeds;
and (C) a seed pod in section.

Peas

Peas, along with lentils, barley, and wheat, form the oldest complex of cultivated foods yet discovered. Fossil seeds unmistakably identified as peas (*Pisum sativum*) have been recorded from excavations in the Near East and Europe and dated as being 9,000 years old. The original domestication of peas was, therefore, probably Mediterranean, although Ethiopia and Central Asia have been proposed as centers of domestication.

The nutritive value of pulses must have been known (at least subconsciously) since early biblical times. In the Book of Daniel (Daniel I), Daniel was taken into the court of Nebuchadnezzar with several other young Jewish boys so that the Israelites could learn the ways of the ruling Babylonians. The young Daniel and a few of his fellow Jews refused to eat the meat served them because it had not been prepared according to Hebrew specifications and the eating of it would have been a violation of their dietary laws. Daniel asked that they be allowed to eat pulses (probably peas) rather than meat. The story continues that, despite the caretakers' fears that the boys would show the effects of this "poor" diet (and hence bring punishment from the king), they granted the request. However, when they were finally brought for an audience at the royal court, Daniel and the other youths who had refused to eat the king's meat had fatter and more beautiful faces than those who had forsaken their religion.

FIGURE 7-9
A wood carving of peas from the
herbal of Clusius.

During the Middle Ages in Europe, dried peas were the mainstay of the peasant population (Fig. 7-9). The familiar nursery rhyme

Pease porridge hot
Pease porridge cold
Pease porridge in the pot
Nine days old . . .

refers to a thick broth or steaming pudding made from dried peas that, if we are to believe the text of the rhyme, could be served from the same pot for over a week.

Peas were not eaten as a fresh green vegetable until the seventeenth century, when a Dutch horticulturalist bred and offered in trade the first "garden" varieties. Green peas were enthusiastically accepted by the members of the court of King Louis XIV, even though, at this early time, fresh peas would probably have been much inferior to our modern fresh peas. Peas are still one of the most important pulses on a worldwide basis today, ranking fourth behind soybeans, common beans, and peanuts (Table 7-3). Breeding in recent times has been for more pods per node so that the fruit crop will mature synchronously, facilitating mechanical harvesting. Chinese snow peas and the new sugar snap peas are pea varieties eaten when the pods are still immature and tender (Fig. 7-2).

Broad Beans

Today, broad, or fava, beans (*Vicia faba,* Fig. 7-10) are associated with the Mediterranean region believed to be the original home of the species. Cultivation was widespread in this region in prehistoric times, and writings attest to the cultivation of broad beans by Egyptians, Greeks, and Romans. From southern Europe, broad beans spread to Asia (China is now the world's largest producer, Table 7-3) and the New World. The second governor of Colombia brought the beans with him when he arrived in South America in 1543 from Spain, but as indicated in Table 7-3, broad beans are primarily a crop of cool regions. In North America, Canada is a larger producer than the United States. In the

FIGURE 7-10
Foliage, flowers, and pod of a broad bean.

FIGURE 7-11
Chick-peas have pods which
contain two or three seeds each.
(A) pod (B) plant.

A   B

United States, production has so declined in recent years that most of the canned and dried beans offered in supermarkets are imported.

The seeds of broad beans are said to produce favism, a disease involving hemolytic anemia. The illness, resulting from the breakdown of red blood cells, is most commonly found in people of Mediterranean origin. Although exposure to the beans was long thought to cause the disease, it is now known that the disease is the result of a genetic disorder that results in a lack of an enzyme (glucose-6-phosphate dehydrogenase). When oxidative agents such as the fava bean alkaloids are ingested by individuals with this disorder, the anemia is aggravated. Oddly enough, before the advent of modern medicine, the disease was often advantageous to individuals who had it because it provided a resistance to malaria in much the same way that sickle-cell anemia conferred malaria resistance to blacks in Africa.

Chick-peas

East of the Mediterranean, another legume replaces peas and broad beans and becomes, along with lentils, a basic component of the diet of literally millions of people. Chick-peas (*Cicer arietinum*, Fig. 7-11) have a record of cultivation that is only slightly younger than that of peas and lentils. The first certain records, from Turkey, are dated to be 7,400 years old. The area of origin of the species is assumed to be in northeast Africa. By 2000 B.C., chick-peas were introduced to India, which now produces 79 percent of the world's crop (Table

7-3). Cultivation also spread around the Mediterranean, and today chick-peas form a part of the cuisines of Italy, Spain, Morocco, and Algeria. Brought to Mexico by the Spanish, who call them garbanzos, they now constitute, along with beans, a Mexican nonmeat source of protein.

Soybeans

Of all the legumes, soybeans (*Glycine max*) must be crowned the king (Fig. 7-12). In China, soybeans are referred to as poor man's meat, indicating their importance in this populous country. While all of the legumes we are discussing can be considered as partial substitutes for animal protein, the soybean has more protein (Table 7-2), and greater versatility, than any of the others. Soybeans also have less carbohydrate per unit weight than any of the other pulses except peanuts. The amino acid content of soybeans, especially in the sulfur-containing compounds, is particularly impressive. These factors combine to make soybeans the most important legume crop in the world (Table 7-3). In the western hemisphere, soybeans are used primarily for oil extraction and animal food. In the East, the use of the beans is quite a different story.

Soybeans are thought to be native to northeastern China, where the domestication is hypothesized to have occurred about over 7,000 years ago. The soybean was the plant Shen Nung (see Fig. 3-6) was supposed to have given human beings when he introduced people to the art of cultivation. Brought from the Orient to France by missionaries, soybeans never became popular outside of China and Japan until after 1890. Now, soybeans are considered by many to be one of the principal plants that will help feed the world's future population. Included in the repertoire of possible foodstuffs from soybeans are curd, cheese, drinks, sauces (soy), "greens" (sprouts) for salad, and oil. The dried or fresh seeds of the soybean can, of course, be eaten directly, but they are more often processed, especially in the Orient, to provide a variety of nutritive products. Of these, tofu, okara, soy milk, and soy sauce are the most important (Fig. 7-13).

In making tofu, dried soybeans are soaked, rinsed, and then crushed in water. The slurry is heated to near boiling and the liquid decanted. This liquid, known as soy milk, can be used as a drink and is the basis of nondairy infant formulas used in the United States. The solid portion left after decanting the soy milk is called okara, a spongy mass that can be eaten like cottage cheese. To make tofu, the soy milk is used. The liquid is boiled with magnesium salts which initiate the production of curds (coagulated protein). Once curd formation has started, the mixture is allowed to stand until almost all of the protein has coagulated. In a manner similar to that of fresh cheese making, the curds are scooped from the now thin liquid and drained as much as possible. After sufficient draining, the curds settle into firm, smooth-textured tofu cakes. Tofu is extremely nutritious, very digestible, and quite bland, qualities that allow it to be used in a variety of dishes.

Soy sauce is another product made from the beans. In making the sauce, okara, sometimes mixed with wheat, is formed into cakes. The cakes are placed in a cool, dark environment for 3 or more months, during which they become encrusted with fungi. The cakes, with the crusted part of the fungi removed, are then soaked in salty water. The filtered, now flavored salt water is soy

FIGURE 7-12
A soybean plant bears its pods tightly clustered on the lower part of the stem. (Courtesy of the USDA.)

FIGURE 7-13
The range of items produced from soybeans shows why it is considered to be the king of legumes. From the left clockwise: tofu, soy sauce, texturized vegetable protein, miso or soy bean paste, soy milk used in infant formulas, and soybeans themselves

sauce. Before you recoil from the thought of pouring juice from moldy soybean paste on your food, remember that most cheeses are made by allowing fungi to act on milk curds (in Roquefort, blue Danish, and Stilton cheeses, the blue veins of fungi are still visible). In fact, the fermented cakes, after they have been soaked and rinsed, are eaten in China like cheese. American soy sauce is often made by flavoring salt water rather than by soaking fermented soy cakes.

One of the conspicuous nutrients that most beans lack is vitamin C. The Chinese, so dependent on soybeans, have managed to overcome this shortcoming by sprouting the beans. Soy sprouts contain abundant vitamin C and are eaten in salads. It is little wonder that soybeans have been held in such high esteem in the Orient for so long and are now coming into their own in the west as well.

Soybeans were not commercially planted in the United States until 1920. Since then, a meteoric rise in soybean production has made the United States the world's largest soybean supplier and led to the legume's being called a "Cinderella crop." Before the twentieth century, soybeans were not used as livestock food because the raw vegetable contains a trypsin inhibitor. Trypsin is an enzyme that is necessary for animal digestion of protein. In the early part of this century, it was discovered that heating destroyed the inhibitor. As a result of this finding, soybeans became a major component of animal feeds, and worldwide demand for the crop soared.

One of the first Americans to become interested in soybeans was Henry Ford, who thought that they had as much potential as a raw material for manufactured goods as they had as a source of protein. The Ford Motor Company built three factories in the 1930s where soybean oil was extracted and used in the production of paints and plastics. Ford himself ate soybeans at every meal and had a suit made from "soy fabric" that he wore to a convention. His company also sponsored a 16-course soybean dinner at the 1934 "Century of Progress" show in Chicago.

Despite Ford's efforts, little of the huge American soybean crop is consumed directly. Over half of our production is exported, and most of the rest is fed to animals, processed into oil, plastic, paints, and adhesives. This pattern may change as a result of the newly discovered process of spinning soy proteins into texturized vegetable protein (TVP). Soy protein, processed in this way (Fig. 7-14), can be flavored to imitate various meats and utilized as a meat substitute or added as a filler to canned or processed meats. Many dry dog foods produced in America contain processed soybeans as well as grain by-products.

Pigeon Peas

While the name pigeon pea (*Cajanus cajan*) might not seem like a familiar one, you have probably eaten them without realizing it. In the United States, pigeon peas are often used as the basis of split pea soup (although true peas are also used). In other parts of the world, however, they constitute a common and important part of the diet. There is a dispute about the place of origin of the crop, with most botanists favoring India because of the prominence of the pea there. Other botanists suggest that the pea was first cultivated in northern Africa because there are 4,000-year-old fossils thought to be pigeon peas from this region. Whatever the origin, the legume is now of major importance in India, which produces 95 percent of the world's crop. Pigeon peas will grow on poor soil, which makes them a good crop for farmers utilizing marginal agricultural land. In India, the dried peas are used like lentils in the preparation of dals. Since pigeon peas have remained a crop grown on a small scale, and one of poor farmers, few large-scale efforts have been made to produce new varieties and better yields.

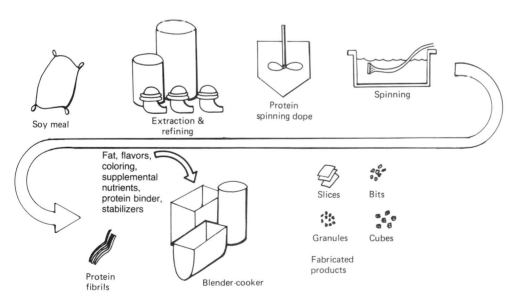

Soy meal   Extraction & refining   Protein spinning dope   Spinning

Fat, flavors, coloring, supplemental nutrients, protein binder, stabilizers

Slices   Bits   Granules   Cubes

Fabricated products

Protein fibrils   Blender-cooker

FIGURE 7-14
The steps in the manufacture of texturized vegetable protein. (After a diagram in R. I. Mateles and S. R. Tannenbaum (eds.). 1968. *Single Cell Protein.* Cambridge, Massachusetts, MIT Press.)

Black-Eyed Peas (Cowpeas)

Africa (Ethiopia) is usually considered to be the original area of domestication of black-eyed peas, *Vigna unguiculata* (Fig. 7-15). There is still an extraordinary amount of diversity in the peas from this region. Spread of their cultivation seems to have occurred both westward within Africa and eastward to India. The majority of the crop, however, is still grown in Africa where seeds, sprouts, and leaves are eaten. Black-eyed peas were introduced to the United States by slaves brought from western Africa. Today, they are part of the regional cooking of the American south. In parts of the southern United States, hoppin' John, a mixture of rice, black-eyed peas (or cowpeas as they are often called), and salt pork is the traditional dish of New Year's Day. Superstition says the stew must

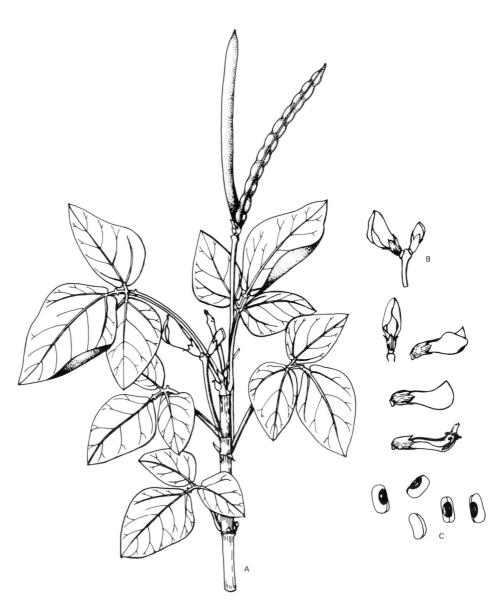

FIGURE 7-15
Black-eyed peas. (A) plant (B) flowers in profile and section (C) shelled peas.

be eaten before noon on January first in order to ensure good luck during the new year.

Lima Beans

The lima bean, *Phaseolus lunatus,* is generally considered to be native to Central America and to have naturally spread northward in North America and southward into South America. However, the oldest fossil material, dated between 7 and 10 thousand years before present, is from Peru. Domestication of the bean may therefore have occurred independently in Central and South America. The common name of the bean, lima, comes from the place from which Europeans first described the species, Lima, Peru. As in the case of peas, lima beans have historically been used in dried form. In many tropical areas, shelled and dried limas are still the primary product for which the crop is grown. In the United States, dried limas are used in soups, but most lima beans are consumed as shelled, immature, fresh (including frozen and canned) seeds.

Some cultivars of lima beans contain cyanogenic compounds which release cyanide when the beans are chewed or ground. Fortunately thorough cooking destroys the enzymes that liberate the cyanide. Moreover, the cultivars commonly used in the continental United States (not Puerto Rico) have very low recorded amounts of cyanogenic compounds. Modern selection has tended to be for disease resistance, modifications of the growth form into bush types, and increases in the numbers of pods per bush.

Common Bean

In the United States, we can find a wide variety of beans such as kidney beans, navy beans, shell beans, pinto beans, pea beans, Mexican beans, black beans, runner beans, green beans, string beans, wax beans, and snap beans (Fig. 7-16). Despite the apparent differences in these commonly encountered beans, all belong to the same species, *Phaseolus vulgaris.* The common bean may have been independently domesticated in both Central and South America. Early dates of fossil cultivated material from Central America are about 7,000 years ago. Recently uncovered material from Peru is as old as, or older than, the Mexican material. By the time of European arrival in the New World, common beans were an important dietary item for native peoples throughout North, Central, and South America. Today, it is the most widely cultivated species of *Phaseolus,* and the second most important legume in the world (next to soybeans, see Table 7-3) in terms of tons produced per year.

FIGURE 7-16
All of these dried beans belong to the same species, *Phaseolus vulgaris.* From the left: yellow wax beans, mature tender green stringless beans, navy beans, pinto beans, great northern beans, black beans, dwarf horticultural beans, kidney beans.

Because plants of the common bean are naturally vines (bush forms are recent selections), they were historically often grown with another crop such as corn to provide the vines with a support. As emphasized early in the chapter, this mixed cropping was also advantageous for the corn, since the beans are associated with nitrogen-fixing bacteria. Native Americans, without realizing the reason, must have found that growing these two crops together increased production of both. The Indian dish succotash reflected the method of mixed agriculture and was made from dried corn and dried common beans. Our modern idea of succotash as being a mixture of corn and lima beans comes from the name on packages of frozen vegetables.

Refried beans, mashed precooked beans fried in oil, commonly associated with Mexican cooking today, came into existence only after the Spanish had introduced domesticated animals to Central America. The Aztecs appear not to have had much cooking oil, probably because Central American wild game had little fat, and native oil seed sources were rare in Central America. Moreover, outside of the llama, dog, guinea pig, and turkey, New World Indians had no domesticated animals and thus no ready source of lard. Frying, therefore, was presumably an innovation brought from Europe and very successfully used with this native American legume.

Peanuts

Peanuts are legumes which are more often thought of as nuts or as an oil seed crop (see Chapter 10) than as a pulse, but in parts of the world, the seeds are cooked and eaten much as any other grain legume. The species from which we obtain peanuts, *Arachis hypogaea,* is native to central South America, perhaps eastern Peru. Domestication probably occurred first in southern Bolivia and northwestern Argentina. By the time Columbus reached the New World, peanuts were cultivated throughout the warm regions of the Americas. The Portuguese took peanuts to Africa where their cultivation was quickly adopted. Peanuts are now an important dietary item in west African countries. Peanuts were also taken to southeast Asia via the Philippines by the Spanish.

Peanuts are called by different names in various parts of the world. The British name groundnut, or ground pea, refers to the way in which peanuts bear their fruits (Fig. 7-17). Like other legumes of the subfamily Faboideae, peanuts bear pea-type flowers above the ground. Self-pollination occurs within the flowers. After fertilization, instead of producing a crop of legumes on the

FIGURE 7-17
Virginia peanuts. (Courtesy of the
USDA.)

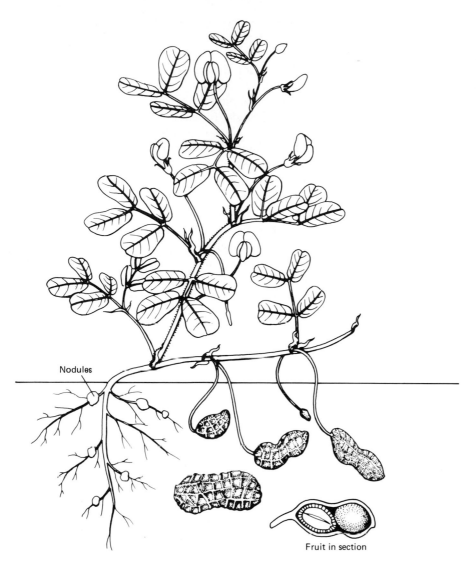

Nodules

Fruit in section

FIGURE 7-18
After fertilization, the flower pedicels of the peanut curve downward, and the developing fruit is forced into the ground by the proliferation and elongation of cells under the ovary. The legume subsequently develops underground.

stems, the pedicels or flower stalks curve downward. The cells under the ovary begin to divide, producing a "peg" that forces the ovary into the ground. As the peg develops, a cap of cells forms next to the withered style. This cap protects the ovary as it is pushed into the soil in much the same way as the root cap protects a root. After the expanding ovary is pushed a few centimeters into the soil, growth downward ceases. The ovary turns sideward and matures underground. Many people unfamiliar with the taxonomy or flowering process of peanuts think the fruits are roots or tubers (Fig. 7-18).

Peanuts are also called goobers, a name that was brought with the peanut back to the New World by African slaves. The widespread production of peanuts in the American southeast can be attributed, in fact, not to its being native in the New World, but to the popularity of the crop in black Africa and its secondary introduction to North America by slaves of southern plantations. With the decline of the cotton empire, peanuts have partially replaced cotton

FIGURE 7-19
George Washington Carver is often credited with establishing the peanut industry in the south after the Civil War had destroyed the cotton empire. Courtesy U.S.D.A.

as a major crop in that part of the country (Fig. 7-19). The popularity of peanuts now extends to all parts of the American population.

Most of the peanut consumption in the United States today is in the form of peanut butter, although a substantial part of the crop is used for hog feed. South American Indians ground peanuts and produced a product similar to modern peanut butter, but "development" of the creamy paste now used is often attributed to a physician, thought by some to be John Harvey Kellogg. The doctor, a health food advocate, was searching for an easily digestible, highly nutritious food that required little effort (chewing) to eat. In 1890, salted peanut paste was being produced. The name Kellogg is probably known by most people as the name of a modern corporation that has gone beyond the sphere of health foods. The Kellogg Company was founded by W. K. Kellogg, John Harvey's brother. The two brothers worked together, and the company originally manufactured flaked cereals that had been developed in the Battle Creek Sanatorium run by John Harvey Kellogg.

Peanuts have been the subject of much agricultural research because they are a good protein source, grow on pool soils, and thrive in tropical regions. Yet, they are susceptible to infestation by a fungus which secretes aflatoxins, chemicals deadly poisonous to humans. Deaths from eating contaminated peanuts have been greatly reduced by instituting proper methods of handling and storage which can virtually eliminate the danger of infestation.

TAMARIND AND CAROB

Tamarinds (*Tamarindus indica*, Fig. 7-20) have been used in Africa, where wild plants grow in the tropical dry savannas, and in southern Asia for thousands of years. The ripe, long brown pods borne on tall, spreading trees are used primarily for their tart, rather sticky pulp, but the seeds can also be roasted or boiled and eaten after the removal of the seed coat. Tamarinds are rarely grown in plantations. Plants are usually grown from seed and pods harvested from semiwild trees. In the United States and Mexico, tamarinds are used primarily as a flavoring in sauces. In India, the fruit pulp is an integral part of chutneys and sauces.

Carob is another caesalpinoid legume that produces pods used primarily for their pulp. The species (*Ceratonia siliqua*) is native to the Mediterranean region. Its common name St. John's bread comes from the fact that it constituted the "locust" on which John the Baptist fed. In ancient times, carob seeds were used as weights for small quantities of precious substances such as gold because they are extremely uniform in size. Our modern unit the "carat" used for gold and jewels is a reflection of this former use.

Traditionally, carob pods were gathered from wild trees and the sweet pulp chewed is the mesocarp. The seeds have also been used to make a coffee-like beverage. Today, carob trees are propagated by seed or by grafting. Since the species is primarily dioecious, grafting assures a large proportion of female trees. Production is still highest in the Mediterranean region, with the tiny island of Cyprus the world's largest producer.

The two most important uses of carob in the United States are as a chocolate substitute and as a source of locust gum, a polymer extracted from the endosperm of the seeds (Chapter 11).

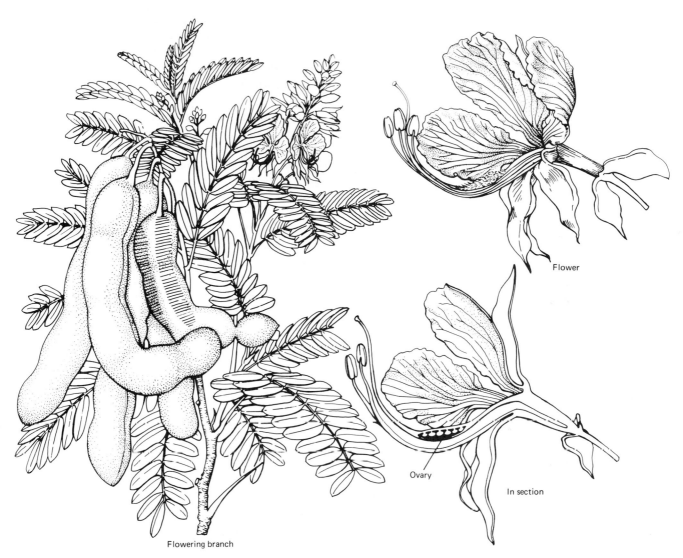

Flower

Ovary

In section

Flowering branch

FIGURE 7-20
Tamarind pods hanging from a branch (left), an individual caesalpinoid flower (above, right) and a cross section of a flower (below) showing the simple ovary that develops into the legume. (After Baillon.)

FORAGES

Just as legumes complement grains for human consumption, legume forages supplement forage grasses and feed grains for domesticated herbivores. In addition to affording the best mixture of foods for grazing animals, legume-grass herbage associations are beneficial for other reasons. Research has shown that legumes alone do not build soil as well as herbaceous mixtures. One of the best ways to improve soil is to plant a grass-legume mixture and use the area for prudent pasturage. Simply harvesting the hay from such a field removes much of the potentially available nitrogen and the plowing under of the cover without cutting precludes any direct profit. Grazing animals on the field allows conversion of part of the vegetable biomass into a cash crop (meat), permits the roots of the grass and legumes to build up the soil, and provides the addition of considerable nitrogen in the form of animal urea. Where conditions permit, such a system is superior to letting the fields go fallow, or using a simple crop rotation system.

The growth of grasses with legumes also circumvents problems presented by pure fields of forage legumes. Ruminants do poorly on a diet consisting principally of legumes because fermentation of large amounts of fresh leguminous material in the rumen causes bloat. Forage legumes also generally do not grow well in pure stands because they cannot compete successfully with weedy grasses and forbs. The growing of legumes with desirable grasses alleviates the problem of weed invasion. If the crop is to be cut for hay, growing a suitable mixture of grass and legumes permits direct baling of the quality of hay desired.

Several hundred legumes are used for forage on a worldwide basis, but a few predominate, especially in this country. Unlike grasses, native legumes are used less for forage in the United States than introduced, domesticated species. Table 7-4 lists some temperate forage legumes which are important in the United States. Of these, alfalfa and the clovers are by far the most important. We therefore discuss these two kinds of forage in some detail and then mention some of the less important American legume forages.

TABLE 7-4

Major Legume Forage Plants of North Temperate Regions

COMMON NAME	SCIENTIFIC NAME	HABIT	NATIVE TO
Alfalfa	*Medicago sativa*	Perennial	Near East
Bird's-foot trefoil	*Lotus corniculatus*	Perennial	Mediterranean
Clovers	*Trifolium*		
Alsike	*T. hybridum*	Perennial	Sweden
Arrowhead	*T. vesticulosum*	Annual	Italy
Crimson	*T. incarnatum*	Annual	Southeastern Europe
Egyptian (berseem)	*T. alexandrinum*	Annual	Near East
Kura	*T. ambiguum*	Perennial	Near East
Persian	*T. resupinatum*	Perennial	Near East
Red	*T. pratense*	Perennial	Asia Minor
Strawberry	*T. fragiferum*	Perennial	Near East
Subterranean	*T. subterraneum*	Annual	Mediterranean
White	*T. repens*	Perennial	Near East
Lespedezas	*Lespedeza*		
Korean	*L. stipulacea*	Annual	East Asia
Sericea	*L. cuneata*	Perennial	East Asia
Striate	*L. striata*	Annual	East Asia
Sweet clover:			
Melilotus officinalis		Biennial	Eurasia
M. alba		Annual	Eurasia

Alfalfa (*Medicago sativa*), or lucerne as it is called in most of the Old World, is king of the forage legumes. Alfalfa was probably the only forage crop cultivated in prehistoric times. It is now extensively grown on every continent except Antarctica, and it has an area of cultivation that exceeds 33 million hectares. This acreage is a little larger in size than the amount of land to which the common bean is planted (ca. 24 million hectares in 1981). Although cultivated alfalfa is tetraploid, primitive diploids occur in Iran and other parts of the Near East believed to be the area of original domestication. It appears likely that the spread of alfalfa cultivation followed the adoption of horse husbandry, extending to the north, east, and west from the Near East after 1300 B.C. Introduction to North America occurred about 1850, although it had been cultivated earlier in Mexico and South America. An initial stumbling block to the use of alfalfa in north temperate areas was its lack of winter hardiness. After 1900, breeding efforts overcame this problem, and the acreage devoted to alfalfa in the northern United States and Eurasia escalated rapidly.

Alfalfa is a perennial that is grown from seed. It can be used as a pasture crop or harvested and used in hay. In warm regions with an adequate moisture supply, as many as six to nine cuttings can be made per year. In dried form, alfalfa makes an excellent fodder, but pure alfalfa hay is comparatively expensive. Consequently, alfalfa is usually grown with forage grasses so that the bailed harvest will contain a proper mixture of the two kinds of forages.

True clovers, species of the genus *Trifolium,* are the next most important group of forage legumes. In the United States, red clover (*Trifolium pratense*) and white clover (*T. repens*) are the most widely grown (Fig. 7–21). Both are perennials and both are native to the Old World (Table 7–4). The two clovers were independently introduced to North America by European colonists. Red clover is the most extensively grown of all of the clovers on a worldwide basis, but white clover is the most important in many temperate regions. Like alfalfa, clovers are often planted with grasses and can be used as pasturage or cut for hay. The other three major clovers grown in the United States are alsike clover (*T. hybridum*), arrowhead clover (*T. vesiculosum*), and crimson clover (*T. incarnatum*).

Sweet clovers, primarily *Melilotus officinalis* and *M. alba,* are the other important ''clovers'' grown in the United States. One reason for the popularity of sweet clovers is their superior ability to improve soils. Both species produce long roots that aerate the soil well during growth. Once senescence begins, their roots rapidly decay, releasing precious nutrients at considerable depths in the soil.

The other major group of pasture and hay legumes grown in the United States are the lespedezas (*Lespedeza* spp.). All are native to eastern Asia and were successfully introduced into this country only after 1919. Despite this rather late start, their cultivation rose rapidly. Annual lespedezas make excellent pasture forage, and plants can be harvested for hay at the end of the summer.

A last forage legume that has become increasingly popular is bird's-foot trefoil, *Lotus corniculatus.* This species is extremely adaptable to a wide range of climate and soil conditions and it is very persistent. It has proved to be especially useful in the northeastern United States, where heavy soils predominate.

FIGURE 7-21
Clover. On the left is an entire and a sectioned flower. The clover inflorescence, often mistaken for a "flower," is made of a cluster of these small papilionaceous flowers. (After Baillon.)

ADDITIONAL READING

Duke, J. A. 1981. *Handbook of Legumes of World Economic Importance.* New York, Plenum. Excellent summaries of the taxonomy, chemistry, uses, and cultivation of most of the legumes used in the world today.

Isley, D. 1982. Leguminosae and *Homo sapiens. Economic Botany* 36:46–70.

Kaplan, L. 1981. What is the origin of the common bean? *Economic Botany* 35:240–254. A discussion of the possible regions of domestication of the New World's most important native legume.

Lappe, F. M. 1971. *Diet for a Small Planet.* New York. Ballantine Books. A layman's discussion of the nutritional components of vegetable foods, primarily legumes and grains.

Purseglove, J. W. 1974. Leguminosae, in Purseglove J. W. 1974. *Tropical Crops. Dicotyledons.* New York, Wiley, pp. 199–332. A thorough discussion of legumes that are important in tropical regions. Several species not included here are discussed.

Zohary, D. 1972. The wild progenitors and the place of origin of the cultivated lentil: *Lens culinaris. Economic Botany* 26:327–332.

Zohary, D., and M. Hopf. 1973. Domestication of pulses in the Old World. *Science* 182:887–894. (peas, lentils, broad beans)

Chapter 8

Foods from Leaves, Stems, and Roots

*P*lants have been compared, in general terms, to animals turned inside out, with the leaves representing animal lungs, and the roots with all their ramifications, the intestinal system. This analogy of the Baron Justus von Liebig in his 1829 publication *Natural Laws of Vegetation* reflected the views of some of the early investigators of plant physiology. While such quaint notions of the roots as the digestive organs of a plant seem ridiculous to us today, they are not much more erroneous than the common modern impression that roots are any underground portion of a plant and that all of the above-ground green or woody portions are stems or leaves. In this chapter we discuss each of these organs in more detail than in Chapter 1 and relate their diverse natural structures and functions to the uses we make of them as food crops.

THE STRUCTURE AND FUNCTION OF STEMS AND LEAVES

In Chapter 1, we mentioned that stems and leaves were parts of the shoot system, but we discussed them separately. Functionally, it is useful to recognize a distinction between the two: stems generally provide support for the above-ground plant parts and house the conduction system, whereas leaves are the usual site of photosynthesis. When we look at stems and leaves from the point of view of their anatomy, or cellular structure, the distinctions between the two become less clear.

The difficulties in defining stems versus leaves arise from the fact that, to many plant anatomists, stems appear to be collections of leaf bases. In monocotyledons such as palms (Figs. 8-1, 8-2), this type of stem organization

FIGURE 8-1
In this palm as in other monocotyledons, the stems are formed by the bases of the leaves.

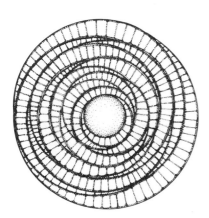

FIGURE 8-2
A cross section of a banana tree shows the overlapping leaf bases surrounding the flowering stalk which, together, form a false stem.

is particularly evident. The vascular system is composed of separate bundles, each with several cells of xylem and phloem, that reflect the pattern of the collected leaf bases (Fig. 8-3). In contrast, most dicotyledonous plants do not have sheathing leaf bases. The vascular strands of herbaceous dicotyledons are in discrete bundles like those of monocotyledons, but rather than being scattered throughout the stem, they are arranged in a ring near its periphery (Fig. 8-3). The impression produced in this case is that the leaves are emergences from the stem rather than its primary elements. In woody perennials, the vascular traces of the leaves merge with one another in the stem to form cylinders of conductive tissue (see Fig. 17-1). Perhaps the most natural way to conceive of the leaves and stems, therefore, is as components of the same system, the shoot system. Within this system, growth can take place at a number of rather precise places. Unlike animals, in which almost all living cells can divide, only certain plant cells (under natural conditions) retain the ability to divide and produce new tissue. These cells occur in regions called *meristems,* from the Greek word meaning "divisible," or able to divide.

Meristems can be classified in several different ways. One way is to group them into primary and secondary meristems. Primary meristems are regions of new growth that contain unspecialized cells like those which began to divide when the seed germinated (Fig. 8-4). Secondary meristems develop in areas of mature tissue such as bark (Chapter 17). Another way to classify meristems is into terminal or lateral meristems, depending upon where they occur in the plant. Terminal capical meristems occur at the ends of stems, branches, and roots. They produce mainly growth in length. Lateral meristems are bands or

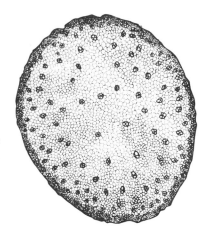

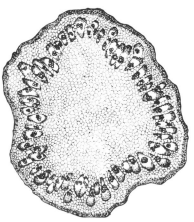

FIGURE 8-3
The vascular bundles are scattered in monocotyledonous stems such as corn (top), whereas they are arranged circularly in the stems of dicotyledons such as this sunflower (bottom).

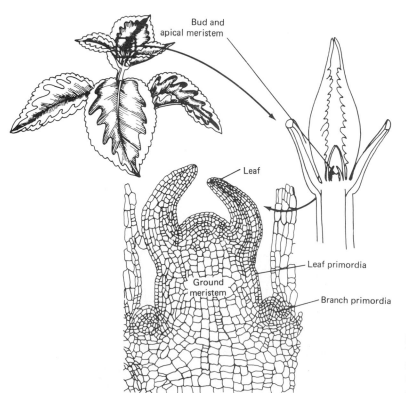

FIGURE 8-4
The apical meristem of a *Coleus* seen under different degrees of magnification has a precise arrangement of cells.

patches of tissue located parallel to the sides of the plant parts in which they occur. During cell division, lateral meristems cause growth in width or circumference (Fig. 8-5). We will discuss lateral meristems in more detail when we treat fibers (Chapter 16) and wood (Chapter 17). Here, we are concerned with shoot terminal meristems.

Terminal meristems are found in the buds at the tips of branches and in buds located in the axils of branches or leaves. Knowledge of the location of axillary buds allows us to ascertain where a leaf "begins" since there is always a bud between the stem and the petiole of a leaf, but not at the juncture of a leaflet and the axis of a compound leaf (Fig. 8-6). When a leaf is attached to a branch, the axillary buds are usually dormant, but once the leaf has been shed,

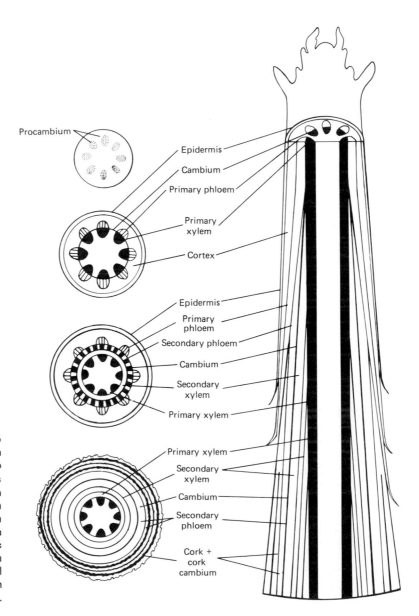

FIGURE 8-5

A diagrammatic representation of a plant's vascular system. In the top section, newly formed cells are differentiating to form a procambium. Further down the stem these cells form the primary xylem and phloem separated by a cambial layer which is responsible for subsequent production of xylem and phloem cells. This cambial activity results in lateral growth in woody plants.

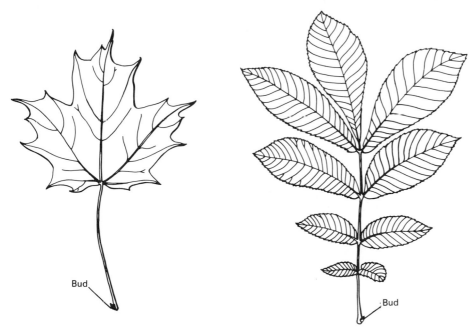

Bud

Bud

FIGURE 8-6
Both the simple leaf of a sugar
maple, (left) and the compound leaf
of a hickory (right) have axillary
buds only where the leaf emerges
from the stem. It is easy to tell that
the leaflets of hickory leaf are
parts of a compound leaf because
they do not have buds at their
bases.

the buds can become active and produce a branch with leaves, each of which will have, in turn, an axillary bud. Axillary buds can also produce flowers or flowering shoots. In contrast to leafy branches, flowers or flowering shoots usually appear while the leaf is still attached. Axillary buds of underground stems are generally found above crescent-shaped scars that are the remnants of aborted leaves. In fact, the presence of such buds on organs that superficially resemble roots indicates that the organ is a stem growing underground rather than a true root. In natural situations, the sprouting of new shoots from the buds of underground stems constitutes a method of vegetative reproduction.

At this point you might ask how leaves and flower parts expand if the only areas of cell division are the meristems we have mentioned. These structures expand as they reach maturity by means of cell elongation and cell division in localized regions of a leaf or flower. Cells in these patches lose the capacity to divide once the leaf or flower part has fully expanded.

THE STRUCTURE AND FUNCTION OF ROOTS

The fallacy of defining roots simply as the underground parts of a plant is evident when we consider that rhizomes, tubers, corms, and bulbs, all modified stems, and even some fruits such as peanuts, can develop underground. Moreover, all roots are not restricted to the soil. Some roots can grow underwater or even in the air. Commercially important roots are often lumped together with tubers and rhizomes into "root crops" because all of these structures can be similarly shaped, fleshy, and full of stored nutrients. However, roots lack the telltale leaf scars and axillary buds of underground stems. While this superficial difference in the two kinds of underground organs is the easiest way to

FIGURE 8-7
Each root hair is formed by the extension of a single epidermal cell. In cross section we can follow the path a water molecule might take from the root hairs into the xylem. There are numerous intercellular spaces between the cortical cells allowing water movement and oxygen exchange, but water can also move across the cortex cells after it is absorbed by the root hairs. In either case, however, water must traverse the endodermis before it enters the vascular cylinder. The Casparian strips of the cell walls of the endodermis are impregnated with a waxy substance, suberin, which aids in the selective screening of ions carried in the water.

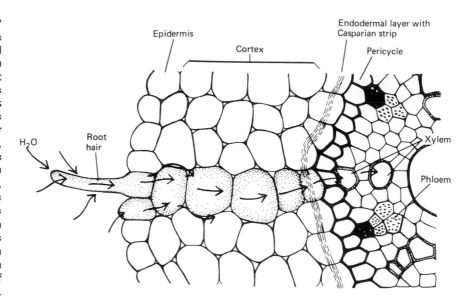

distinguish roots from stems, the two actually differ in many other aspects of their anatomy and physiology. Since roots do not carry out photosynthesis, they lack stomates. Because they must absorb water and dissolve nutrients, they have root hairs, each an extension of a single cell but capable of probing between soil particles as it takes up aqueous solutions (Fig. 8-7).

Like stems, roots have apical meristems, but the pattern of growth is different from that of the shoot system. When the root grows downward, cells of the root cap slough off as they protect the meristematic region where primary cambium, xylem, and phloem are produced (Fig. 8-8). Instead of occurring as discrete bundles or rings inside the phloem (as in stems), the primary xylem in roots forms radial strands. If the strands merge in the center of the root, the xylem forms a core with radiating arms. Clusters of phloem cells occur in the spaces between the xylem arms (Fig. 8-8). In roots of woody plants, the cambium can eventually form a cylinder by growing laterally around the arms of the xylem. In some cases, primarily among the monocotyledons, the xylem strands do not extend to the center of the root. In such situations, there is some pith inside the xylem.

Encircling the conductive tissues is a layer of cells called the *pericycle* that is important in the formation of new roots. Since roots have no axillary buds, formation of lateral roots must proceed by a process different from that of branch formation. Lateral roots arise from specific patches of cells within the pericycle that retain the ability to divide. As they divide, these cells form protuberances that push through the cortex and epidermis (Fig. 8-9) as they elongate. The initiation of lateral organs from near the middle of a root thus contrasts with stem development where lateral branches arise from the peripheries of shoots. Once they are sufficiently mature, lateral roots can themselves give rise to lateral roots, eventually forming large, ramified systems.

Surrounding the pericycle is a ring of cells called the *endodermis* (Fig. 8-7). Outside of the endodermis is the cortex, a relatively thick layer of unspecialized cells. These cells can serve as storage sites by accumulating starch in their vacuoles. Commercially important edible roots have particularly large accu-

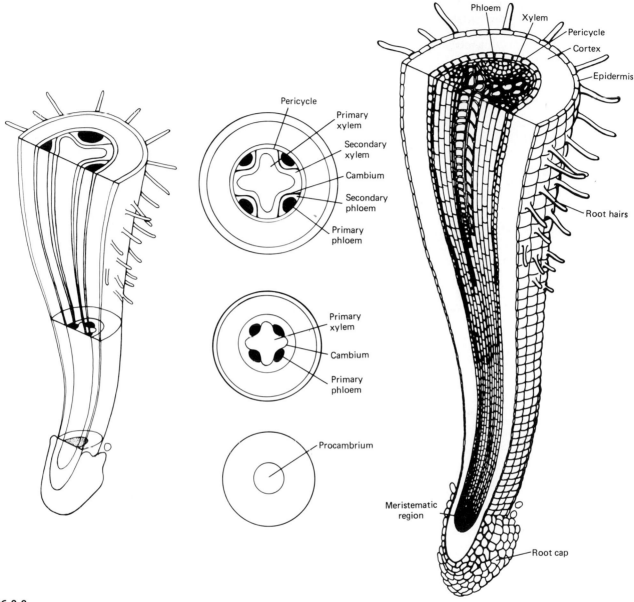

FIGURE 8-8
The apical meristem of the root lies protected beneath the root cap. The youngest, least-differentiated cells produced by the apical meristem lie just above it. The oldest root cells are farthest from the root tip. Differentiated cells lie some distance above it. The root of a dicotyledon typically has a central mass of xylem with projecting arms between which lie clusters of phloem cells. The cambial cells lie between the two kinds of conductive tissue. In old roots, the cambium can grow around the xylem arms and, eventually, form a ring of meristematic tissue.

mulations of starch in the cortex. Surrounding the entire root is the epidermis. Just behind the root tip, epidermal cells can produce root hairs. As the root grows downward, new absorbing regions are continually produced and the hairs of older, distal portions wither. Figure 8-8 gives an indication of the arrangement of progressively older tissues behind the apical meristem of a root.

FIGURE 8-9
Lateral roots are initiated by cells
of the pericycle near the middle of
the roots.

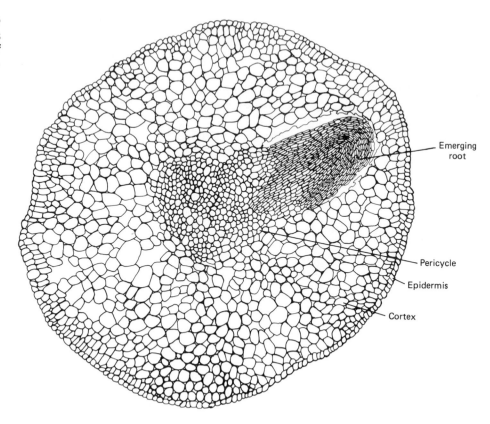

FIGURE 8-9
Lateral roots are initiated by cells
of the pericycle near the middle of
the roots.

Emerging root

Pericycle

Epidermis

Cortex

The delicate appearance of growing root tips belies the rapidity with which they grow and their ability to push through rock and concrete. It has been estimated that at the end of only 4 months of growth, the root system of an annual rye plant has a total length of 11,000 km (6830 miles) and a surface area of 630 m² (6780 ft²). Since only the tips of either primary or lateral roots carry out absorption, continuous root growth is essential. The size of the root system does not expand indefinitely, however, because not all roots are maintained from year to year. In temperate regions, there is extensive root elongation and initiation of lateral roots throughout the spring and summer. During the winter, many roots die back.

HOW WATER AND NUTRIENTS MOVE INTO PLANTS

If nutrients were drawn with water into the plant solely by capillary action or by suction caused by evaporation from the leaf and stem surfaces, chemicals would move through the plant body in the same concentrations in which they occurred in the soil water. Exhaustive studies have shown that most compounds and ions are selectively moved into the xylem for transport. Consequently, the mechanism of nutrient absorption is not strictly passive. Once water is absorbed by a root hair, it flows toward the center of the root by moving either between or across cells (Fig. 8-7). In either case, the first real barrier to movement is the endodermis. The thickened areas (*Casparian strips*) of the radial and cross walls of this single layer of cells provide barriers that all of the absorbed material must traverse. The endodermis thus carries out an initial screening of the nutrients that ultimately reach the xylem. A second active mechanism is also involved in determining which ions are finally secreted into the xylem. It is believed that this mechanism uses protein molecules as carriers to transport ions across cell membranes. Since this is an active mechanism, ions can move in a direction that is against a concentration gradient.

SHOOTS, ROOTS, AND PLANT HABIT

While all roots serve the same basic function, they can differ considerably in size and shape. In many cases, the kind of root a species possesses is correlated with its life history.

In many temperate biennial species, storage in roots is so pronounced that their taproots have become fleshy (Fig. 8-10). Humans have taken advantage

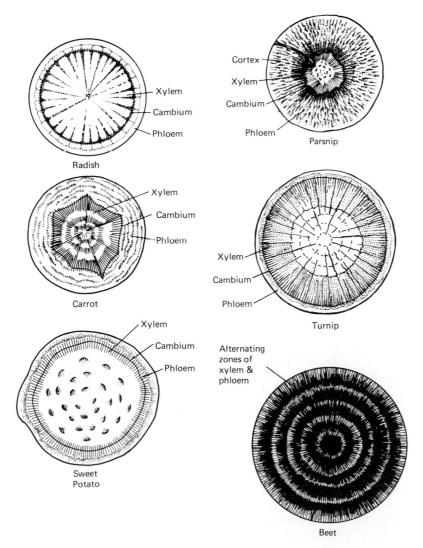

FIGURE 8-10
Cross sections of storage roots.

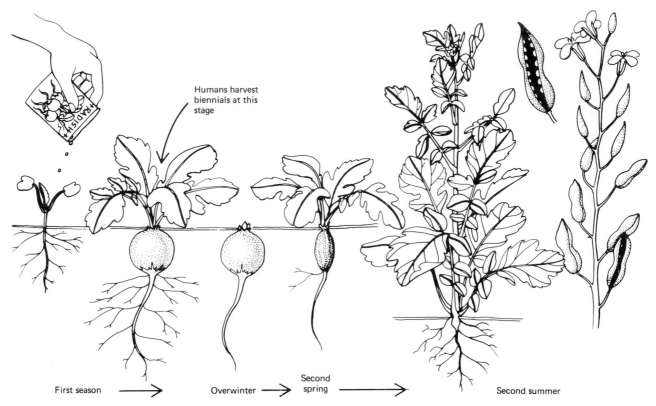

Humans harvest biennials at this stage

First season ⟶   Overwinter ⟶   Second spring ⟶   Second summer

FIGURE 8-11
Humans take advantage of the biennial habit of many species by harvesting the roots after storage is complete at the end of the first year's growth.

of this characteristic of the biennial life history pattern and have learned to harvest the roots at the end of the first year of growth after storage is complete (Fig. 8-11). Sometimes these biennial species are sources of both a root and a leafy crop, but it is difficult to obtain both from the same plant. For leafy crops, farmers want mature, but tender leaves or immature shoots, so plants are harvested when they are relatively young. For root crops, farmers wait as long as possible before harvesting so that the greatest amount of the nutrients, often including those in the leaves themselves, will be translocated to the roots. The roots swell as parenchyma tissue in the xylem proliferates or because secondary cambia are formed that produce additional layers of parenchyma tissue. The swollen roots are dug at the end of the year, usually after the leaves have withered. In general, in those biennial species that are cultivated for both a leaf and a root crop, there has been divergent selection leading to one set of cultivars producing the leafy crops and another set that produces the root crops.

Annual species have roots that are less elaborated than those of either biennials or perennials. Since all growth and reproduction occurs in less than a year, there is no appreciable storage of nutrients in the roots. Food crops that come from annuals are therefore restricted to stems, leaves, fruits, and seeds.

In many regions of the world, woody plants are deciduous, because of either freezing winters or strong seasonal aridity. Most perennial plants that grow in such regions transport photosynthetic products from the leaves to the stems and roots during the end of the growing season and store them in parenchyma cells in the secondary xylem. In temperate woody species, these nutrients, stored as starch, are broken down to sugar at the end of the winter to provide energy for the production of new leaves that will subsequently manufacture the sugar needed for growth and maintenance. In arid areas, particularly tropical arid areas, herbaceous perennials store appreciable amounts of nutrients in modified roots. The major sources of starch in many tropical areas are the roots of such species.

TABLE 8-1

Composition of Leafy, Root, Tuber, Rhizome, and Bulbous Crops Based on 100 g Edible Portion

CROP	WATER, %	CALORIES	PROTEIN, g	FAT, g	CARBOHYDRATE, g	VITAMINS	
						A, units	C, mg
LEAFY CROPS							
Brussels sprouts	85	45	5	0.4	8	550	102
Cabbage	92	24	1	0.2	5	130	47
Cauliflower	91	27	3	0.2	5	60	78
Celery	94	17	1	0.1	4	240	9
Chard	91	25	2	0.3	5	6,500	32
Collards	87	40	4	0.7	7	6,500	92
Endive	93	20	2	0.1	4	3,300	10
Kale	88	38	4	0.1	6	8,900	125
Kohlrabi	90	29	2	0.1	7	20	66
Lettuce	96	13	1	0.1	3	330	6
Spinach	91	26	3	0.3	4	8,100	51
BIENNIAL ROOT CROPS							
Beets	87	43	2	0.1	10	20	10
Carrots	88	42	1	0.2	10	11,000	8
Parsnips	79	76	2	0.5	18	30	16
Radish	95	17	1	0.1	4	10	26
Rutabaga	87	46	1	0.1	11	580	43
Turnips	91	30	1	0.2	7	Trace	36
PERENNIAL STEM CROPS							
Artichokes	86	ca 20	3	0.2	11	160	12
Asparagus	92	26	3	0.2	5	900	33
BULBS							
Leeks	85	52	2	0.3	11	40	17
Onions	88	45	1	0.2	11	Trace	25
STARCHY RHIZOMES, TUBERS, ROOTS							
Manioc (tapioca)	13	352	1	0.2	86	0	0
Potato, sweet	71	114	2	0.4	26	8,800	21
Potato, white	80	76	2	0.1	17	Trace	20
Taro	73	98	2	0.2	24	20	4
Yam	74	101	2	0.2	23	Trace	9

Source: Data from *Composition of Foods.* USDA handbook no. 8. 1975. USDA, Washington, D.C. Values for vitamin A are in international units.

EDIBLE STEMS, LEAVES, AND ROOTS

In our discussion of edible products from stems, leaves, and roots, we first treat crops from biennial species. Many of these produce both a leafy and a root crop. Leafy vegetables have few calories and little protein (Table 8-1) and are consumed primarily in developed countries. In cultures which suffer from food shortages, they are luxuries used primarily to flavor dishes or stretch a meal. The fleshy roots of biennials have higher caloric values than their leafy counterparts, but they are low in carbohydrates compared to the storage roots, rhizomes, and tubers of many perennial species (Table 8-1). After discussing biennials, we treat a few important green vegetable crops obtained from annual and perennial species and then turn to the starchy crops obtained from the rhizomes, tubers, and roots of tropical perennials. These crops constitute some of the world's most important food plants because of their high food value and impressive yields. Last, we treat the species from which we obtain sweeteners, primarily sugar. Table 8-2 gives the worldwide production of most of the crops discussed in this chapter. Table 8-3 lists the plants discussed in this chapter.

TABLE 8-2

Production of Major Leaf, Stem, and Root Crops

CROP	TOP 5 COUNTRIES	AMOUNT, 1,000 t	TOP 5 CONTINENTS	AMOUNT, 1,000 t	WORLD TOTAL, 1,000 t
Artichoke	Italy	521	Europe	967	1,216
	Spain	303	Africa	92	
	France	96	South America	91	
	Argentina	80*	North & Central America	52	
	United States	52*	Asia	14	
Cabbage	U.S.S.R.	10,150*	Asia	14,802	35,794
	China	6,240*	Europe	7,484	
	Japan	3,270*	North & Central America	1,914	
	Korea (Republic of)	3,056	Africa	730	
	United States	1,532*	South America	609	
Cassava	Brazil	2,096	Africa	48,251	123,153
	Thailand	17,000*	Asia	45,929	
	Zaire	14,600	South America	27,839	
	Indonesia	13,770	North & Central America	898	
	Nigeria	9,950*	Oceania	237	
Carrots	U.S.S.R.	2300*	Europe	3,575	11,379
	China	2,085*	Asia	2,997	
	United States	1,010	North & Central America	1,407	
	Japan	627*	South America	554	
	United Kingdom	617	Africa	397	
Cauliflower	China	918*	Europe	2,136	4,776
	India	660*	Asia	1,974	
	France	500*	North & Central America	326	
	Italy	480	Africa	146	
	United Kingdom	280*	Oceania	105	

TABLE 8-2 (Continued)

Production of Major Leaf, Stem, and Root Crops

CROP	TOP 5 COUNTRIES	AMOUNT, 1,000 t	TOP 5 CONTINENTS	AMOUNT, 1,000 t	WORLD TOTAL, 1,000 t
Onion	China	2,843*	Asia	10,391	22,024
	India	2,700*	Europe	4,079	
	U.S.S.R.	2,030*	North & Central America	1,909	
	United States	1,689	Africa	1,734	
	Japan	1,200*	South America	1,683	
Potato, white	U.S.S.R.	83,000	Europe	95,155	286,472
	China	50,033*	Asia	74,464	
	Poland	34,473	North & Central America	18,559	
	United States	14,774	South America	8,813	
	India	10,108	Africa	5,434	
Potato, sweet	China	95,700*	Asia	106,315	114,842
	Indonesia	2,120	Africa	5,081	
	Viet Nam	1,700*	South America	1,396	
	India	1,560	North & Central America	1,289	
	Japan	1,400*	Oceania	629	
Sugar beets	U.S.S.R.	82,000	Europe	135,508	271,002
	France	23,955	Asia	28,497	
	United States	19,152	North & Central America	20,322	
	Germany, federal Republic of	16,500	Africa	2,833	
	Poland	16,358	South America	1,842	

*Indicates an estimated or unoffical value.
Note: Only crops of major importance in international trade are included. Russia is treated as a country only. Its values are not added into those for either Europe or Asia.
Source: Data from *FAO Yearbook for 1983*, vol. 37. 1984. FAO, Rome.

TABLE 8-3

Plant Names in Chapter 8

COMMON NAME	SPECIES	FAMILY	CHROMOSOME NUMBER
Arrowroot	*Maranta arundinacea*	Marantaceae	$2n = 18, 48$ diploid
Artichoke	*Cynara scolymus*	Asteraceae	$2n = 34$ diploid
Asparagus	*Asparagus officinalis*	Liliaceae	$2n = 20$ diploid
Beet	*Beta vulgaris*	Chenopodiaceae	$2n = 18$ diploid with triploid cultivars
Broccoli	*Brassica oleracea* var. *botrytis*	Brassicaceae	$2n = 18$ diploid
Brussels sprouts	*Brassica oleracea* var. *gemmifera*	Brassicaceae	$2n - 18$ diploid
Cabbage	*Brassica oleracea* var. *capitata*	Brassicaceae	$2n = 18$ diploid
Canna	*Canna edulis*	Cannaceae	$2n = 18, 27$ diploid, triploid
Carrot	*Daucus carota*	Apiaceae	$2n = 18$ diploid
Cauliflower	*Brassica oleracea* var. *cauliflora*	Brassicaceae	$2n = 18$ diploid
Celery	*Apium graveolens*	Apiaceae	$2n = 22$ diploid
Chicory	*Cichorium endivia*	Asteraceae	$2n = 18$ diploid

TABLE 8-3   (Continued)

Plant Names in Chapter 8

COMMON NAME	SPECIES	FAMILY	CHROMOSOME NUMBER
Chive	*Allium schoeno-prasum*	Liliaceae	$2n$ = 16, 24, 32 diploid, triploid, tetraploid
Cocoyam	*Colocasia esculenta*	Araceae	$2n$ = 28, 42 diploid triploid
Collard	*Brassica oleracea* var. *acephala*	Brassicaceae	$2n$ = 18 diploid
Corn	*Zea mays*	Poaceae	$2n$ = 20 diploid
Dasheen	*Colocasia esculenta*	Araceae	$2n$ = 28, 42 diploid triploid
Date palm	*Phoenix dactylifera*	Arecaceae	$2n$ = 36 diploid
Eddo	*Colocasia esculenta*	Araceae	$2n$ = 28, 42 diploid triploid
Endive	*Cichorium intybus*	Asteraceae	$2n$ = 18, 36 diploid tetraploid
Garlic	*Allium sativum*	Liliaceae	$2n$ = 16 diploid
Kale	*Brassica oleracea* va. *acephala*	Brassicaceae	$2n$ = 18 diploid
Kohlrabi	*Brassica oleracea* var. *gongyloides*	Brassicaceae	$2n$ = 18 diploid
Leek	*Allium ampeloprasum*	Liliaceae	$2n$ = 32 tetraploid
Lettuce	*Lactuca sativa*	Asteraceae	$2n$ = 18 diploid
Manioc	*Manihot esculenta*	Euphorbiaceae	$2n$ = 36 diploid
Maple, sugar	*Acer saccharum*	Aceraceae	$2n$ = 26 diploid
Onion	*Allium cepa*	Liliaceae	$2n$ = 16 diploid
Parsnip	*Pastinaca sativa*	Apiaceae	$2n$ = 22 diploid
Potato,			$2n$ = 90 hexaploid
sweet	*Ipomoea batatas*	Convolvulaceae	
white	*Solanum tuberosum*	Solanaceae	$2n$ = 24, 36, 48; primarily 48 tetraploid
Radish	*Raphanus sativus*	Brassicaceae	$2n$ = 18 diploid
Rutabaga	*Brassica napus*	Brassicaceae	$2n$ = 38 amphidiploid
Sorghum	*Sorghum bicolor*	Poaceae	$2n$ = 20 diploid
Spinach	*Spinacia oleracea*	Chenopodiaceae	$2n$ = 12 diploid
Swiss chard	*Beta vulgaris* var. *ci-*	Chenopodiaceae	$2n$ = 18 diploid
Sugar cane	*Saccharum officinarum*	Poaceae	$2n$ = 80 octoploid
Tannia	*Xanthosoma sagittifolium*	Araceae	$2n$ = 26 diploid
Turnip	*Brassica campestris*	Brassicaceae	$2n$ = 20 diploid
Yam	*Dioscorea alata*	Dioscoreaceae	$2n$ = 30–80 polyploids
	D. rotundata		$2n$ = 40 tetraploid
	D. cayenensis		$2n$ = 40, 60, 140 polyploids
	D. esculenta		$2n$ = 40, 60, 90, 100 polyploids
	D. trifida		$2n$ = 54, 72, 81 polyploids
Yantia	*Xanthosoma sagittifolium*	Araceae	$2n$ = 26 diploid

BIENNIAL AND ANNUAL STEM, LEAF, AND ROOT CROPS

Turnips and Their Relatives

The array of vegetables produced by modifications of the shoots and roots of members of the Brassicaceae or mustard family (Fig. 8-12) exceeds that of any other plant group. The food crops of the Brassicaceae, including mustard (Chapter 9) and rapeseed oil (Chapter 10), share a pungent flavor imparted by a class of compounds known as the mustard oil glycosides, or glucosinolates. While not toxic to humans in the quantities in which we consume them, mustard oils have been shown to be toxic to insects.

In America the most commonly consumed vegetables in the mustard family belong to the genus *Brassica*. These are grouped together as the cole crops. Interestingly, almost all of the leafy vegetables in this genus belong to one species, *Brassica oleracea*. Each of these vegetables, kale, Brussels sprouts, cabbage, kohlrabi, broccoli, and cauliflower, is produced by a different modification of the leaf or shoot system, and each has its own history of domestication.

Fossil evidence for the cultivation of *Brassica oleracea* is lacking, but historical documentation shows that it was cultivated by the Greeks as early as 650 B.C. By the time of the early Greek writings, there is mention of a type of leafy vegetable with a head which might have been cabbage, and another which sounds from the description rather like kohlrabi. Selection within the species was already advanced at an early date, and the train of selection of various modern morphological forms shown in Fig. 8-13 was underway. Presumably, the plants first taken into cultivation were rather loose-leaved, rank herbs. Forage kales [the European black cabbage (*Brassica oleracea* var. *acephala*)] are probably the modern cultivated forms which most closely resemble wild ancestral types (Fig. 8-13). These early sea kales were adapted to coastal habitats where they were frequently exposed to salt spray. The waxy layer that protected these plants from salt damage is still evident in our modern cole crops and contributes to their resistance to drought and cold.

The common cabbage (*Brassica oleracea* var. *capitata*) includes red and green cabbages with tight heads and smooth leaves (Fig. 8-14). Savoy cabbage, considered by many to be a distinct variety (*B. oleracea* var. *bullata*) has looser, crumpled leaves and is less common in the United States than the more compact type. Although a headed cabbage was described by ancient Greeks (e.g., Pliny the Elder, 23 to 79 A.D.), the modern variety (*B. oleracea* var. *capitata*) is believed to have been developed in Germany where both red and white (= green) cabbages were grown by 1160 A.D. Headed cabbages are formed when the terminal meristem of the primary shoot does not elongate and the "inner" leaves do not expand. The terminal bud (Fig. 8-15) remains buried deep inside the closely appressed leaves.

Cabbages have always been an important item in the diets of European peasants. They could be harvested late in the fall and stored for considerable periods of time. Moreover, since about 200 B.C., the process of packing shredded cabbage leaves and salt into earthenware crocks to make sauerkraut has been used to preserve the vegetable throughout the winter. However, the cold tolerance that has made cole crops so popular in the north temperate zone has also dictated their limits of successful cultivation. These hardy vegetables should

FIGURE 8-12
The flowers of wild mustard (top) and cabbage (below) are typical of the flowers of all of the species in the Brassicaceae. The four petals form a cross, and the six stamens are arranged in two groups with four of them long and two short.

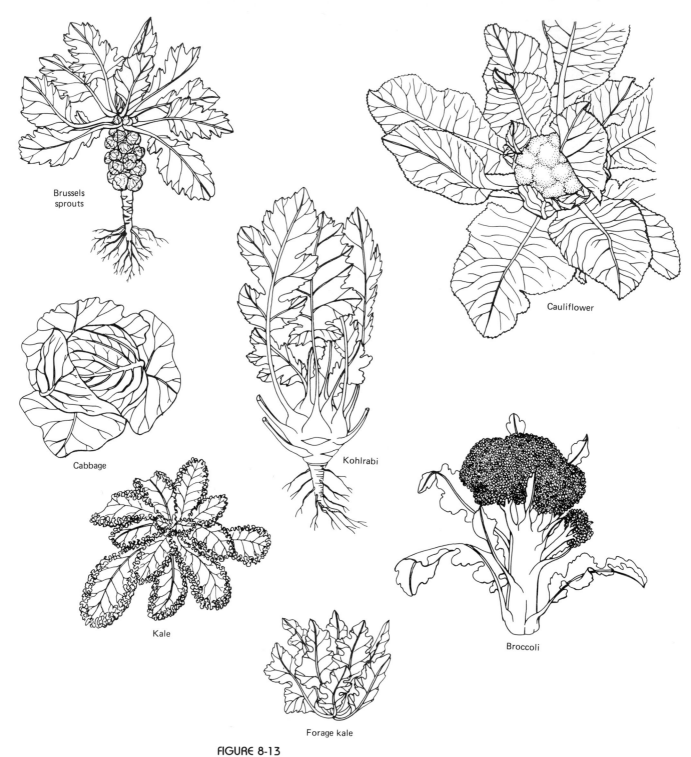

FIGURE 8-13
The postulated proliferation of the varieties of edible brassicas. The ancestral type of the modern crops is presumed to have been similar to the forage kales shown at the bottom.

FIGURE 8-14
Cabbage harvest in Egypt.
(Courtesy of the FAO.)

be planted in the early spring or fall because they mature best under cool temperatures. In addition, if cabbages are not picked at the proper time, they can bolt, i.e., produce a flowering shoot. The expression "gone to seed" comes from the uselessness and tattered appearance of a bolted, leafy crop. In the case of cabbage, the leaves become bitter following bolting, and the plants assume the appearance of a wild or weedy mustard.

Brussels sprouts (*Brassica oleracea* var. *gemmifera*) look like miniature cabbages but are formed when axillary buds form lateral "heads." Brussels sprouts were apparently selected from a mutant plant that appeared in a European garden about 1750. Because lateral shoots are smaller and more tender than the primary stem, Brussels sprouts are eaten in their entirety. The

FIGURE 8-15
A cabbage cut in half clearly shows that the head is formed by the suppression of the terminal bud. Lateral buds remain undeveloped in the leaf axils.

"core" (main axis) of a cabbage is usually removed because it is woody and fibrous.

Kohlrabi (Fig. 8-13) is a bizarre form of the cabbage plant often given a separate varietal name (*Brassica oleracea* var. *gongyloides*). Some workers, however, consider it a distinct species. While a sketchy early Greek description in Pliny's works would seem to apply to kohlrabi, a unequivocal description appears only as late as the sixteenth century. Nevertheless, Pliny's account of a cabbage with an expanded, edible portion just above the roots would seem to correspond to few other vegetables.

Both cauliflower and broccoli are later selections than cabbages or kales. Cauliflower was recorded by the Arabians in the twelfth century, and broccoli appears in old European texts about the middle of the seventeenth century. Despite the fact that both vegetables are derived from inflorescences of the cabbage plant, they are considered to be different varieties. Broccoli (*Brassica oleracea* var. *botrytis*) is a mass of fertile flower buds (Fig. 8-13) that will turn into an inflorescence of yellow flowers if not picked soon enough. Cauliflower (*B. oleracea* var. *cauliflora*), on the other hand, is formed primarily of sterile, abortive flowers. In order to produce a pure white head of cauliflower, the leaves borne under the inflorescence are gathered and tied around it to prevent exposure to the sun which promotes chlorophyll synthesis and a green color.

The two most important cole root crops are turnips and rutabagas. They are often confused, even at the produce counter in grocery stores because, once harvested, they can look quite similar. Nevertheless, they are produced by different species. The parts of turnips and rutabagas we eat actually contain more than just the taproots of the plants. The upper part of the globular structures is the *hypocotyl,* or the portion of the plant where the stem and root intergrade. Most of the mass is root tissue, however, and the vegetables are more properly considered roots than stems.

Turnips, or turnip rapes, are the roots of *Brassica campestris,* thought to have grown originally wild in Europe and Asia (Fig. 8-16). There are references to a plant that appears to have been a turnip in Indian writings of 2000 B.C. European cultivation appears to have begun only in the thirteenth century. Different cultivars of this species yield Chinese cabbage. The English name *turnip* comes from the same etymological base as the verb "to turn" because turnips are so smooth and perfectly formed they appear to have been turned on a lathe. The flesh of the turnips sold as vegetables is usually white. The roots themselves are flat on the top and generally tinged with purple. Yellow-fleshed varieties are grown but are not commonly found in stores.

Turnips have for some reason always been held in low esteem. In Roman times they were spoken of in derogatory terms, and they were a favorite item to throw at miscreants. The Aryans disdained them because they were eaten by Indian races. Young German women in some areas would present suitors that they wanted to reject with a plate of boiled turnips. The majority of Europeans, however, consumed great quantities of turnips throughout the Middle Ages and still include them as a part of many winter meals. Yet this dependency has not seemed to change the opinions of most people about the vegetable. Even in the United States, children are led to believe that nothing could be worse than having to eat turnips. Most people believe that they have a strong flavor, but they have perhaps never had a good turnip or have mistakenly confused the vegetable they ate with its relative, the rutabaga.

FIGURE 8-16
These freshly dug turnips clearly show the flat top of the hypocotyl that differs from the pointed top of a rutabaga. Of turnips Thomas Elyot said, "Boiled it nourysheth moch, augmenteth the sede of man and provoketh carnal lust."

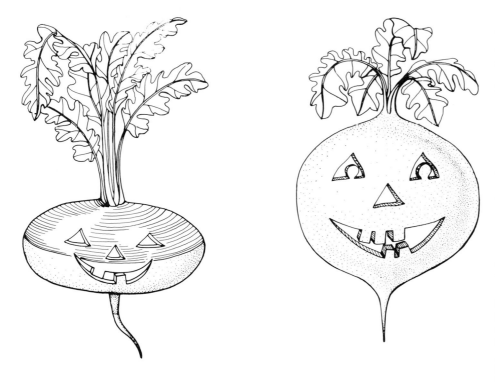

FIGURE 8-17
The first Jack O'Lantern was carved out of a turnip. The Irish, who started our Halloween tradition, explain it this way. One night Satan came to a local pub to claim the soul of Jack, a drunken miserly fellow. Jack suggested that they "have one for the road," but when the Devil turned himself into a sixpence to pay the barkeeper, Jack grabbed the coin and put it in his wallet. After some bickering, Jack agreed to set the Devil free with the promise that he be left alone for another year. Each year thereafter, Jack managed to outwit the Devil. Finally, Jack grew old and died of natural causes. Heaven refused to admit him, and the Devil was not about to offer his adversary a resting place so Jack's soul was stuck between Heaven and Hell. Satan gave Jack a burning coal to light his way, and Jack fashioned a lantern from a turnip he was eating to hold it. The Irish have used turnips (left), rutabagas (right), and potatoes to make lanterns for All Saints' or All Hallow's Day. Americans introduced the practice of using pumpkins for Jack O'Lanterns.

Rutabagas, or Swedish turnips, come from *Brassica napus* and are more nutritious, but have a more pronounced flavor, than their smaller cousins (Fig. 8-17). Rutabagas are believed to have resulted from hybridization and polyploidization between cabbages and turnips because their chromosome number and complement is double that of these two species. The history of rutabaga cultivation has paralleled that of turnips except that cultivars of this species have been selected for use as an oil seed crop producing rape, or colza, oil (Chapter 10). In Europe, rutabagas are often used as food for domesticated animals. In U.S. markets, rutabagas are yellow and larger than turnips. The tops are cut off before marketing, obscuring the fact that, unlike turnips, they are pointed at the upper end. Shippers often coat rutabagas with wax to help prevent drying.

Radishes (*Raphanus sativus,* Brassicaceae) are an important root vegetable on a worldwide basis, but in the United States, they have been relegated to the position of a garnish, carved to resemble roses and to be observed, not eaten, on relish trays and platters. Other cultures hold the peppery roots in much higher esteem. Inscriptions from Egyptian tombs dated to be almost 4,000 years old show that radishes were important in this ancient civilization. Egyptians not only ate the roots, but also valued radish seed oil. In China, selected cultivars are still grown as an oil seed crop. In Japan and many other oriental countries, radishes are among the most important vegetables. In the orient, however, the roots are not the small red spheres that are consumed in the United States, but rather, elongate white, or giant black-skinned forms known as *daikons.* Although the ancestor of the domesticated radish is not known with certainty, the large straight-rooted types are probably similar to the wild types from which modern variants have been selected. The globular form was developed only in the eighteenth century and was initially white. The bright red color was the result of a later mutation that was perpetuated by man.

Lettuce and Its Relatives

Lettuce, chicory, and endives belong to the Asteraceae, one of the largest angiosperm families, containing over 13,000 species. The members of this family are easily recognized by the characteristic heads of many small flowers (see Fig. 2-13). Yet, despite its large number of species, the family has not contributed many plants to the human diet. There are, however, a few notable exceptions. The most important of these is lettuce (*Lactuca sativa*). In ancient times, it was thought that the milky juice exuded from a cut lettuce possessed medicinal virtues similar to those of opium latex. Although such ideas were quickly dispelled, Linnaeus chose the generic name *Lactuca* derived from the Latin *lac*, or milk, because of the milklike sap. Some forms of lettuce may have been cultivated as early as 4500 B.C., and lettuce is depicted on Egyptian tombs. The Romans later enjoyed various greens tossed with olive oil and vinegar as preludes to their renowned feasts. Like many domesticated species, there is no wild plant that can be definitely associated with lettuce, although *L. serriola* and some other species of the Mediterranean region appear to be very close relatives.

Throughout history, lettuce has been subjected to divergent selection. By the fourteenth century, salads had become the rage of England, and housewives had over 50 different types of greens, many of them varieties of lettuce, at their disposal. A virtuous homemaker arranged her assortment of buttery, crisp, sweet, and bitter greens with snippings of herbs, violets, and other flowers so as to give them an aesthetic appeal as well as an interplay of tastes. The salad greens in our modern grocery stores seem banal in contrast.

The lettuce types we do encounter today can be categorized into heading, loose-leaf, and cos types. Iceberg lettuce, a heading type, has become the most common form sold in the United States because its tight, round heads ship well. Other forms include loose-leaf types such as salad-bowl, red-tipped, and oak-leaf lettuces. The primary cos type with stiff, rather elongate leaves, is romaine (Fig. 8-18). Like cabbage, lettuce heads are formed by the suppression of the terminal bud, and they will also bolt if not picked soon enough. Flowering lettuce is a rank-looking plant about 1-m tall with small yellow heads resembling those of dandelions.

Endives and chicory have graced salad bowls as long as lettuce, but both belong to a related genus *Cichorium*, not *Lactuca*. In the United States, the salad delicacies known as Belgian endives come from the perennial species *Cichorium intybus*. To produce these small, tender heads, seeds are sown in the spring and the plants are allowed to grow freely until the first frost, after which they are dug and topped. The thick root stocks are then planted in sand and blanched (forced to initiate new shoots in the absence of light). Under these conditions, chlorophyll synthesis is prevented or greatly reduced, and mild-flavored, torpedo-shaped heads are produced. In Europe, these forced, blanched heads are called witloof. If the plants are allowed to leaf naturally, they produce large, rather bitter leaves that are also used as salad greens. As an escaped wild plant in the United States, *C. intybus* is known as chicory. Likewise, the taproot and products made from it are called chicory. In Europe and in parts of the American south,

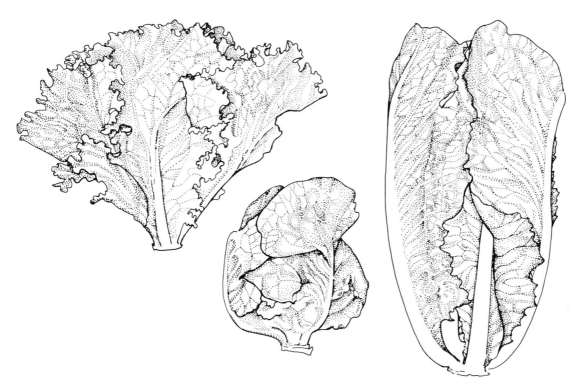

FIGURE 8-18
Different forms of the heads of lettuce are illustrated by (from the left) leaf lettuce, a leafy variety, Boston lettuce, a head-forming variety, and romaine, a cos type.

the roots are roasted and ground. The powder is used mixed with coffee or as a coffee substitute (see Figs. 14-8 and 14-9). Adding to the confusion is the fact that in American supermarkets, the salad green known as chicory or escarole comes from another species, *C. endiva*. The heads of this annual species can also be blanched by tying the large, outer leaves around the inner "heart" of the head.

Carrots and Their Relatives

The Apiaceae, or carrot, family is known primarily for the herbs it contains (Chapter 9). The seeds of celery (*Apium graveolens*) are used as an herb like those of many other members of the family, but the species is most commonly grown as a vegetable. That the edible parts of the plant we use as a vegetable are the swollen leaf bases becomes apparent when we examine the stalks (Fig. 8-19). On the inside, at the bottom of each stalk is an axillary bud, indicating the juncture of a leaf and a stem. What are commonly called celery leaves in cookbooks are, therefore, simply the partially expanded leaf blades. Celery occurs wild today in temperate Eurasia. The species, cultivated since ancient Greek and Roman times, has a variety that produces celeriac (Fig. 8-20), a vegetable consisting of the swollen basal stem of a celery plant. Celeriac is rather popular in Europe where it is eaten raw or cooked as a side dish.

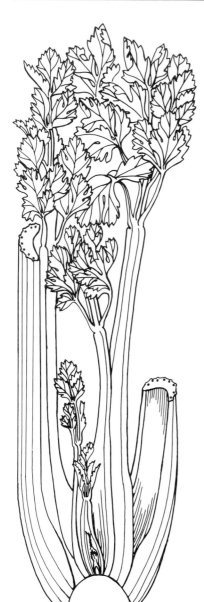

In addition to culinary herbs and celery, the Apiaceae yield two root vegetables of some importance, carrots (Table 8-2) and parsnips. Of the two, carrots (*Daucus carota*) are the more popular in this country. Like most biennials, carrots store reserves in a main taproot during the first summer. As a result, we seldom see flowers on harvested carrots. If we did, we would recognize that the weed known as Queen Anne's lace is nothing more than a carrot that has escaped from cultivation (Fig. 8-21). Originally, carrot roots must have been rather woody and branched. Human selection led to the conical, unbranched versions we now consume. A study of the cultivars grown in Asia Minor and information from historical writings have revealed that the original domesticates were purple, although a yellow mutant that was devoid of the red anthocyanin commonly occurred. Since the purple pigment was water-soluble, the use of carrots in soups and stews often produced a purple-brown, rather unappetizing dish. It was soon seen that the yellow forms tasted the same as the purple and circumvented the bad color problem. Consequently, they eventually became the most commonly grown type. Later selection intensified the color. Carrots are now eaten not only cooked, but as a raw vegetable.

The pigment responsible for the orange color of carrots is carotene (listed as vitamin A in Table 8-1). Chemically, this pigment consists of two vitamin A

FIGURE 8-19
The unopened leaf blades of celery are often discarded. The portion we usually eat is the expanded petiole.

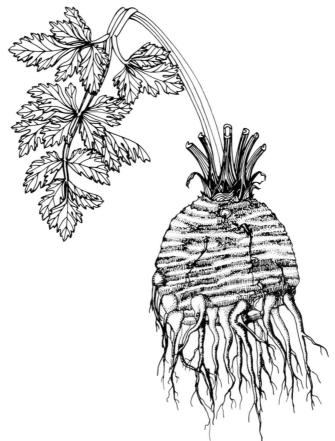

FIGURE 8-20
Celeriac, also called celery root, celery knob, and turnip-rooted celery, is highly prized in Europe for its swollen root.

FIGURE 8-21
Carrot plants that have escaped from cultivation are commonly known as Queen Anne's Lace. At the left is a whole and a sectioned flower. Many of these small flowers make up a flat-topped inflorescence characteristic of the Apiaceae. (After Baillon.)

molecules joined end-to-end (Fig. 8-22). During digestion, carotene is immediately broken down to vitamin A by the mucus lining of the intestine and by the liver. Vitamin A, in turn, is almost identical to the chemical receptor retinal present in the rod cells of the retina of the human eye (Fig. 8-22). Rod cells are important in vision because they gather light at low intensities. Thus, among its other functions, vitamin A supplies the precursor for the important pigment of the eye's rod cells. Consequently, there is some truth to the old wives' tale that eating carrots can help you see in the dark. However, eating carrots merely helps to ensure that an individual receives enough vitamin A. Beyond a threshold, additional vitamin A cannot improve vision.

In general appearance, parsnips (*Pastinaca sativa*) resemble carrots except that they are pale yellow instead of orange. Although they are sweeter than

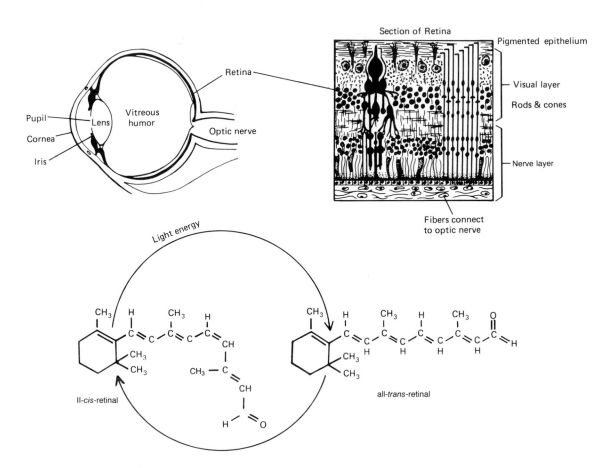

FIGURE 8-22

The relationship between beta carotene, the orange pigment in carrots and sweet potatoes, vitamin A, and the receptor pigments of the eye. Beta carotene is converted to vitamin A in the intestine, so we generally give the vitamin A rather than the beta carotene content of vegetables.

carrots, they are not as popular in this country. They seem to fall into the category of "old-fashioned" vegetables. Like carrots, they are native to the eastern region of the Mediterranean and were eaten throughout the Middle Ages in Europe. The Pilgrims brought them to the New World where they became one of the few European crops enthusiastically adopted by North American Indians.

Beets

Beets (*Beta vulgaris,* Chenopodiaceae Fig. 8-23) might seem to be an odd vegetable to number among the most important crops in the world (see Fig. 19-8), but this apparent anomaly is due to the fact that most of us think of beets only as the small, round, deep red roots that turn up as a side dish, pickled at salad bars, or as a primary ingredient of Russian borscht soup. In addition, the species provides three other crop plants, the *mangel-wurzel,* Swiss chard, and the sugar beet. The first is a cultivar with a yellowish-white root that is used as a cattle feed. Chard (Fig. 8-23) is a leafy vegetable. The sugar beet (see Fig. 8-41), the most economically important of all of the products obtained from the species, is discussed later along with sugar cane.

It is believed that chard (*Beta vulgaris* subsp. *cicla*), or chardlike greens, has been eaten by humans since prehistoric times in the Mediterranean region where beets are native. Early Greeks mentioned varieties of chard in their writings. The roots of Swiss chard plants are stringy and more branched than those of the subspecies which yields edible beets and are consequently not generally eaten.

Spinach, An Annual Leafy Crop

Spinach (*Spinacia oleracea*), like beets, is a member of the goosefoot family (Chenopodiaceae). Spinach is now often substituted for lettuce in salads, but it

FIGURE 8-23
Swiss chard and beets come from different cultivars of the same species. (Courtesy of Garry Fox.)

would not have been included among the salad greens of the Romans because the species is native to western Asia and appears to have been domesticated after the fall of the Roman Empire. The cartoon character Popeye the Sailorman (Fig. 8-24) was originally devised to cajole children into eating spinach because of the large amounts of iron it contains. Unfortunately, it is now known that much of the iron in the vegetable cannot be absorbed because calcium oxalate in the leaves forms a chemical complex with the iron, hindering its absorption.

THE PERENNIAL GREEN VEGETABLES: ARTICHOKES AND ASPARAGUS

The Perennial Habit

When farmers plant a perennial crop, they use procedures quite different from those followed by farmers cultivating annual or biennial plants. Great care is taken in the preparation of the soil for perennial crops because once the plants start to grow, they remain in place for many years. Unlike annual or biennial species, harvesting of crops from perennial vegetables is delayed until 2 or more years after planting in order to allow the plants to become well established and to store up reserves in the root system. However, as the years progress, the farmer reaps increasingly larger harvests until the plants begin to senesce. Perennial temperate vegetable crops are generally replanted every 15 to 25 years.

Artichokes

The Arabic word *ardischauki*, meaning "earth thorn," is the origin of the name artichoke given to one of our most distinctive vegetables. Artichokes (*Cynara scolymus*) are members of the Asteraceae, but unlike lettuce and the other leafy family members, they are cultivated for their immature flowering heads (Fig. 8-25). When we eat an artichoke, we eat the carbohydrates stored in the tender portions of the fleshy bracts surrounding the undeveloped flowers (Table 8-1). The undeveloped flowers and the chaff intermingled with them form the inedible "choke" of the artichoke. The savored artichoke "heart" is the swollen part of the stem (receptacle) on which the "choke" is borne. If allowed to mature, an artichoke becomes a beautiful, large, blue thistle (Fig. 8-26). Artichokes are generally planted using suckers, or basal branches, in order to ensure uniformity and to hasten plant establishment.

Asparagus

Asparagus (*Asparagus officinalis*, Liliaceae) is a dioecious perennial (Fig. 8-27) native to the temperate scrub communities of southern Europe, western Asia, and northern Africa. These habitats are subject to periodic burning, and many of the plants which occur natively within them possess the ability to sprout rapidly from a robust underground system of rhizomes. Humans have made use of this resprouting habit by growing plants in habitats where plants die back in the winter and send up new shoots in the spring. The vegetable consists

FIGURE 8-24
This statue of Popeye stands in Crystal City, Texas, reputed to be the spinach capital of the world.

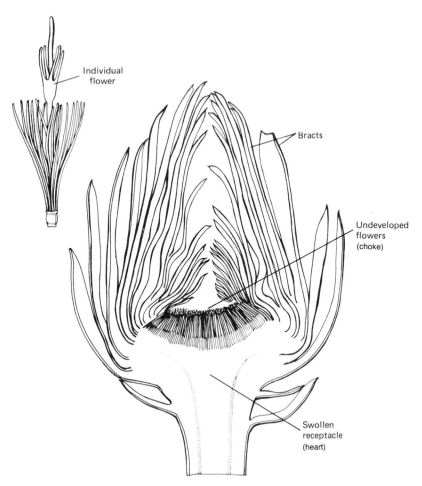

Individual flower

Bracts

Undeveloped flowers (choke)

Swollen receptacle (heart)

FIGURE 8-25
An artichoke cut in half shows that the vegetable is really an immature flowering head.

FIGURE 8-26
If allowed to bloom, the artichoke will produce a beautiful purple or blue thistle head.

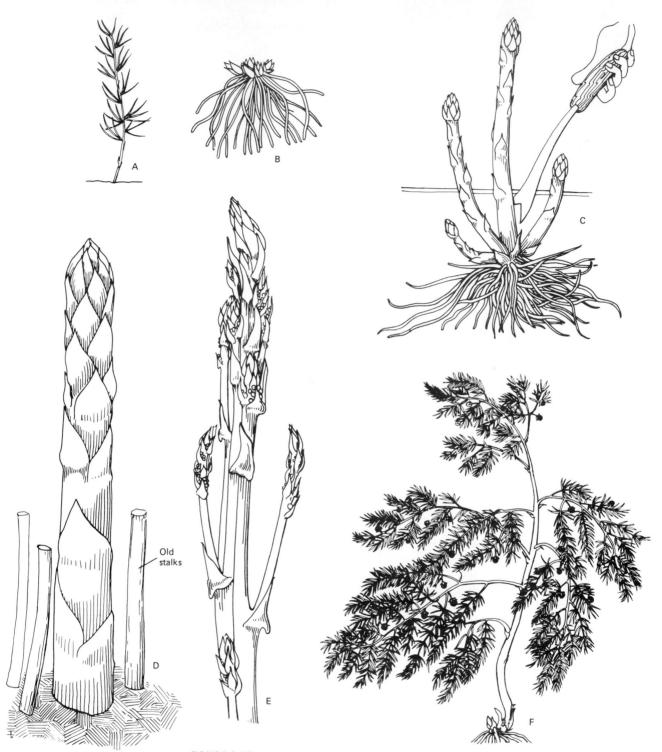

Old
stalks

FIGURE 8-27
Asparagus is a perennial herb that produces a crop for many years. Because of the long
life of individual plants, planting and harvesting procedures differ from those of annual
crops. (A) Seedlings are established and transferred to permanent beds, or (B) rootstocks
(crowns) are purchased and set into permanent beds. (C) After a plant is 3 years old,
some of the shoots produced in the spring can be harvested using an asparagus knife.
(D–F) Some shoots are left to develop into the fernlike branches that will produce
photosynthetic products, some of which will be transported to the rootstock for storage.

of young, unexpanded shoots that are cut where they emerge from the ground. Asparagus is expensive because shoot production occurs for only a limited period in the spring and only a certain proportion of the new shoots can be harvested without damaging the plants. Shoots that are allowed to mature expand into ramified branch systems, or "cladophylls," which resemble feathery leaves. In the horticultural trade, some varieties of asparagus plants are called "ferns" because of their green, highly dissected vegetative systems. The true leaves of an asparagus are reduced to scales at the base of each cladophyll. Asparagus is usuallly propagated using pieces of rhizomes called crowns because seed set is generally poor.

VEGETABLES FROM BULBS:
ONIONS, LEEKS, GARLIC, AND SHALLOTS

The lily family's reputation for ornamental plants is challenged only by its contribution to the culinary arts, principally the contributions made by members of the genus *Allium* (Fig. 8-28), which includes the onion and its relatives. Most of these species were cultivated, or at least eaten, before recorded history. Like most lilies, these pungent herbs are bulbous plants. People have made use of the stored nutrients found in their bulbs and in all cases (except chives) consume the bulb and sometimes the basal portions of the flattened leaf blades. Onions (*A. cepa*, Fig. 8-29) and garlic (*A. sativum*) probably originated in central Asia and the leek (*A. ampeloprasum*) in the Near East. All were cultivated in Egypt, however, by the year 3200 B.C. Onions are grown for their single, large bulb, and garlic for its cluster of bulbs, each of which is called a clove. Leeks produce a bulb that is ony slightly differentiated from the basal portions of the leaf blades. Consequently, a substantial portion of the leaf blades of leeks as well as the swollen basal portions are eaten. Another cultivated member of the genus, the chive (*A. schoenoprasum*), is eaten for the leaves alone.

Almost all members of the *Allium* group have been reputed to have medicinal properties. At one time or another, various species have been prescribed for bad eyesight (Hippocrates), voice improvement (Nero), baldness (Gerarde), or as a cure for the common cold, acne, arthritis, hypertension, or poor digestion. General Ulysses S. Grant was so convinced that a supply of onions was needed to protect his troops from dysentery that he declared "I will not move my army without onions." There is still controversy about the real effects of onions and garlic with many health food enthusiasts advocating their use and others denying any therapeutic benefit.

The pungent quality common to all of the alliums is linked with the compounds that makes us cry when cutting onions. These are volatile sulfur compounds (including methyl di- and trisulfides, and *n*-propyl di- and trisulfides) which are released when the cells of an onion are ruptured. When these compounds dissolve in the fluid covering our eyes, they form sulfuric acid. The tears brought on by the burning sensation only provide more water into which the compounds can dissolve. Since the chemicals are water-soluble, their effects can be diminished by submerging an onion in water while it is being sliced or by freezing the onion before slicing.

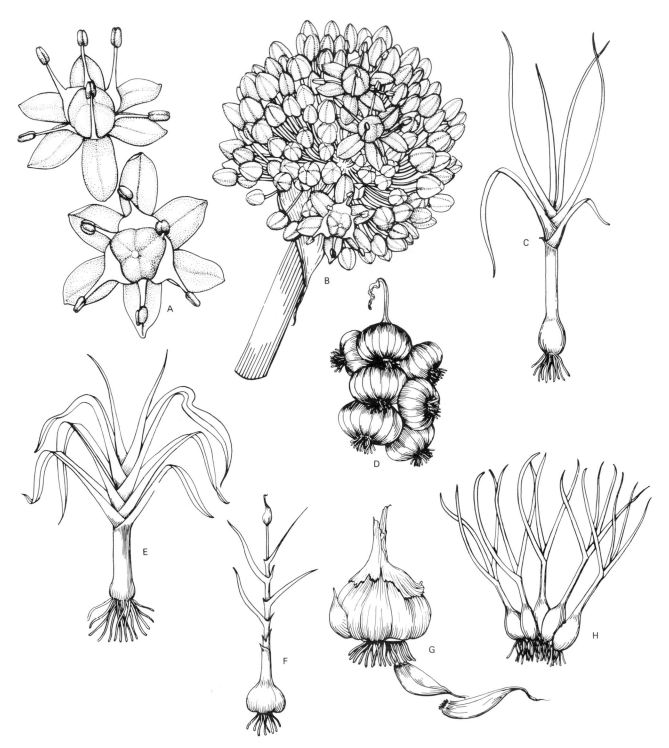

FIGURE 8-28
The lily family provides us with the onion and its relatives. (A) onion flowers, (B) onion inflorescence, (C) young onion, (D) dried onion bulbs, (E) leek, (F) garlic, (G) garlic bulbs, (H) shallots.

Starches from Tubers, Rhizomes, and Roots

Tubers and Rhizomes

As we saw in Chapter 1, one important modification of stems has been for underground growth. Underground stems serve as overwintering organs and/or for asexual reproduction. Particularly adapted for human use are those underground stems that store nutrients. A seedling draws the nutrients it needs for initial growth from either the cotyledons or the endosperm as we saw in Chapters 4 to 7. If a plant reproduces vegetatively from underground stems, the newly emerging shoot can derive nutritive material directly from the "parent" plant's vascular system until it has established roots of its own, or it can pull reserve nutrients from an underground storage organ such as a fleshy rhizome or tuber (see Fig. 1-16).

People have made abundant use of storage rhizomes and tubers since prehistoric times. Undoubtedly, wild tubers along with fleshy fruits were among the principal foods gathered by preagricultural people. They are still an important item in the diets of hunter-gatherer tribes that exist today. Humans have also taken advantage of the characteristics of rhizomes and tubers in their use of them for propagation. People can use a part of the storage stem, or a few "seed" rhizomes or tubers, to produce a plant which, when mature, provides an abundant supply of food plus enough material for further propagation.

White Potatoes

Although the vast majority of the world is fed primarily by grains (Chapter 6), one nongrain food item, the potato, is of comparable importance as a food staple in temperate regions of the world. Considering the modern importance of potatoes (*Solanum tuberosum*, Solanaceae), it is surprising that they have been known outside of the Americas for only a few hundred years. The commonly

cultivated potato and several less well known tuber-bearing members of the same genus are natives of South America.

Spanish explorers found potatoes being grown by Indians all along the Andes from Colombia to Chile. In high elevation areas of the central Andes, Indians preserved potatoes by a primitive method of freeze-drying (Fig. 8-30). Temperatures in these areas can fluctuate each 24 h from below freezing to almost 15°C (84°F) during the day. In addition, the climate is dry for most of the year. These combined climatic factors allow Indians to spread harvested potatoes on the ground and leave them overnight to freeze. The next day, the Indians stomp on them to squeeze out as much water as possible. This process yields, after a few days, a light, dry mass of cellulose and starch known as *chuño*. Since this method is not as rapid as modern freeze-drying (and ruptures the cells which is avoided in modern techniques), chuño often tastes moldy, but once completely dry, it keeps for months.

While chuño never became a favorite of the Europeans, the Spaniards quickly saw that potato tubers had a potential as a food source and they introduced them into Spain and used them for food on return voyages to Europe. The rest of Europe learned about potatoes from an independent introduction to England. Whether Sir Walter Raleigh or Sir Francis Drake first brought the potato to England is the subject of some debate, but evidence seems to favor the latter captain. A folktale says that Queen Elizabeth I accused Sir Francis of trying to poison her when her cooks served a salad of potato leaves rather than a dish of cooked tubers from the plant delivered to them by Drake. Nevertheless, the potato was destined to become a great success in England and, soon, an even greater success in Ireland.

Ireland's enthusiastic reception of the potato was related not only to its growing well in the country's cool, moist climate, but also to the fact that a small amount of land could produce enough calories to meet the needs of an entire family. In addition, Ireland was in the midst of a rebellious conflict with England, and potatoes could be buried to hide them from enemy troops that periodically swept into Irish towns. By the 1840s, Ireland was not only fond of potatoes, she was completely dependent upon them. The entire country had essentially adopted a monoculture of the tuberous plants. A week's food for an average family of five typically consisted of about 40 lb of potatoes, several pounds of meat (probably mutton), a few loaves of bread, a few quarts of beer, bits of butter, sugar, and tea. About 1845 disaster struck. Potato blight, caused by a fungus (*Phytophthora infestans*) reached Europe, and within 5 years, virtually all of the Irish (and British) potato crops were destroyed. It is estimated that during these 5 years at least 1 million people (some say 2 million) died of starvation and over another million emigrated, many to the United States. After this lesson, farmers began to pay attention to selecting for disease resistance and to the dangers of dependency on a single crop. Luckily, at about the same time, a fungicide (Bordeaux mixture) was developed which helped in controlling the blight.

Today, potatoes are grown throughout temperate parts of the world (Table 8-2) and in upland tropical areas. Planting is almost exclusively by seed tubers or by planting portions of the tubers which contain dormant buds (Fig. 8-31). While we generally see only a few potato varieties in the United States, hundreds of varieties have been developed.

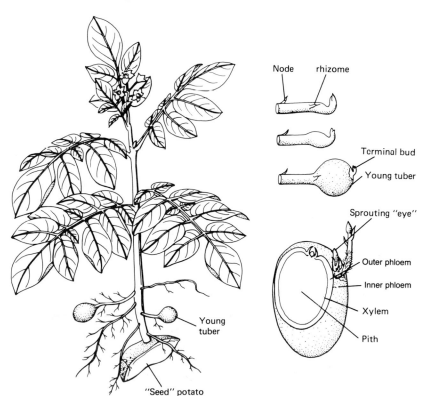

FIGURE 8-31
Tuber formation in potatoes.

Potatoes have traditionally been grown from pieces of tubers because the seeds are almost always nonfertile, but using tubers creates problems because they carry pathogens, particularly nematodes and fungi. Gardeners can help reduce infestations in their crop by buying only "Certified disease-free" seed tubers. Recently, a new variety of potato called "Pioneer" has been produced which can be grown from seed. This variety has caused some excitement because using seeds will not only reduce pest problems but also allow new breeding programs to be undertaken.

Yams

The second most important tuber in terms of world production is the yam (Fig. 8-32). Although the sweet potato (*Ipomoea batatas* Convolvulaceae) is often called a yam in the United States, true yams belong to the genus *Dioscorea* (Dioscoreaceae). There is some debate about whether a yam is derived from stem or root tissue; most authorities consider that the cultivated species have tubers. However, it is possible that some of the species have storage organs that are derived from roots rather than from stems. Different species of yams are native to Africa, Asia, and South America, and domestication appears to have occurred independently on each continent. The species traditionally cultivated in the three native areas are *D. rotundata* and *D. cayenensis* in Africa, *D. alata* and *D. esculenta* in Asia, and *D. trifida* in the New World. Today, however, *D. rotundata* is grown throughout the world in the largest quantities.

In many parts of Africa and Asia, yams are important in all aspects of the culture. In both New Guinea and Melanesia, there are ritual yam cults. Among the practices of these cults is the planting of ceremonial yams in special gardens. The size of these yams, which can reach up to 50 kg (122 lb), is thought to reflect the grower's status in the community. The ceremonial yams are used for gifts in ritualized exchange. Yams are held in such high esteem that only men are allowed to cultivate both eating and ceremonial yams. A yam festival is held at harvest time, during which tubers are dressed or covered by masks (Fig. 8-33).

Propagation of yams is by asexual means using pieces of tubers with buds. Many of the cultivated species are poisonous or semipoisonous because of the presence of oxalic acid (or oxalates) in cells just below the skin of the tubers. Peeling and boiling eliminates most of the potential problems caused by the crystals. Yams are higher in protein than many of the other starchy tuber crops (Table 8-1), but unfortunately, yam cultivation has been replaced in many areas of the tropics by that of different tuberous crops which are easier to grow.

Other tubers grown on a more limited scale include the taro, eddoe, dasheen, or "old" cocoyam (all *Colocasia esculenta*, Araceae), malangas, yautias or tannias (*Xanthosoma sagittifolium*, Araceae), arrowroot (*Maranta* spp., Marantaceae), and canna (*Canna edulis*, Cannaceae) (Fig. 8-32). Only the first two of these are important crop plants. Dasheens are staple foods in the tropical Pacific areas and widely cultivated in west Africa. The genus *Colocasia* is native to east Asia and Polynesia but had spread eastward by the time of the Egyptian empire. Tannias, in contrast, are native to the New World, where they have been cultivated since ancient times. Arrowroot has a limited area of cultivation

FIGURE 8-32
Many tubers and rhizomes such as (A) yam, (B) arrowroot, (C) taro or dasheen, and (D) yautia are food staples in tropical areas.

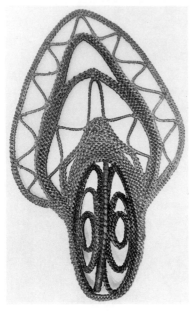

FIGURE 8-33
This Tumbuan mask from the Maprik District in the northeastern part of New Guinea was made for a yam to wear during the harvest ceremonies. Like most yam masks, the figure depicted is animistic and has two prominent eyes separated by a nose or beak. Woven masks are also worn by young male dancers during the yam festivals. (Mask a gift of Molly Whalen.)

in the West Indies and northern South America where the genus is native. The tubers are generally not eaten but used to make a fine, easily digested, powdered starch. Adulterants, starch from other tuberous crops, are often sold under the name "arrowroot," but they are inferior in quality.

Cannas are also native to South America, but their propagation has now spread throughout the tropics. Canna plants are also grown as ornamental canna "lilies." The tuber is somewhat purplish and when dried and powdered, produces a starch known as purple arrowroot or Queensland arrowroot.

STARCHY ROOT CROPS

Manioc

While most of the root crops of temperate regions are vegetables that are used as side dishes during a meal, the major root crops of tropical regions often constitute the entire meal. The most important of these tropical root crops has many names, manioc, manihot, or cassava (*Manihot esculenta*, Euphorbiaceae, Fig. 8-34). While practically unknown in temperate regions, manioc is the staple food of over 500 million people. It contributes about 37 percent of the calories consumed in Africa and 11 percent of the caloric intake in Latin

FIGURE 8-34
The manioc, or cassava, plant. According to a Tupi legend, there was once a mother with no food who had to watch her starving child die. Sadly, she buried the child under the floor of her hut. That night, a wood spirit, or "mani," came and transformed the child's body into the roots of a plant that grew up to feed future generations of Indians. The plant was called "mani" "oca" (root) for the root that the wood spirit brought.

America. As is true of all of the world's staple foods, manioc is almost pure starch. The storage roots are produced by monoecious perennial plants (Fig. 8-34) native to the tropical areas of the New World. The species is known only as a cultigen and has been grown for so many thousands of years by small tribes of Indians that there are hundreds of local varieties or land races.

The variability of the species and its widespread cultivation has created problems in interpreting the evolutionary history of manioc and determining the area of initial cultivation. It appears likely that it was domesticated independently in two regions, the semiarid areas of southern Mexico and adjacent Guatemala, and northeastern Brazil. Archaeological evidence from Mexico dates from only the first millennium B.C. Griddles for baking manioc bread dated to the year 2000 B.C. have been uncovered in South America. Still, the wild species which appear to be most closely related to manioc grow in southern Mexico.

Manioc seems like an ideal crop for tropical regions. It grows well in both arid and wet climates, produces storage roots in poor soil, is relatively resistant to insect and fungal pests, requires a minimum of agricultural labor, and has a high yield per unit area. Planting consists simply of clearing the ground and inserting into it 30-cm-long pieces of the stem which have several axillary buds. Each plant produces several storage roots which can be harvested in about 18 months or left in the ground until needed.

Unfortunately, manioc has its drawbacks. Despite their high carbohydrate content (about 35 percent by weight, Table 8-1), the roots contain very little protein (1 percent) and essentially no vitamins. Moreover, manioc roots often contain poisonous compounds that must be removed before the roots can be eaten by humans. The poisonous compounds are two cyanogenic glycosides, linamarin and lotaustralin. These hydrolyze in the presence of the enzyme linamarase to form, among other compounds, hydrogen cyanide. Any bruising, cutting, or grating of the roots brings the enzyme and the glycosides into contact and leads to cyanide production. Roots that have little or none of the glycosides are known as sweet maniocs. Those with appreciable amounts of the glycosides are called bitter maniocs. If present, the glycosides are located primarily in the outer cortical cells of sweet forms, whereas the toxic precursors are dispersed throughout the roots of bitter varieties.

In addition to the bitter and sweet forms, there are two major morphological types of manioc. The first type has smooth-skinned tubers and plants with silver-gray stems. The second group has rough-skinned tubers and red or purple-tinged stems. Unfortunately, the morphological groups do not coincide with the groups based on cyanogenic glycoside content. In fact, it is impossible to tell without prior knowledge of the kind of manioc that was planted whether it is of the bitter or sweet type. This lack of certainty may account for the fact that both forms are often processed similarly, particularly in South America. In Africa, where sweet types were primarily introduced, and are most extensively grown, the roots are often simply peeled, boiled, or dried. During the boiling process, the few cyanogenic glycosides are broken down and cyanide gas is driven off. The roots are then sold as dry, heavy, starchy lumps. These detoxified lumps can be ground into manioc flour to be used in breadlike products, or they can be reboiled to yield a rather gelatinous, tasteless, starchy vegetable reminiscent of boiled potatoes.

FIGURE 8-35
The cassava root is grated to rupture the cells and to release the toxic compounds that must be removed before the manioc can be safely eaten. (Courtesy of the FAO.)

In Brazil and Venezuela, both important countries of manioc production and consumption, the roots are often processed by a mechanized version of an ancient hand process. The roots have traditionally first been shredded (Fig. 8-35) and pounded to ensure all of the cells are ruptured and all of the enzymes and glycosides have been released (Fig. 8-36). The soggy mass is drained and the resultant wet fibrous residue allowed to sit overnight. Chemical studies have

FIGURE 8-36
Pounding cassava root is another method of rupturing the cells. (Courtesy of the FAO.)

shown that right after shredding and pounding, there are still large amounts of toxic compounds in the mass. After standing, the concentrations are lessened since some of the volatile hydrogen cyanide gas has had time to escape. The wet, day-old pulp can be used directly to make a kind of flatbread by spreading it on a large circular griddle about a meter in diameter. The circular manioc pancakes that are produced do not contain cyanide because the cooking drives off the last of the gas. These flat manioc breads, which taste something like shredded wheat, can be stored for long periods of time (Fig. 8-37). The drained manioc pulp can also be dried and powdered. If the powder is toasted, it is known as fafaro. Fafaro often constitutes the only food eaten during an entire day by many natives of the Brazilian tropical lowlands. To those unaccustomed to it, the powder tastes like a cross between plain corn meal and flour. Yet, large numbers of people eat it simply by the handful.

Where factory processing is not available, various other primitive methods are still employed. The roots are often ground by hand, and the wet pulp poured into basketry tubes 2 or more meters tall. The baskets, which are hung from the top and closed at the bottom, are constructed in such a way that they act like Chinese finger puzzles. The weight of the wet manioc causes the tubes to constrict and squeeze out the moisture which runs down the baskets into vessels below.

Although the juice squeezed from the grated manioc mass is usually discarded, it contains a modest amount of starch that can be recovered by letting it settle out of the solution and decanting the liquid. The starch cake left behind can be dried and eaten raw, baked into a breadlike product, or used commercially like corn or potato starch. Purified manioc starch is a better sizing material (Chapter 11) than either of these two other starches, but it is not produced on a large enough scale to challenge them commercially. If the starch is heated on a hot metal plate or tossed in a hot metal drum, it becomes gelatinized into pellets known as tapioca. Because tapioca is such a good thickening agent, it is a common ingredient in fruit pies and puddings.

The water solution drained from grated or pounded manioc can also be fermented into a beer that is perfectly safe to drink because the fermenting yeasts break down any traces of toxic compounds left in the liquid.

Sweet Potatoes

The thought that humans crossed the Pacific from the New to the Old World long before Columbus discovered America has intrigued people for years. Thor Heyerdahl built his balsa raft, the Kon-Tiki, fashioned after the reed rafts of the Oru Indians living on Lake Titicaca in Bolivia (Fig. 8-38), and sailed westward from the coast of Peru in 1947 in order to prove that such a journey was possible. His success heightened the controversy as to whether prehistoric exchanges had occurred between South America and the islands of Oceania. Among the evidence mustered to support this hypothesis is the cultivation of sweet potatoes (*Ipomoea batatas*, Convolvulaceae) by native peoples of Malaysia and Polynesia as well as by those of South America.

There is little doubt that the species is native to South America. Fossilized sweet potatoes from the Andes have been dated as being 8,000 and 10,000 years old. Archaeological evidence has shown that they were cultivated in South America by at least 2500 B.C. The unsolved question is when and how

FIGURE 8-37
The large, flat circles of cassava bread store well but are a little tough and dry to eat by themselves. Courtesy J. L. Neff.

FIGURE 8-38
The reed rafts of the Oru Indians in Lake Titicaca, Bolivia, provided a model for the Kon Tiki. Thor Heyerdahl sailed from Peru in a craft similar to this in order to prove that there could have been pre-European exchanges between South America and Oceania.

the sweet potato reached the Old World. There are no fossil or archaeological remains from the Pacific Islands, but the sweet potato forms part of an agricultural complex in Polynesia thought to date from 1200 A.D. Ancient pits believed to have been used to store sweet potatoes have also been found in New Zealand, and the roots figure prominently in the mythology of primitive New Zealand peoples in much the same way as true yams (*Dioscoreaceae*) do in Melanesia and New Guinea. Was the sweet potato really carried across the Pacific by humans before Europeans began to traffic the oceans? The answer currently seems to be no. This conclusion will undoubtedly stand unless fossil evidence of sweet potatoes establishes their presence in Oceania prior to the seventeenth century. Even if such evidence is found, it will not provide the answer as to how the roots crossed the ocean.

Today, the sweet potato is cultivated throughout the world. In temperate areas it is grown as an annual, and in the tropics it persists as a perennial. The plants themselves are trailing vines (Fig. 8-39) with morning glory flowers. Although sweet potatoes resemble tubers and were confused in the eighteenth century with white potatoes, they are true roots as evidenced by the fact that sweet potatoes cannot be propagated by planting pieces of the root in the ground. A whole sweet potato would have to be planted to get a new plant.

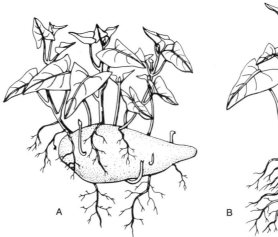

FIGURE 8-39
Sweet potatoes sprout from a storage root but most shoots sprout from the "top" end (A). A seedling (B).

Plantings are therefore made using pieces of the aerial stems. The process of swollen root formation in sweet potatoes has been studied in some detail. Regular roots become tough and hard as they mature because of lignin deposition in the cell walls. Lignification appears to be prevented in the storage roots by hormones that also initiate the swelling process. During swelling, the storage roots produce additional cambium layers that make cortical cells in which starch is stored.

Unlike manioc, sweet potatoes contain a modest amount of protein (about 2 percent by weight), some sugar, and almost as much carotene (tabulated as vitamin A) as carrots (Table 8-1). They are less extensively grown than manioc in the tropics because they are much less resistant to rot and insect attack. Improvements resulting from breeding programs have been slow because of the difficulty of producing flowers under temperate summer conditions.

In the United States, sweet potatoes are eaten like white potatoes or "candied" (cooked with sugar). In the southern and western portions of the country, the roots are often called yams, which causes some confusion with the true yams. Varieties of sweet potatoes with moist yellow-orange flesh are usually eaten in the south. In the northern part of the country, varieties with drier, lighter yellow-colored flesh are preferred. In other parts of the world such as Japan, over half of the annual crop is used for starch, wine, and alcohol. A relatively large percentage is also used for animal food. In view of the fact that the species is native to the Americas and to tropical regions, it is interesting that China is now the world's leading producer (Table 8-2).

SWEETS FROM STEMS AND ROOTS

When we eat candy or pour sugar into our coffee, we seldom think about the fact that sugar comes from the stem or root of a plant. Historically, most sweeteners used by humans (other than honey) have been syrups or crystallized sugar obtained by boiling down the sap or juice expressed from the stems of various plants. These include most table sugar and sweet syrups from cane, sorghum, date palms, and trees such as the sugar maple. Since sugar is by far the most important sweetener in the modern world, we will discuss it in some detail. What we commonly call sugar is sucrose, a disaccharide of glucose and fructose (see Fig. 1-3). Sucrose tastes sweeter to humans than either of its constituent monosaccharides. It is for this reason that sugar tastes sweeter than corn syrup, which is a mixture of glucose and fructose. Most sugar is obtained from cane, but in some areas, sugar beets are used for its production. The final product, refined (white) sugar, is identical regardless of whether cane or sugar beets are used. Consequently processing of both is discussed together.

Sugarcane

Sugarcane is a perennial grass (*Saccharum officinarum*, Poaceae) believed to have been selected from wild species of *Saccharum* by primitive peoples of the Far East, probably New Guinea. The process of producing sugar from cane has been known in India since 3000 B.C. Alexander the Great brought back to Europe stories of a honey from reeds, but since the Far East was the only area of the

FIGURE 8-40
Sugarcane harvest in the West Indies over 100 years ago. (From *Histoire et Legends des Plantes*, J. Rambosson, 1868.)

Old World where cane grew well, very little arrived in Europe. Honey was the primary sweetener in Europe until almost 1500. Sugar was initially so scarce in Europe that it was used primarily to make medicines more palatable (a practice still used for childrens' medicines). The discovery of the New World provided Europeans with land that could grow tropical and subtropical crops such as sugar cane. Columbus brought cane to the New World on his second voyage (1493) and successfully established its cultivation in the West Indies (Fig. 8-40). Sugar rapidly became an important American export, and its production was intimately linked with the history of the New World.

The famous American sugar triangle of the eighteenth century involved New England, Africa, and the West Indies. Raw sugar or molasses produced in the West Indies was shipped to Connecticut, where it was used to make rum. The rum was sent to Africa to buy slaves, who were brought to the West Indies to provide labor for the cane fields. Payment for the slaves was in cane products that again went to New England. In 1764, the British Parliament imposed a tax on sugar with the passage of the sugar taxes. As a result, people in Connecticut started to smuggle sugar into the colonies. When British customs vessels began to patrol the waters, one, the *Gaspee*, was burned and sunk. This overt act of aggression, resulting from the Sugar Act, predated the Boston Tea Party (1773) as the initial piece of violence that finally led to the American Revolution.

One of the reasons for the successful establishment of cane, once suitable areas were found, is the method by which it is grown. Virtually all sugar cane is propagated vegetatively by means of *setts*. A sett is part of the stem which includes lateral buds and a circle of cells which give rise to a number of adventitious roots. The initial adventitious roots absorb nutrients until new roots are formed. Once established, the cane puts out tillers which give rise to new plants. This method of growth allows the harvesting of multiple crops per year. After several years, yields per plant decrease and a new field is planted.

Sugar Beets

Beet greens and beet roots (*Beta vulgaris*, Chenopodiaceae, Fig. 8-41), were eaten by the Romans, but it was not until the eighteenth century that the Germans discovered that the roots contained appreciable amounts of sugar. In 1747, Andreas Margraff developed the first process to extract the sugar, but a truly productive extraction process did not become available until 130 years later. At the same time, artificial selection began to lead to beets with higher and higher sugar contents. Napoleon I realized the value of a domestic source

FIGURE 8-41
A white sugar beet. (Courtesy of the USDA.)

of sugar and ordered research to be conducted on the plant. Subsequent work so increased the sugar content that the French achievements began to draw the attention of other temperate countries. In the 1890s, sugar beet cultivation expanded rapidly in the United States and throughout Europe.

The sugar beet represents one of the greatest triumphs of plant selection (see Fig. 19-4). Under the encouragement of artificial selection, the sucrose content of sugar beets has risen from a lowly 2 percent in the eighteenth century to over 20 percent today. Although tough and basically inedible for humans, sugar beets now provide most of the sugar for many European countries. In the United States, they are important in some regions, but production is limited by the availability and capacity of local processing plants. Since the United States has a ready supply of cane sugar from Hawaii, Puerto Rico, and the Gulf Coastal States, there has not been any great pressure to expand sugar beet production. The predominant use of cane sugar in the United States is, therefore, a matter of economics, and not because of any inherent superiority of cane over beet sugar.

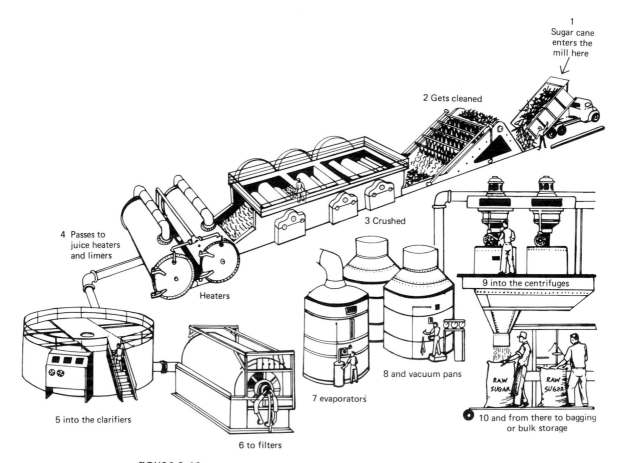

FIGURE 8-42
The major steps in sugarcane processing. (Redrawn by permission from *Sugar*, 1949, Hawaiian Sugar Plantation Association.)

Sugar Refining

Crystallized sugar is made by processing the juice of cut sugarcanes (Fig. 8-42) or harvested sugar beets. Cane is ready to harvest 22 to 24 months after planting the setts. The canes can simply be cut (Fig. 8-43), or the fields can be burned to remove the leaves and evaporate much of the water in the stems, thereby reducing the time needed later to concentrate the expressed juice. Sugar beets are harvested in the fall of the year they are planted. When they are harvested, they are often "topped" (the top of the root and the rosette of leaves are severed from the root). The tops are used as animal food. Once canes reach the mills, they are fed through rollers that crush them and express the juice. Sugar beets are washed, shredded, and crushed. From this point on, the process of sugar production from the sweet juice of either crop is essentially identical.

The expressed juice is usually heated with calcium hydroxide and water and allowed to cool and settle. The upper layer of clarified juice is siphoned off and sent to an evaporater. The muddy lower portion is often filtered and the clear filtrate added to the liquid in the evaporater. Heat and vacuum evaporaters are used to thicken the syrup until it has begun to crystallize. This partially crystallized sugar syrup is called the *massecuite*. When crystallization has reached a maximum, the crystals of raw sugar are separated out by centrifugation. The brown liquid separated from the crystals is molasses. Most sugar-producing areas ship raw sugar, or use part of it to make rum. Consequently the modern rum-producing countries are also major cane sugar producers (Table 8-2). Raw cane sugar is crystalline, but it is brown and not free-flowing. The refining of the raw sugar into white table sugar is accomplished in refineries located in the sugar-importing countries.

FIGURE 8-43
Cutting cane in Louisiana.
(Courtesy of the USDA. Photo by
J. M. Cross.)

The refining of raw sugar involves several steps. First, the crystals are redissolved into a heavy syrup and centrifuged to remove some of the remaining unwanted chemicals. Next, a filtering agent is added to the syrup which clarifies it as it passes through the liquid. Another filtration, usually through charred bone, produces enough clarity to allow final crystallization and production of the various forms of white sugar. Granulated sugar is made by crystallization of the syrupy mass in large pans, followed by spinning the rough crystals in hot, revolving drums. Powdered sugar is made by finely grinding the crystals. Because it prevents clumping, corn starch is added to powdered sugar to produce confectioner's sugar. Sugar cubes are formed by slightly moistening the sugar and pressing it into forms.

Despite the claims that raw sugar is "better" for you than refined sugar, the major ingredients used in the refining process, diatomaceous earth and bone char, are both innocuous substances. Most of the unhealthy effects of sugar are caused by the ingestion of too much sucrose. Raw sugar still contains many chemicals present in the cane stalks, none of which have ever been shown to have any beneficial effects. Actually, what is often sold as "raw sugar" is refined sugar to which molasses has been readded.

Other Sources of Sweeteners

The making of sorghum "molasses" or corn syrup is similar to that of making sugar from cane except that the sugar is not crystallized from the syrupy juice. Sorghum (*Sorghum bicolor*) and corn (*Zea mays*) are both grasses like cane, but they are known primarily for their grains (Chapter 6). Compared to sugarcane, both of these grasses are relatively little used as producers of sweeteners.

In the Old World, primitive people of the Mediterranean region often collected the sap from palm trees (*Phoenix dactylifera*, Arecaceae) and condensed it by boiling off part of the water in order to produce a sweet syrup. Likewise, early American settlers used the sap of native plants for syrup production. Importation of sugar was costly, and honeybees were not present in North America until they were established by Europeans. Learning from the native Indians, the settlers began to slash the bark of various trees, particularly the sugar maple (*Acer saccharum*, Aceraceae) (Fig. 8-44) and to collect the sap in the spring as it exuded from the cut. The sap was then boiled down until it reached the desired consistency. The amount of sap of temperate trees needed,

FIGURE 8-44
In the spring, as the ground thaws, the sap rising in maple trees is tapped and evaporated down into maple syrup. (Photo courtesy of USDA.)

however, is enormous compared to the amount of juice from crushed cane because the sugar content of sap is very much lower (maple sap has 8 percent sugar) than that of cane juice (about 22 percent). It takes approximately 40 gallons of sugar maple sap to produce a gallon of syrup.

ADDITIONAL READING

Ayensu, E. S., and D. G. Coursey. 1972. Guinea yams. The botany, ethnobotany, use and possible future of yams in West Africa. *Economic Botany* 26:301–318.

Banga, O. 1976. Carrot, in N. W. Simmonds (ed.). *Evolution of Crop Plants*. New York, Longman, pp. 291–293.

Boswell, V. R. 1949. Our vegetable travelers. *National Geographic* 96(2):145–217. An interesting account written for the lay population of the spread of our common vegetables around the world. Some of the statements about specific crops are out of date, but the article is still worth reading.

Buck, P. A. 1956. Origin and taxonomy of broccoli. *Economic Botany* 10:250–253.

Cobley, L. S. 1976. *An Introduction to the Botany of Tropical Crops*, 2d ed. Revised by W. M. Steele. New York, Longman. An excellent and concise summary of the major crops of tropical regions.

Correll, D. S. 1962. The potato and its wild relatives. Section *Tuberarium* of the genus *Solanum*. Texas Research Foundation. Renner, Texas. An authoritative treatment of the group of solanums to which the potato belongs. The introduction provides interesting background.

Kahn, E. J., Jr. 1984. The staffs of life. 11. Man is what he eats. *New Yorker*. November 12 (1984):56–106.

Lancaster, P. A., J. S. Ingram, M. Y. Lim, and D. G. Coursey. 1982. Traditional cassava-based foods: Survey of processing techniques. *Economic Botany* 36:12–45.

Lathrap, D. W. 1973. The antiquity and importance of long-distance trade relationships in the moist tropics of pre-Columbian South America. *World Archaeology* 5:170–186.

Liebig, Baron J. von. 1863. *Natural Laws of Husbandry*. New York, Appleton.

Rogers, D. J. 1963. Studies of *Manihot esculenta* Crantz and related species. *Bulletin of the Torrey Botanical Club* 90:43–54.

Schery, R. W. 1947. Manioc—A tropical staff of life. *Economic Botany* 1:20–25.

Seigler, D. S., and J. F. Pereira. 1981. Modernized preparation of cassave in the Llanos Orientales of Venezuela. *Economic Botany* 35:356–362.

Simmonds, N. W. (ed.) 1976. *Evolution of Crop Plants*. New York, Longman.

Ugent, D. 1970. The potato. *Science* 170:1161–1166. The origins of the cultivated potato and the dangers involved in modern practices of cultivation are discussed.

Vaughan, J. G. 1977. A multidisciplinary study of the taxonomy and origin of *Brassica* crops. *BioScience* 27:35–40.

Wilson, C. M. 1968. *Roots: Miracles Below*. Garden City, New York, Doubleday.

Wilson, L. A. 1970. The process of tuberization in sweet potato [*Ipomoea batatas* (L.) Lam], in D. L. Plucknett (ed). *Proceedings of the Second International Symposium on Tropical Root and Tuber Crops*. Honolulu, Hawaii, University of Hawaii, pp 24–26.

Yamaguchi, M. 1983. *World Vegetables*. Westport, Connecticut, AVI Publishing Company. Although this text deals with many different kinds of vegetables, the treatments of starchy tubers and roots are particularly good. Cross sections show the different tissues that are starch-filled.

Yen, D. E. 1974. The sweet potato and Oceania. *Bulletin of the Bernice P. Bishop Museum* 236:1–389.

Chapter 9

Spices, Herbs, and Perfumes

*S*pices, herbs, and perfumes are inextricably linked to one another historically, chemically, and in terms of their physiological effects on humans. Initially, there was little distinction between them. Aromatic woods and resins were burned to produce scented smoke that would carry offerings or entreaties to the gods (Fig. 9-1). When mummified, bodies were scented with spices and herbs, and ancient perfumes often consisted of fragrant herbs, crushed into oil or wine. The same kinds of compounds are responsible for the distinctive qualities that have placed spices, herbs, and perfumes among our most coveted botanical products. Although they essentially have no nutritive value, herbs, spices, and perfumes add variety and interest to life and attest to the strength of our acquired food habits and social mores.

In this chapter we first consider the chemistry and ecological properties of the compounds for which we use plant-derived scents. We then discuss the history of spices and herbs and treat those of importance today. We last deal with perfumery from ancient times to the present. Table 9-1 lists the plants discussed in detail in this chapter.

CHEMISTRY AND ECOLOGY

Plants used as herbs or spices or in perfumery are chosen because they produce, in small quantities, characteristic flavors and/or odors when added to food or other objects. Most of the chemicals responsible for these distinctive tastes and smells are compounds known as *essential oils*. These chemicals, also called *volatile oils*, are usually terpenes, but they can also be aldehydes, ketones, alcohols, or

FIGURE 9-1
The king offers the falcon-headed god Selchmet burning incense and a libation that he pours into a bowl of lotus flowers. In return for these offerings, he hopes to receive bravery and strength. (Redrawn from an Egyptian wall sculpture at Abydos, ca. 1300 B.C.)

TABLE 9-1

Plants Discussed in Chapter 9

SPICE OR HERB	SCIENTIFIC NAME	FAMILY	CHROMOSOME NUMBER
Allspice	*Pimenta dioica*	Myrtaceae	Unreported
Anise	*Pimpinella anisum*	Apiaceae	$2n = 18, 20$ diploid
Basil	*Ocimum basilicum*	Lamiaceae	$2n = 48$
Bay	*Laurus nobilis*	Lauraceae	$2n = 42, 48$
Caper	*Capparis spinosa*	Capparaceae	$2n = xx$
Caraway	*Carum carvi*	Apiaceae	$2n = 20$ diploid
Cardamom	*Elettaria cardamomum*	Zingiberaceae	$2n = 48, 53$
Cassia	*Cinnamomum cassia*	Lauraceae	Unreported
Cayenne	*Capsicum annuum*	Solanaceae	$2n = 12, 24$ diploid, tetraploid
	C. fructescens		$2n = 24$ tetraploid
Celery seed	*Apium graveolens*	Apiaceae	$2n = 22$ diploid
Chervil	*Anthriscus cerefolium*	Lamiaceae	$2n = 28$ diploid
Cinnamon	*Cinnamomum zeylandica*	Lauraceae	$2n = 24$ diploid
Cloves	*Syzygium aromaticum*	Myrtaceae	Unreported
Coriander (cilantro)	*Coriandrum sativum*	Apiaceae	$2n = 22$ diploid
Cumin	*Cuminum cyminum*	Apiaceae	$2n = 14$ diploid
Dill	*Anethum graveolens*	Apiaceae	$2n = 22$ diploid
Fennel	*Foeniculum vulgare*	Apiaceae	$2n = 22$ diploid
Ginger	*Zingiber officinale*	Zingiberaceae	$2n = 22$ diploid
Horseradish	*Armoracia rusticana*	Brassicaceae	$2n = 32$ tetraploid
Lemon thyme	*Thymus citriodorus*	Lamiaceae	$2n = 42$
Mace	*Myristica fragrans*	Myristicaceae	$2n = 42$
Marjoram	*Origanum majorana*	Lamiaceae	$2n = 30$
Mustard	*Brassica nigra*	Brassicaceae	$2n = 16$ diploid
	B. alba (hirta)		$2n = 24$ tetraploid
Nutmeg	*Myristica fragrans*	Myristicaceae	$2n = 42$
Oregano	*Origanum vulgare*	Lamiaceae	$2n = 30, 32$
Parsley	*Petroselinum crispum*	Apiaceae	$2n = 22$ diploid
Pepper:			
Black	*Piper nigrum*	Piperaceae	$2n = 48, 52, 104, 128$
Red	*Capsicum annuum*	Solanaceae	$2n = 12, 24$
	C. fructescens		$2n = 24$
Peppermint	*Mentha piperita*	Lamiaceae	$2n = 36, 48, 64, 65, 72, 84, 122, 144$
Rosemary	*Rosmarinus officinalis*	Lamiaceae	$2n = 24$
Saffron	*Crocus sativus*	Iridaceae	$2n = 24$ triploid
Sage	*Salvia officinalis*	Lamiaceae	$2n = 14, 16$
Spearmint	*Mentha spicata*	Lamiaceae	$2n = 36, 48, 64$
Tarragon	*Artemisia dracunculus*	Asteraceae	$2n = 18, 36, 54, 72$
Thyme	*Thymus vulgaris*	Lamiaceae	$2n = 30$
Turmeric	*Curcuma domestica*	Zingiberaceae	Unreported
Vanilla	*Vanilla planifolia*	Orchidaceae	$2n = 28\text{–}32$

Note: Plants included in Table 9-3 are not included. For several species, the base chromosome number has not been determined so the level of ploidy is difficult to ascertain.

various phenol-based compounds (Fig. 9-2). The adjective "volatile" refers to the property of diffusing readily into the air where the oils may be detected by our scent and taste receptors.

Volatile oils are found in specialized plant cells, glands, or vessels that can occur in any or all parts of a plant (see Fig. 9-10). Many of the essential oils

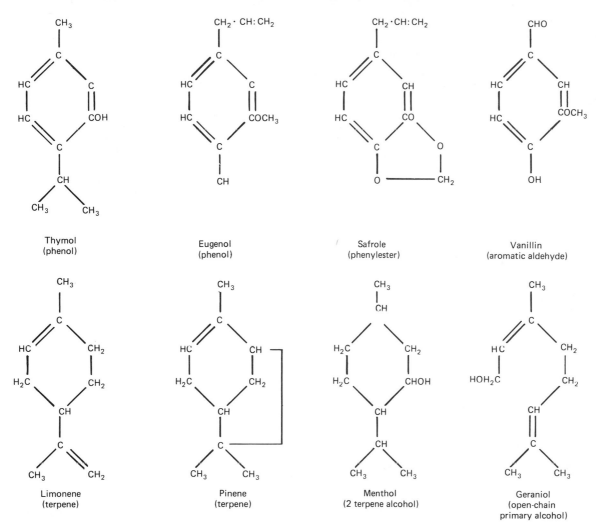

FIGURE 9-2
Structures of some common volatile oils responsible for the flavors and aromas of herbs, spices, and perfumes. Note that several classes of compounds, aldehydes, alcohols, phenols, and terpenes are used.

used in perfumes are produced by plants to attract animals that will serve as pollinators or fruit dispersers (see Chapters 1 and 4). Other perfume oils, such as thymol, come from glands on plant herbage. The role of oils found in leaves, seeds, and even roots is less obvious than the function of such compounds in flowers and fruits and has therefore been the subject of some discussion.

One possible explanation for the presence of volatile oils in stems and leaves and roots is that they are waste products which are sequestered to prevent their interfering with normal plant processes. However, while some volatile compounds are produced as parts of metabolic processes, others are not. Instead, they seem to be synthesized via specialized pathways. This suggests that there has been selection for their production, perhaps as deterrents to the growth of competing plants. This effect, called *allelopathy*, assumes that essential oils in the leaves are washed onto the soil around the plant. The chemicals in the soil might subsequently prevent germination of seeds, or cause the death of encroaching plants, thereby reducing competition for water, nutrients, and sunlight. Recent studies do not, however, support the idea that allelopathy is an ecological factor in most plant populations. Instead, there is growing evidence that the chemicals responsible for the pungent flavors and odors of spice and herb plants serve to deter predation by insects or infestation by pathogens. Fortunately, there is no evidence that spices, in the quantities in which we consume them, have any adverse effect on humans. In small amounts, some spices or herbs may even be beneficial because they can aid digestion. In large doses, a few spices (e.g., mace and nutmeg) can have harmful effects.

HERBS AND SPICES

To the epicure, spices and herbs bring to mind delightful smells and evoke thoughts of delicious foods. Spices are associated with feistiness, piquancy, and pungency, while herbs somehow make one think of pastoral scenes, old-fashioned settings, home remedies, and subtle flavors. Yet, it is difficult to draw any real distinction between spices and herbs. As we found with fruits and vegetables, there is a discrepancy between the botanical and common definitions of the terms. To a botanist, the word *herb* refers to a nonwoody plant. It can be an annual or a perennial, as long as the stems do not form any appreciable wood. The word *spice*, in contrast, has no botanical definition. Instead, it is rather loosely applied to an assortment of dried barks, roots, seeds, fruits, and flower parts, used for their scents and flavors. Some people define herbs and spices in terms of their places of origin, herbs from temperate regions and spices from tropical areas. In practice, this way of grouping plants used for flavorings and condiments matches well with a grouping based on the parts of the plants used. Herbs do tend to come from temperate regions while most spices come from tropical regions. The main exceptions to the notion that herbs come from leaves are found in an important group of temperate herbs including dill, coriander, caraway, and cumin, which yield aromatic fruits used to flavor foods. Similarly, saffron, usually called a spice, comes from flower parts and is native to the temperate areas of Eurasia. Spices come from a variety of plant parts, but comparatively rarely from leaves.

Spices and Herbs in History

We take for granted the large repertoire of spices and herbs commonly found in our supermarkets and used in most American kitchens. Today, the average American consumes almost 1 kg (2 lb) of herbs and spices annually. Yet, only a few hundred years ago, many of our common spices would have been rare or exceedingly expensive. Several of them would have even been unknown in Europe. While they have lost their former importance as items of commerce, the value placed upon spices and herbs was once enough to motivate exploration and to set into motion a struggle for power between European nations.

We do not know when humans first began to use spices and herbs as flavoring agents, but there are records of the use of garlic and onions as far back as 4,500 years ago. Before the advent of refrigeration, humans used spices

FIGURE 9-3
Anubis, a jackal-headed god who brought the dead to judgment, preparing a mummy. (Redrawn from an Egyptian wall fresco.)

and herbs to help in the preservation of food, to cover up the flavors of spoiling food, and to make otherwise monotonous meals more interesting. Spices were also used at a very early time for religious ceremonies, for embalming, and to produce fragrant smoke during ritualized cremation of the dead. The practice of embalming goes back 5,200 years in Egypt (Fig. 9-3). Herodotus provided a description of the mummification process and the use of myrrh, cassia, cinnamon, cumin, anise, and marjoram (among other plants) as materials for stuffing body cavities. Many of the spices and herbs used in ancient embalming are not native to Egypt or even the Mediterranean region. Demand for them in Egypt eventually led to an important set of trade routes to southeast Asia and China that crisscrossed the Mid East, Arabia, and India by 1400 B.C. (Fig. 9-4).

The Greeks expanded the spice trade routes, establishing Alexandria (Cairo) as the trade center for the Mediterranean. By the times of Hippocrates (ca. 400 B.C.) and Theophrastus (ca. 300 B.C.), herbs and spices were commonly used in medicine as well as for other purposes, and many were accurately described botanically (Fig. 9-5). Theophrastes even noted a correlation between semiarid and arid areas and the predominance of aromatic herbs.

During the Roman Empire, spice trading reached a peak because of the widespread use of herbs and spices in everything from wine, lamp oils and perfumes, to incense and food. The fall of the Empire carried down with it the network of the spice trade routes. Nevertheless, the Arabs who assumed control of Alexandria were themselves users of spices. Mohammed, the spiritual leader of the Arabs, had even been a spice trader as a young man.

Europe was, however, cut off from a supply of spices during the Dark Ages (641 to 1096 A.D.). The few herbs available were generally cultivated in monasteries. Only with the advent of the Crusades (after 1096 A.D.) did many spices reappear in Europe. The crusaders fighting in Palestine and Syria were exposed to new and exotic flavors and acquired a taste for them. Upon their return to Europe, they helped regenerate a demand for imported spices. Venice and Genoa gradually became centers of trade linking Europe to the Near and the Far East. The wealth brought to these Italian cities from their trading activities helped foster the cultural rebirth we know as the Renaissance. By the end of the Crusades (ca. 1300), Europe was a changed continent. Spice

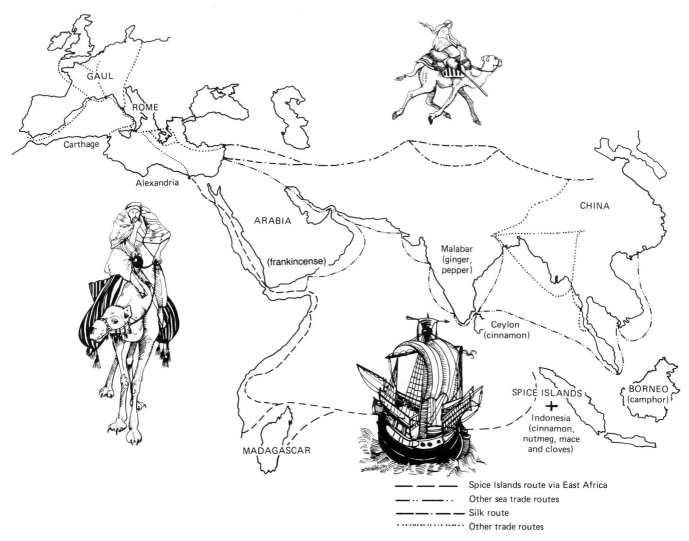

FIGURE 9-4
A map of the ancient trade routes which provided the Mediterranean region with silk and spices from about 4000 B.C. until after 1498 when Vasco da Gama reached India by sailing around the African Cape of Good Hope.

merchantry had become a legitimate profession, and spicers were given a charter by King Henry in 1429 to sell spices *en gros* (from which our word grocery is derived), or wholesale.

After the Crusades, the search for new trade routes and sources of spices escalated. In 1453, Constantinople (Istanbul) was captured by the Turks, and, once again, Europe was severed from Eastern spice sources. A new way to reach India and the Orient was needed. This time, Europe was technologically ready to search for it. The Portuguese were among the first explorers. In 1498, one of their ships, captained by Vasco da Gama, managed to reach India by sailing around the southern tip of Africa. Columbus, of course, had set out earlier on his Spanish-sponsored voyage to find a route to the East Indies. Instead, in

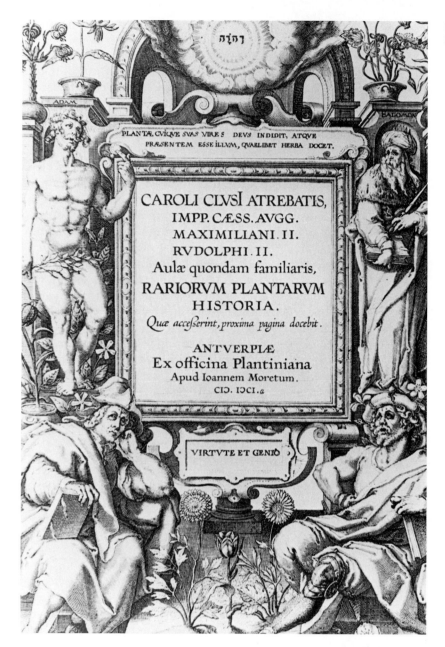

FIGURE 9-5
Frontispiece from Clusius' herbal.

1492, he inadvertently set the stage for the influx of a hitherto unknown set of spices from the New World (Fig. 9-5). Both the navigation of a direct sea route to India and the eventual reopening of the overland routes to the Orient reestablished the flow of Asian spices into Europe by 1560.

Because of their early explorations, the Portuguese dominated the spice trade for some time, but the Dutch eventually managed to take over the Spice Islands (islands in the Malay Archipelago now known as the Moluccas) between 1605 and 1621. Holland then virtually controlled the spice trade for over 200 years until overproduction of some spices and the development of the British Navy brought an end to the Dutch monopoly in about 1796.

The New World introduced few spices into commerce, but even so, the United States had a small place in the history of spice trading. Salem, Massachusetts (usually associated with whaling), served for many years as a center for black pepper trading between Sumatra and England. Still, spices never played as important a role in American history as they had in that of Europe, in part because by the time America was settled by Europeans, travel was much improved from that of the Middle Ages. In addition, plantations of Old World spices were eventually established in the West Indies, reducing some of the necessity for long, expensive trips to procure them. Increased availability served to diminish the value of most spices and the importance of spice trading gradually faded. Since 1900, there has been an increase in the use of artificial flavorings, which has caused a further reduction in the trading of spices. Nevertheless, spices and herbs give the foods of various countries their distinctive characteristics, and the use of both seems to be increasing in the United States.

It is interesting to reflect upon the flavors we now associate with different cuisines, and the native homes of the spices and herbs which produce those flavors. For example, we associate the hotness of red peppers with both Indian and Szechwan Chinese cooking, but red peppers could not have been used in Asia before 1492. Likewise, cumin and coriander are integral parts of Mexican cooking, but both were brought to the New World by Europeans. Table 9-2 gives the areas of the world in which the spices and herbs we discuss in this chapter are produced today. In our discussion of these spices and herbs, we treat first the classical herbs of Europe, then the spices of the Old World, and, lastly, those spices and herbs native to the New World.

Herbs and Spices of the Mediterranean Region

In terms of the definition of herbs as flavorful plant leaves, the mint family (Lamiaceae or Labiatae, Fig. 9-6) stands out. Almost all of the members of this family have fragrant herbage, and many are dominant species of the type of scrub vegetation that occurs around the Mediterranean Sea and the coast of California. Almost all of the green herbs that readily come to mind—rosemary, thyme, marjoram, oregano, basil, sage, and mint—belong to this family and are native to the northern and eastern edges of the Mediterranean Sea. These herbs have been used for millennia, and during the course of history many have acquired symbolic meaning (see Table 18-5). In cooking, the leaves of any of these can be used fresh, but most commonly they are dried and sold in whole or powdered form.

Rosemary (*Rosmarinus officinalis*, Lamiaceae) comes from a shrubby plant with needlelike leaves (Figs. 9-6, 9-7). Like many herbs in this family, it has long been used to brew a tea thought (without substantiation) to have calming properties. The oil responsible for its flavor in cooking can also be extracted and is sometimes used as a constituent of perfume or hair conditioners. When plants of rosemary are grown for commercial leaf production, the plants are continually divided and pruned to maintain a low, dense mat that frequently sends up new, harvestable shoots.

TABLE 9-2

Modern Areas of Production of Herbs and Spices

HERB OR SPICE	WHERE GROWN
Allspice	Jamaica, Guatemala, Honduras, Mexico
Anise seed	Turkey, Spain, Syria
Basil	United States, France, Hungary, Belgium
Bay leaves	Turkey, Greece, Portugal, Yugoslavia
Capers	Mediterranean
Capsicum peppers	Japan, Turkey, Africa, Mexico, United States
Caraway seed	Netherlands, Denmark, Poland, Russia, Syria
Cardamom	Guatemala, India, Ceylon
Celery seed	India, France
Chervil	France, United States
Cinnamon	Indonesia, South Vietnam, China, Ceylon
Cloves	Madagascar, Tanzania
Coriander	Morocco, Rumania, Argentina, France
Cumin	Iran, Morocco, Lebanon, Syria
Dill	India
Fennel	India, Argentina
Ginger	Nigeria, Sierra Leone, Jamaica, India
Horseradish	Widely cultivated in temperate regions, in United States primarily in Missouri and Illinois
Nutmeg and mace	Indonesia, Grenada
Marjoram	France, Portugal, Greece, Rumania, United States
Mint	Belgium, France, Germany
Mustard	Canada, Denmark, United Kingdom
Oregano	Greece, Mexico, Dominican Republic, Turkey, Italy, Sicily
Paprika	United States, Spain, Hungary, Yugoslavia, Morocco, Bulgaria
Parsley	United States
Pepper	Indonesia, India, Brazil, Ceylon, Malaysia
Poppy seeds	Netherlands, Poland, Denmark, Sweden, Turkey, Balkan countries, Argentina
Rosemary	France, Spain, Portugal, Yugoslavia, United States
Saffron	Spain, Portugal
Sage	Yugoslavia
Tarragon	United States, France, Yugoslavia
Thyme	Spain, France, Portugal
Turmeric	India, Haiti, Jamaica, Peru

Plants of thyme are generally smaller than those of rosemary. Although there are several hundred species of the genus *Thymus* (Lamiaceae), *T. vulgaris* (Fig. 9-6) is the one most commonly grown as an herb. The major part of the world's production is still centered around the Mediterranean, but both the common thyme and lemon thyme, *T. citriodorus,* are also grown in California. Thymol, a crystallized phenol constituent of thyme oil has some antiseptic and fungicidal properties, and is used in mouth washes and cough drops.

Although they are distinctly different in flavor, oregano and marjoram are closely related (Fig. 9-6). Both are perennials belonging to the genus *Origanum* (Lamiaceae). Plants of oregano tend to be robust and about 1 m tall. Marjoram (*Origanum majorana*) plants are much more delicate in appearance and about 30 cm tall. The modern fondness for oregano (*Origanum vulgare*) in the United

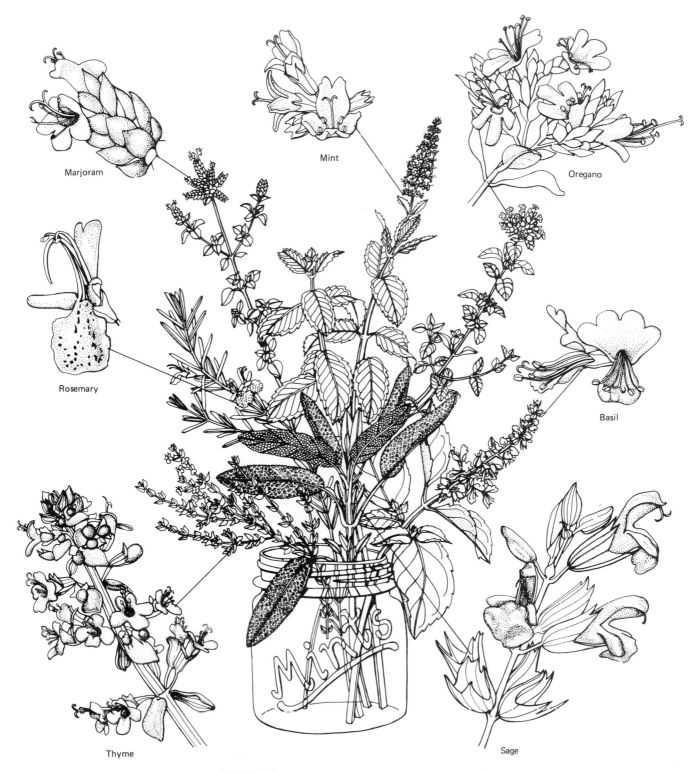

FIGURE 9-6
Herbs of the Lamiceae (mint family) are characterized by flowers with bilabiate (two-lipped) corollas, square stems, and opposite leaves.

FIGURE 9-7
A rosemary sprig. A poem by Sir
Thomas Moore (1779–1852) shows
how herbs were used to express
sentiments in nineteenth-century
England and Ireland:

> As for rosemary, I lette it runne
> all over my garden walls,
> not onlie because my bees
> love it,
> but because it is the herb
> sacred to remembrance
> and to friendship, whence a
> sprig
> of it hath dumb language.

States can be attributed to the rise in popularity of Italian food—particularly pizza—after World War II. Soldiers fighting in Italy learned to like the flavor of oregano, and, as a result it is now one of the most commonplace herbs in the United States. Marjoram has a milder flavor than oregano and has had a more consistent history of use at a relatively low level.

Another herb often associated with Italian cooking is basil, *Ocimum basilicum* (Lamiaceae, Figs. 9-6, 9-8). Part of its use in Italian food reflects its particular affinity for tomatoes, a component of many pasta sauces. Before tomatoes were introduced to the Old World, the herb was used in a variety of dishes and beverages. It is an important ingredient of Chartreuse, a green or yellow sweet cordial manufactured by Carthusian monks in France (see Fig. 15-29).

Of all of the members of the Lamiaceae used as herbs, sage (*Salvia officinalis,* Fig. 9-6) was the one most commonly used in medicine from classical Greek times through the Middle Ages. Nevertheless, the results of modern testing have not found it to have medicinal value. The leaves were the most popular dried

FIGURE 9-8
Basil has played various roles in different cultures. In India, the herb is considered
sacred, and in time, it earned a reputation as a mood elevator. The ancient Greeks
thought the plant grew best if it was addressed in insulting tones while being tended.
Others associate the plant with female purity because the leaves were said to wilt if
handled by an unchaste woman. Perhaps the most interesting use of basil was
developed by the wives of Genoese sailors who ground the leaves into olive oil with
garlic and cloves, mixed in parmesan cheese and pine nuts, producing the classic pesto
sauce.

cooking herb in the United States until the spectacular rise of oregano after 1945. Sage is considered an indispensable part of "stuffing," which usually accompanies a Thanksgiving turkey or other poultry dishes. Since sage oils mix particularly well with fats, sage is commonly used in processed meats and sausages.

Mint is a name that can be applied to a number of species of the genus *Mentha* (Lamiaceae). In ancient times, spearmint, *M. spicata,* was probably most commonly used, but *M. peperita,* peppermint (Fig. 9-6), is the herb now most frequently cultivated. The genus is named for Menthes, a young woman of classical mythology who apparently provoked Persephone's jealousy and was consequently turned into a mint plant. Today, mint is extremely important in home cooking and as a commercial flavoring for mint jelly, candies, medicines, toothpastes, and mouth washes. Mint is cultivated in the United States in California, Washington, Oregon, and Indiana. Like other herbs of the Lamiaceae, the mints grown in the United States are native to the Mediterranean, where they are also commercially produced. Although various mints are widely used today in herbal teas and to flavor medicines, there is no substantiation of the claims that they have antiseptic or antispasmodic properties.

Second in importance to the Lamiaceae, in terms of the number of herbs that it contains, is the Apiaceae, or carrot family. Members of this family are easily recognizable because of their flat-topped clusters of flowers (see Fig. 8-21, Fig. 9-9). The older family name, Umbelliferae, is derived from the word "umbel," the name of this type of inflorescence. Some members of the family are used for their leaves or roots (see Chapter 8), others for their fruits, and a few for both parts of the plant. Parsley, *Petroselinum crispum,* and chervil, *Anthriscus cerefolium,* are used exclusively for their herbage. Parsley, the often neglected garnish laid aside on our plate, is the most commonly used culinary herb in the United States today. Unfortunately, it is more often used as a decoration than as a flavoring agent. Rather than ignoring it, we would do well to eat it because it is rich in vitamins A and D and supposedly dispels odors of garlic and onions if chewed after a meal containing them. Enormous quantities of dehydrated parsley leaves (estimated to be over 0.5 million pounds per year, which would require more than 18 million pounds of fresh parsley) are produced annually in the United States. Chervil, a much less common herb, is one of the few members of this group of herbs that is not native to the Mediterranean. The species occurs naturally in Russia and western Asia.

The other two members of the Apiaceae used for their leaves are dill, *Anethum graveolens,* and cilantro, *Coriandrum sativum* (Fig. 9-9). Dill "weed" is indispensable as a flavoring in pickled items. Cilantro leaves, which come from the same plant as coriander, have had a recent surge in popularity because they are a component of many Mexican dishes.

Features common to almost all of the herbs of the Apiaceae are a musky flavor and aroma. While rather hard to define, these features are characteristic of both the leafy herbs and those used in the form of dried fruits. Included in this last group are coriander, dill, fennel (*Foeniculum vulgare*), cumin (*Cuminum cyminum*), anise (*Pimpinella anisum*), celery seed (*Apium sativum*), and caraway seed (*Carum carvi*). Although the part of the plant used in each of these cases is generally called a seed, it is in fact half of a fruit (see Fig. 9-9) composed of both the pericarp and the seed. The oils which give the flavor to the fruits are

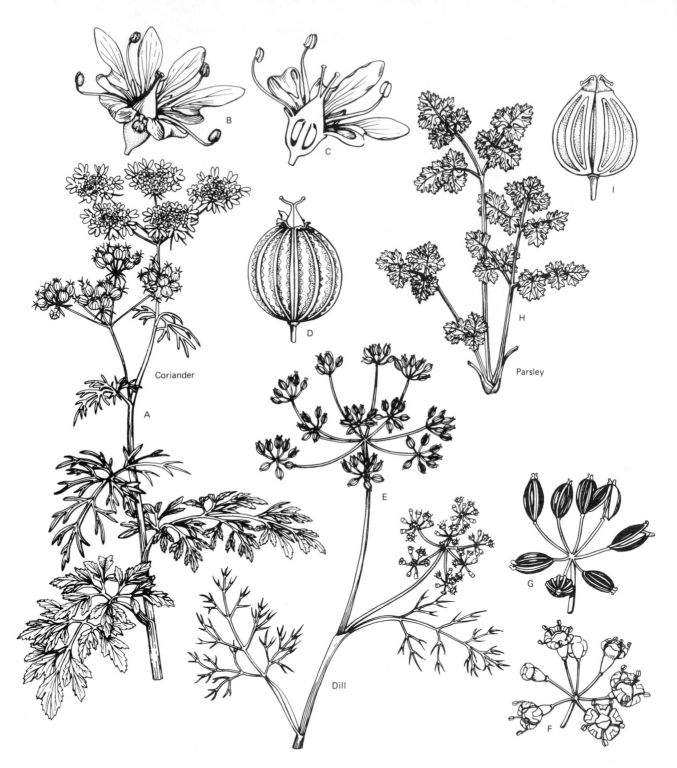

FIGURE 9-9

Coriander and cilantro come from the fruit and foliage of the same member of the Apiaceae. (A) Flower and flowering plant; (B) close-up of a flower; (C) cross section of a flower; (D) fruit. Dill has the same kind of two-parted fruit (usually called a seed in cookbooks) as other herbs in the family. (E) Dill flower and fruit; (F) inflorescence; (G) infructescence. We do not often see the fruits (I) of parsley (H) since it is grown primarily for its foliage.

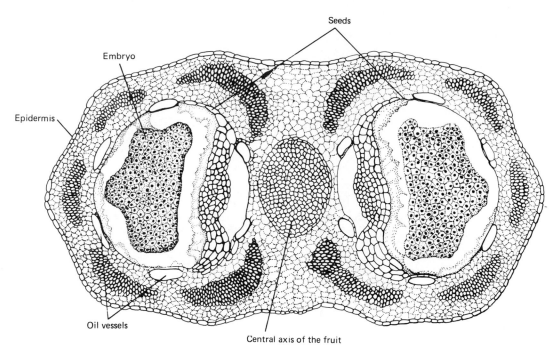

Seeds

Embryo

Epidermis

Oil vessels

Central axis of the fruit

FIGURE 9-10
Oil vessels which contain the volatile compounds found in many members of the
Apiaceae appear as a ring of empty spaces in this prepared fruit cross section.

usually located in vessels in the fruit wall (Fig. 9-10). These oils can be extracted
and used as flavorings for medicines and candies.

Because of two genera in the mustard family (Brassicaceae), Europeans
were not deprived of piquant flavors during the Middle Ages despite their lack
of access to exotic spices. One of these is mustard itself (*Brassica*) and the other
is horseradish. The traditional "hot dog" mustard we use as a condiment is a
mixture of the dried seeds of two *Brassica* species, *B. nigra* and *B. alba*
(Brassicaceae). The first, known as black mustard, is native to Europe and
Africa. Black mustard has been used since Greek and Roman times, and frequent
references to it are found in the Bible. The second species, white mustard, is a
European species that probably came into use at a later date. In Europe today,
usage of the two species differs in various countries. The French and the
Germans prefer mustard made from black mustard seeds, whereas the British
prefer the milder yellow or white species. The difference in the strength of the
flavors of these two species is due in part to the fact that each produces a
different oil when mixed with water.

Horseradish (*Armoracia rusticana*) is another European member of the
mustard family (Fig. 9-11) enjoyed by early Scandinavians and Germans as a
condiment for meats and fish, but it was not until the seventeenth century that
the rest of Europe began to appreciate the pungent flavor of the grated root of
this plant. Although the roots are yellowish-white, the English refer to the plant
as "red cole," perhaps alluding to the hot sensation one gets when ingesting
even small pieces. The root itself is only about 2.5 cm (1 in) in diameter but
may reach a length of 1 m (3 ft). The smell and flavor of horseradish come

FIGURE 9-11
Horseradish plant. (After Fuchs.)

from the glycoside sinigrin, which decomposes in the presence of water to form a mustard oil.

Other Herbs from Temperate Europe and Asia

Tarragon (*Artemisia dracunculus*, Asteraceae), a native of southern Russia and neighboring areas, is a newcomer as a flavoring agent compared to many of the herbs of the Lamiaceae and Apiaceae. It is mentioned in the literature of the Middle Ages but was apparently unknown to the Greeks and Romans. Now, however, it is one of the most popular herbs in Europe. In the United States it is perhaps best known as the flavoring in tarragon vinegar.

Bay leaves are often considered to be a spice rather than an herb, even though they are obviously leaves. One reason for the frequent inclusion of bay leaves with the spices may be that they come from a tree, *Laurus nobilis*, which belongs to the same family (Lauraceae) as cinnamon. The classical species of bay used since antiquity is native to the Mediterranean region. Other species which occur in Central America are sometimes substituted. Laurel (or bay)

FIGURE 9-12
A flowering branch of bay from
Clusius' herbal published during the
15th century.

sprigs were, of course, used in Rome to crown the winners of athletic events
(Fig. 9-12). Our word baccalaureate comes from the Latin word (*bacca*) for
berrylike and laurel. Today a baccalaureate signifies the completion of one's
studies, just as in Roman times being crowned with branches of laurel celebrated
the successful completion of an event.

A last noteworthy flavoring agent from the Mediterranean region is saffron
(Fig. 9-13), the most costly of all herbs and spices. The reason for this costliness
is that saffron comes from the stigmas of a crocus (*Crocus sativus,* Iridaceae). No
other part of the plant is used. An acre planted solely with saffron will yield

FIGURE 9-13
Because only the red stigmas of
Crocus sativus are used to produce
saffron, its cost greatly exceeds
that of any other spice.

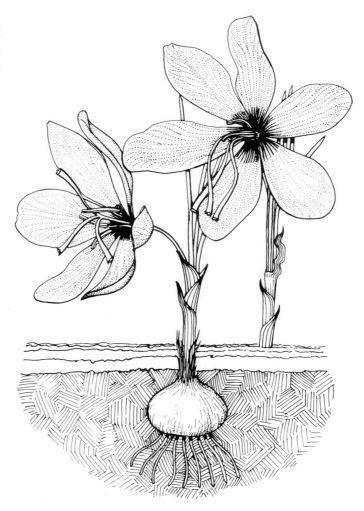

only 3.5 to 5.5 kg (8 to 12 lb) of dried spice per year: 462,000 stigma branches representing 154,000 flowers are needed for each kilogram (about 210,000 stigmas per pound) of dried spice. Nevertheless, there is always a demand for saffron because of its amazing taste and ability to produce an intense yellow color. The robes of ancient Irish kings were dyed with saffron, and classic Spanish and Indian dishes such as paella are flavored and colored with saffron. Often, a much less expensive spice, turmeric, is substituted for saffron, but with a corresponding loss in flavor.

Spices of the Old World Tropics

The exotic spices which fostered European explorations of the world came primarily from India and southeast Asia. Two of the most important of these spices from an historical point of view are cinnamon and cassia (Fig. 9-14). While the name cassia may seem unfamiliar to you, most of what you have eaten that was labeled cinnamon, was probably cassia. True cinnamon, *Cinnamomum zeylanicum* (Lauraceae, Fig. 9-15), is considered to have a more delicate, finer flavor than cassia, which comes from *C. cassia*. Under the guidelines

FIGURE 9-14
The cassia quills (left) are thicker than those of true cinnamon (right) because they are made from the entire bark. Cinnamon quills are made from only the inner layers of bark.

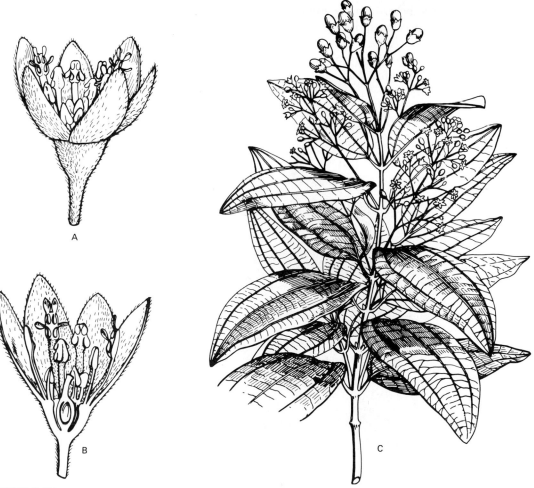

FIGURE 9-15
True cinnamon comes from the ground bark of *Cinnamomum zeylanicum*. Like other members of the laurel family, cinnamon has simple leaves, and its flowers have anthers with pores covered by flaps from which pollen is shed. (A) Flower; (B) flower cross section; (C) branch. (After Baillon.)

FIGURE 9-16
Workers stripping bark from a
cinnamon tree during harvest. Photo
courtesy American Spice Trade
Association.

of the U.S. Federal Trade Commission, either can be sold under the name of
cinnamon. Both are among the few spices of the world obtained from bark. To
produce the quills, or curls, of cinnamon that are sold whole or ground, the
bark is stripped from the shoots of trees that have been forced to root-sprout
(send up shoots from the roots around the trunk) (Fig. 9-16). The outer bark
is then rotted from the inner bark. The cleaned inner bark is dried, causing it

FIGURE 9-17
Cloves are the unopened flower
buds of *Syzygium aromaticum*.
(After Baillon)

FIGURE 9-18
Harvesting cloves in Zanzibar.
(Courtesy of the American Spice
Trade Association.)

to curl into tight, red-brown quills. Cassia is obtained in the same way, except that the outer bark is not stripped away. Cassia is therefore a much coarser product than true cinnamon (see Fig. 9-14). Ceylon and several countries of southeast Asia are the major suppliers of both cinnamon and cassia.

Other spices which come from tropical trees of the Old World are cloves, nutmeg, and mace. Cloves are rather unique in that they come from unopened flower buds (Fig. 9-17). The only other "spice" that comes from flower buds is capers (*Capparis spinosa*, Capparaceae), a relatively minor crop from the Mediterranean.

Cloves (*Syzygium aromaticum*. Myrtaceae) are probably native to the Spice Islands (Fig. 9-18) or to nearby areas in Indonesia. They were one of the major spices imported into Alexandria for distribution throughout the Greek Empire. The flower buds are harvested when they are just turning pink. If a flower has opened before the clove is picked, it produces an "empty" clove when dry. An empty clove, considered inferior in commercial trading, has lost the corolla, stamens, and style. Round, fat, unopened cloves have superior flavor and are particularly attractive compared with empty cloves. The collected buds are dried, cleaned, graded, and packed for export. About half of the world production of cloves is used in the manufacture of a particular brand of Indonesian cigar. Clove oil used to be commercially extracted from the buds, but it is now synthetically produced.

Nutmeg and mace are two differently flavored spices which come from separate parts of the fruit of *Myristica fragrans* (Myristicaceae, Fig. 9-19). The fruit can be considered a drupe which splits open at maturity, exposing the stony endocarp surrounded by a red, slightly fleshy network called an aril. Once peeled off and dried, the aril becomes mace (Fig. 9-20). The endocarp with the

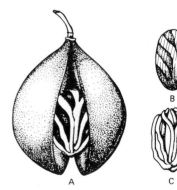

FIGURE 9-19
(A) Nutmeg and mace come from different parts of the same fruit. (B) Nutmeg is the dried endosperm of the seed and (C) Mace is produced from drying the netlike aril around the seed.

FIGURE 9-20
Women in Grenada separating mace from nutmegs. (Courtesy of the American Spice Trade Association.)

seed inside is also dried. Removal of the hard endocarp leaves the round nutmeg. Unlike true nuts, the nutmeg is composed primarily of a mass of endosperm, not cotyledonous tissue. Nutmeg trees are dioecious and produce flowers only after they are about 6 years old. Consequently, it is difficult when planting an orchard to know if the trees will be male or female. Once the sex of the trees has been determined, male trees are often decapitated and female scions grafted onto the rootstocks. A few male trees are retained in the population to ensure fertilization and fruit production. Nutmeg, toxic in large quantities, will produce hallucinogenic effects in intermediate doses (see Chapter 19). The mace used in modern riot control is a synthetic compound unrelated to the spice we get from *Myristica fragrans.*

Two other spices of great importance in ancient cultures were cardamom and ginger, both members of the ginger family, or Zingiberaceae. Cardamom (*Elettaria cardamomum*) comes from the dried seeds of an exotic-looking plant (Fig. 9-21). Oils extracted from cardamom seeds were formerly used for a

FIGURE 9-21
Harvesting cardamom in India. (United Nations photo by John Isaac.)

variety of purposes including medicines. Ginger is obtained from the fresh or dried rhizome (see Fig. 1-15D) of *Zingiber officinale*. Both cardamom and ginger are native to southern Asia but were introduced to Europe by Arabs several hundred years B.C.

Ginger is usually propagated vegetatively by planting pieces of the rhizomes because the plants rarely produce viable seeds (Fig. 9-22). Ginger yields an oleoresin that is used to flavor soft drinks and ginger beer. Turmeric (*Curcuma domestica*) is an Asian member of this family that produces rhizomes that are dried and ground and incorporated into curry powder, or used as a source of a yellow dye.

In terms of quantities traded, black pepper is the most important spice in the world today. Like cinnamon, nutmeg, cloves, cardamom, and ginger, pepper was an important item in early East-West trade. Black pepper comes from *Piper nigrum*, a viney member of the Piperaceae (Fig. 9-23) native to the jungles of southwestern Asia. Either white or black pepper can be made from the drupes. To produce black pepper, the drupes are stripped from the clusters green and allowed to ferment a few days before they are dried. The entire dried fruit is then ground, or left whole as pepper corns. White pepper is obtained from ripe drupes which have become somewhat fleshy. The fruits are soaked and lightly crushed to remove the outer flesh. The fruit with the white endocarp exposed is then dried. Most black pepper is still produced in southeastern Asia, primarily India and Indonesia, but Brazil is now also a major world producer. The United States is the world's largest consumer of black pepper.

Spices from the New World

The New World has contributed only three major spices to the repertoire of those commonly used, allspice, capsicum peppers (red peppers), and vanilla. The last two of these have assumed very prominent places in the world spice trade.

Allspice, *Pimenta dioica* (Fig. 9-24), comes from the same family (Myrtaceae) as cloves but is native to Central America and the West Indies. The name allspice refers to the flavor of the dried green berries, which is rather like a combination of cinnamon, cloves, and nutmeg. Allspice was among the few treasures that Columbus was able to present to the court of his sponsors (Fig. 9-25). Of all the American spices, it has the somewhat dubious distinction of being the only one which is still exclusively grown in the New World. Almost all of the modern production is in Jamaica.

In contrast to allspice, capsicum peppers, species of *Capsicum* (Solanaceae), are now grown in many areas of the world. One species, *Capsicum annuum* (Fig. 9-26), produces both sweet edible bell peppers and a variety of other peppers of variable shapes and degrees of hotness. Paprika, the flavor base of the paprikashes of the Balkan countries, is obtained from dried, powdered peppers of this species, as is the chiltecpin, a tiny, very hot pepper occasionally found in United States supermarkets. Other pungent kinds of peppers are often obtained from *C. fructescens* or, occasionally, from one of the other less common members of the genus. Tabasco sauce is made from processed fruits of *C. fructescens*.

There is still disagreement about exactly how many species are involved in the vast array of capsicum peppers. The situation is complicated further by the

FIGURE 9-22
A ginger plant showing the tangy rhizome used in many cuisines.

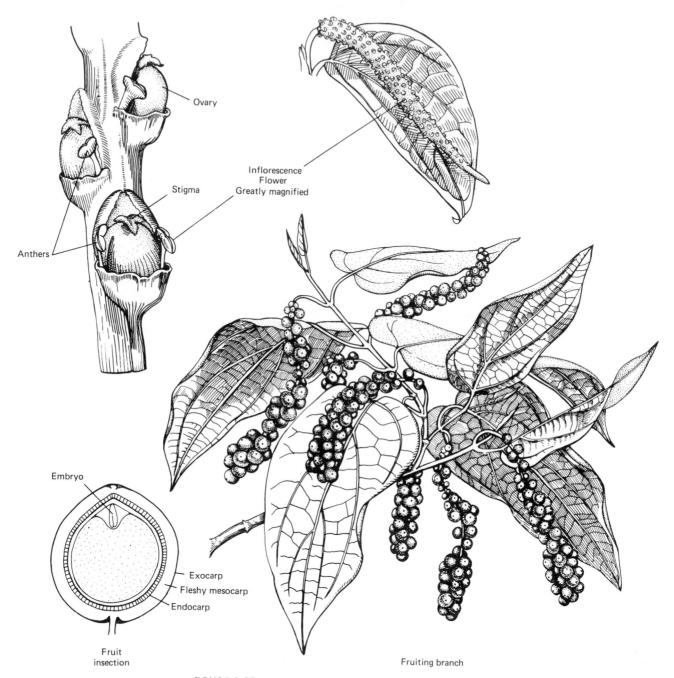

Ovary

Inflorescence
Flower
Greatly magnified

Stigma

Anthers

Embryo

Exocarp
Fleshy mesocarp
Endocarp

Fruit
insection

Fruiting branch

FIGURE 9-23
A fruiting pepper vine, inflorescence, flower, and peppercorn in cross section. (After Baillon, in part.)

fact that the common names applied to powdered, dried peppers or mixtures in which they occur are somewhat arbitrary. For example, the name "cayenne" refers simply to dried, pungent red peppers (regardless of the species involved). Red pepper (as a label on a can of spice) can be anything from cayenne to paprika in terms of hotness. The coarsely ground red pepper flakes seen in the

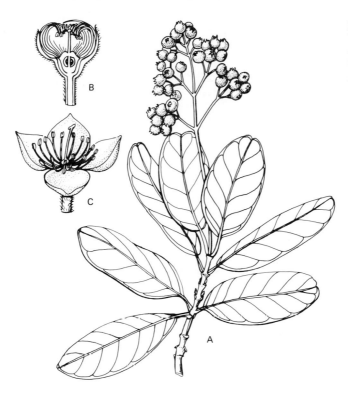

FIGURE 9-24
The clovelike flowers with a whorl of stamens show that allspice belongs with cloves in the Myristicaceae. The spice itself comes from the dried berries.

shakers of fast food restaurants are often a mixture of species of *Capsicum*. Finally, the chili powder that is sold in cans or jars is actually a mixture of oregano, cumin, garlic powder, and dried red peppers.

The compound in some red peppers which causes the burning sensation is capsaicin, a crystalline substance when pure. The compound is concentrated in the placenta of the fruit on which the seeds are borne. Removal of the seeds and the placenta can render an otherwise scorching red pepper relatively harmless. Since their introduction to Europe, Asia, and Africa, capsicum peppers have become integral parts of the cooking of Italy, Spain, Hungary, India, China, Indonesia, and many parts of Africa. Peppers thus rank with corn and tobacco as a crop of the Americas adopted around the world.

Vanilla must certainly be considered one of the strangest of spices. It comes from one of the largest families of flowering plants, the Orchidaeae (Fig. 9-27), but, except for horticultural flowers, vanilla is the family's only crop. Vanilla

FIGURE 9-25
Allspice was one of the New World treasures that Columbus brought back to his sponsor, Queen Isabella of Spain. [This photogravure is from *Great Men and Famous Women* by V. Brozik (1894) edited by C. Howe.]

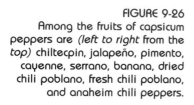

FIGURE 9-26
Among the fruits of capsicum
peppers are *(left to right* from the
top) chiltecpin, jalapeño, pimento,
cayenne, serrano, banana, dried
chili poblano, fresh chili poblano,
and anaheim chili peppers.

FIGURE 9-27
The flower of the orchid that will
produce a vanilla pod after
pollination.

plants are perennial vines native to Central and South America. The Aztecs used vanilla (*Vanilla planifolia*) as a flavoring for chocolate, and the Spanish carried it to Europe, where it was first used in the same way. The recognition of vanilla as a delicious flavoring in its own right came somewhat later. The part of the plant used to produce the flavoring is the fruit, usually called a vanilla bean. However, strictly speaking, the fruit is a false berry (the ovary is inferior), despite its similarity in appearance to a legume pod (Fig. 9-28). Inside the fruit are millions of small seeds embedded in a slightly pulpy mass. The tiny black specks in "real" vanilla ice cream are vanilla seeds. Their presence is supposed to indicate that the manufacturer used real vanilla fruits to flavor the ice cream.

Like many orchids, the flowers of vanilla are intricately shaped and adapted for pollination by specialized insects (Fig. 9-27). Consequently, for commercial production of vanilla, particularly when it is grown in areas of the world where it is not native, the flowers are hand-pollinated by workers who carefully transfer pollen to about 1,500 flowers per day. The fruit takes 9 months to mature. The mature pods are then picked and placed on tables to sweat (Fig. 9-29) and dry for several months. During this process, which is called fermentation, though it is basically a chemical and not a microbial alteration, the pods are exposed to the air each morning to warm up and then wrapped in cloth for the remainder of the day. This laborious process is carried out daily for up to 6 months. During this time, a flavor-producing crystalline substance, vanillin, is produced. Unfermented pods lack vanillin and do not have a vanilla flavor.

Faster methods of maturing the vanilla pods have been tried, but not widely adopted. It is little wonder that vanilla ranks second to saffron as the most expensive spice. Vanilla extract is made by chopping the fermented beans and continually percolating an ethanol-water mixture over them to dissolve out the

FIGURE 9-28
Women in Madagascar picking mature vanilla beans. (Photo courtesy of the Vanilla Information Bureau.)

FIGURE 9-29
Spreading vanilla beans out to dry, one of the steps in the lengthy curing process. (Photo courtesy of the Vanilla Growers Association.)

pure vanillin and other flavoring agents. Any product labeled pure vanilla extract must come from vanilla beans. Vanillin can be made synthetically from a number of products including wood pulp. Artificial vanilla is extensively used as a flavoring but for most people synthetic vanilla lacks the subtle qualities of real vanilla.

PERFUMES

Of all the products obtained from plants, perfumes have played the most provocative role in human affairs. It is impossible to tell when people began to use perfumes. Prehistoric humans probably tried many ways to capture for themselves the pleasing scents of flowers and herbs before true perfumery techniques, as we know them today, were developed. Our word for perfume comes from the Latin "per," meaning through, and "fumus," air or smoke, referring to the early recognized tendency for fragrances to diffuse through the air. Over 5,000 years ago, the Egyptians had become skilled perfumers who taught the art of perfumery to the Hebrews. The frequent references to perfumes and incenses in the Torah reflect the importance that perfumes achieved in Judaic culture. In Eastern cultures as well, perfumes were used in temples and private homes. The Chinese used scents in ceremonies and even developed an incense clock that indicated the hour as it burned. The Japanese modified the Chinese clock by making sticks of different odors. As they burned, these "joss sticks" provided scents appropriate for the different parts of the day.

The wealthy Greeks employed perfumes in many aspects of their lives, but it was during the Roman Empire that perfumery reached its pinnacle. Incenses were commonly used in temples and homes. Perfumes were added to baths, (Fig. 9-30) oils for anointing the body, and even to wine. The value of scents at this time is reflected in the story of the birth of Christ. Two of the three gifts presented to Jesus were used as incense (see Chapter 11).

After the fall of Rome, perfumery remained important in the Near East and Asia, but Europe seemed to have forsaken the pleasures of sweet smells. As in the case of spices, it was the returning crusaders who reintroduced perfumes to Europeans. Shortly thereafter, the popularity of perfumes soared. The French led in the recognition of the importance of perfumery as a trade by granting a charter to perfumers as early as 1190. To this day, the French are still considered to be the leaders in the perfume industry, although they have recently been severely challenged by American firms.

How Perfumes Are Made

Originally, all perfumes were derived from natural sources, flowers, fruits, leaves, roots, wood, resins, and, occasionally, animal secretions. Unless a substance was to be used directly, as in the case of a resin, the trick of the perfumer was to extract fragrant substances from plants or animals and to present them in a usable form. The most successful extraction procedures are those which destroy or alter the natural fragrances as little as possible. Since the compounds responsible for fragrances are oils, they are not water-soluble. Consequently, extraction methods must use substances into which volatile oils

FIGURE 9-30
Lavender and the generic name
Lavandula come from the Latin
lavare meaning to wash, because
in Roman times the herb was used
to add fragrance to wash water.
Lavender is among the plants
cultivated commercially in the
perfume region in the south of
France. (Redrawn from Matthiolus,
1565.)

dissolve. In the manufacture of a perfume, substances known as *fixatives* are added that help retard rapid dissipation of the volatile compounds.

The basic ingredients that are used in the perfume trade are known as *odorants*. These fall into five groups: concretes, absolutes, distilled and fractionally distilled oils, expressed oils, and tinctures. These classes of odorants are based on the way in which the oils are extracted. A sixth group of odorants includes balsams and resins that are often used intact or to "fix" other oils. *Concretes* are considered to be the purest of the natural odorants. They are obtained by immersing fragrant plant products in a hydrocarbon solvent which penetrates the tissues and dissolves out the oils. The liquid is then evaporated from the oils under reduced pressure. *Absolutes* result when a concrete is extracted to a more concentrated state (because waxes and glycerides are left behind) with alcohol. The alcohol is then evaporated from the oil under conditions of low heat and reduced pressure. Most perfumes consist of mixtures of absolutes. The alcohol used in the extraction can be subsequently added to colognes and lotions.

There are two other, more traditional, ways of making concretes and absolutes. One is to macerate fragrant substances in hot fat or oil and subsequently extract the fat or oil with alcohol (Fig. 9-31). The second is a process called *enfleurage* (Fig. 9-32). In this procedure, flower petals are pressed onto a coating of pure lard (usually pig lard). Each day the petals are changed until the fat is saturated with absorbed oils. The essential oils are then extracted from the pomade with alcohol. Once extracted, the fat is practically odorless and is often later used for soap manufacture. Enfleurage is a time-consuming and expensive operation because it requires extensive hand labor. However, it produces extremely true fragrances and is able to capture very delicate scents that elude other techniques.

Distillation and *fractional distillation* are the most common methods employed today for the extractions of natural fragrances (Fig. 9-33). Both involve exposing plant parts to steam, often superheated steam. The volatile oils are carried off in the steam. Because volatile oils are insoluble in water, they rise to the surface when the steam is condensed into water and are easily removed by skimming. Fractional distillation utilizes the fact that different volatile oils vary in their solubilities in steam. The most soluble are carried away first and the less soluble, later. By condensing different fractions of the steam, the fragrant compounds can be separated from one another and individually collected. Distillation has an advantage over other extraction methods in that it is inexpensive and rapid. However, heat is detrimental to many fragrances.

Expressing oils directly from plant parts is a straightforward process. Although

FIGURE 9-31
One way to capture plant fragrances is to macerate scented parts in hot fat or oil. The mixture is usually allowed to stand in copper pots for a period of time before the oil or fat is extracted with alcohol.

FIGURE 9-32
Each day this woman removes
faded jasmine flowers from a layer
of fat and replaces them with fresh
flowers. This process, called
enfleurage, is used to capture
particularly delicate scents.

it is an easy technique, its use is limited, because most plants do not have localized areas of high oil concentration. Expression of oils is practiced primarily with citrus peels, which contain appreciable quantities of oil in pockets dotting their skins. The oils are generally obtained by scoring the rinds and absorbing the exuded oils, but they can also be simply pressed from the peels. Since other substances are forced out with the oils when extensive pressure is applied, oils obtained in this way are less pure than those recovered after scoring the peels.

Tinctures have been used since ancient times and are still used to extract medicinal compounds as well as fragrant oils. They are produced by macerating (chopping) a substance in 95 percent ethanol (the ancient Egyptians used wine). If the macerated plant parts are heated in the alcohol, an infusion results. While tinctures may seem very similar to absolutes, it should be noted that absolutes are obtained by the extraction of concretes, whereas tinctures are obtained by extracting plant or animal material directly. The important animal substances in perfumery, ambergris, civet, castoreum, and musk, are generally used in the form of tinctures.

While all of these techniques are used in modern operations, about 60 percent of the fragrant compounds involved in modern perfumery are synthetically produced today. Usually, synthetic fragrances are based on natural compounds. The chemical structure of a natural fragrance is determined and then procedures are devised for its synthesis. It is often cheaper to synthesize a compound than to extract it from a natural source. In addition, synthetic fragrances are pure, whereas natural extracts are mixtures of volatile oils. If some of the compounds of an absolute are undesirable, the mixture must be fractionally distilled to separate them from the desirable components. In some cases, however, it is so inexpensive to use natural sources that there has been little attempt to produce a synthetic compound. This is the case with cedar (*Juniperus*) oil. In other cases, attempts by chemists to reproduce a natural fragrance have failed, with the result that natural sources must be used.

FIGURE 9-33
Distilling fragrances (Photo
courtesy the Parfumerie Fragonard,
Grasse, France.)

FIGURE 9-34
The master perfumer uses his nose
to guide him in combining scents
into harmonious mixtures. (Photo
courtesy of Parfumerie Fragonard,
Grasse, France.)

FIGURE 9-34
The master perfumer uses his nose to guide him in combining scents into harmonious mixtures. (Photo courtesy of Parfumerie Fragonard, Grasse, France.)

From Oils to Perfumes

Early in the profession of perfumery it was discovered that the blending of scents provided new, sometimes superior fragrances. The mixing of tinctures with fixatives such as resins, ambergris, musk, etc., also tended to slow degradation, retard diffusion into the air, and balance different scents. The creation of a perfume is thus an art that requires the careful blending of different fixatives and odorants. Today, perfume houses employ not only chemists, but also at least one perfume master whose job can be compared to that of a musical composer (Fig. 9-34). This analogy is carried through the fragrance lexicon, where different scents, called *notes*, are used to create masterpieces. As in music, the same notes are recognized all over the world. The various notes in the perfume industry and the series into which they are grouped are given in Table 9-3. Table 9-3 also gives examples of plants which provide the repertoire of notes. As can be seen, the same note can be obtained from quite unrelated sources.

Once a perfume has been produced by a skillful blending of diverse fragrances, it is classified by its dominant series, or note, if one is particularly evident. In choosing a perfume, it is often a good idea to consider the personality of the potential wearer in terms of the dominant note. Also, there is truth in the adage that a person should "try on" a perfume. The fragrances in a perfume are chemicals that blend and react differently with different body chemistries.

It takes about 2 to 3 million dollars to develop and launch a successful major perfume. A very small part of this cost is the odorants, however. The development and marketing of the perfume are the expensive parts of modern perfume making. Nevertheless, if a great perfume is developed and becomes popular, manufacturers can expect to gross over $50 million a year in sales.

TABLE 9-3

The Series and Notes of Perfumers and Their Sources

SERIES: NOTES	PLANT SOURCES
	FLORAL SERIES
Rose notes	Rose de Grasse; Bulgarian rose; oil of rose geranium (*Pelargonium graveolens*); rose leaf oil
Jasmine notes	Oil of jasmine (*Jasminum odoratissimum*); absolute of ilang-ilang (*Cananga odorata*)
Hyacinth notes	Absolute of oil of *Hyacinthus orientalis*
Lilac notes	Oil of lilac, *Syringa vulgaris*
Lily notes	Oil of lily of the valley (*Convallaria majalis*)
Orange blossom notes	Oil of orange flowers; oil of mock orange (*Philadelphus coronarius*)
Tuberose notes	Absolute of tuberose (*Polianthes tuberosa*); Oil of narcissus (*Narcissus poeticus*); oil of jonquil (*Narcissus jonquilla*); oil of champaca (*Michelia champaca*); oil of honeysuckle (*Lonicera caprifolium*); oil of lily (*Lilium candidum*)
Violet notes	Absolute oils of *Viola* flowers; oil of boronia (*Boronia megastigma*); oil of cassie (*Acacia farnesiana*); oil of mimosa (acacia and mimosa flowers); oil of iris or orris (*Iris pallida*)
Mignonette notes	Oil of mignonette (*Reseda odorata*)
	WOODY SERIES
Sandal notes	Solvent extracts of sandalwood (*Santalum album*)
Peppery notes	Oil of thyme (*Thymus vlugaris*)
Carnation notes	Absolute oil of carnation (flowers of *Dianthus caryophyllus*); oil of cloves (buds of *Syzygium aromaticum*); oil of tobacco flowers (*Nicotiana* spp.)
	RURAL SERIES
Herbaceous notes	Absolute oil of flouve, a grass; oil from flowers of *Lavandula angustifolia*; oil of lavandin, *Lavandula* hybrids; tea leaf absolute (*Camellia sinensis*)
Green notes	Oak moss, various species of lichens; oil of fern, usually male fern; oil of ivy, leaves of *Hedera helix*
	BALSAMIC SERIES
Resinous notes	Absolute oil of cypress, needles and twigs of *Cupressus sempervirens*; oil of fir (*Abies alba*); everlasting absolute (*Helichrysum angustifolium*)
Vanilla notes	Absolute oil of vanilla, pods of *Vanilla planifolia*; oil of heliotrope (*Heliotropium arborescens*)
Fruity series	Absolute oil of fig leaves, *Ficus carica*
	ANIMAL SERIES
Amber notes	Ambrette seed absolute (*Abelmoschus moschatus*); angelica absolute (*Angelica archangelica*); cumin absolute (*Cuminum cyminum*); labdanum absolute (*Cistus* spp.)
Maritime notes	Absolutes of several seaweeds
Musk notes	Costus absolute, rhizomes of *Saussurea alpina*
	EMPYREUMATIC SERIES
Tobacco notes	Maté absolute, leaves of *Ilex paraguariensis*; melilot (*Melilotus* spp.); tobacco leaf absolute, leaves of *Nicotiana* spp.; tonka bean absolute, seeds of *Dipteryx odorata*

Source: Adapted from F. V. Wells and M. Billot. 1975. *Perfume Technology. Art, Science, Industry.* 2nd ed. New York, Wiley. In their discussion of series and notes, these authors characterize the fragrances listed above and comment on their importance.

ADDITIONAL READING

American Spice Trade Association. 1966. *A Glossary of Spices*. New York.

American Spice Trade Association. 1966. *A History of Spices*. New York.

Andrews, J. 1984. Peppers. The Domesticated Capsicums. Austin, University of Texas Press. A beautifully illustrated and scholarly work on America's most famous spice.

Grieve, M. 1931. *A Modern Herbal*. Originally published by Harcourt, Brace and Co. 1971. Dover Edition in two volumes.

Morris, E. T. 1984. Fragrance. The Story of Perfume from Cleopatra to Chanel. New York, Scribner's. A very readable account of the use of perfumes in Eastern and Western cultures from 2800 B.C. to the present.

Rosengarten, F., Jr. 1969. *The Book of Spices*. Philadelphia, Livingston. An extensively illustrated book that discusses the botany, chemistry, history, and modern production of almost all herbs and spices.

Stuart, M. (ed.). 1979. *The Encyclopedia of Herbs and Herbalism*. New York, Grosset and Dunlop.

Chapter 10

Vegetable Oils and Waxes

*T*he expression "100 percent vegetable oil" has recently been popularized in the United States through television commercials. The advertisements seem to imply that pure vegetable oil is a recent innovation which is now being produced because it is healthier than animal-derived oils and fats. Contrary to the implication of these ads, the use of vegetable oils dates from several thousand years before Christ. The importance of such oils in ancient cultures is highlighted by the story of the naming of the capital of Greece. The legend says that the capital of the empire was to be named for the deity who gave the Greeks the most valuable tribute. As her offering, Athena threw her spear into southern Greece. An olive tree sprang up at the place where it touched the ground. This tree with its oil-containing fruits was declared the most valuable of any of the gods' gifts, and the capital was named Athens. The Greeks held the olive in high esteem because it provided food, their most important cooking and lubricating oil, and a popular cleansing agent (Fig. 10-1).

The olive notwithstanding, vegetable oils come almost exclusively from seeds. Within the seeds, oils are stored in the endosperm (castor oil, coconut); the cotyledons of the embryo (peanut, soybean, cotton); or the scutellum (corn). In most cases, the stored oils are the primary, or sole, food material for the germinating embryo. Since oil contains a large number of calories relative to starch (9 vs. 4 cal/g, respectively), oils are excellent sources of energy for emerging seedlings. Usually, seeds with large amounts of oil do not contain large amounts of starch.

In some cases, such as olives and palm fruits, oils are pressed from the fruit pulp. It is not easy to explain why fruit pulps should be rich in oil. The pulp is not used by the embryo for nutrition, but by animals that disperse the fruits. While these animals benefit from the rich food sources, the plant must expend considerable energy in their production. Nevertheless, humans have capitalized on several plants that do produce oil-rich fruit pulps, and we have incorporated them into the repertoire of edible plants since early times.

While correlations have been made between the consumption of saturated fats and oils, and an increased incidence of heart disease, the prevalent belief that all vegetable oils are unsaturated and therefore healthier than animal fats, is incorrect. Vegetable oils such as palm and coconut oil are more saturated than either butter or lard. Saturation is a chemical property of an oil that affects human nutrition, the ways in which the oil can be most profitably used, and, to some extent, how the oil is extracted, refined, and processed. Before we explore the ancient and modern uses of oils (Fig. 10-2) and the ways they are obtained and processed, we briefly describe the chemistry of vegetable oils. Table 10-1 lists the plants discussed in this chapter.

THE COMPOSITION OF SEED OILS

While the word "oil" can refer to any of a number of substances that are insoluble in water, only one kind, glycerides or acylglycerides, is expressed in large quantities from plants. Glycerides do not diffuse into the air like volatile oils, and they are consequently called fixed, neutral, fatty, or nonvolatile oils.

FIGURE 10-1
The cleaning properties of vegetable oils, palm and olive in particular, were known to the ancient Greeks as this 1915 advertisement for Palmolive soap emphasizes.

These oils are constructed from two fundamental kinds of compounds: glycerol, a three-carbon alcohol, and fatty acids (Fig. 10-3). The glycerol "backbone" is the same for all glycerides. Differences between oils are the result of the various combinations of fatty acids attached to the three available sites along the backbone.

A *fatty acid* is basically a hydrocarbon chain with a carboxyl (COOH) group

FIGURE 10-2
During this century, there has been an increase in the consumption of fats and oils in the United States. Much of this increase has been attributed to the rise in popularity of fast-food outlets. Note that the biggest increase is in salad and cooking oils, shortening and margarine all of which are plant derived. (Adapted from Rizek, R. et al., 1974, Journal American Oil Chemist's Society 51:244–250.)

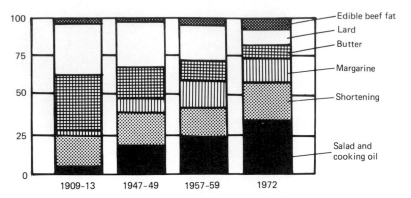

at one end that causes it to interact chemically as an acid. When such an acid combines with an hydroxyl group of the glycerol molecule, water is released. If only one acid is attached to the backbone, a monoglyceride results. If two acids are attached, a diglyceride is formed. In a triglyceride, fatty acids are attached at all of the hydroxyl positions on the glycerol molecule. Correspondingly, fatty acids can be cleaved from a glyceride by hydrolysis, or the addition of an OH group to the glycerol backbone and some sort of positive ion to the fatty acid. Alkalis, or "bases," are often used to effect this separation. The most commonly used alkalis, sodium and potassium hydroxide, produce glycerol and sodium or potassium salts of the fatty acids. These metal salts of fatty acids are commonly called *soaps* (Fig. 10-4). As their name implies, they are the primary ingredients

TABLE 10-1

Plants Discussed in Chapter 10

COMMON NAME	SCIENTIFIC NAME	FAMILY	CHROMOSOME NUMBER
Bayberry	*Myrica pensylvanica*	Myricaceae	Unreported
Candellila wax	*Euphorbia antisyphilitica*	Euphorbiaceae	Unreported
Carnauba wax	*Copernicia cerifera*	Arecaceae	$2n = 36$ diploid
Castor	*Ricinus communis*	Euphorbiaceae	$2n = 20$ diploid
Coconut	*Cocos nucifera*	Arecaceae	$2n = 32$ diploid
Corn	*Zea mays*	Poaceae	$2n = 20$ diploid
Cotton, upland	*Gossypium hirsutum*	Malvaceae	$2n = 52$ tetraploid
Jojoba	*Simmondsia chinensis*	Simmondsiaceae	$2n = 52$
Linseed	*Linum usitatissimum*	Linaceae	$2n = 30$ diploid
Olive	*Olea europaea*	Oleaceae	$32n = 46$ diploid
Palm, palm kernel	*Elaeis guineensis*	Arecaceae	$2n = 32$ diploid
Peanut	*Arachis hypogaea*	Fabaceae	$2n = 20$ diploid
Rape	*Brassica spp.*	Brassicaceae	$2n = 20, 38$, diploid, amphiploid
Safflower	*Carthamus tinctorius*	Asteraceae	$2n = 24$ diploid
Sesame	*Sesamum indicum*	Pedaliaceae	$2n = 26$ diploid
Soybean	*Glycine max*	Fabaceae	$2n = 40$ diploid
Sunflower	*Helianthus annuus*	Asteraceae	$2n = 34$ diploid
Tung oil	*Aleurites fordii*	Euphorbiaceae	$2n = 22$ diploid (usually)

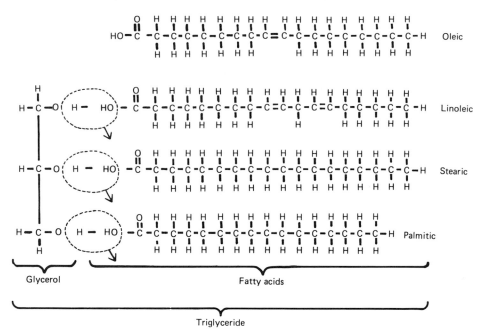

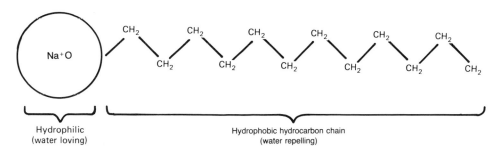

Glycerol Fatty acids

Triglyceride

FIGURE 10-3
Three molecules of water are released in the formation of triglycerides from glycerol and fatty acids. In this process, the carboxyl groups of each of three fatty acids react with the alcohol groups of the glycerol molecule. Palmitic, stearic, and oleic are the fatty acids most commonly occurring in vegetable oils. Oleic is considered unsaturated because there is a double bond between the ninth and tenth carbon atoms. The other two are saturated. Linoleic acid is one of the fatty acids required by humans because it cannot be manufactured in the human body.

Hydrophilic (water loving) Hydrophobic hydrocarbon chain (water repelling)

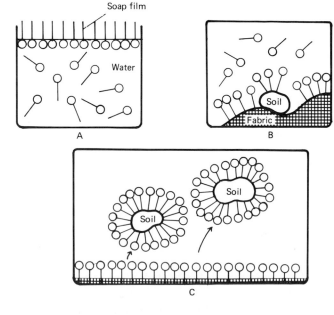

FIGURE 10-4
Soaps are basically fatty acid metal salts. The metal, usually sodium or potassium, carries a positive charge and acts hydrophilically, while the long hydrocarbon chain (top) is hydrophobic. This duality of the soap molecule causes soap to form a film on water (A), but also allows soaps to attract oily dirt particles with their hydrocarbon tails (B) and then disperse them into the water (C) so that they can be rinsed away.

of common soap. Soap is thus one of the major products derived from seed oils. Other metal ions besides sodium and potassium can be used to manufacture specialty soaps. Lead and zinc are often used to make medicinal soaps, and aluminum is used in soaps that have waterproofing properties.

Since fatty acids are long-chain organic compounds, there are several ways in which the chains can be constructed. The structure of fatty acids determines their chemical properties and, ultimately, those of a vegetable oil itself. The primary ways in which the fatty acids differ from one another are in the number of carbon atoms (the length of the chain), the position of the carbon atoms relative to one another (straight chain or variously branched), and in the number of double bonds between carbon atoms. In general, the longer the hydrocarbon chain of a fatty acid, the higher the melting point. Triglycerides with predominantly long-chain fatty acids will have higher melting points than those composed of shorter-chain acids with the same saturation levels.

In commercial vegetable oils, most of the fatty acids have straight chains. The fatty acids most commonly found in vegetable oils are palmitic, stearic, and oleic (see Fig. 10-3). The first of these acids has a chain length of 16 carbons. The second two both have 18 carbon atoms, but differ from one another because stearic acid has no double bonds and oleic acid has one. In vegetable oil jargon, fatty acids with only single bonds are considered *saturated*. When adjacent carbon atoms share two electrons and produce a double bond, the bond between the carbons is called *unsaturated* because each carbon atom is capable of accepting another atom with which it can share a pair of electrons. A given fatty acid can have none, one, or several double bonds. In the first case, it is called a *saturated fatty acid*. If it has one double bond, it is called *monounsaturated*, if two, *diunsaturated*. For more than two double bonds, the acid is usually simply referred to as *polyunsaturated*. However, since all vegetable oils are mixtures of triglycerides and some free (not attached to an alcohol) fatty acids, the expressions "saturated" or "polyunsaturated" refer in common usage to the relative proportions of saturated or unsaturated fatty acids in the mixture of glycerides that make up an oil.

The double bonds of an unsaturated fatty acid can be broken by the addition of reactive atoms such as those of hydrogen, iodine, and chlorine. The degree of unsaturation is standardly measured by allowing iodine to combine with an oil. Iodine ions are incorporated into the oil's fatty acid chains at positions where there were double bonds. After all chemical reaction has ceased, the amount of iodine left can be subtracted from the amount initially added, giving a value called the *iodine value*. Fatty oils can be classified by their iodine numbers. The iodine values of edible oils range from about 7 to over 200 (Table 10-2). Oils with values below 70 are usually referred to as fats because they are solid at room temperature.

Another way of grouping vegetable oils which reflects the degree of saturation is into drying, semidrying, and nondrying oils. The name "drying oil" refers to the fact that oils with many double bonds tend to link together forming polymers. If these oils are spread in a thin film over a substance, they dry into impervious coatings. Nondrying and most semidrying oils will not form such coatings. A general rule of thumb is that drying oils have iodine values higher than 150, semidrying oils between 100 and 150, nondrying oils between 70 and 100, and fats below 70 (see Table 10-2).

The presence of double bonds in many fatty acids allows another kind of

TABLE 10-2

Saturation Indexes and Oil Protein Content of Vegetable Oils

CATEGORY/VEGETABLE OIL	IODINE NUMBER	OIL, %	PROTEIN, %
		DRYING	
Linseed	165–204	35–45	33
Tung	160–175	16–18	Not used*
		SEMIDRYING	
Safflower	140–150	25–37	16–22
Soybean (shelled)	103–152	12–24	30–40
Sunflower	113–143	22–36	37–38 (meal)
Corn	103–133	2–5	10
Sesame	103–118	44–54	40
Cottonseed	90–117	19	21
Rapeseed	94–105	2–49	20
		NONDRYING	
Peanut	84–100	45–55	25–28
Olive	78–88	75	**
Castor	81–89	50	Not used
		VEGETABLE FATS	
Palm (fruit)	46–60	50	**
Palm (kernel)	14–22	44–53	17
Coconut	7–10	65–68	21

Note: Iodine values from E. A. Weiss. 1971. *Castor, Sesame, and Safflower.* New York, Barnes and Noble; and D. Swern. 1979. *Bailey's Industrial Oil and Fat Products*, vol. 1, 4th ed. New York, Wiley. Percentage composition figures based on dry weights.
*"Not used" indicates that the meal is toxic and therefore the protein content is not important.
**Indicates fruit pulps that are low in protein and not used for feed.

chemical reaction, oxidation, to occur. A fatty acid can absorb oxygen at positions where carbons are linked by double bonds. The resultant oxidation of the fatty acids leads to a change in the taste of the oil. We perceive this change as obnoxious and describe the oils as rancid. To thwart rancidity, antioxidants such as BHT, BHA, and polysorbate 80 can be added to oils or to processed oil-containing foods, such as crackers, cookies, cereals, margarine, etc. These antioxidants are called "preservatives."

Natural antioxidants also occur in association with fats and oils. Some of these, such as vitamin K, are necessary for human nutrition. Although they are not antioxidants, the vitamins A, D, and E are oil-soluble and absorbed by the body only when they are dissolved in fats and oils. Oils also aid in the accumulation of the B vitamins, provide linoleic acid (see Fig. 10-3), and are a major source of calories in the United States (Fig. 10-2).

NONFOOD USES OF VEGETABLE OILS

Although most vegetable oils have traditionally been, and are still, used for food-related purposes, large quantities are consumed by industry for the manufacture of nonedible products. In some cases they are used as lubricants, but in most cases they are used as the basis of soap or polymer production. We mentioned earlier how "soaps" are made from triglycerides by the addition of alkali. The old-fashioned method of soap making involved boiling animal fat (lard) and ashes together. When combined with water, the ashes formed an alkali that cleaved the fatty acids from the triglycerides of the lard. Soap has cleaning properties because the metal salts of the fatty acids have an affinity for water, while the remaining parts of the molecules are attracted to grease or "dirt" (Fig. 10-4). Today, vegetable oils are used more extensively than lard in soap making.

Oils, particularly drying or semidrying oils, are also extensively used to manufacture paints and varnishes. Linseed oil or tung oil polymerize easily and can simply be applied to surfaces without processing. To form a decorative coating, color can be added to the oil. The Flemish were the first people to combine pigments with vegetable oils and produce what we now call oil paints. They perfected a technique in the fifteenth century of glazing paintings with translucent colors over an undercoating (Fig. 10-5). The result was an illusion of texture and dimension never before achieved in painting. Oils are now used in paints as the binder in which pigments are dispersed. In the manufacture of paints, oils are first boiled with compounds containing heavy metals such as magnesium, cobalt, or lead. These help the oils to absorb oxygen after they are applied resulting in hard films. Varnishes are made by mixing boiled oils with

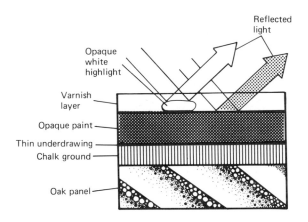

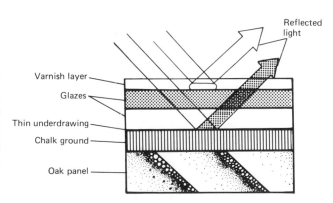

FIGURE 10-5
Flemish artists added a new dimension to painting when they developed a technique of mixing pigments with vegetable oil, usually walnut or linseed, to produce translucent glazes. As is shown in the lower diagram, light penetrates the glaze layer and is reflected to create an illusion of depth that was not possible with previous opaque paints (shown in the upper diagram).

resins or gums (e.g., damar, copal, etc.; see Chapter 11). Enamels are made by mixing pigments with varnishes. Paints differ from varnishes and enamels because they do not contain gums or resins. They have traditionally been prepared from mixtures of boiled oils, pigments, and thinning agents such as turpentine. Modern latex paints are made from alkyd resins, which are polymer resins manufactured from fatty acids cleaved from vegetable oils. Because of their ease of application and clean up (they are water-soluble), latex paints now make up about 40 to 80 percent of the modern commercial paint production.

Linoleum, as implied by the name (from *linum*, flax, and *oleum*, oil), is another product for which seed oils have been used. The tough, yet pliant, material is made from a mixture of oils, gums, and synthetic resins plus various pigments. To prepare oils for use in linoleum manufacture, they are first "blown," or subjected to a current of air, oxygen, or ozone which causes them to become thick. This oxidized oil is mixed with ground cork or wood, various resins and pigments, pressed onto a backing (usually felt or burlap), and heated for a few days. Blown oils are soluble in petroleum oils (regular oils are not), which allows them to mix better with synthetic resins. The manufacture of true linoleum has, for the most part, given way to plastic floor coverings in this country. However, even these are often made from polymers derived from seed oils.

Vegetable oils are now being increasingly used as starting points for the synthesis of a number of organic compounds formerly obtained from petroleum. Fatty acids separated from a glycerol backbone can be converted chemically into alcohols, amines, esters, and other compounds that can be used to manufacture detergents, resins, and special types of industrial oils. To a limited extent, vegetable oils have been used in medicine and in the cosmetics industry. Today, only castor oil is employed medicinally. Finally, some seed oils have been proposed as substitutes for diesel oil.

METHODS OF EXTRACTING VEGETABLE OILS

Examination of a sesame seed or a castor bean gives little indication that over half of the seed's weight (Table 10-2) is oil held in special cellular organelles. Humans have devised various methods of removing oils from seeds and fruits. Among the earliest methods of extraction was grinding with stones to crush the tissues and release the oils (Fig. 10-6). Sometimes the seeds or the basal stones were heated to facilitate the flow of the oil from the crushed debris. Because heating increases the fluidity of oil, more oil is extracted when heat is used than when the pressing is done under ambient conditions. Both a mortar and pestle arrangement, often with a deep, inverted cone and a revolving pestle, and a press with a wedge that was manually pounded, have been used for oil expression since early times. In many undeveloped countries, various types of manual or animal-driven presses are still employed. Almost all of the oil extracted by these kinds of methods is consumed locally.

The first mechanized presses, produced in the seventeenth century, were steam-driven and operated on the same principle as the manual press. In 1795, the hydraulic press was developed in England and, in 1900, the first screw

FIGURE 10-6
These Tunisian workers are using a
traditional stone press to extract
oil from olives. (WPF/FAO photo by
F. Botts.)

presses were used for seed oil extraction. Before the widespread adoption of the screw press, pressing was done either under ambient or elevated temperatures. In the first case, the procedure was called *cold pressing*, and in the latter, *hot pressing*. Although cold pressing extracted less oil than hot pressing, the oil obtained was purer than that extracted with heat because high temperatures promoted the extraction of other compounds along with the glycerides. Today, these distinctions mean little because the process of extracting with a screw press itself generates heat that can reach temperatures between 65 and 72°C (149 to 162°F).

The screw press was readily adopted in industrial countries because it made large-scale oil production economically feasible. A screw press, or expeller (Fig. 10-7), allows the continuous feeding of seeds (usually with seed coats removed) into one end of a constantly turning screw. The seeds are crushed by the increasing amounts of pressure exerted as they are forced toward the distal end of the screw. The oil released during the crushing flows out of the press through slots along the cylinder enclosing the screw. The residue, or cake, is expelled through the end of the press, which is fitted with a conical choke. The size of the opening through which the residue is ultimately forced is controlled by varying the extent to which the choke is fitted into the screw cylinder. Because the residual cake is often very high in protein (Table 10-2), it is frequently used as animal food. Today, oil seed meal constitutes an important commercial by-product.

Because expeller pressing leaves a residue that still contains 2 to 4 percent oil, a further refinement using organic solvents was developed. Solvent extraction, which leaves a scant 0.5 to 1 percent of the oil in the cake, has now been adopted in most large operations. During solvent extraction, the husked seeds or fruits are crushed and agitated in the solvent, usually hexane. The solution containing the oils is separated from the residue and then the solvent itself is driven from the mixture by distillation. Hexane boils off at temperatures between

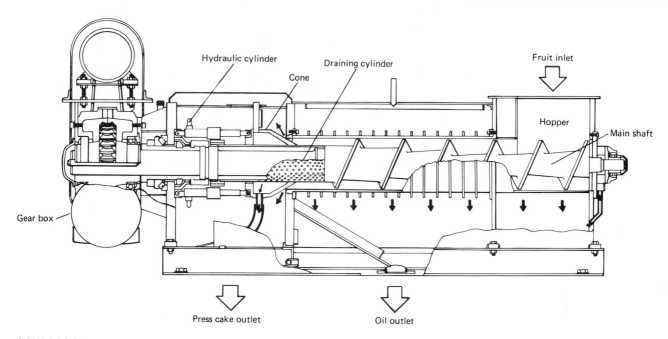

FIGURE 10-7

A screw press, or expeller used to express oil from palm fruits. After a diagram supplied by the French Oil Mill Machinery Co.

63 and 65°C (145 to 149°F), leaving the crude oil behind. The solvent is recovered by chilling, and reused.

Figure 10-8 diagramatically shows the steps followed in most modern commercial extraction operations. Any procedures that are peculiar to individual seed or fruit crops are discussed below under the treatments of the crops. Usually, seeds are cleaned and dried before they are sold to oil refining companies or just after they arrive at the mill (see Fig. 10-8). The seeds or fruits are then recleaned, and dehusked or decorticated. "Husk" is a general term referring to the fruit wall, the seed coat, or both. In the case of sesame seed or cottonseed, the seed coat is removed. Sunflower and safflower seeds are actually entire, single-seeded achenes and the "shell" is composed of the ovary wall. The seed itself has a thin, membranous coat. Both the shell and the seed coat are removed in the decortication of these two oil crops.

The kernels separated from all covering material are then pretreated by breaking and flaking them with a series of rollers. The flakes are cooked or subjected to high temperatures to reduce their moisture content. This process, known as *conditioning*, also facilitates oil release and helps to coagulate proteins so they will not be extracted with the oil. The mass of prepared, flaked, and conditioned seeds is then sent to the presses. The meal may be pressed once or several times or sent directly to tanks for solvent extraction. In many operations, the meal is subjected to an initial pressing before it is solvent extracted. The wet meal left after the solvent-oil mixture has been removed is heated with steam to evaporate the toxic solvent. The solvent is recovered by condensation, and the meal is sold for animal feed.

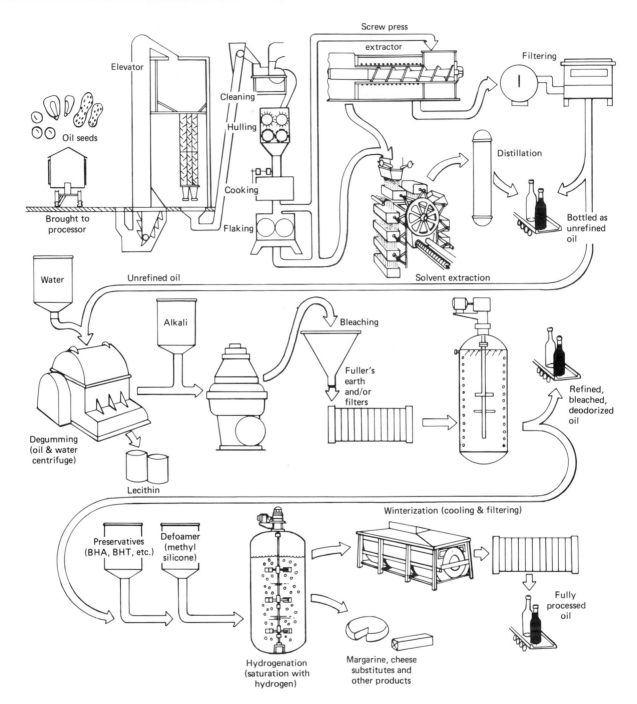

FIGURE 10-8
A diagrammatic representation of the extraction and processing of vegetable oils. After the addition of preservatives and defoaming agents, the oil can either be winterized or hydrogenated.

OIL REFINING AND PROCESSING

Following extraction, oils are usually refined and then subjected to one or more additional procedures. Refining is most often accomplished by mixing the oil with caustic soda to remove any unbound (free) fatty acids. Refining leads to the production of a neutral oil. In addition to being refined, the oil can be degummed, bleached, deodorized, and/or winterized. *Degumming* is effected by mixing the oil with water at 32 to 49°C (90 to 120° F) and centrifuging the mixture. The process removes any mucilaginous material that might have been extracted with the oil. Lecithin, a phosphoglyceride sold in health food stores, is recovered from the water fraction obtained during the degumming of soybean and other seed oils. Triglycerides themselves are usually odorless and colorless. Both the taste and colors of vegetable oils are caused by other compounds that are extracted with the oil. These include volatile oils and pigments such as gossypol and the carotenoids. *Bleaching*, or the removal of pigments, is carried out by adding fuller's earth (diatomaceous earth) or charcoal to the oil and filtering the mixture. Most oils are also *deodorized* by heating them with steam under vacuum conditions. *Winterizing* is a process that attempts to circumvent the clouding of oil under low temperature conditions by chilling the oil to about 7°C (45°F) and filtering out all of the particles that condense.

An oil can undergo a number of additional processing steps, depending on its eventual use. One of the most common is *hydrogenation*. During hydrogenation, hydrogen gas is bubbled under pressure through an oil in the presence of a catalyst, usually nickel. The hydrogen binds to carbon atoms previously linked by double bonds and causes the unsaturated fatty acid components of the oil molecules to become completely or partially saturated. Hydrogenation thus lowers the iodine value of an oil and raises its melting point. Most hydrogenated oils are therefore solid at room temperature. Since its discovery in the early 1900s the process of hydrogenation has opened up an entirely new industry that now produces vegetable lards, margarines, and cheese substitutes.

Refining and additional processing tend to produce oils and fats that have little individual character. While connoisseurs relish the flavor of olive, sesame, or peanut oil, oils such as cottonseed, soybean, and palm oil are purposely treated to render them virtually tasteless and identical. The Food and Drug Administration allows the substitution of several such oils in food manufacture (Fig. 10-9). This interchangeability permits manufacturers to purchase whichever oil is least expensive at any one time.

Using this background of vegetable oil chemistry, extraction, and processing, we can turn to the different plant species from which oils are now commercially extracted. Table 10-2 lists the seed and fruit oils in the order we discuss them and gives their iodine numbers and classification according to their drying properties. Table 10-2 also shows the relative amounts of oil and protein per dry weight of the different vegetable oil crops we discuss, and Table 10-3 gives recent worldwide production figures.

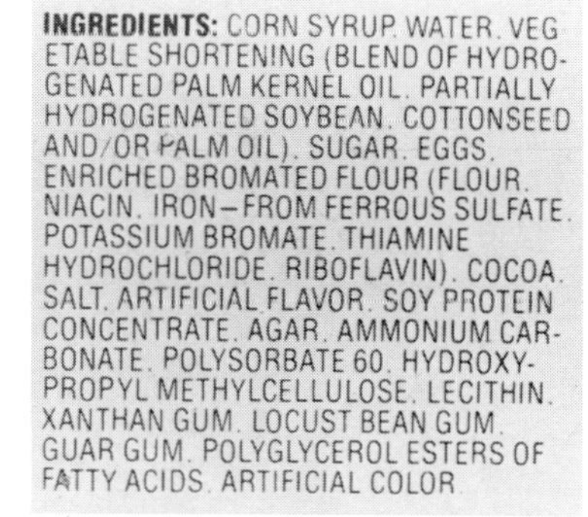

INGREDIENTS: CORN SYRUP, WATER, VEGETABLE SHORTENING (BLEND OF HYDROGENATED PALM KERNEL OIL, PARTIALLY HYDROGENATED SOYBEAN, COTTONSEED AND/OR PALM OIL), SUGAR, EGGS, ENRICHED BROMATED FLOUR (FLOUR, NIACIN, IRON—FROM FERROUS SULFATE, POTASSIUM BROMATE, THIAMINE HYDROCHLORIDE, RIBOFLAVIN), COCOA, SALT, ARTIFICIAL FLAVOR, SOY PROTEIN CONCENTRATE, AGAR, AMMONIUM CARBONATE, POLYSORBATE 60, HYDROXYPROPYL METHYLCELLULOSE, LECITHIN, XANTHAN GUM, LOCUST BEAN GUM, GUAR GUM, POLYGLYCEROL ESTERS OF FATTY ACIDS, ARTIFICIAL COLOR.

FIGURE 10-9
Because they are indistinguishable once refined, the Food and Drug Administration considers cottonseed, soybean, and palm oils interchangeable in packaged foods. This ruling leads to the common statement on food packages that the product may contain any or all of the three oils.

TABLE 10-3

World Production of Vegetable Oil Crops

CROP	TOP 5 COUNTRIES, 1,000 t		TOP 5 CONTINENTS, 1,000 t		WORLD, 1,000 t
	COUNTRY	t	CONTINENT	t	
Castor	India	345	Asia	629	936
	China	195*	South America	195	
	Brazil	172	Africa	42	
	U.S.S.R.	60*	North & Central America	5	
	Thailand	33*	Europe	5	
Coconut (copra)	Philippines	1,930*	Asia	3,835	4,548
	Indonesia	1,070*	Oceania	306	
	India	350*	North & Central America	192	
	Malaysia	204*	Africa	177	
	Mexico	145*	South America	38	
	Sri Lanka	145*			
Cottonseed	China	9,282*	Asia	14,587	27,498
	U.S.S.R.	5,615*	North & Central America	3,439	
	United States	2,817	Africa	2,142	
	India	2,540*	South America	1,689	
	Pakistan	1,040	Europe	362	
Linseed	Argentina	670*	South America	692	2,374
	India	476	North & Central America	658	
	Canada	465	Asia	609	
	U.S.S.R.	220*	Europe	128	
	United States	187	Africa	58	
Olive oil	Italy	670	Europe	1,226	1,564
	Spain	268	Africa	186	
	Greece	259	Asia	133	
	Tunisia	132	South America	17	
	Turkey	70	North & Central America	3	
Palm kernels	Malaysia	840	Asia	1,059	2,147
	Nigeria	360*	Africa	733	
	Brazil	250*	South America	300	
	Indonesia	154	Oceania	37	
	Benin	75*	North & Central America	18	
Palm pulp	Malaysia	3,000	Asia	4,194	5,870
	Indonesia	950	Africa	1,351	
	Nigeria	710*	South America	176	
	China	200*	Oceania	108	
	Zaire	140*	North & Central America	41	
Peanuts (in shell)	India	7,300*	Asia	13,410	19,792
	China	4,036*	Africa	4,099	
	United States	1,485	North & Central America	1,625	
	Sudan	900*	South America	598	
	Indonesia	760	Oceania	35	

TABLE 10-3   (Continued)

World Production of Vegetable Oil Crops

CROP	TOP 5 COUNTRIES, 1,000 t		TOP 5 CONTINENTS, 1,000 t		WORLD, 1,000 t
	COUNTRY	t	CONTINENT	t	
Rapeseed	China	4,288*	Asia	7,172	14,342
	Canada	2,681	Europe	4,340	
	India	2,472	North & Central America	2,684	
	France	969	Africa	23	
	Germany, Federal Republic of	580	Oceania	22	
Safflower seed	Mexico	545	North & Central America	645	1,120
	India	372	Asia	373	
	United States	100*	Oceania	49	
	Australia	49	Africa	32	
	Ethiopia	32*	Europe	15	
Sesame seed	India	590*	Asia	1,391	2,076
	China	350*	Africa	477	
	Burma	204	North & Central America	142	
	Sudan	200*	South America	65	
	Mexico	99	Europe	1	
Soybeans (part for oil)	United States	43,421	North & Central America	45,023	78,566
	Brazil	14,582	South America	19,334	
	China	9,770*	Asia	12,352	
	Argentina	3,750	Europe	757	
	Paraguay	780*	Africa	391	
Sunflower seed	U.S.S.R.	5,300*	Europe	3,690	15,766
	Argentina	2,300	Asia	2,463	
	United States	1,447	South America	2,349	
	China	1,370*	North & Central America	1,518	
	France	837*	Africa	354	
Tung oil	China	62*	Asia	62	91
	Argentina	14*	South America	28	
	Paraguay	13*	Africa	1	
	Brazil	1*	—		
	U.S.S.R.	1*	—		
	Malawi	1*	—		

*Indicates an estimated or unofficial figure.

Note: Only crops of importance in international trade are included. The U.S.S.R. is considered a country only. Its values are not included in those of Asia or Europe.

Source: Data from *FAO Production Yearbook for 1983,* vol. 37. FAO, Rome, 1984.

Drying Oils

Linseed Oil

Linseed is probably the oldest domesticated oilseed crop and forms part of the complex of wheat, barley, peas, lentils and chickpeas that were early domesticates in the Near East (Fig. 10-10). Fossil linseeds clearly show signs of human selection by 6000 B.C. and, combined with evidence from other crops, indicate that a rather well developed agriculture had been established by this time. The same species, *Linum usitatissimum* (Linaceae), that produces seeds for linseed oil is the source of flax (see Chapter 16), but oil extraction predated fiber use by several thousand years.

The application of linseed oil to surfaces to produce a water-repellent glaze has been practiced since at least classical Egyptian times when coffins were coated with mixtures of linseed oil and resin. Today, linseed oil is incorporated into paints or used alone to protect natural wood surfaces such as shingles, decking, and fences. Linseed oil is usually blown if it is to be used for coating purposes. The high level of unsaturation (Table 10-2) that gives linseed oil such good drying properties, fosters rapid oxidation. As a result, linseed oil quickly becomes rancid and acquires a pungent, unpleasant taste.

Linseed plants are erect annual herbs about 1 m tall (see Fig. 16-15). They have blue "flax" flowers that give way to capsules containing about 10 elliptic seeds. The cultivars used for oil extraction are shorter and more branched than those used for fiber. The dark amber oil extracted from the seeds has an acrid odor and taste. Even bleached and refined, linseed oil has a flavor too strong for most people. The decline in use of linseed oil as an edible oil has resulted from a combination of this unpleasant taste, the rapidity with which the flavor further degenerates due to oxidation, and the increased availability of other, more palatable oils. In the United States, linseed oil is used almost exclusively for nonedible products, although bottles of cooking oil can be found in health or "natural" food stores.

FIGURE 10-10
Linseed oil was used extensively in ancient Egypt. A copy of a scene on an Egyptian tomb shows the harvesting of flax from which both fiber and linseed oil were obtained.

Tung Oil

Most people have never heard of tung oil since it is never sold in grocery stores, never added to commercial food products, and rarely sold under its own name. Tung oil is not eaten because it is poisonous, but its high level of unsaturation makes it an excellent oil for many other purposes. The oil is extracted from the seeds of several species of the Old World genus *Aleurites* (Euphorbiaceae), a family with many species that contain poisonous alkaloids. Species of the genus are tall (15 m, or 49 ft), deciduous trees that begin to bear fruit only after the age of 3 years. Each lobed fruit contains three, oil-rich seeds. Most of the oil produced comes from the seeds of *A. fordii* (Fig. 10-11), a native of China which has been employed for hundreds of years as a source of oil for paints, waterproof coverings, and caulking. Tung oil is an ingredient of "India" ink, an important item in China's traditional calligraphy. It is also used in varnish, paint, and linoleum manufacture, but it has to some extent been replaced by other oils or synthetic materials. The so-called teak oil sold for fine furniture is usually refined

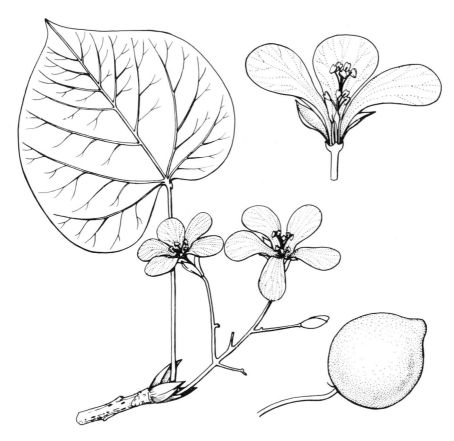

FIGURE 10-11
A flowering branch of *Aleurites fordii*, the species that provides the seeds most commonly used for the extraction of tung oil. A cross section of the flower and a whole fruit are shown on the right.

tung oil, which is perfect for conditioning wood and providing a thin protective covering.

Tung oil was formerly a relatively important oil in the United States, but all of the oil used was imported. When the U.S. supply of tung oil was cut off during World War I, attempts were made to start tung plantations in the southern portions of the country, primarily in southern California, Texas, South Carolina, and southern Florida. By 1950, production reached a level of almost 50,000 tons of oil per year. Subsequent crop losses caused by frost damage and commercial difficulties led to an abandonment of virtually all of the plantations. As a result, no significant quantities of tung oil are now produced in this country. Most tung oil today comes from China.

Semi-drying Oils

Safflower Oil

Safflower oil has recently received considerable attention in the United States because it is the most unsaturated of the commonly used edible oils. It seems likely that the safflower (*Carthamus tinctorius*, Asteraceae) was originally domesticated in the eastern Mediterranean region, but this is difficult to prove

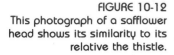

FIGURE 10-12
This photograph of a safflower
head shows its similarity to its
relative the thistle.

because the species is known only in cultivation. Safflowers were originally grown for their flowers, which produce a deep red-yellow dye. Ancient writings suggest that the dye was in use about the time of Christ, but Hebrew writings mention it only in the second century A.D. The Indians seem to have been the first to have extracted oil from the achenes, but they initially employed the oil only for medicinal purposes.

Safflower plants are spiny annuals (Fig. 10-12) with colorful, thistlelike heads. The achenes are large (6 to 7 mm, or 0.25 in, long) and reminiscent of those of the safflower's distant cousin, the sunflower. Until recently, the major use of safflower oil in the developed countries was as a commercial oil for paint, varnish, and alkyd resin production. Safflower oil has the highest linoleic acid content of any known seed oil and is now exploited as a source of this acid. Linoleic acid (Fig. 10-2), one of the few fatty acids that cannot be synthesized by man, is a nutritional requirement in the human diet.

Today, safflower oil is commonly found in cooking oils and manufactured products with an oil base. Since the oil tolerates cold well, it is especially suited for use in the manufacture of salad dressings and margarines that are kept at low temperatures to preserve their flavor. Safflower oil must be used quickly after pressing unless antioxidants are added to it because it readily oxidizes. The cake remaining after pressing is used for feed. If the shells are removed before pressing, the cake can have up to 52 percent protein.

Soybean Oil

Since the origin of soybeans (*Glycine max*, Fabaceae see Fig. 7-12) and their history and cultivation have been discussed in Chapter 7, we will concentrate here on soybeans as an oilseed crop. Although soybeans are not widely used for oil in the Orient, Europeans were pressing them for oil and meal in the 1700s. Prior to 1940, the principal North American use of soybeans was for forage, but by 1947, the acreage devoted to it had greatly increased and 85 percent of the crop was processed for oil. By 1978, this figure had risen to 98 percent. Sixty percent of all of the edible oil consumed in this country today comes from soybeans.

Before processing, the pods are shelled and the beans are cleaned and dried. Like peanuts, soybeans can serve as hosts to *Aspergillus flavus*, which excretes deadly toxins (aflatoxins). With proper handling, fungal infestations can practically be eliminated. After recleaning and hulling, the oil-filled cotyledons are flaked and, in the United States, extracted with organic solvents. As might be expected, the cake is exceedingly high in protein and can be used as animal feed, or further processed and used as an additive in human food. Unfortunately, texturized vegetable protein cannot be made from the cake.

Soybean oil is always highly refined, yielding a tasteless, stable oil that constitutes the majority of the products that go under the label of "vegetable oil." Processing consists of degumming, neutralizing, and bleaching. Liquid forms of the oil are winterized. Because soybean oil contains natural antioxidants (primarily tocopherols), it stores well despite its high level of unsaturation. Most soybean oil is used for salad or cooking oils and in artificial fluffy or creamy products such as whipped toppings, mixes for icings, or instant cheesecakes.

Sunflower Oil

Of all of the plants used for seed oils, only the sunflower can claim North America as its original home. Wild, weedy populations of the species, *Helianthus annuus* (Asteraceae), still occur throughout Canada, the United States, and Mexico. Wild plants are branched and bear numerous, relatively small heads. Domesticated sunflowers usually have one large head atop a single flowering stem (Fig. 10-13). The oldest fossil achenes that indicate sunflower domestication have been found in eastern North America. Native Americans ate the seeds and crushed them for oil.

There is some confusion about the time and date of the introduction of the sunflower into Europe, especially since the first description of the species stated that the plant came from Peru where the species did not occur naturally and where it was introduced only much later. The first firm dates of sunflower cultivation in Europe are in the middle of the sixteenth century when specimens were grown as horticultural novelties. In the last part of the nineteenth century, Russians began to cultivate sunflowers. Throughout the twentieth century, the U.S.S.R. has dominated the world's production of sunflower seeds (Table 10-3). Although fossil evidence indicates that Americans had already selected for large-headed, large-seeded sunflower plants before the arrival of Europeans in the New World, the Russians continued to select for giant inflorescences and now have varieties that can produce up to 1,000 seeds per head. Selection has also produced dwarf oil-yielding varieties that are relatively easily harvested by machine.

The achene coat of sunflowers can vary in color, but two types predominate: black, and gray with brown or black stripes (Fig. 10-14). These two color variants are maintained in cultivation and generally indicate whether the seeds are to be used for confectionary purposes (striped shells) or for oil (black shells). Achenes for eating or confectionary use are larger than those used for oil and have a lower oil content (about 30 vs. 40 percent oil on a dry weight basis).

FIGURE 10-13
In the course of domestication, the sunflower, naturally a much branched annual with numerous rather small heads (A), became a plant with unbranched stems each topped by a single large head (B & C). Each head is composed of numerous flowers (florets) (E). The sequence of blooming and fruit maturation of a single flower is shown (D).

Sunflower seeds are processed in the usual way and yield an oil that is subsequently moderately refined. The oil has been used without refining in a few areas of the northern United States mixed with diesel fuel (75:25 ratio, sunflower to diesel) in farm machinery. This mixture performs as well as pure diesel fuel and studies indicate that if the oil is neutralized, the mixture will outperform diesel fuel. Refined sunflower oil is used as a salad, cooking, and commercial oil in prepared foods. In Russia, it is an important source of oil for paints, varnishes, and synthetic resins.

Corn Oil

Like soybeans, corn (*Zea mays,* Poaceae) is a plant for which humans have found innumerable uses, among them the production of oil. Unlike soybeans, however, oil production constitutes a minor use for the grain (see Chapter 6 for the history of corn use). Corn oil production is basically a by-product of the corn milling industry. Corn can be milled wet or dry. In the wet process, the cleaned corn is soaked or steeped in water and then lightly macerated to separate the germ from the endosperm (Fig. 10-15). The pressed, wet grains are washed into flotation tanks where the lighter germs are removed as they float to the top. The water containing valuable soluble proteins is evaporated down and sold for use as a medium for the culture of yeasts or other microorganisms. The dry milling process, now the most commonly used, is often called the TD system

FIGURE 10-14
Sunflower seeds used for oil extraction are usually solid black, (above) whereas those used for confectionary or eating purposes are striped. (below)

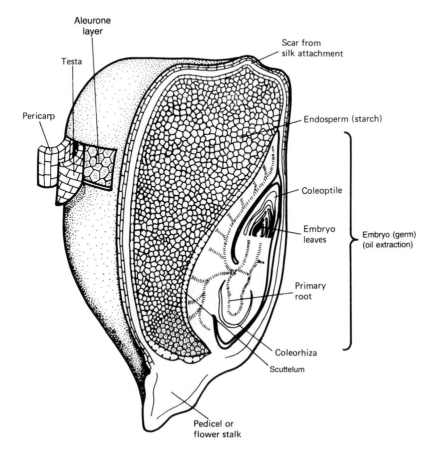

FIGURE 10-15
Corn oil is extracted from the embryo or germ, which constitutes a small portion of the kernels compared to the endosperm.

because it involves tempering and degerming. The grain is tempered by adding just enough moisture to dry corn to make the grains pliant. The germ and cap of the grains are then removed by friction. The oil-containing germs recovered from the milling operations are dried, preexpelled, and extracted with solvents. Starch recovered from the endosperm of either milling operation is used in brewing (see Chapter 15) and in the textile and paper industries (Chapters 11 and 17).

Since corn is used for both oil and starch production, and starch is the principal product, there has been little selection for high oil content in corn destined for commercial use. For feed, however, there is a premium on high oil content since oil contains many more calories than starch. As a result of selection for better feed corns, oil contents as high as 10 percent of the grain (dry weight) have been produced. In general, values range from 2 to 5 percent of the dry weight of the kernels.

Most of the refined corn oil that is marketed is used for salad dressings and margarine. The oil contains natural antioxidants that give it good stability, but it is not a very good frying oil because it smokes when heated to high temperatures.

Sesame Oil

While less important than the other edible oils included here, sesame has historically been an important source of oil and still ranks among the major seed oil crops. There has been some suggestion that the use of sesame (*Sesamum indicum*, Pedaliaceae) as a source of oil predated the use of linseed oil, but since the earliest archaeological finds of seeds from Palestine and Syria date from only 3000 B.C., there is little firm evidence for the hypothesis. Indications of its early use come from Sumerian writings made about 2350 B.C. Asian dates from West Pakistan are about the same age. Although the early records are all from the Mediterranean area, botanists have proposed that the species, now known only as a domesticated crop, was initially taken from the wild in Ethiopia.

Sesame oil is not mentioned in the Bible, but it appears to have been important in non-Hebrew cultures during the few thousand years before Christ. Loans were negotiated in both silver and sesame seeds. The expression "open sesame" made famous by the tale of Ali-Baba and the 40 thieves in the *Tales of One Thousand Arabian Nights* is probably based in some way on the sesame seed. Some authors have suggested that the expression was adopted by the writer of the tales because the capsules must be precisely tapped in order to open them without crushing the valuable seeds.

Sesame plants are herbaceous annuals with attractive white, pink, or purple flowers (Fig. 10-16). In the United States, they are sometimes grown as ornamentals. The fruit is a flat capsule (Fig. 10-17) which contains a variable number of white, oval seeds. The seeds can be pressed with or without the seed coats removed. Most sesame oil extraction is accomplished by relatively unsophisticated cold pressing methods. The yellow oil pressed from the seeds is simply filtered before local consumption. In large-scale operations, a combination of pressing and solvent extraction is employed, and the oil is generally refined. Sesame is considered a semidrying oil because of its high iodine number, but it is one of the oils most resistant to oxidation because it contains the

FIGURE 10-16
Flowers of the sesame plant, usually grown for seed production, are sometimes cultivated in this country as ornamentals. A cross section of a flower (above, left), a sesame branch (middle) and a fruit (right) are shown.

powerful natural antioxidants sesamolin and sesamium. The cake remaining after pressing or extraction is an excellent livestock food and, in times of famine, is eaten by people as well.

Sesame oil is most frequently used as an edible oil, and most of it is consumed at, or near, areas of production in Africa, the Middle East, India, and China. The oil has a characteristic flavor that is preferred by people accustomed to it. The seeds themselves can be eaten or crushed and sweetened to make the Turkish candy known as halva.

FIGURE 10-17
Wild sesame plants (right) have dehiscent capsules that throw out their seeds when mature. Human selection led to the domesticated sesame plant with indehiscent capsules (left) that hold their seeds. (Courtesy of the U.S.D.A.)

Cottonseed Oil

It is doubtful if cottonseed oil would have ever become a major product if cotton (*Gossypium* spp., Malvaceae, Chapter 16) were not so important a crop. The ginning of cotton produces tons of seeds every year that have only recently been commercially exploited on a large scale. The ancient Hindus extracted oil from Asian cottonseeds by pounding them and boiling the pounded meal. The oil was subsequently skimmed from the surface of the water and used in medicine and as lamp oil. Yet, until almost 1900, cottonseed oil was considered essentially inedible because it contained, among other things, gossypol, a bitter

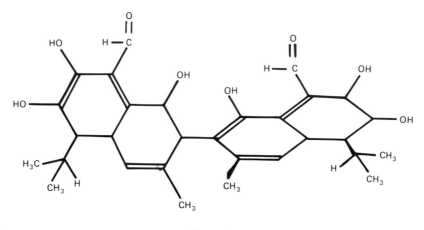

Gossypol

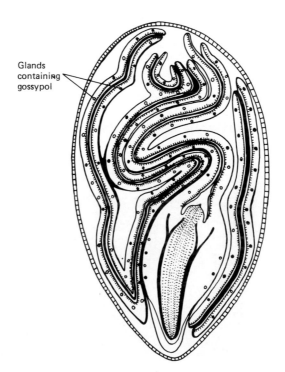

FIGURE 10-18
A cross section of a cottonseed reveals the gossypol glands dispersed throughout the embryo.

pigment produced in glands located throughout the cotyledons (Fig. 10-18). Consequently, most cottonseed was historically used for fertilizer.

In 1899, David Wesson began using caustic soda and fuller's earth to purify seed oils. By 1900, the young chemist had perfected the process by using steam with the decoloring agent (fuller's earth). In this year, the first edible cottonseed oil was marketed under the name of Wesson oil. The procedures developed for this oil essentially started the modern vegetable oil industry that today includes cottonseed oil, soybean oil, rapeseed oil, and palm oil.

Cottonseed oil experimentation lead to another landmark discovery. In 1911, Proctor and Gamble adopted and improved upon a British technique for hydrogenation that led to the first American vegetable shortening, Crisco, which was made from cottonseed oil.

Cottonseeds present a unique problem in the oil extraction process because bits of fiber called *linters* adhere to the seed coats (Fig. 10-19). In modern factories, the linters are removed by machines with fine rasping teeth. The removed linters are used in paper making and for the production of semisynthetic fibers such as rayon and acyl celluloses (see Chapter 16). The delinted hulls are used for cattle food or for mulch. Modern refining includes filtering, neutralizing, washing, bleaching, winterizing, deodorizing, and refiltering. This processing changes the dark-colored, semitoxic oil into a colorless, tasteless, and stable product. The refined oil can be used in liquid form or hydrogenated to produce various types of shortenings and margarines. The cake remaining after pressing is often used as cattle food. While cottonseed residues seem to have little or no adverse effect on cattle, they can cause death in rabbits, rats, guinea pigs, swine, and, presumably, humans. The toxic agent has been identified as gossypol, a substance that has been tested in China as a potential birth control agent. The adverse effects of the chemical, however, have thus far outweighed its potential medicinal value.

Rapeseed Oil

Rape (or colza) oil is relatively rare in the United States but is an important product in Europe and Canada. The oil is extracted from the seeds of *Brassica napus*, rape, or *B. campestris* the turnip (both Brassicaceae), (see Chapter 8). Historically, the oil has been used both for edible purposes and as a lubricant. In recent years, its use as a food item has been declining because of its acrid taste and the suspicion that it might be harmful if eaten in large quantities. Both the taste and the presumed toxic quality are due to erucic acid, one of the unusual fatty acids that occur in seed oil triglycerides. Even after deodorizing, the oil often reacquires an unpleasant taste. Extensive breeding programs have been carried out and have developed strains with a low erucic acid content and properties similar to those of soybean oil. However, a high level of crucic acid is the feature of rapeseed oil that makes it a very good lubricant. While other oils can replace it as an edible oil, none can match it as an industrial oil. In addition to being used for lubrication, the oil is used to quench steel plates after they have been forged, an indication of its ability to withstand high temperatures.

The procedures for extracting the oil are similar to those used for sesame, safflower, and sunflower oil. The cake can be used as livestock food only in limited quantities.

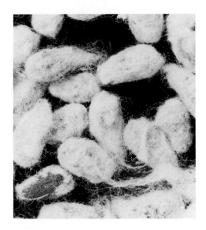

FIGURE 10-19
After the long fibers have been removed, upland cottonseeds are still covered with fine hairs that must be removed before processing.

NONDRYING OILS

Peanut Oil

In Chapter 5, we discussed the origin, dispersal, and rise of the peanut (*Arachis hypogaea*, Fabaceae) to its modern position as a major world crop. Less well known is the beginning of its use as an oilseed crop. Peanuts were not used for their oil until about 125 years ago when a French firm began to crush peanuts imported from Africa in a factory in Marseilles. After this humble beginning, oil extraction spread to other areas of the world. In the United States today, one-fourth of the crop goes into oil production and the majority of the crop worldwide is used for oil.

Before pressing, shelled peanuts must be carefully dried. The thin seed coats (skins) are removed and the oil expressed by crushing and solvent extraction. Although the oil will solidify if kept in a refrigerator, it is considered to be a premium cooking oil because it does not smoke when heated to high temperatures and it imparts a pleasant flavor to food. The cake is a valuable livestock feed as would be predicted by the high protein content of the seeds (Table 10-2).

Olive Oil

As we saw earlier, olives (*Olea europea*, Oleaceae) have sustained cultures nutritionally and spiritually since ancient times. Yet, despite the Greek legend, the first records of olive cultivation are from Crete where there is archaeological evidence of olive culture dating from 3500 B.C. (Fig. 10-20). By early Egyptian times (ca. 3100 B.C.), olive oil was used as a cleanser (see Fig. 10-1), to anoint bodies, as a lamp oil, medicine, foodstuff, and religious accoutrement. It has been suggested that the Egyptians slid the enormous blocks used in the construction of the pyramids over one another by placing olives or olive oil between them.

Olive oil is not one of the major commercial oils today, but it is one of the few that is rarely refined to tastelessness and thus it remains an important minor crop and a delight for gourmets. Its high level of saturation also ensures a long shelf life and retention of its subtle flavor. Olive oil is also one of the few oils that is obtained from a fruit pulp rather than from seeds. The seeds yield little or no additional oil. Because the pulp rather than the seed is used, the procedures for extraction differ from that of most oils. The fruits are macerated, the seeds removed, and the pulp pressed. The first press, a cold press, yields olive oil labeled "virgin" olive oil. Subsequent pressings, usually with heat, yield oils of lower grades.

Castor Oil

Although it is rarely used medicinally today, castor oil (*Ricinus communis*, Euphorbiaceae) was forced down the throats of many children in the past by their well-intentioned mothers. The laxative action for which the oil was administered is caused by ricinoleic acid, the predominant fatty acid in castor

FIGURE 10-20
This drawing copied from a classical Greek vase shows how olives were picked in ancient Greece.

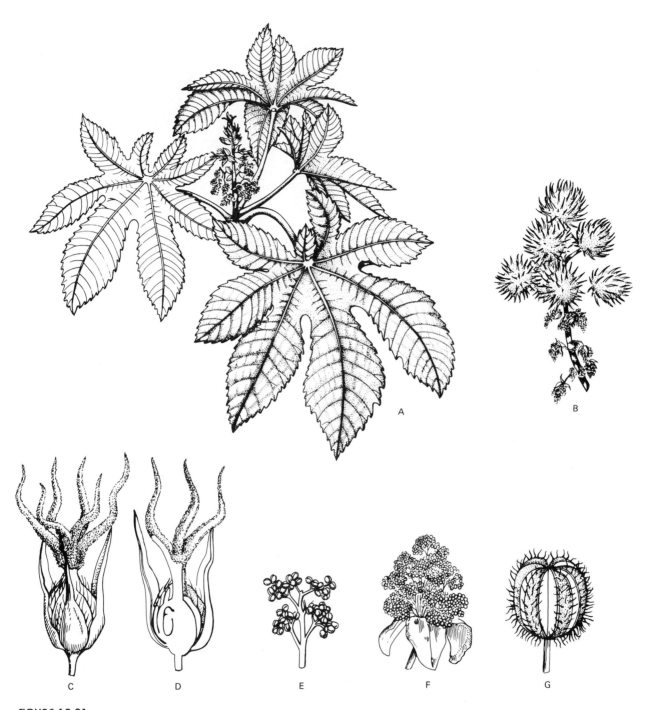

FIGURE 10-21
Because of its attractive foliage, (A) the castor bean plant is sometimes grown as an ornamental. The female flowers (C and D, in section) of the monoecious plants occur at the tops of the inflorescences (B) with the male flowers (E and F) below them. The mature fruits (G) are spiny and three-parted with a single seed in each locule.
A—B from nature, C—G after Baillon.

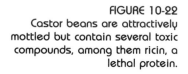

FIGURE 10-22
Castor beans are attractively mottled but contain several toxic compounds, among them ricin, a lethal protein.

oil. The seeds themselves, the hulls, or even the unrefined oil taken in large quantities are toxic to humans. Three compounds are responsible for the poisonous effects: ricine, a mildly toxic alkaloid; ricin, a highly toxic protein; and a protein-polysaccharide mixture called CB-A, which causes violent allergenic reactions in people sensitive to it.

As yet, there is no agreement among botanists about the original area of distribution of the species. By the time it was domesticated, it appears to have already spread across Africa and southern Asia. Investigations of the taxonomy of the genus suggest that the species arose in tropical Africa, but the oldest records of its use are from Egypt where seeds in tombs have been dated to be almost 6,000 years old. Ancient uses of the oil seem to have been as a lamp oil, in medicines, and for religious ceremonies. Brazil is now the world's largest producer, processing almost three times as much as India (Table 10-3).

Castor plants (Fig. 10-21) are monoecious perennials that can reach 13 m height in tropical areas. In temperate areas, the plants are grown as ornamental annuals. Both kinds of flowers usually occur on the same inflorescence with the males at the bottom and the females at the top (see Fig. 10-21). The fruits are three-lobed, usually spiny capsules that split open at maturity to release the three seeds (Fig. 10-22) which are dried, pressed and/or extracted. If the oil is to be used for medicinal purposes, the seeds are pressed at temperatures lower than 50°C so that ricin will not be removed with the oil. The oil is rarely refined as such, but it is often bleached. Because of its toxicity and high level of saturation, the oil can be stored for long periods with no change.

Today castor oil primarily goes into soaps, paints, and Turkey red oil, a sulfated form of the oil that is used as a wetting agent to promote color adherence during dyeing operations. Hydrogenated castor oil is used as a lubricant for airplanes and rocket engines. The cake left after extraction contains enough toxic compounds to render it unsuitable as a livestock food, so it is consequently used as a fertilizer.

VEGETABLE FATS

Palm and Palm Kernel Oil

Oil palms (*Elaeis guineensis*, Arecaceae) differ from all of the other important oil crops in that distinct oils are obtained from the fruit pulp and the seed. The fruits of the palm (Fig. 10-23) are 2.5 cm (1 in) long with an oil-rich mesocarp surrounding a hard endocarp. In commerce, the seed and fruit pulp are extracted separately and the oils kept separate because the two oils differ from one another in chemical composition (Table 10-2).

As is common with cultivated species, there is uncertainty about the original home of the oil palm. Some authors have argued for a New World origin since palm, or dende, oil is such an integral part of the culture of coastal Brazil and because early explorers reported finding the palm in South America. In addition, almost all of the species to which the oil palm is related are New World natives. However, only in Africa is there a variety of indigenous names for the palm and more recent studies can find no records of the species in the American tropics before a possible introduction by humans. Finally, a species of vulture

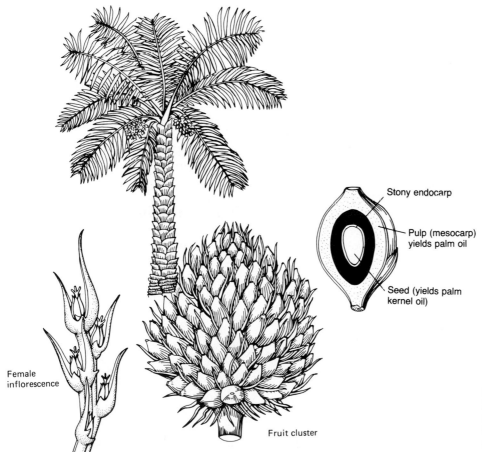

Female inflorescence

Fruit cluster

Stony endocarp

Pulp (mesocarp) yields palm oil

Seed (yields palm kernel oil)

FIGURE 10-23
Palm oil is extracted from the fleshy mesocarps of the palm fruits. Another oil, palm kernel oil, is extracted from the seeds of the same fruits.

FIGURE 10-24
Palm fruits being sold in an African market. (Photo courtesy of B. L. Turner.)

native to Africa and called, appropriately, the palm oil vulture (*Gyphohierax angolensis*), is adapted for a diet that consists primarily of the fruits of the oil palm. Current opinion thus favors the hypothesis that the species originally came from the rainforest of west Africa and was brought to the New World by slaves from this region.

In many areas where they are grown, the palms are only semicultivated. The fruits are picked when ripe and sold in local markets (Fig. 10-24). Customers press out their own oil by fermenting the fruits for 2 to 4 days, boiling them, and pounding the mass. The oil is recovered by stirring the pulp in water and skimming off the oil as it floats to the top. The recovered oil, which has a characteristic orange-red color, caused by the presence of carotenes, is heated to drive off any remaining traces of water. The famous cooking of Bahia, Brazil, liberally uses unprocessed palm oil.

In commercial operations, palm fruits are processed as soon as possible after picking. The fruit pulp is mashed from the seeds and heated into an homogeneous mass. The oil is extracted rapidly once the fruits are picked because the fruits contain natural lipases that begin to break down the oil after they are removed from the trees. Therefore, once picked, the fruits are sterilized with steam, pressed with hydraulic presses, and the oil filtered and stored for possible future processing. The endocarps containing the seeds are dried, graded, and cracked. The kernel, or seed, is then dried and sacked for large-scale extraction operations.

Palm oils have primarily been used for soap and candles, but their importance in margarine and solid shortenings is rising. Palm kernel oil is so much like coconut oil in composition (Table 10-2) that the two are often substituted for one another.

Coconut Oil

Coconuts seem to have limitless uses, among them the production of an oil, or perhaps more correctly, a fat, since the triglyceride is solid at room temperatures (below 24°C). We have little documentation of the early uses of coconut oil,

FIGURE 10-25
Coconut cultivation in Indonesia.
(United Nations photo by S.
Bunnag.)

and we lack fossil, or even historical, evidence with which to assess the origin and spread of the coconut (*Cocos nucifera*, Arecaceae, Fig. 10-25). However, since scrapers used for processing coconut endosperms were common tools in Polynesia when the Europeans arrived, coconuts were presumably used there as oil sources. Oil constitutes such a high percentage of the coconut endosperm that it can be extracted by simply pressing the dried meat.

Until the nineteenth century, almost all coconut oil produced was consumed locally. In 1841, a patent was issued for a process that used coconut oil in soap manufacture. Since that time, coconut oil production has risen dramatically as new uses were found. Soon after the discovery of the process of hydrogenation, coconut oil found its way into margarine. The oil is naturally so saturated that little additional hydrogenation is needed. The flavor of coconut oil also mixes well with a variety of flavors, both sweet and salty, with the result that it is used in many processed foods. Macadamia nuts are roasted in coconut oil to give them an additional, subtle, flavor. In the Asian tropics where most coconuts are grown, the cake is used as cattle and poultry food. Compared to other oils it is little used in paint and varnish manufacture, but it is a common constituent of detergents and resin products.

In recent years, coconut oil has become prominent in cosmetics and nondairy, "dairy" products. Coconut oils are often claimed to enhance the effectiveness of shampoos, hand lotions, and suntan creams. Fast-food restaurants offer their clients nondairy creamers and artificial whipped toppings made from coconut oil. Preference for the nondairy substitutes stems from their low price and resistance to spoilage, not because they are less saturated than natural cream (see Table 10-2).

WAXES

Jojoba (pronounced ho-ho-ba) (*Simmondsia chinensis*, Simmondsiaceae) has recently received much publicity as an important future oilseed crop. The excitement about jojoba stems from several of its properties. First, the shrubby

FIGURE 10-26
The belief that jojoba oil could duplicate the lubricating qualities of sperm whale oil seemed to offer a way to save the mammal from extinction.

plants grow naturally in the Sonoran and Great Basin deserts of North America. Consequently, it is a crop that can be grown on land unsuitable for other purposes. Second, jojoba oil is chemically different from any other seed oil because it is composed of esters (linkages of a straight-chain alcohol and a fatty acid, see Fig. 1-5) rather than triglycerides. The chemistry of the oil is like that of sperm whale oil, and the hope was that this renewable source of oil would replace the almost depleted supply of sperm whale oil (Fig. 10-26) as a lubricant for fine machinery. It is now clear that jojoba oil cannot tolerate high enough temperatures to be a substitute for sperm whale oil, but the oil does seem to be particularly effective in penetrating the outer layers of human skin making it useful in medicines and cosmetics. Jojoba oil today commands high prices because of its widespread use in cosmetics.

The chemical nature of jojoba oil emphasizes the similarity between oils and waxes. "Seed oils" are combinations of an alcohol, glycerol, and fatty acids. Waxes are made from simple alcohols and fatty acids. All flowering plants manufacture waxes which cover their aerial surfaces and prevent desiccation or injury. Yet few plants have pronounced accumulations, or readily obtainable amounts, of wax. Those which produce substantial quantities have been exploited

FIGURE 10-27
Candellila stems are stacked ready to be boiled for wax extraction. (Photo courtesy of J. L. Neff.)

as sources of wax, but only three are used to any extent in the United States today.

The first, carnauba wax, is obtained from the leaf surfaces of *Copernicia cerifera* (Arecaceae), a native of northeastern Brazil. The wax is obtained by collecting new or immature leaves of wild palms and allowing them to dry. Once dry, the leaves are beaten to dislodge the wax. The wax is shipped in particulate form and processed in importing countries. Carnauba wax is used primarily in car waxes and shoe polishes. Although extraction methods are laborious and crude, carnauba wax is still collected because it is harder than beeswax or many synthetic waxes.

Another wax, often substituted for carnauba wax, is candellila wax, obtained from the succulent stems of *Euphorbia antisyphilitica* (Euphorbiaceae). The stems are gathered (Fig. 10-27) and boiled in water. The wax which melts and floats to the surface is scooped off and allowed to harden. Candellila plants are native to the Chihuahuan Desert along the United States–Mexican border. Plants collected for wax extraction are all wild. Entire plants are uprooted by bands of nomadic candellila workers who leave behind a trail of devastation. In the United States, collection of the plants is forbidden because the species is in possible danger of extinction.

The last natural plant wax that is still available in America is bayberry. Early settlers in the northeast collected the fruits of the bayberry (*Myrica pensylvanica*, Myricaceae) and boiled the berries to melt off the wax. The wax, scooped from the surfaces of the boiling cauldrons, was cooled and later remelted for use in candles. Bayberry is still used for novelty candles because of the pleasant fragrance produced when they burn.

ADDITIONAL READING

Bailey, A. E. (ed.). 1948. *Cottonseed and Cottonseed Products*. New York, Interscience.
Beard, B. H. 1981. The sunflower crop. *Scientific American* 244(5):150–161.
Blackmon, G. H. 1943. The tung-oil industry. *Botanical Review* 9:1–40.

Corner, E. J. H. 1966. *The Natural History of Palms*. Berkeley, University of California Press. A most readable account of this economically very important group of plants.

Heiser, C. B., Jr. 1976. *The Sunflower*. Norman, University of Oklahoma Press. Heiser is known for his delightful extended essays on important groups of crop plants. In this book, he explores the evolution and botany of North American seed oils.

Inglett, G. E. (ed.). 1970. *Corn: Culture, Processing, Products*. Westport, Connecticut, AVI Publishing Company.

Norman, A. G. (ed.). 1978. *Soybean Physiology, Agronomy, and Utilization*. New York, Academic.

Purseglove, J. W. 1975. *Elaeis*, pp. 479–510, and *Cocos*, pp. 440–470, in J. W. Purseglove. *Tropical Crops. Monocotyledons*. New York, Wiley.

United Nations. 1977. *Guidelines for the Establishment and Operation of Vegetable Oil Factories*. U.N. Industrial Development Organization, New York. A guide written for individuals interested in starting or upgrading an oilseed processing plant. The style and diagrams are easy to follow, and the major oil seed crops are discussed.

Weiss, E. A. 1971. *Castor, Sesame, and Safflower*. Boston, Barnes and Noble. An extremely thorough discussion of the history, biology, and utilization of these three oilseed crops.

Chapter 11

Hydrogels, Elastic Latexes, and Resins

*T*he various plant products we discuss in this chapter (Table 11-1) do not come from the same plant organs, are not produced in similar ways, and do not even contain the same kinds of chemicals. The one quality that they have in common is that they are all sticky substances exuded or extracted from plants. Because of this single feature, they are often confused with one another. For example, many people chew gum, but chewing gum is not a gum, it is a latex. Likewise, pine pitch or pine gum is actually a resin. By discussing true gums with latexes and resins, we can highlight the differences between them. While all of these kinds of plant products have been used for thousands of years, natural resins and latexes are used today less frequently than in former times because similar or superior synthetic products are now

TABLE 11-1

Plants Discussed in Chapter 11

COMMON NAME	SCIENTIFIC NAME	FAMILY	CHROMOSOME NUMBER
Balata rubber	*Manilkara bidentata*	Sapotaceae	Unreported
Castilla	*Castilla elastica*	Moraceae	$2n = 28$
Ceara rubber	*Manihot glaziovii*	Euphorbiaceae	$2n = 36$
Chicle	*Manilkara zapota*	Sapotaceae	$2n = 26$
Copal	*Copaifera* spp.	Fabaceae	$2n = 24$
	Hymenaea spp.	Fabaceae	$2n = 24$
	Agathis spp.	Araucariaceae	$2n = 26$
Dammar	*Shorea* spp.	Dipterocarpaceae	Variable
	Bursera spp.	Burseraceae	$2n = 22, 24$
Frankincense	*Boswellia carteri*	Burseraceae	Unreported
Guar	*Cyamopsis tetragonolobus*	Fabaceae	$2n = 14$, diploid
Guayule	*Parthenium argentatum*	Asteraceae	$2n = 36$–144, diploid to octoploid
Gum arabic	*Acacia senegal*	Fabaceae	$2n = 26$
Gum ghatti	*Anogeissus latifolia*	Combretaceae	$2n = 24$
Gutta-percha	*Palaquium gutta*	Sapotaceae	$2n = 24$
Hevea or para rubber	*Hevea brasiliensis*	Euphorbiaceae	$2n = 36$
Indian rubber	*Ficus elastica*	Moraceae	$2n = 26, 39$
Karaya gum	*Sterculia urens*	Sterculiaceae	$2n = 40$
Kauri resin	*Agathis australis*	Araucariaceae	$2n = 26$
Lacquer	*Rhus verniciflua*	Anacardiaceae	$2n = 22$
Landolphia	*Landolphia* spp.	Apocynaceae	$2n = 22$
Larch	*Larix occidentalis*	Pinaceae	Unreported
Locust gum	*Ceratonia siliqua*	Fabaceae	$2n = 24$, diploid
Mastic	*Pistacia lentiscus*	Anacardiaceae	$2n = 24, 30$
Myrrh	*Commiphora myrrha*	Burseraceae	Unreported
Turpentine	*Pinus palustris*	Pinaceae	$2n = 24$, diploid
	P. pinaster		$2n = 24$, diploid
	P. sylvestris		$2n = 24$, diploid
	P. taeda		$2n = 24$, diploid
Tragacanth	*Astragalus gummifer*	Fabaceae	$2n = 16$

Note: For many of the species listed, the base chromosome number, and therefore the ploidy level, is unknown.

available. The use of hydrogels, in contrast, has exploded in the last 50 years, and the demand has only been partially filled by synthetically produced compounds.

HYDROGELS

Hydrogels, defined as water-modifying substances, include all the products that we collect, extract, or synthesize for the purpose of altering the behavior of water. A familiar illustration of how a hydrogel works is provided by the procedures used in making gravy. The cook starts with a paste made from flour and fat. She or he then adds a thin water-based stock to the hot paste while stirring continuously. After a few minutes, the liquid becomes thick and the stock turns into "gravy." The thickening occurred because the starch molecules of the flour associated with the water molecules of the stock. The water molecules, once they became associated with starch molecules, could no longer move freely. Their sluggishness was manifested as a thickening of the solution (Fig. 11-1). Starch, pectin, and the animal-derived counterpart, gelatin, are the

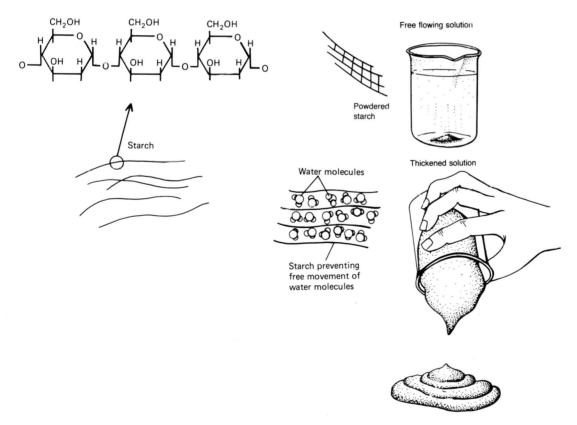

FIGURE 11-1
When hydrogel molecules associate with water molecules preventing their moving freely, the mixture becomes thick.

most common household hydrogels. Commercial operations use a much wider array of water-modifying agents

Plant-derived hydrogels fall into three main classes: gums, pectins, and starches. True gums are plant exudates produced only at specific times in response to injury or wounding. It is thought that they result from the breakdown of compounds in injured cells and serve to seal wounds and help prevent invasion by fungi and bacteria. Natural gums, therefore, are not always present in the plants that can produce them and there are no special cells, tissues, or organs for their production. While natural gums are all plant exudates, many of the modern commercially important "gums" are either extracted from plant tissues (primarily seeds and wood) or synthetically produced using cellulose as a base. Such substances are not breakdown products of cellular components following injury, but they are like true gums in that they have similar chemical and biological effects. Two additional important gums, agar and carrageenan, derived from algae, are not included in our discussion.

Chemically, all gums are polysaccharides of acid salts of sugars other than glucose (primarily galactose, fucose, xylose, arabinose, rhamnose, and galacturonic acid; Fig. 11-2). The calcium, magnesium, and potassium ions of the salts cause the gum molecules to associate with water. The polymers themselves

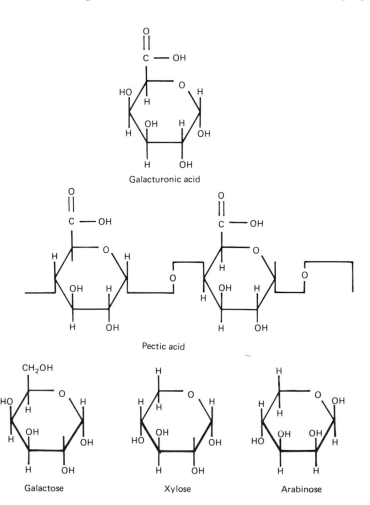

FIGURE 11-2
True gums are polysaccharides generally composed of acids salts of sugars such as galactose, xylose, and arabinose. Pectic acid is, in part, a polymer of galacturonic acid molecules.

Galacturonic acid

Pectic acid

Galactose   Xylose   Arabinose

can be linear, branched, or cross-linked. In this chapter, we will employ the term gum for an exudate, extractive, or synthetic polymer that has this sort of chemical structure.

All gums are only partially digested by humans and, with a few exceptions, have no adverse effects when ingested. From the point of view of human metabolism, gums can thus be considered inert substances. For this reason, they are particularly suited for use in foods, diet products, and medicines. The other major uses of gums are in the paper, textile, and petroleum industries.

In the food industry, gums are used to texturize products, provide "body," improve the feel of foods in the mouth, stabilize emulsions, retain moisture, thicken liquids, and suspend particles. Most dairy products other than milk, cream, and butter use gums to disperse fat and protein molecules evenly in a water base. In frozen products, gums help to prevent the formation of ice crystals (Fig. 11-3) and in sauces and syrups, they produce a thick, rich consistency. Gums also increase a product's shelf life because they prevent solid particles from settling out of suspensions. Powdered "crystals" that are used to make instant beverages are often sprayed with a thin film of gum to prevent water absorption before use and to help disperse the particles once they are mixed with water or milk. The minute amounts of gum used also thicken the beverage slightly, causing it to seem smoother and richer. This smoothness improves "mouth feel" and has led to the widespread use of gums in ice cream, whipped toppings, commercial frostings, and cream fillings. Finally, gums are often added to sandwich spreads and luncheon meats to bind together the processed protein.

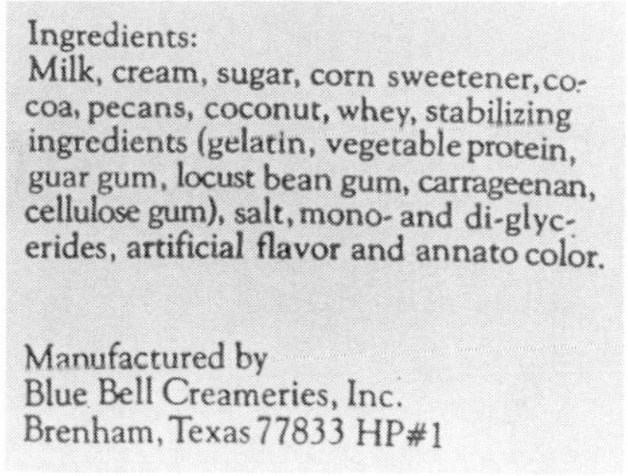

In medicine, gums hold tablets together and keep the particulate matter in oral antibiotic solutions (such as penicillin) dispersed in liquid. Some gums can act as laxatives and have been used alone or in combination with other products. Gums also make your toothpaste a "paste" (or a gel) and allow the fats and oils in hand and body lotions to maintain their smooth, creamy consistency.

The paper and textile industries are the primary consumers of many gums because of their use as sizing agents. *Sizings,* or sizes, are substances spread on cloth or paper to fill in the pores and irregularities. In the textile industry (Chapter 16), sizing stiffens and strengthens threads during weaving and, occasionally, during garment manufacture. After weaving or stitching, the sizing is usually washed out of the fabric and flushed into a stream or river flowing near the mill.

Starch has traditionally been the most popular sizing substance because it is inexpensive and readily available. However, it has a few serious drawbacks: it must be used in large quantities, and it is highly digestible by many organisms. Disposal of starch sizing wastes into waterways has upset the ecology of many aquatic systems. Starch does not kill organisms and is not toxic to humans. Instead, it leads to *eutrophication,* or nutrient enrichment, of the water. This unnatural abundance of nutrients causes population increases of algae. When the algae die, they provide abundant food for bacterial decomposition. As populations of the bacteria balloon, they cause a depletion of the oxygen in the streams and, ultimately, the death of most aquatic plants and animals.

Gums have an advantage over starches as sizes because only small quantities are required. As previously mentioned, they are also less readily metabolized by organisms than starch. Consequently, gums have relatively minor effects when, or if, they are washed into waterways.

Ingredients:
Milk, cream, sugar, corn sweetener, cocoa, pecans, coconut, whey, stabilizing ingredients (gelatin, vegetable protein, guar gum, locust bean gum, carrageenan, cellulose gum), salt, mono- and di-glycerides, artificial flavor and annato color.

Manufactured by
Blue Bell Creameries, Inc.
Brenham, Texas 77833 HP#1

FIGURE 11-3
The gums in this "German chocolate cake" ice cream prevent the formation of ice crystals and make the ice cream feel smooth in the mouth.

In the paper industry, gums are spread on paper products to produce a smooth texture and a surface that gives a crisp printed image. Gums are also applied to the surfaces of butcher and freezer papers to give them water repellency.

The petroleum industry was once a large gum consumer because gums have characteristics that facilitate two of the complex operations involved in modern drilling procedures. Soil particles through which the drill is moving are potentially very abrasive, but act like lubricants if gums are added to the water-soil mixture surrounding the bit. As in the case of food products, gums keep the particles well dispersed in the water and give the mixture a smooth consistency. The other major application in the petroleum industry involves secondary recovery operations. Gums are added to the water or brine that is pumped into the ground to maintain enough pressure to keep the gas and oil rising to the surface. If pure water were pumped in, it would quickly penetrate through the rock and flood the well. Gums thicken the water and slow its movement. While plant gums were originally used, almost all gums used in drilling operations today are synthetically produced.

Pectin is the name given to a special group of plant polysaccharides which form gels under particular conditions. The saccharides involved are arabans, galactans, and methyl esters of galacturonans made up primarily of D-galacturonic acid. The number of methoxyl groups is important in determining the properties, and thus the uses, of different pectins. Pectins are found between cells and as components of the primary cell wall. Pectin found in the cell walls is associated with calcium and forms molecules of extremely high molecular weight. This form of pectin is highly insoluble. During fruit senescence, or in the presence of acid or alkali, the large molecules break down into smaller, water-soluble units. Pectin also occurs in many species in a layer of the epidermal

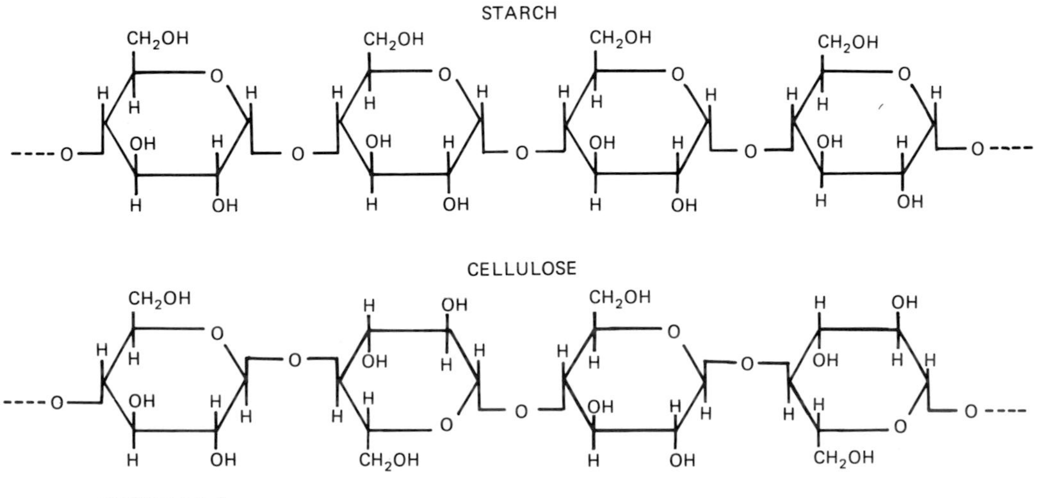

FIGURE 11-4
Cellulose is composed of glucose molecules joined by beta bonds. The exposed OH groups on either side of the cellulose chain facilitate cross-links between cellulose chains. Starch is also composed of glucose molecules, but the molecules are linked with alpha bonds. The exposed OH groups all lie on the same side of the starch polymer, which causes the molecules to coil with the OH groups to the inside.

cells between the cuticle and the inner portion of the cell wall. Some species have a particularly pronounced layer in the epidermal cells of their fruits. Peels of such fruits therefore constitute a convenient source of pectin. Humans are incapable of digesting pectins, but the bacteria which live in human intestines are capable of doing so. Although these bacteria catabolize pectins and use the products as carbohydrate sources, their human hosts are still unable to receive any nutritional value from them.

Starch constitutes the last category of commercially important plant-derived hydrogels. In contrast to gums and pectins, starches are linear or branched polymers of glucose (Fig. 11-11). Starch is easily broken down by enzymes known as amylases because the molecules of glucose in the starch polymer are linked in such a way that the enzyme can readily hydrolyze the bonds. Glucose molecules in starch are linked by alpha bonds, in cellulose they are linked by beta bonds (Fig. 11-4). Beta bonds, unlike alpha bonds, are exceedingly difficult to break.

Starch is found in almost all plant parts, but it is commercially extracted only from seeds and roots that store large quantities of it. While some commercially processed starch is used in the food industry, most is used in the manufacture of paper and cardboard. Starch is spread on the surfaces of these products once they are made. Large rollers then squeeze off all but a very thin layer, giving the final products glossy, smooth surfaces.

Important Sources of Natural Gums

The classic, and still most widely employed, natural gum is gum arabic. This gum, exuded from wounded trees of *Acacia senegal* (Fabaceae, Fig. 11-5) has been used since at least Egyptian times. The small spindly trees of this acacia are native to western Africa, primarily the Sudan, but because the gum was sent to Arabian ports for shipment to Europe, it acquired the English name of gum arabic. Ninety percent of the world's supply is still obtained from wild trees that are purposefully slashed or punctured in order to induce a wound reaction. Natives then collect the dried globs or beads of gum that exude from the wounds (Fig. 11-6). Most of the collecting takes place between October and June during the dry season. Particles of dry gum are carried to a central locality, graded, and further dried before shipping to processing plants around the world (Fig. 11-7). Final processing consists of regrading the particles and cleaning and blending them, if necessary, before they are milled or dissolved and spray-dried into powder.

Gum arabic is almost ubiquitous in our daily lives. The "lace curtain" left on the sides of a freshly drained beer glass is created by the gums added to the beer to stabilize the foam. Chances are, every time you lick a postage stamp, the water-soluble glue is gum arabic. Gum arabic is the gum most often used to coat flavor particles, and small quantities of it are added to candies and confections with high sugar contents in order to prevent sugar crystallization. This gum also emulsifies fats in many foods, hand lotions, and liquid soaps. The Egyptians mixed gum arabic into their paints to suspend the pigments. The gum is still a component of fine water colors. Gum arabic is also the most important gum used in the manufacture of ink, but it is less used in paper (except tissue paper) and fabric manufacture than other gums.

FIGURE 11-5
A branch of *Acacia senegal*, the
small tree that provides us with
gum arabic. (After Baillon.)

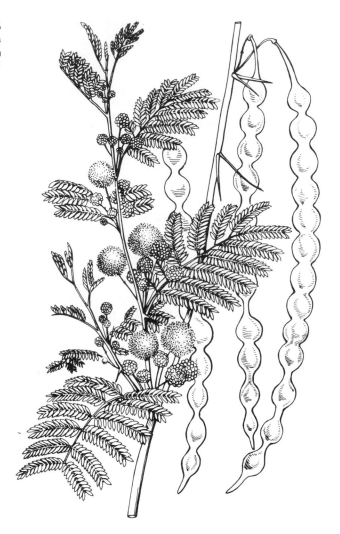

Gum tragacanth is another silent partner in our lives. Tragacanth is one of the substances first recorded as an emulsifier. Its use predates the Christian Era. The most important characters of gum tragacanth are its ability to avoid degradation in acid solutions (in contrast to most other gums), and its ability to emulsify oil without a surfactant (a wetting agent). It is also one of the few gums to which some people have an allergenic reaction.

The gum comes from any of several species of *Astragalus* (Fabaceae), but primarily *A. gummifer* (Fig. 11-8), native to the Near East and Asia Minor. To obtain the gum, shrubby bushes of wild *Astragalus* species are tapped by carefully making incisions in the top part of the root or on the bases of the larger branches. The name "tragacanth" comes from the appearance of the exuded gum, which tends to form ribbons similar in appearance to a goat horn (*tragos* meaning goat, and *akantha,* horn, in Greek). The gum is collected as partially dried flakes or ribbons, which are brought to a trade center and sold. During processing, the irregular pieces are cleaned and ground as needed. Although

FIGURE 11-6
The tapping and collecting of gum arabic. (Photo courtesy of The Gum Arabic Co., Ltd., Khartoum.)

FIGURE 11-7
Cleaning and grading gum in Port Sudan on the Red Sea. (United Nations photo issued by FAO.)

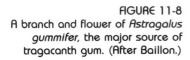
FIGURE 11-8
A branch and flower of *Astragalus gummifer*, the major source of tragacanth gum. (After Baillon.)

gum tragacanth stores well, it can be sprayed with ethylene oxide or a similar agent in order to deter microbial growth.

Gum tragacanth is primarily used in food preparations such as mayonnaise, sandwich spreads, and pre-prepared milkshakes. It is also used in toothpaste and hand lotions and medicinally as a binder for tablets or as a suspending agent for oral penicillin.

Gum karaya, or sterculia gum, is unique among the gums because it can form a strongly adhesive gel when mixed with only small quantities of water. Although it is the least soluble of the commercial gums, its adhesive properties combined with its resistance to bacterial and enzymatic breakdown have led to its use as a dental adhesive and as a binder for the fibers in bologna and other luncheon meats.

Gum karaya is collected from trees of *Sterculia urens* (Sterculiaceae), a widespread species native to the rocky hills and plateaus of India. The trees are tapped or blazed during the dry season and the exuded gum collected before the monsoons arrive. Some karaya gum is obtained from trees grown on commercial plantations which are tapped year-round. The collected tears of gum are sorted, cleaned, blended, and granulated or powdered.

Karaya gum was little used in the United States before World War II, but the war interrupted the American supply of gum tragacanth. Karaya gum constituted an available, inexpensive substitute. Granulated karaya gum sometimes serves as a bulk laxative. The powdered, dried gum is also used throughout the food industry, but primarily in salad dressings, ice creams, cheese spreads, and whipped toppings. Hair setting gels owe their texture to karaya gum, not gelatin. The gum often replaces gum arabic in the tissue paper industry and was formerly used in drilling operations for thickening mud or plugging wells.

The last of the important natural gums is gum ghatti, named because it was originally brought to ports in India from the interior over mountain passes, or *ghats*. The plants which yield this gum are trees of *Anogeissus latifolia* (Combretaceae), which grow natively in the dry deciduous forests of India and Sri Lanka. As in other gum collection operations, the trees are wounded and the exudate collected. The particles of gum are sun-dried, sorted by hand, and shipped, usually from Bombay. Inferior grades of the gum are powdered and used in India as a stabilizer in whitewash. Gum ghatti maintains a constant place in the spectrum of available gums because it has properties intermediate between those of arabic and karaya gum, but it is a better oil emulsifier than either since it has a high viscosity. Consequently, it is used as an emulsifier in liquid and paste waxes and for fat-soluble vitamins. Like many other gums, it has also been used in drilling operations.

While the true gums come from wounded woody tissue, similar substances can be obtained from the endosperm of some legume species. The two most important of these seed gums are locust and guar gums. Locust gum, or carob seed gum, was used by the Egyptians as an adhesive for mummy bindings, but the tree from which the gum-producing seeds are obtained (*Ceratonia siliqua*, Fabaceae, Fig. 11-9) is best known in the West as a chocolate substitute (see Chapter 7).

To obtain locust gum, the pods are shaken from the 3–7 m (6–23 ft) tall trees. The seeds are removed from the pods, and the seed coats plus the hard, yellow-green embryo are abraded from the translucent endosperm. The endosperm, which constitutes about one-third of the seed, is dried and ground into powder. Locust gum is used almost exclusively in the food industry, primarily in ice creams, salad dressings, and pie fillings (see Fig. 11-3).

Guar gum is extracted from the seeds of *Cyamopsis tetragonolobus* (Fabaceae) a species known only as a cultigen. Guar gum was apparently domesticated in India, although all of the wild species of the genus are native to Africa. Guar is an herbaceous perennial which can be used as a cattle feed. It was originally introduced to the United States in 1903 as a cover crop for grazing. During World War II, when locust gum was difficult to obtain, alternative sources of gum were sought. Domestic production of guar escalated, and by 1953, guar gum was replacing locust gum in many operations, particularly in the paper and textile industries.

Guar shrubs are 1 to 2 m (3 to 6 ft) tall with vertical stalks bearing clusters of pods. The way in which the pods are borne led to the common name of cluster bean in Texas and Oklahoma, where almost all of the guar in the United States is grown. Guar is the only commercially exploited source of gum that lends itself to mechanized agriculture. Its pods can be harvested with a grain harvester, a definite advantage over the laborious hand wounding and collection of true gums. Once collected, the seeds are separated from the pods and the seed coats removed. The endosperm is then dried and ground to a fine powder.

Guar gum is principally used in the paper industry. The petroleum industry used to consume a large amount. Guar gum was added to the water in drilling operations to fracture rocks and to brines used to maintain pressure in pumping operations. A recently publicized use of guar gum is its addition to water

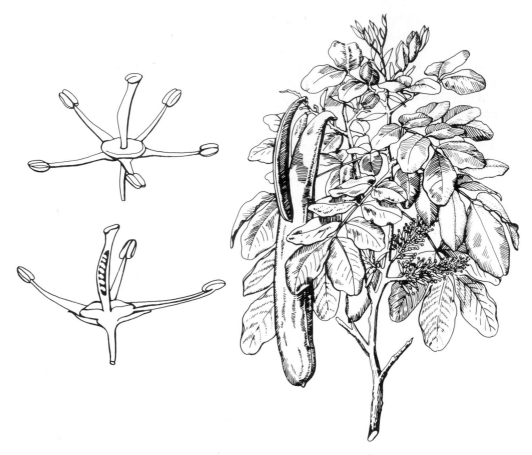

FIGURE 11-9
The seeds of the carob tree are the source of locust bean gum, while the powdered, sweet mesocarp of the fruit is used as a chocolate substitute. Shown here is a branch (right) and an individual flower (left) whole and in section. (After Baillon.)

pumped through fire hoses. Water with traces of guar flows faster through the hoses than regular water because the gum reduces friction between the water and the sides of the hose. Guar gum is still less extensively used than gum exudates, particularly in the food industry, but its use is steadily growing.

Larch gum is unique in that it is obtained neither from exudates nor from seed endosperm. Although it can be exuded, it is usually extracted today from wood chips of the American larch, *Larix occidentalis* (Pinaceae). It has been known for almost 100 years that the wood of this species housed an abundant supply of a gumlike substance, but only in recent years has it been commercially exploited as a source of gum. In its behavior, larch gum is similar to gum arabic and has replaced it in some lithographic operations. Although approved for use in foods, it has not been used to any appreciable extent in edible products.

A recent and promising development in the gum industry is the production of semisynthetic cellulose gum (carboxymethylcellulose). This gum is made by taking purified cellulose, a polymer of glucose (see Fig. 11-4), and allowing it to react in an alkaline medium with sodium monochloroacetate. The resultant gum is used today in more diverse ways than any other water-soluble polymer. The largest use for carboxymethylcellulose is in detergents. Remember the detergent ad claiming that its product does not leave the wash dingy like other brands? The extra whitening and brightening effect is due to the presence of cellulose gum. Ordinarily, in washing machines, part of the dirt that has been flushed from the clothes can be redeposited on them when they are spun dry. Cellulose gum helps prevent redeposition and thus reduces graying of the fabric.

Cellulose gum is also important in the paper industry and is the major gum replacing starch in the textile industry. Small quantities are indispensable in latex paints because they give the paint the proper viscosity for even flowing from brushes or rollers. Modern processed foods often use cellulose gum rather than, or in combination with, one of the other gums to texturize, stabilize, thicken, and improve their sensory qualities (see Fig. 11-3).

Pectins

Over 75 percent of the world's pectin goes into the manufacture of jams and jellies. While some fruits contain sufficient pectin to "gel" by simply cooking them in water, others do not. Consequently, commercially prepared pectins are added to produce a thick, sticky, or gelatinous consistency. For this reason a small amount of pectin is used in medicines such as Kaopectate.

Since there are many potential sources of pectins, economics has determined which have become commercially important. The most widely used sources are apple pomace (the residue left after pressing apples) and citrus peels (Fig. 11-10). Apple residues contain 10 to 15 percent pectin on a dry weight basis, and lemon, orange, lime, or grapefruit peels about 20 to 30 percent. Because these substances are by-products of the apple and citrus fruit juice industries, pectin extraction constitutes a profitable use of what would otherwise be waste.

Pectins are extracted from the pomace and peels by heating them in water between 60 and 95°C (140 and 203°F) at carefully controlled acidity levels

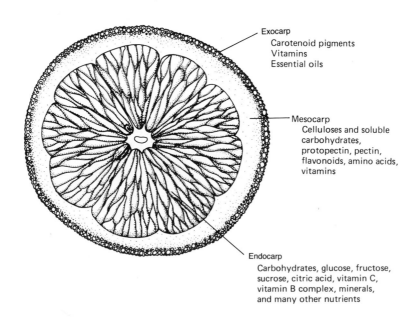

Exocarp
Carotenoid pigments
Vitamins
Essential oils

Mesocarp
Celluloses and soluble carbohydrates, protopectin, pectin, flavonoids, amino acids, vitamins

Endocarp
Carbohydrates, glucose, fructose, sucrose, citric acid, vitamin C, vitamin B complex, minerals, and many other nutrients

FIGURE 11-10
A cross section of a lemon shows the region of the peel richest in pectin. Because of their thick skins, relative to other citrus fruits, lemons and grapefruits are most often used as pectin sources.

(about pH 2.5). The high acidity breaks down the insoluble pectins and allows them to dissolve in the warm water. The pectin is separated from the water by centrifugation and/or filtration. If the peeling and mashings from the juice factories are not processed immediately, they are dried to avoid natural enzymatic degradation.

Starch

We have already discussed most of the major crop plants exploited for their large amounts of starch (Chapters 6 and 8). While our earlier emphasis was on the growing of these plants as carbohydrate food sources, we also use many of them as sources of starches used as hydrogels or as substrates for conversion into sugar used in fermentation (Chapter 15). For hydrogel use, starches are always extracted from the seeds, tubers, or roots in which they were stored. For conversion into sugar, starches are usually not purified.

Starch exists in plants in two forms: linear polymers called *amyloses*, and branched forms known as *amylopectins* (Fig. 11-11). Within plant cells, the

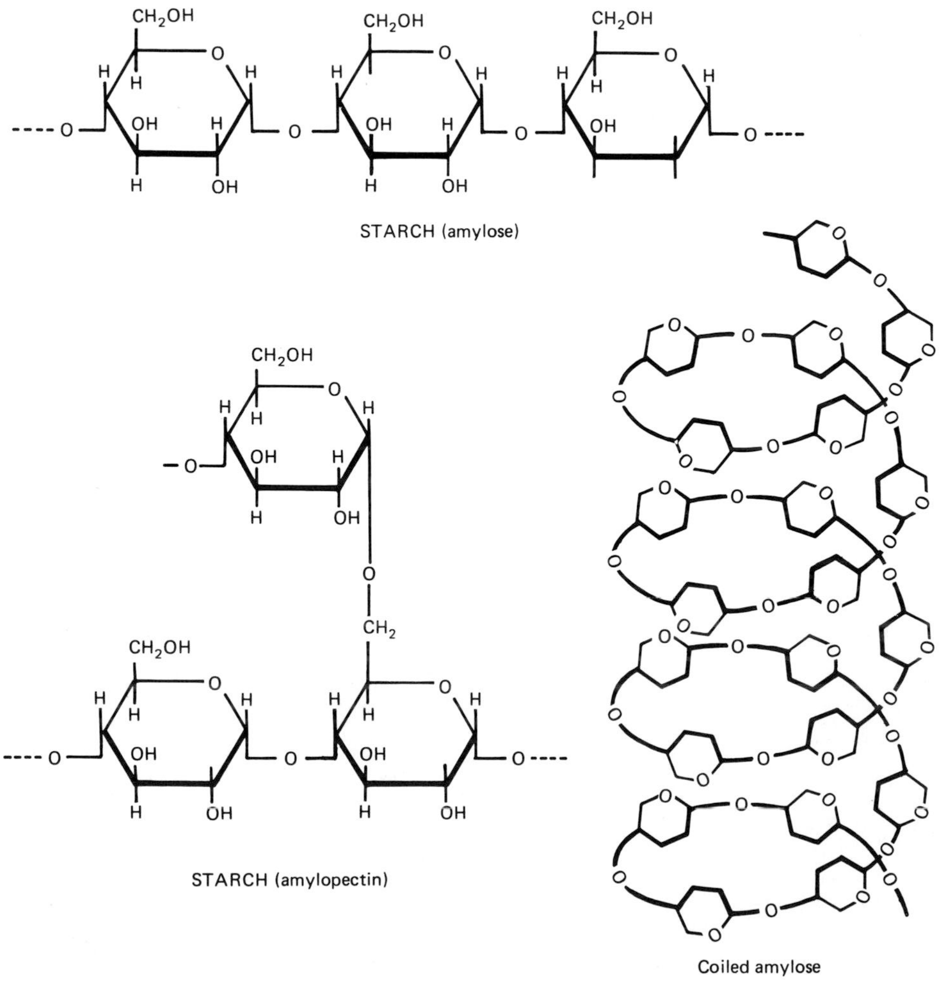

FIGURE 11-11
Starch molecules can be either linear or branched. Amylose contains more than 1,000 glucose molecules in unbranched chains which form coiled molecules. Amylopectin (right) is composed of 48 to 60 glucose molecules in branched chains that do not coil.

STARCH (amylose)

STARCH (amylopectin)

Coiled amylose

polymers are packed into insoluble granules that have to be disrupted before the starch becomes soluble. The starch granules of different plant genera have such characteristic shapes and sizes that a particular starch source such as wheat, corn, rice, or taro can be identified by simply looking at the starch granules (Fig. 11-12).

When starch granules are initially extracted from plant material, they are flushed from the ruptured cells with cool water. The water is then evaporated, leaving a solid residue that is easily dried and powdered. It is in this form that we buy cornstarch, potato starch, or arrowroot. When the starch is used, it is remixed with water, or a water-based solution (such as the meat stock for gravy), and heated. Within a temperature range between 55 and 88°C (131 to 176°F), the starch granules open and absorb water. It is this sensitivity to heat that necessitates the mixing of the flour or cornstarch with cool water (or the preliminary mixing of the flour with hot fat) before starch is added to a solution. If the starch were simply dumped into hot or boiling liquid, it would immediately turn into a gelatinous lump. By first dispersing it in a cool liquid or stirring a hot liquid into the starch-fat mixture, the granules are not clumped when they burst. The starch polymers subsequently complex evenly with the water molecules, resulting in a thick sauce, gel, or paste. Alkali solutions also cause starch to gelatinize, and they are sometimes used instead of hot solutions to produce a hydrogel effect.

The most common sources of commercial starches are corn, wheat, sorghum (see Chapter 6), arrowroot, cassava (Chapter 8), and sago, a palm that has large quantities of starch in its pith. Over 60 percent of the commercially produced starch is used for making cardboard. It adds strength to the paper, holds it together, and produces a finished outer surface. In the food industry, starch is used as a thickener. Starch can also be employed as the basis for adhesives such as library paste and for starching shirts.

RUBBER-PRODUCING LATEXES

While gums come predominantly from Old World species, elastic latexes tend to come from the New World. Strictly speaking, a *latex* is any mixture of organic compounds produced in laticifers. *Laticifers* are single cells (nonarticulated laticifers) or strings of cells (articulated laticifers) that form tubes, canals, or networks in various plant organs (Fig. 11-13). The cell walls between the cells of articulated laticifers can remain intact to varying degrees. In mature laticifers, the cellular organelles have usually disintegrated. Laticifers are not known in gymnosperms but occur sporadically throughout the angiosperms, primarily among dicotyledons.

Latex is an emulsion of a variety of compounds. The particular composition varies between species, but since hydrocarbons predominate, latexes are all insoluble in water. Mixed with these nonpolar compounds can be alkaloids, resins, phenolics, terpenes, proteins, and sugars. Some latexes have elastic properties. Others such as opium poppy latex (see Chapters 12 and 13) or papaya latex (Chapter 5) are essentially inelastic. The "latex" in latex paints, we might mention, is not latex at all. Latex paint consists of synthetic plastic

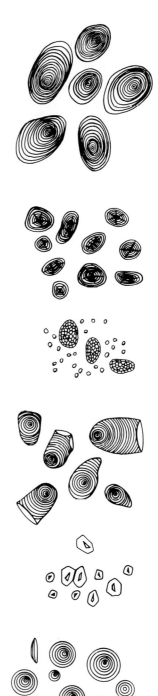

FIGURE 11-12
Starch granules (top to bottom) of potato, bean, rice, sago palm, corn and wheat.

FIGURE 11-13
A cross section of a hevea stem shows the location of the laticifers. Unlike many other rubber sources, hevea has large laticifers that lie just under the inner bark. Because of the location of the laticifers, individual hevea trees can be tapped every 3 to 4 days without harming the cambial layer.

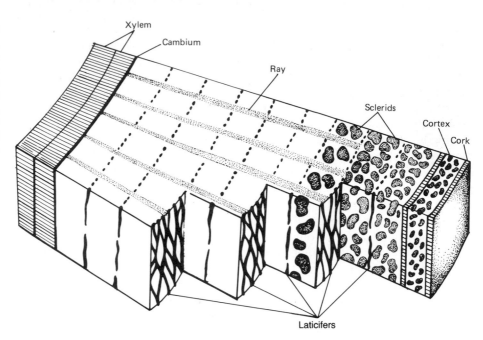

particles dispersed in water with a binding agent. When the water evaporates, the particles fuse, producing a latexlike coating.

The functions of latexes are not known. It is possible that they serve as antiherbivore agents, but many plant anatomists maintain that they are composed of by-products of primary chemical reactions within the plant that are secreted into the laticifers to keep them from interfering with normal cell functions.

In this discussion, we deal only with latexes that exhibit elastic properties. In common language, these are lumped together as "rubber," a name given to hevea latex by Sir Joseph Priestly in 1770 when he discovered that it could be used for rubbing errors from a page. Today we restrict the term rubber to terpenoid polymers (usually based on isoprene units) with elastic properties (Fig. 11-14).

Rubber-containing latexes were independently discovered by native peoples in many parts of the world and employed in various ways. In 1510, a chaplain from Queen Isabel and King Ferdinand's court wrote that the Aztecs played a vigorous game using balls made from the juice of an herb (Fig. 11-15). In all probability, the balls were made from the latex of *Castilla elastica*, a member of the fig family (Moraceae). Likewise, early explorers in the Amazonian rain forest reported that the Indians had the curious custom of dipping their feet in "sap" collected from a tree and then holding them in the smoke of a fire. The tree in this case was *Hevea brasiliensis* (Euphorbiaceae), the species from which almost all natural rubber is now obtained. By holding their feet in the smoke, the natives coagulated the rubber on the bottoms of their feet and instantly produced a pair of perfectly fitted tennis shoes. Coagulated latex was something of a curiosity in Europe, but the Spanish in South America were quick to see that it had certain possibilities. They began to dip their hats and cloaks in latex and smoke them to make them waterproof.

In 1823, Charles Macintosh discovered that hevea (or para) rubber was

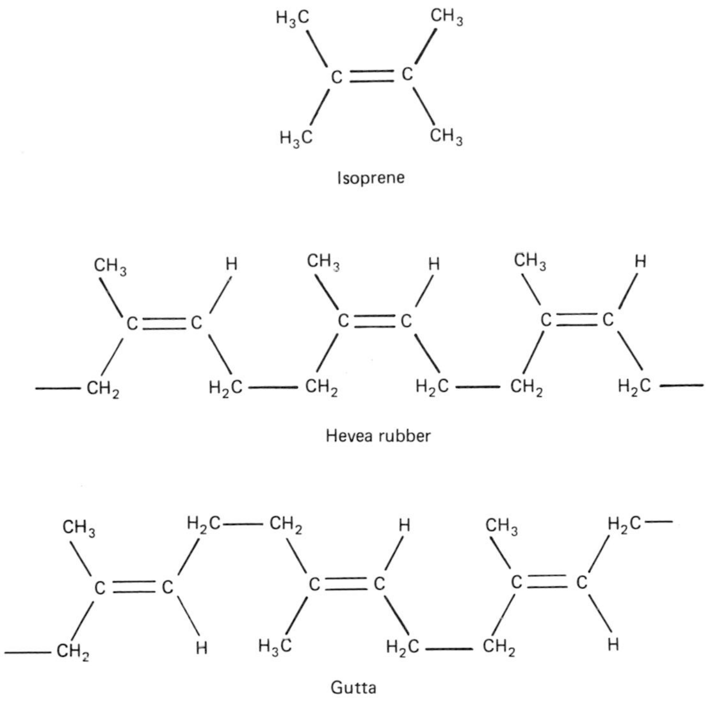

Isoprene

Hevea rubber

Gutta

FIGURE 11-14
The five-carbon isoprene molecule is the basic building block of natural rubber. Both hevea and guayule rubber are composed of up to 6,000 of these units. Less elastic natural rubbers like gutta-percha contain fewer, branched units. Most synthetic rubber is manufactured by polymerizing either butadiene or styrene molecules obtained from coal, petroleum, or alcohol.

soluble in hexane (naphtha). This discovery led to new uses because coagulated latex could be shipped, redissolved, and then applied to specific substances under controlled, industrial conditions. When the solvent was evaporated, a thin coating of rubber was left. The famous mackintosh, a sort of rainproof "slicker," named after Macintosh, was made by applying dissolved rubber onto fabric. Nevertheless, coatings made in this way had numerous problems. In very cold weather, they cracked and in hot weather, they became sticky. These problems were overcome when Charles Goodyear discovered *vulcanization* in 1839 (Fig. 11-16).

Vulcanization is simply the addition of sulfur to rubber, which cross-links the molecules of the isoprene chains. This relatively simple change makes the latex impervious to weather conditions and improves its elasticity. It has recently been discovered that vulcanization can be accomplished without sulfur by irradiating the latex with cobalt-60.

Until the 1890s, all hevea rubber was extracted from wild trees. The latex was collected by natives who were each assigned individual areas within the forest. The hevea trees within this area were tapped by diagonally slashing the bark (Fig. 11-17). Cups placed at the lower end of the slashes were periodically collected, and the day's load of latex brought to a smoking house. There, it was filtered, mixed with water, and poured over a paddle placed above a smoky fire. By continually pouring latex over the paddle, a ball of coagulated latex was eventually produced. The rubber was, and is still sometimes, shipped in this form. Today, dilute acetic acid or formic acid, rather than smoke, is often used to coagulate the rubber, which is formed into sheets instead of balls. At a later time, either sheets or balls are shredded and dissolved in an organic solvent. Complete purification of the latex is rare.

FIGURE 11-15
A Mayan ball player was carved on the walls of the ball court at Chichen Itza in the Yucatan State of Mexico.

FIGURE 11-16
Charles Goodyear had been experimenting with rubber for several years when, in 1839, he accidentally came upon the most important development in rubber history. Goodyear spilled a mixture of rubber and sulphur on a hot stove and noted that when cooled, the rubber lost its stickiness and retained its elasticity. The process was named vulcanization after Vulcan, the Roman god of fire. (Photo courtesy of the Goodyear Tire Company.)

In 1876, seeds of a highly productive population of hevea known to produce large quantities of latex were taken to Kew Gardens in London, and from there to Sri Lanka where they were used to establish plantations. Attempts to grow hevea in plantations in the Amazonian Basin had failed, largely because of the South American leaf blight which invariably destroyed plantations once they were established. Luckily, the seeds taken to the Far East were free of the fungus, and by the turn of the century, rubber production in the Old World began to rival that of South America. By World War II, 90 percent of the world's natural rubber was collected from Asian plantations.

During World War II, fighting operations effectively cut off the supply of rubber to the United States, threatening disaster (Fig. 11-18). The United States began serious experimentation with other sources of rubber and simultaneously discovered a process to produce a synthetic substitute. The first synthetic developed was styrene butadiene rubber. This synthetic rubber still predominates, but several others have been developed, including polybutadiene and polyisoprene, which is chemically similar to the major component of natural rubber (see Fig. 11-14). All of the synthetics are compounds made from the polymerization of dienes. The procedures used for polymerization differ between the types of synthetic rubber, but all of the units, or monomers, used for commercial synthesis come from petroleum.

In view of the inevitable increases in petroleum costs and the impending depletion of petroleum sources, the synthetic rubber industry is faced with

FIGURE 11-17
Tapping a rubber tree. (Photo from the collection of the Field Museum of Natural History.)

FIGURE 11-18
On October 24, 1927, Harvey Firestone (left) received this message: "Suppose war was declared; embargo shuts off rubber—in one year our rubber is exhausted—but still guayule is obtainable. This being the case, can inner tubes and shoes be made of guayule so that a fair mileage would be possible? Yours very truly, Thomas Edison." (right) The two men are shown here examining the results of their research, a tire made from guayule grown in Edison's Florida gardens. Guayule has since been shown to be chemically identical to hevea rubber. (Correspondence and photo courtesy Firestone Tire and Rubber Company.)

having to find new sources of basic compounds to use for synthesis. Suggested possibilities include shale oil, coal, old tires, and plant-derived carbohydrates. As in all business operations, the ultimate consideration will be the profit margin. If synthetics become more expensive to make than natural rubber (the reverse has been true for the past 30 years), there may be a swing back to natural sources that have been abandoned, or sparingly used, in recent years.

For the moment it might appear that synthetic rubber has replaced natural rubber. This is not the case, however. In fact, in the last 10 years, the demand for natural rubber has increased because its properties are still superior to those of the various synthetics. Over two-thirds of all rubber, either synthetic or natural, goes into the manufacture of tires. Until recently, 95 percent of the rubber used was polybutadiene. Radial tires, which have swept the market during the last decade, need more resilience than that afforded by butadiene polymers. Consequently, manufacturers incorporate large amounts of natural rubber into radial tires. This renewed demand for natural rubber has brought into focus the traditional rubber source, hevea, and a new potential source, guayule, or *Parthenium argentatum* (Asteraceae, Fig. 11-19).

Guayule is a shrub native to the Chihuahuan Desert of southwestern Texas and adjacent Mexico. The species was not even botanically described until 1859 and was not used as a source of latex by native peoples, presumably because the latex occurs in individual, thin-walled, single cells dispersed throughout the cortical and ray tissues of the stems and roots of the small, brittle bushes (Fig. 11-20). The stems also contain resin ducts, but the resin has never been considered a primary recovery product.

Guayule was among the plants studied during World War II as a source of rubber. Many wild plants were actually harvested and used as auxiliary rubber sources. There were high hopes for guayule because the rubber is virtually identical to hevea rubber, and the plants from which it can be extracted grow naturally in semiarid regions. Furthermore, shrubs can produce up to 20 percent of their dry weight in rubber. With breeding, this amount could probably be substantially increased. The most important drawback of guayule is that plants

FIGURE 11-19
A chemist with the USDA spraying guayule with chemical bioregulators to increase the plant's rubber production. (Photo courtesy USDA.)

cannot be profitably harvested until they are about 7 years old. Still, Mexico already has a functional guayule rubber production plant and the United States may soon follow its lead.

To obtain the latex, the shrubs are collected, either by cutting them at the base or, more often, uprooting them. They are dipped in hot water to coagulate the rubber and to help in the removal of leaves. The woody tissues are then

FIGURE 11-20
A cross section of a guayule stem showing the distribution of laticifers. Because the latex is produced in cells throughout the cortex of the roots and stems, entire plants are harvested for rubber extraction.

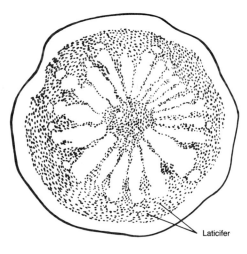

Laticifer

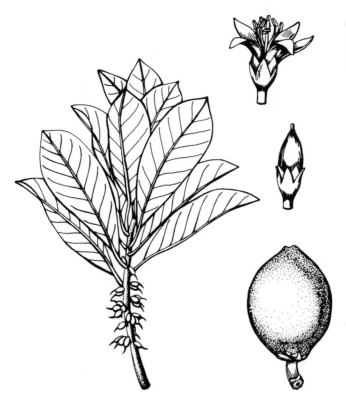

FIGURE 11-21
A branch, flower, and fruit of
Palaquium gutta, the source of
gutta-percha rubber.

pulped in water, and the resins and latex skimmed from the bagasse. The resins are separated from the latex and the latex purified.

Several other kinds of rubber were extracted from New World species before 1900. Among them were ceara rubber from *Manihot glaziovii* (Euphorbiaceae) and Panama rubber from *Castilla elastica* (Moraceae). Gutta-percha (*Palaquium gutta*, Sapotaceae) was extracted from trees in both the Old and New Worlds. Indian fig rubber (*Ficus elastica*, Moraceae) and landolphia (*Landolphia* spp., Apocynaceae) were all extracted from species native to the Old World. The primary source of rubber, however, has always been *Hevea brasiliensis* because of the high percentage of rubber in its latex.

Two other types of rubber have periodically been used for specialty purposes. One, balata rubber, is extracted from trees of *Manilkara bidentata* (Sapotaceae), a native of Trinidad and South America. Balata produces machine belts of very good quality because they stretch less than those made with other rubbers. The second, gutta-percha, is extracted from *Palaquium gutta* (Sapotaceae, Fig. 11-21), a tree native to southeast Asia. Gutta-percha has been widely used for golf balls and coatings on undersea cables. This rubber disintegrates rapidly when exposed to the air, but it is very bouncy and resistant to salt water. The former uses of both these rubbers have, for the most part, been taken over by synthetics.

Chicle

A final latex that must be mentioned is chicle, commonly known as "gum." The Aztecs of the Yucatan Peninsula chewed chicle, a practice to which some

FIGURE 11-22
A chiclero tapping a chicle tree. (Courtesy of Wm. Wrigley, Jr., Co.)

early explorers attributed their flashing white teeth. Our modern chewing gum industry is indebted to Santa Anna (1794 to 1876), the charismatic Mexican from the Yucatan who served several times as president of Mexico and who led the Mexico troops that killed Davy Crockett and James Bowie in the battle of the Alamo (1836). During one of his periods of exile between presidencies, Santa Anna was confined to a small house on Staten Island. While there, he chewed chicle. When he returned to Mexico, he left behind a large chunk of chicle. His secretary, a Mr. Adams, found the mass in a drawer and gave it to his father.

The elder Adams was something of an entrepreneur and, realizing the similarity between the material and hevea latex, tried to vulcanize it. When this failed, he tried to interest dentists in using the substance as an adhesive for dentures. When the dentists showed no interest, he took his wife's rolling pin in frustration, rolled out the latex, cut it into small rectangles which he wrapped in paper, and took it to a local candy store for sale. The new treat was an instant success. After all, this was the heyday of chewing tobacco for adult males. Shortly afterward, Adams began to add sugar and flavorings to his confection and soon patented the first chewing gum machine. The company he started eventually became the American Chicle Company. Chewing gum is still made by mixing chicle latex in giant vats with sugar and flavoring. Chicleros (Fig. 11-22) still tap wild trees of *Manilkara zapota* (Sapotaceae) to provide chicle for chewing gum. The crude latex is collected, molded into blocks, and shipped. Before heating and mixing, it is cleaned.

Resins

Natural resins have played a varied role in the history of many cultures. Early paints, incense for religious services, and caulking for ships all came from resins. Now, almost all resins are synthetically produced, but a demand for natural resins provided the impetus for the development of synthetic replacements.

Resins are actively synthesized and secreted into specialized canals or ducts. The ducts are either simple intercellular spaces or canals formed by the disintegration of a series of adjacent cells. Like latexes, resins seem to have an antiherbivore function in the plant. They are known to deter herbivory by some insects, and some have been shown to have antibacterial properties. Resin canals and production occur in the xylem, phloem, and bark of many gymnosperms and in some dicotyledons. Chemically, natural resins are a rather heterogeneous group of compounds, but all are polymerized terpenes generally mixed with volatile oils. Resins are also insoluble in water. The largest number of natural resins used by humans come from species native to the Old World, but numerous New World species have historically been significant resin sources as well.

One of the most ancient uses of resin was as incense. We have no idea when the practice of burning substances that produced a sweet-smelling smoke began, but the earliest writings attest to its common incorporation into religious services. Unlike flowers, or even most aromatic herbs, incense does not simply smell like burned leaves when ignited. Instead, it gradually releases volatile oils that diffuse outward in the smoke. Two classic resins (or gum-resins since they

FIGURE 11-23
Queen Hatshepsut sponsored what scholars consider to be the world's first plant collecting expedition around 1500 B.C. She sent her expedition to the land of Punt, a spur on the eastern coast of Africa which projects into the Indian Ocean. Besides ivory, gold, ebony, and cinnamon, the party returned with monkeys and baboons for the Queen's zoo and 31 frankincense trees which were planted at the temple of Karnak on the banks of the Upper Nile in Egypt. Carvings, such as the one reproduced here, on the walls of one temple show how frankincense shrubs were loaded onto a ship and transported in wicker baskets.

have gums mixed with them) used in incense are frankincense (*Boswellia carteri*, Burseraceae) and myrrh (*Commiphora myrrha*, Burseraceae). The value placed on these resins in early Egyptian, Greek, and Roman times is underscored by the biblical story of the Wise Men who presented these two resins along with gold to the infant Jesus. Both frankincense and myrrh are natives of Abyssinia and were brought to the Mediterranean region from the interior of Africa (Fig. 11-23). They were collected much like gums, by scoring the shrubs which produced them and collecting the exudate. Neither is an important commodity today, and their use is restricted to some church services.

Resins from these and other sources were also used in embalming (Fig. 11-24). Records from early Egypt indicate that myrrh and frankincense were used in the body cavity or for anointing the head of the deceased. Once coffins began to be used in the third Egyptian dynasty (2780 to 2680 B.C.), resins, probably obtained from pines and junipers, were used to varnish them.

Other resins that have been extensively used include mastic and lacquer. In countries bordering the Mediterranean, mastic (*Pistacia lentiscus*, Anacardiaceae) has been employed for centuries as a sealing material, a masticant to sweeten the breath, and, more recently, as an adhesive for dental caps. Lacquer (*Rhus verniciflua*, Anacardiaceae) is collected by tapping the trunks of trees which grow natively in China and Japan. The liquid resin is filtered and kept in a dark, tightly closed container.

The use of lacquer was developed into an art form in China before the Christian era, but the Japanese perfected its use during the Ming Dynasty (1368 to 1644 A.D.). The production of a good piece of lacquered work took months. The wooden surfaces to be lacquered were smoothed, coated with a thin layer of lacquer, dried, rubbed with charcoal powder, and polished. Up to 300 layers of lacquer would be applied and smoothed in this way. Designs of gold leaf or rice paper were placed on the surface and coated with numerous additional layers of lacquer. The final products have exquisite designs embedded in glasslike surfaces.

Two other groups of resinous substances of limited economic, but of major

FIGURE 11-24
The process of embalming as depicted in Ridpath, 1890, *Cyclopedia of Universal History*.

aesthetic, importance in our lives are copals and dammars. Copal usually refers to recent or fossilized resins, most of which come from species of *Copaifera* or the related genus *Hymenaea* (Fabaceae). To obtain fresh material, the resin-bearing trees are slashed and the resin collected once masses of it have accumulated over the cut. Copal is also collected from gymnosperm species of the genus *Agathis* (Araucariaceae). Despite the disparity of the sources, the resins of all of the species are soluble in non-polar solvents and, when dry, are lustrous and transparent. Dammars are likewise shiny and transparent when dry. They are obtained in the same way as copals, but they come from members of the Dipterocarpaceae, primarily species of *Shorea*. Both copals and dammars have historically been used in many kinds of paints and varnishes, but today their use is practically restricted to artists' paints. When painting with oils, colored pigment is squeezed onto a palette and then taken up in a thinning solution of linseed oil mixed with either copal or dammar. The resins "hold" the paint when dry and provide the luminous depth characteristic of oil paintings.

Because of their importance in caulking and rendering wood and ropes resistant to seawater, pine resin products acquired the name of *naval stores* (Fig. 11-25). The principal products of the naval stores industry are pitch, turpentine, and rosin. The pines most commonly used as sources of the resins used in this industry are *Pinus pinaster* and *P. sylvestris* in Europe, and *P. palustris* and *P. taeda* (all Pinaceae) in the southeastern United States. Pine pitch was well known in Egypt, Greece, and Italy centuries before the birth of Christ as an effective sealant and water-proofing substance. Pine pitch smeared on the inside of Grecian clay wine urns to prevent leakage imparted a resinous or piney taste for which the Greeks developed a fondness. Pine flavoring is still added to reproduce this taste in Greek retsina wine.

In the production of turpentine and rosin, trees are tapped and the resin collected and allowed to stand. Rosin precipitates from the crude resin, and the liquid is distilled to produce turpentine. Rosin is a brittle, friable substance when dry, but it becomes sticky when heated. The rosin bag of baseball players contains powdered rosin that becomes sticky when the pitcher's hand warms it. The slight stickiness helps the ball player to grip the ball, which, hopefully, improves the accuracy of the pitch. The bows of string instruments are drawn across blocks of rosin to make them slightly sticky so as to create more friction between the bow and the strings. Increasing the contact enhances the tone of the music produced. Rosin also finds its way into printer's ink, paper coatings, varnishes, and sealants.

Turpentine is important as a solvent and cleaning agent for oil-based paints. It is mostly used today as a source of organic compounds for further synthesis. Everyday items such as deodorants, shaving lotions, and some medicines contain turpentine or chemicals derived from it. Limonene is a commonly used lemon flavoring made from turpentine precursors.

Natural resins played a role in the development of linoleum in 1860. During his experiments with linseed, the English scientist Frederick Walton discovered that a mixture of linseed oil, resin, and cork particles could be rolled onto fabric to produce a solid, easy to clean, and durable surface. The resin initially preferred was kauri resin obtained from a New Zealand gymnosperm, *Agathis australis* (Araucariaceae). Modern linoleum-like floor coverings primarily contain synthetic resins.

Finally, we mention the only jewel of plant origin, amber. Amber was gathered by peoples of both the Old and New World and used for ornamentation or, in a few cases, for arrow points. It is lightweight, brilliant, easily carved or drilled, and can range in color from yellow to dark brown. Some ambers are opaque, others are transparent (Fig. 11-26). Amber is now known to be fossilized terpenoid resins. While it was originally thought that amber came only from pine resins, the work of Jean Langenheim showed that other trees, especially species of legumes that now provide copal, were important sources of amber in the New World.

FIGURE 11-25
Pine pitch, obtained by tapping several species of pines, is the basis of the naval stores industry. Rosin and turpentine are the primary products of the industry. (Courtesy of the Georgia Forestry Commission.)

FIGURE 11-26
A crane fly trapped in Baltic amber.
(Photo from the collection of the
Field Museum of Natural History,
Chicago.)

ADDITIONAL READING

Davidson, R. L. (ed.). 1980. *Handbook of Water-soluble Gums and Resins*. New York, McGraw-Hill.

Furia, T. E. (ed.) 1972. *Chemical Rubber Company Handbook of Food Additives*, 2nd ed. Cleveland, Chemical Rubber Co.

Howes, F. N. 1949. *Vegetable Gums and Resins*. Waltham, Massachusetts, Chronica Botanica.

Howes, F. N. 1950. Age-old resins of the Mediterranean region and their uses. *Economic Botany* 4:307–316.

Hymowitz, T. 1972. The trans-domestication concept as applied to guar. *Economic Botany* 26:49–60.

Langenheim, J. 1969. Amber. A botanical inquiry. *Science* 163:1157–1169.

Larsson, S. G. 1978. *Baltic Amber. A Palaeobiological Study*. Entomograph 1. Klampenborg, Denmark, Scandinavia Science Press.

Mantell, C. L. 1947. *The Water Soluble Gums*. New York, Reinhold.

Mantell, C. L. 1950. The natural hard resins. Their botany, sources and utilization. *Economic Botany* 4:203–242.

National Academy of Sciences. 1977. *Guayule. An Alternative Source of Natural Rubber*. Washington, D.C., National Academy of Science.

Schultes, R. E. 1977. The odyssey of the cultivated rubber tree. *Endeavour* 1(New Series):133–138.

Schultes, R. E. 1976. The taming of wild rubber. *Horticulture* 54:(11)10–21.

Whistler, R. L., and T. Hymowitz. 1979. *Guar. Agronomy, Production, Industrial Use and Nutrition*. West Lafayette, Indiana, Purdue University Press.

Chapter 12

Plants in Medicine

Our modern sweet pink and yellow medicinal liquids and neatly packaged pills seem far removed from the bitter extracts of roots and herbs used in many other cultures. Yet many of our modern medicines are based on chemicals initially leached from natural sources. Of the 100 most prescribed American medicines in 1979, 22 contained active compounds derived from flowering plants. Nine of these drugs relied on plant-derived steroids, the rest contained primarily alkaloids. Fungi, not discussed in this text, were the source of another 15 of the major prescription drugs. Not only are plant and fungal natural products important medicines in their own rights, but many have also provided blueprints for synthetic or partially synthetic drugs.

In this chapter, we provide a brief history of man's search for plants that heal and we discuss the chemical nature of the medicinally important secondary plant compounds. We also indicate how many of these natural products led us to the discovery of some of our most effective modern drugs. The plants discussed in this chapter are listed in Table 12-1.

TABLE 12-1

Plants Discussed in Chapter 12

COMMON NAME	SCIENTIFIC NAME	FAMILY	CHROMOSOME NUMBER
Agave	*Agave sisalana*	Agavaceae	$2n = $ ca. 138–148 pentaploid
Aloe	*Aloe barbadensis*	Liliaceae	$2n = 14$ diploid
Belladonna	*Atropa belladonna*	Solanaceae	$2n = 50, 72$
Chaulmoogra	*Hydnocarpus* spp.	Flacourtiaceae	$2n = 22, 24, 48$
Cinchona	*Cinchona officinalis*	Rubiaceae	$2n = 34$ diploid
Coca	*Erythroxylum coca*	Erythroxylaceae	$2n = 24$ diploid
Duboisia	*Duboisia myoporoides*	Solanaceae	$2n = 60$
	D. leichtardtii		$2n = 60$
Ephedrine	*Ephedra sirica*	Ephedraceae	unreported
Foxglove	*Digitalis purpurea*	Scrophulariaceae	$2n = 56$
Hellebore	*Veratrum viride*	Liliaceae	$2n = 32$, diploid
Henbane	*Hyoscyamus niger*	Solanaceae	$2n = 34$
	H. muticus		$2n = 28$
Ipecac	*Cephaelis ipecacuanha*	Rubiaceae	$2n = 22$
Mayapple	*Podophyllum peltatum*	Berberidaceae	$2n = 12, 16$
Opium	*Papaver somniferum*	Papaveraceae	$2n = 22, 44$ di, tetraploid
Papaya	*Carica papaya*	Caricaceae	$2n = 18$ diploid
Periwinkle	*Catharanthus roseus*	Apocynaceae	$2n = 16$
Plantain	*Plantago ovata*	Plantaginaceae	$2n = 8, 10$
	P. psyllium		$2n = 12$
Quinine	*Cinchona officinalis*	Rubiaceae	$2n = 34$ diploid
Snakeroot	*Rauwolfia serpentina*	Apocynaceae	$2n = 20, 22, 44$
	R. tetraphylla		$2n = 66$
Willow, white	*Salix alba*	Salicaceae	$2n = 76$
Wormwood	*Artemesia anuua*	Asteraceae	$2n = 18$
Yam	*Dioscorea floribunda*	Dioscoreaceae	$2n = 30, 36, 54, 72, 144$
	D. composita		$2n = 30, 36, 54$

For many species listed, the ploidy level is hard to determine because the base number of the group is unknown.

HISTORY

Humans must have searched from early times for any substance that would relieve pain and cure loved ones. Since illness was thought to be caused by mystical agents underlying the natural world, cures for mental and physical diseases were sought among plants and animals. Curative agents must have been discovered by the treacherous processes of trial and error. Medical knowledge accumulated slowly as it was painstakingly passed on by word of mouth from generation to generation. Since communication between tribes was poor, remedies were probably independently discovered several times.

Unmistakable knowledge of the use of medicinal herbs comes only after the advent of recorded history. Sumerian drawings of opium from 2500 B.C. suggest a good knowledge of medicinal plants, but substantial records of the use of herbs in medicine comes first from the Code of Hammurabi, a series of tablets carved under the direction of the king of Babylon about 1770 B.C. These tablets mention plants such as henbane, licorice, and mint that are still used in medicines. Later, the Egyptians recorded their knowledge of illnesses and their cures on temple walls and on the Ebers papyrus (1550 B.C.) which contains over 700 medicinal formulas. Many of these recipes contained plant substances from species now known to have therapeutic value. These included *Cannabis*, aloe, castor, mandrake, and numerous gum- and resin-producing shrubs (Fig. 12-1).

The Golden Age of Greece was a time of great advancement in medicinal and biological knowledge. Several prominent figures of this era stand out for their contributions to pharmacology (Fig. 12-2). Hippocrates (ca. 460 to 377 B.C.) earned his reputation as the father of medicine by being the first chronicler to discuss illnesses and their treatment in a rational way. Unlike most of his predecessors, Hippocrates did not believe that sickness was caused by evil spirits but by corporal problems. He therefore prescribed sound nutrition, purgatives, and, in certain cases, botanical drugs. The number of effective medicinal plants he discussed came to between 300 and 400 species. The philosopher Aristotle (384 to 322 B.C.) also compiled a list of plants of medicinal value a few years later, and his best pupil, Theophrastus (372 to 287 B.C.), started the science of botany with his detailed descriptions of the species growing in the botanical gardens in Athens. Theophrastus also provided the first thorough account of opium and its effects. The most significant Greek contribution, however, was made by Dioscorides whose legacy was a five-volume work entitled *De materia medica*. This encyclopedic work described the preparation of about 1,000 simple drugs. Although poorly organized, and often inaccurate, it became the prototype for future pharmacopoeias and was accepted without question by Europeans until the fifteenth century.

Historians are often quick to say that the writings of Dioscorides and his contemporaries set back medicinal science for the next 1,500 years because Europeans slavishly followed their detailed, but often incorrect, ideas about diseases and their cures. It is probably more accurate to ascribe the stagnation of medicine during the Middle Ages to the dormancy of intellectualism in Europe rather than to the stifling effects of the works of the early Greeks.

Actually, during the Middle Ages, the studies of botany and medicine

FIGURE 12-1
An Egyptian queen holding a mandrake flower. Mandrake preceded ether as an anesthetic by 2,000 years. Because atropine and scopolamine are present throughout the plant, mandrake extracts can kill pain and produce a dreamlike sleep.

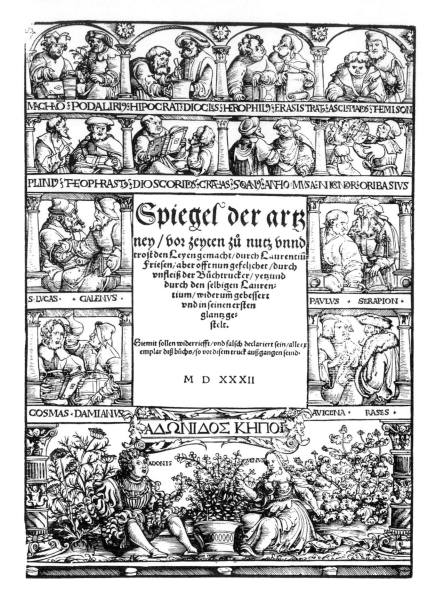

FIGURE 12-2

When Hans Weiditz designed his frontispiece for a medical handbook (*Spiegel del Artzney*, Strassbourg, 1532), he included 24 portraits of the most significant contributors to medical knowledge up to that time. Although the drawings do not necessarily provide true likenesses of these famous men, they do provide an outline of medical history. For this reason, and because their names are abbreviated in the drawing, we provide an annotated translation beginning at the *top left*:

Machaon and Podalarius, mentioned by Homer in the Iliad as "healers both and skillful."
Hippocrates (460–377 B.C.), the father of medicine for whom the Hippocratic oath, the traditional code of physicians, was named.
Diocles, considered Greece's second best physician.
Herophilus (300 B.C.) introduced dissection as a part of the study of anatomy.
Erasistratus (310–250 B.C.), founder of the science of physiology.
Asclepiades (ca. 124. B.C.), a Greek physician who moved to Rome where he espoused that medical treatment be given *cito, tuto, et iucunde* (promptly, safely, and pleasantly).

became more closely linked than ever before in history. Because of the lack of any central power in Europe during this period, society became more and more provincial. Virtually all reading and writing were carried out in monasteries where monks laboriously copied and compiled manuscripts. Following a Greek precedent, monks produced herbals (see Fig. 9-5), manuals for the identification and preparation of plants of supposed or real medicinal value. During this period, information in Europe about botany was essentially restricted to writings in such herbals. But until the invention of the printing press in 1439, herbals were available to only a handful of people. Once the printing press allowed the wide circulation of ideas and recipes for medicines, remedies could be explained, compared, and discarded if they were found to be ineffective. The general availability of printed herbals with information about useful plants and their habits, collection, and preparation was thus the prelude to the discovery (or rediscovery) of some truly effective plant-derived medicines, many of which are still in use today (Fig. 12-3).

Finally, in the fourteenth century, the Renaissance arrived, bringing with it a new desire for knowledge. Suddenly, studies of the human body were renewed, and surgical procedures were duly improved. The widespread extent of the search for knowledge during the next three centuries is shown by drawings made by Leonardo da Vinci (1452 to 1519 A.D.). Da Vinci believed that the task of the artist, like that of the scientist, was to observe one's subjects as carefully as possible and to transmit these visual perceptions faithfully.

A contemporary of da Vinci's, Paracelsus (1493 to 1541), exhibited his break with tradition by publicly burning the works of Theophrastus, Dioscorides,

FIGURE 12-3
This old lithograph entitled "The Apothecary" shows a pharmacy at the end of the sixteenth century. At this time, pharmacists often sold spices as well as medicines. Shown in the back on a shelf are cones of raw sugar. The sugar was used to make medicines more palatable. From Hans Sach's "Beschreibung aller stande," Frankfort, 1574. [Reproduced from *Medicine and the Artist (Ars Medica)* by permission of the Philadelphia Museum of Art, Carl Zigrosser, Dover Publications, Inc., 1970.]

Themison (123–43 B.C.), pupil of Asclepiades, founded the "Methodist" school of medicine in Rome.

Pliny (43–79 A.D.), author of *Natural History*, a well-known compendium of ancient scientific knowledge.

Theophrastus (372–287 B.C.), considered Aristotle's finest pupil and the founder of botany with his *Enquiry into Plants*.

Dioscorides (ca. 40–90 A.D.) compiled known medical knowledge into his *De materia medica*.

Crateuas (120–63 B.C.) introduced the concept of polyvalent drugs and serums.

Soranus (98–138 A.D.) produced the text *Diseases of Women* that laid the foundation for obstetrics and gynecology.

Nicander (185–135 B.C.) wrote a book about poisons called *Theriaca*.

St. Luke, the biblical apostle who was both a physician and artist.

Galen (138–201 A.D.) established the science of experimental physiology.

Paul of Aegina (625–690 A.D.), author of a comprehensive work on surgery.

St. Cosmas and St. Damian practiced medicine and were killed in the Christian persecutions of 303 A.D. As a result, they became patron saints of physicians and pharmacists respectively.

Avicenna (980–1037 A.D.), acclaimed Arabian physician who produced a classic medical text that was used in medical schools into the seventeenth century.

Rhazes, a Persian physician (865–925 A.D.), who studied smallpox and is credited with introducing animal gut as suture material.

Last, but not least, Adonis and Venus are pictured to represent regeneration and fertility. As they sit clutching potted herbs beside them, we are reminded of the close ties between botany, medicine, and mythology until more scientific ways of curing disease were adopted.

[Reproduced from *Medicine and the Artist (Ars Medica)* by permission of the Philadelphia Museum of Art, Carl Zigrosser, Dover Publications, Inc., 1970. Museum of Art.]

and Galen while proclaiming that he had superior insights into the prevention and cure of illness. Plants, he explained, were placed on earth by God for human use. Consequently, God had provided signs embodied in the plants themselves to indicate their potential uses (Fig. 12-4). For example, if a plant has red sap, it is a sign that the plant is intended for the treatment of blood disorders. The brainlike convolutions of a walnut are an indication that walnuts are effective for treating brain ailments. This idea, called the Doctrine of Signatures, seems absurd to us now, but received great acclaim when proposed. Luckily, it was soon displaced by less subjective and more secular methods of determining a plant's medicinal efficacy. Paracelsus did, however, contribute to

FIGURE 12-4
The doctrine of signatures holds that a plant's appearance contains clues to its use. Thus, the heart-shaped flowers such as these of the bleeding heart (B), fruits, or leaves of plants indicate their usefulness in treating heart problems. Dutchman's breeches (A) was thought useful for treating syphilis. The leaves of hepatica (C) were used to treat liver ailments, while the convoluted surface of walnuts (D) gave a sign that the nuts were useful for head and brain problems.

the advance of medicine by introducing the concept of chemical drugs in the place of vegetable remedies.

By the seventeenth and eighteenth centuries, science and philosophy advanced to the stage of hypothesis testing. The experimental approach to medicine led to an improved understanding of physiology and provided a framework for the careful testing of medicines that was manifested in the work of Edward Jenner (1796), who discovered the process of vaccination, and that of Dr. William Withering, who experimented with foxglove extracts as remedies for heart problems. During the nineteenth century significant progress was made in surgical procedures. Anesthesia was introduced, and Joseph Lister promoted the use of chemicals to prevent anasepsis.

The first half of the twentieth century was a period of tremendous advancements in medicine as causes of diseases were uncovered and new "miracle" drugs were isolated and synthesized. Initially, many modern medicines were isolated products from traditional, plant-derived extracts. These included morphine, quinine, and ephedrine. Medical chemists took as their job the identification of the active compounds in natural extracts and their subsequent synthesis. In addition, once the natural active principle was determined, it could serve as a "lead" for the synthesis of numerous chemically related compounds which were potentially better medicines than the original product. A synthetic counterpart might retain the active site on the molecule, but have different side groups that facilitate entry into the system or negate some of the adverse effects of the original chemical.

Today, natural products are comparatively less studied for potential use in medicine than in former times for several reasons. Medical chemists believe that over the years people have sorted out and tested most of the plants that might have some medicinal value. To uncover a new, potentially useful plant would necessitate massive screening and testing programs. Many companies feel that the chance of return is increasingly diminishing. In addition, as we become more knowledgeable about the biochemistry of disease, it becomes easier to design completely synthetic drugs that address specific biochemical problems. Nevertheless, plants are still being screened as potential antitumor agents, and drug companies realize that many plants of purported use in Africa, Asia, Australia, and Latin America have never been tested. These still hold promise for new compounds of medicinal importance.

THE CHEMISTRY OF PLANT-DERIVED MEDICINES

Fatty acids and essential oils, gums and resins, and alkaloids and steroids are all plant products that find their ways into modern drugs. Oils and gums are used as purgatives and as carriers or emulsifiers in many drug preparations. Volatile oils and resins are often used to help penetrate tissues and as antiseptics. Because we have discussed these kinds of products in Chapters 9, 10 and 11, we concentrate here on alkaloids and steroids, the two major classes of plant-derived compounds used in medicine today. Compounds in both these classes can occur in forms with one or more sugar molecules attached. Such forms, called *glycosides,* are often the medicinally active form of the compound.

Steroids are complex chemical compounds which all have the same fun-

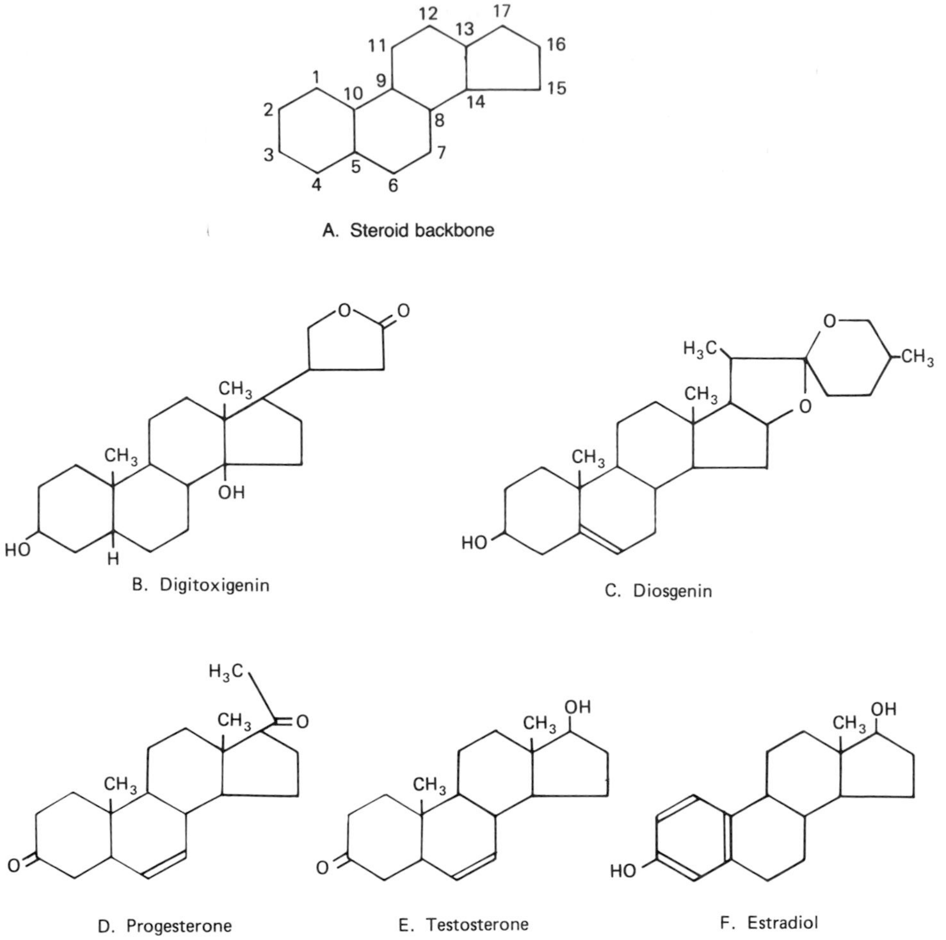

FIGURE 12-5
With the addition of particular side chains or extra rings to the steroid backbone, various cardiac glycosides and steroid hormones are produced. Cardiac glycosides (B) have a unique ring attached to the seventeenth carbon of the steroid backbone. Steroidal saponins (C) are common in the Liliaceae, Agavaceae, and Dioscoreaceae. They differ from other steroids because they have a spiroketal complex attached to the steroid skeleton. This complex makes them especially useful as precursors of human steroid hormones (D, E, and F).

A. Steroid backbone

B. Digitoxigenin

C. Diosgenin

D. Progesterone

E. Testosterone

F. Estradiol

damental structure of four carbon rings (Fig. 12-5) called the steroid backbone. The addition of diverse chemical moieties at different places of the backbone (the positions are numbered as shown in Fig. 12-5) leads to the production of a variety of different steroidal compounds. Additions of sugar molecules to the steroidal backbone produce steroidal glycosides.

Both steroids and steroidal glycosides occur in many unrelated groups of angiosperms. No direct physiological functions for steroids have been found in plants, so these compounds are classified as secondary products. However, steroids often have a pronounced effect on animals, particularly vertebrates. Consequently, many biologists believe that the production of these compounds is for the purpose of deterring herbivory. Supportive evidence for this view comes from studies of the monarch butterfly, which feeds in the larval stage (as a caterpillar) on milkweeds (Fig. 12-6). Milkweeds, members of the genus *Asclepias* (Asclepiadaceae), are toxic to humans because they contain abundant steroidal glycosides. Monarch larvae are not poisoned by the glycosides they ingest when they eat milkweed leaves. Instead, they store the compounds in parts of their bodies. When the caterpillars metamorphose into butterflies, these stored glycosides are carried into the adult stage where they occur primarily in

FIGURE 12-6
Glycosides ingested by the caterpillar of this butterfly while feeding on milkweed are stored in the body and, after metamorphosis, appear in the adult monarch's body. These sequestered chemicals act as a defense against predation.

the wings. Because the butterflies now contain these compounds, the butterflies themselves are toxic to vertebrate predators such as birds. A taste of one such butterfly causes a bird to become ill and/or to eject the bite of the insect. Birds quickly learn to avoid monarchs. Hence, the butterflies have made a secondary use of a plant product which may have been produced by the plant itself to ward off predation.

The second major group of medicinally important plant products are *alkaloids*, a diverse group of heterocyclic chemicals lumped together because they contain nitrogen and exhibit an alkaline reaction. Some chemicals now classified as alkaloids have an acid reaction because of component parts of the molecule. Alkaloids were formerly considered secondary products like steroids, but unlike steroids they have recently been shown to enter into the primary metabolism of plants. The plant is continually making new alkaloid molecules and breaking down old ones. Fig. 12-7 gives examples of some familiar alkaloids.

Many alkaloids are extremely poisonous to humans, and many, as we shall see in Chapter 13, have been used as poisons. Perhaps a familiar example of a plant toxic to animals because of its alkaloid constituents is locoweed. Others include deadly nightshade and oleander. Some natural insecticides such as nicotine are alkaloids. Many alkaloids that are toxic to humans in large or moderate amounts can be effective in alleviating physiological problems if ingested in small quantities. Consequently, one of the most important steps in the development of medicines from alkaloids (and steroids) is the determination of standardized, safe doses. Without proper administration of the medicines, patients in former times were as likely to die from the cure as from the disease.

PLANTS OF FORMER IMPORTANCE IN MEDICAL TREATMENT

Among the worst diseases that have historically afflicted humans are leprosy and malaria. The first of these was one of the most dreaded diseases of ancient times because of the terrible disfiguration it caused and the slow and painful way in which it led to death. The second, malaria, has been called the world's greatest killer because it has been the cause of more human deaths throughout recorded history than any other disease or all our wars combined. The medicines first used with any success in treating these two diseases came from plants.

The horror with which people react to leprosy is still evident in the expression "to treat someone like a leper." For centuries leprosy was considered an incurable disease, and lepers were shunned or confined to colonies designated for them. We now know that leprosy is caused by a bacterium which is apparently transmitted by contact between susceptible individuals. Although the Vedas mentioned over 2,000 years ago an oil called chaulmoogra that helped in curing leprosy, Europeans did not learn of it until the middle of the nineteenth century.

Once chaulmoogra reached Europe, its effectiveness was evident, but the source of the seed oil was not. Europeans and Americans received fruits without any indication of the plants from which they had come. Joseph Rock, a naturalized American with a reputation for being able to do the impossible, was sent to Asia in the 1920s to locate the species which produced the fruits (Fig. 12-8). Following a journey of thousands of miles in all types of conveyances, Rock found the source of chaulmoogra growing on trees of *Hydnocarpus* spp. (Flacourtiaceae) native to India and surrounding countries. Rock sent viable seeds to Hawaii where cultivation of *Hydnocarpus* began. From shortly after the time of Rock's discoveries until the production of sulfa drugs in 1946, chaulmoogra oil provided the only effective treatment of leprosy.

The drastic effects of malaria as a cause of death have been documented since ancient times. Examination of skulls from the Bronze Age of Greece have shown that malaria was a major cause of mortality. During World War I, more people died from malaria than from enemy fire. As late as 1950, an estimated 1 million people in India alone died each year from malaria. Even today, entire populations are infested with the disease. Before the cause of malaria was known, people believed that it was transmitted through the air. The name

FIGURE 12-8
Joseph Rock had taught Arabic in Austria and Chinese in Hawaii before being chosen to lead the U.S.D.A.'s search for the source of chaulmoogra oil. Knowing only that the fruits appeared in native markets in China, Burma, and India and that description indicated the plant was a member of the tropical family Flacourtiaceae, Rock journeyed thousands of miles through the wilds of Asia. Following linguistic and botanical clues, he finally tracked down the wooly fruits on 25- to 35-m-tall trees in Burma. He is shown here in 1925 in an Hawaiian plantation of *Hydnocarpus* established with the seeds he collected. (Photo courtesy of the U.S.D.A.)

''malaria'' comes from the French words *mal*, bad, and *aire*, air, reflecting this idea of aerial transmission. Since we now know that the disease is transmitted by mosquitos, the guess that malaria moved through the air does not seem so far-fetched.

Malaria is caused by a sporozoan belonging to the genus *Plasmodium*. When bitten by a mosquito that has previously had a blood meal from a malaria victim, an individual becomes inoculated with the microorganism. The disease is characterized by fever, chills, anemia, and spleen enlargement. Usually the person who has contracted the disease has spells that come and go. Attacks occur when large numbers of blood cells simultaneously rupture, releasing a form of the sporozoan that had been multiplying within the cells.

For countless centuries, there was no way to control the effects of malaria. Finally, in the middle of the seventeenth century, Jesuits in South America discovered a native remedy for the disease consisting of an infusion made by boiling pieces of bark of cinchona (*Cinchona officinalis*, Rubiaceae, Fig. 12-9) in water. The Indian name for the tree from which they obtained the bark was *quina*, or *quina quina*, (hence quinine). The Jesuits proclaimed that they had found a cure for malaria, but the religious order was so hated and feared in Europe that large segments of the population, believing promotion of the drug to be a conspiracy to kill Protestants, would not try it. Oliver Cromwell died of

FIGURE 12-9
A flowering and fruiting branch of *Cinchona officinalis*, the bark of which yields quinine, a remedy for malaria.

malaria rather than take the "Jesuits' powder." Not until 1681 was cinchona universally accepted as an efficacious treatment for malaria.

It was soon recognized that individual trees of cinchona differed in the quality of the alkaloids they contained. Knowledge of particularly potent trees was a carefully guarded secret. The Dutch were finally able to acquire seeds taken from a high-yield plant near Lake Titicaca, Bolivia, from an Indian who died in prison as a result of his trading secrets about potent trees. After overcoming many agronomic problems, the Dutch finally managed to establish productive plantations from these seeds in Java. Their efforts quickly gave them a monopoly over the world supply of quinine. With the onset of World War II, Europe and the United States were cut off from their supplies of the drug, causing the United States to send expeditions to Bolivia to discover new sources. While one of these missions was successful, its accomplishment was somewhat overshadowed by the synthesis of quinine in 1944. Soon afterward, chemically similar alkaloids were also synthesized, reducing the need for natural quinine.

For many years, virtually all of the malaria drugs have been synthetically produced. Nevertheless, strains of *Plasmodium*, particularly in the Far East, have recently become resistant to many of the synthetic quinine analogs. As a result, there is renewed interest in natural quinine, which seems still to retain its effectiveness against the new strains. One heritage of the former use of natural quinine to prevent malaria is the tonic of a gin and tonic, a drink originated by the British whose tours of duty included tropical countries where the threat of malaria caused them to take prophylactic doses of quinine.

Recently, a potent antimalarial compound has been isolated from *Artemesia anuua* (Asteraceae). This species, a wormwood related to the plant from which absinthe is made, has been used in China for various medicinal purposes since 168 B.C. The compound, artemisinin, is not an alkaloid and does not even contain nitrogen.

While their impact was less dramatic, three other drugs initially derived from plants, and now produced synthetically, are worth mentioning. One is *ephedrine*, originally obtained by soaking the dried stems of *Ephedra sinica* (Gnetaceae, a gymnosperm, Fig. 12-10). Infusions of the plant stems have been prescribed in China for centuries as a stimulant and in the treatment of high blood pressure, asthma, and hay fever. It was only after 1920 that westerners accepted *Ephedra* extract as a decongestant and isolated the alkaloid ephedrine responsible for its action. Synthetic ephedrine or similar compounds are now used in many medicines.

The second plant of note is the white willow (*Salix alba*, Salicaceae). Since the time of Dioscorides, solutions made by soaking willow leaves (decoctions) were often placed on areas of the body which throbbed or ached. The active ingredient in willow that alleviated pain was isolated in 1827 and called *salicin*. Salicin could not be taken internally, but a derivative, acetylsalicylic acid produced in Germany in 1899, could be ingested and provided relief for all types of pains. This compound, named *aspirin* by its original producers, is now probably the most widely used medicine in the world.

We seldom think of cocaine as a medicine, but it has historically been used as a calmative and as a local anesthetic. Coca (*Erythroxylum coca*, Erythroxylaceae, Fig. 12-11) is a native of the South American Andes where Indians have chewed

FIGURE 12-10
Ephedra, from which ephedrine is obtained, is a gymnosperm as indicated by the presence of cones rather than flowers. (A) Staminate cones; (B) ovulate cones; (C) plant.

FIGURE 12-11
A branch and flower of a coca plant, which yields the alkaloid cocaine.

the leaves mixed with lime for countless ages. The alkaloids extracted during the chewing process reduce feelings of hunger and pain.

The potential use of cocaine in medicine was first discovered in 1884 when an assistant to Sigmund Freud placed a solution of cocaine, first isolated from coca in 1858, on his tongue and found that it produced a numbing sensation. A later series of experiments showed that a cocaine solution could be used as a local anesthetic in eye surgery, dentistry, and other operations where only a part of the body needed to be desensitized. Cocaine has never been synthesized, but similar compounds have. The most important of these similar alkaloids is procaine, better known by its trade name, Novacaine. Synthetic alkaloids similar in action to cocaine have virtually replaced it in medical use in the United States.

PLANTS OF IMPORTANCE IN MEDICINE TODAY

While natural quinine, cocaine, and chaulmoogra oil are rarely used today to treat disease, many plants are still of great importance as sources of medicinal compounds (Fig. 12-12). In our discussion of plants important in modern medicine, we treat first those yielding steroids and then those from which alkaloids are obtained.

Steroids

As pointed out earlier, steroids have a pronounced effect on humans. One of the primary reasons for these effects is that most animal hormones have a

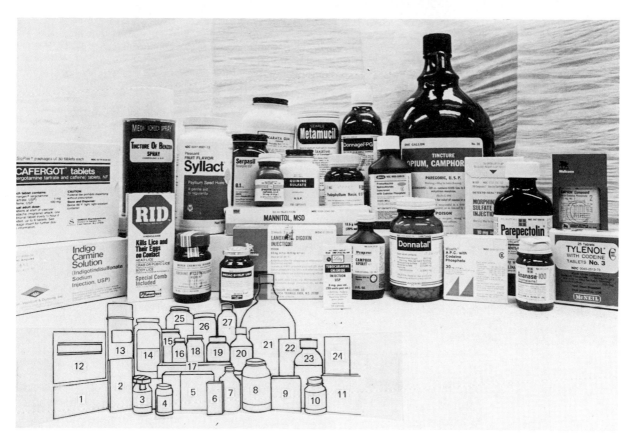

FIGURE 12-12

Drugs from plants. Shown are many medicines currently available in pharmacies. Since many of the labels show trade names, we provide a brief summary of the botanical constituents and the uses of the drugs:

1. Indigo carmine, a histological stain from indigo.
2. RID contains pyrethrins that kill head and body lice.
3. Cocaine hydrochloride, a local anesthetic (for external use).
4. Ipecac syrup made from *Cephaelis ipecacuanha* is used to induce vomiting after poison ingestion.
5. Lanoxin contains digoxin, a foxglove cardiac glycoside that helps to stabilize heart rhythms.
6. Tubocuranine chloride contains curare from *Strychnos* species. It is used as a skeletal muscle relaxant in surgery.
7. Camphor spirits are made from *Cinnamomum camphor* and are used externally as a topical anesthetic.
8. Donnatal contains belladonna alkaloids and barbiturates which relieve cramps and spasms of the stomach, bladder, and intestines.
9. A.P.C., or aspirin with codeine phosphate and caffeine, is prescribed as an analgesic and anti-inflammatory agent.
10. Ananase 100 contains bromelain, a proteolytic enzyme from pineapple used to reduce edema, ease pain, and accelerate tissue repair.
11. Tylenol with codeine is taken for extra pain relief.
12. Cafergot tablets contain ergotamine tartrate (obtained from the fungus ergot) and caffeine. The tablets are used to treat migraine headaches.
13. Tincture of benzoin spray is an example of a plant product used as a topical protectorant.

14. Syllact is one of several brands of laxative that contain psyllium (see no. 26).
15. Serpasil is a trade name of a drug that includes *Rauwolfia* alkaloids that are effective in treating high blood pressure and emotional conditions.
16. Ephedrine sulfate is used as a bronchodilator.
17. Mannitol extracted from the dried exudate of the ash, *Fraxinus ornus* (Oleaceae), is used as a laxative.
18. Quinine sulfate is an antimalarial drug.
19. Mayapple yields podophyllin, which prevents cell proliferation in tumorous growths.
20. Promethazine hydrochloride with codeine phosphate forms the basis of some cough syrups or antihistamines.
21. Tincture of opium has been used to alleviate menstrual cramps, for diarrhea, and in other ailments that lead to painful abdominal spasms.
22. Morphine sulfate is an injectable pain killer that is used only in extreme cases.
23. Parapectolin contains opium extracts and kaolin (clay) to absorb irritants and provide relief in cases of simple diarrhea and colicky cramps.
24. Empirin with codeine phosphate provides stronger relief than aspirin alone.
25. Karaya gum (Chapter 11) is used as a bulk laxative.
26. Metamucil contains psyllium extracted from plantain seed husks. This product acts like Syllact (see no. 14).
27. Donnagel-PG contains hyoscyamine sulfate, hyoscyamine hydrobromide, and powdered opium. It is prescribed in cases of diarrhea, gastritis, colitis, and nausea.

(Photo taken with the help of Dr. James Delucio, Department of Pharmacy, University of Texas.)

---

steroidal skeleton like those shown in Figure 12-5. A *hormone* can loosely be defined as a substance produced in one part of the body that affects the functioning of tissues or organs in other parts of the body. In humans, hormones are produced by the pituitary, adrenal, and thyroid glands, the sex organs, and several other organs. Among the hormones produced by the adrenal cortex and those considered to be sex hormones are cortisone, estrogen, androgen, progesterone, and testosterone (see Fig. 12-5). It is possible to synthesize these hormones or to extract them from the excised glands or the urine of pregnant or postmenopausal women, but such methods are extremely time consuming and expensive. The most cumbersome part of chemical synthesis of these hormones is the steroid backbone. Consequently, if a steroid skeleton can be obtained from a relatively inexpensive, accessible source, the final synthesis of a human hormone by addition of peripheral components to the molecule can be relatively easily accomplished.

During the years between 1936 and 1940 it was discovered that certain members of the yam genus *Dioscorea* (Dioscoreaceae, Fig. 12-13; see also Chapter 8) contain particular kinds of steroids called *saponins* (see Fig. 12-5). The name "saponin" refers to the fact that these compounds make a soaplike foam when shaken with water. The medicinally important saponins have an additional ring added to the steroid backbone (see Fig. 12-5) that makes them particularly similar to human sex hormones. Plant species, such as *Agave sisalana* (Agavaceae, see also Chapter 16), species in the Liliaceae, and several species of *Dioscorea*, contain such saponins (actually in the form of sapinogens, saponins to which one or more sugars are attached), but two Central American species of *Dioscorea*, *D. floribunda* and *D. composita*, have been found to be the most productive sources. Both of these species yield *diosgenin*, a good starting point for chemical

FIGURE 12-13
The roots of yam vines (shown here) are the principal sources of steroid precursors used to produce active compounds in oral contraceptives, and to treat hormone imbalances and heart disease.

synthesis. This sapogenin is extracted from the tubers, which are washed and chopped, or chopped and dried, for later extraction.

The majority of the hormones synthesized from diosgenin are used in birth control pills, for the production of hormones to regulate menstruation cycles, or as a component of fertility drugs. Cortisone and hydrocortisone are two other important hormones that are synthesized from diosgenin. They are used for the treatment of severe allergenic reactions, for arthritis, and for Addison's disease, caused by malfunction of the adrenal glands.

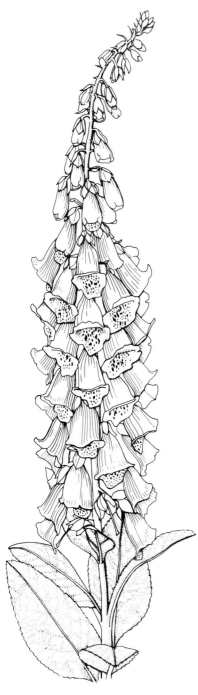

Heart disease is one of the most common causes of natural death in the United States today. The apparent relative unimportance of heart disease as a source of mortality in former times was probably due to the fact that most individuals died of other causes before reaching the ages during which heart disease becomes a problem. It is also possible that heart disease was not formerly recognized as such. For example, dropsy was a common disease 100 years ago, but it is unheard of today. Dropsy is an ailment characterized by the retention of fluid in the tissues (edema). We now know that the problem is caused by the poor circulation associated with congestive heart failure.

It is little wonder therefore that the scientific investigation into the cure for dropsy led to the detection of one of the most important modern treatments for certain heart problems. William Withering, a prominent British physician, took heed when, about 1775, a patient thought to be incurable recovered from dropsy after being administered an infusion of various plant parts. The folk cure included the leaves of the purple foxglove (Fig. 12-14), a common garden ornamental. Foxglove (*Digitalis purpurea,* Scrophulariaceae) is an extremely poisonous European plant (it is rumored that drinking the nectar will kill a child). Withering's first job was, consequently, to ascertain what constituted a nonlethal dose of dried, powered *Digitalis* leaves and to record meticulously the effects of the drug. Withering's careful treatments and documentation of amounts administered eventually led to the widespread use of *Digitalis*.

Many years after Dr. Withering's work, the chemicals which were responsible for the effectiveness of *Digitalis* were isolated and characterized. They proved to be steroids, primarily a type of glycosides known as cardiac glycosides. These steroids foam in water like saponins, but they have different rings attached to the steroid backbone (see Fig. 12-5). Cardiac glycosides are so named because they have a strong effect on the cardiac (heart) muscles. They change the rhythm of the heart beat in such a way that the systolic contraction is lengthened. The net effect is that the time between heart contractions is increased and the ventricle is emptied more completely than without the drug. Cardiac glycosides used medicinally can also improve circulation in general, relieve edema (dropsy) often associated with heart failure, and help renal secretion. The major cardiac glycosides isolated from *Digitalis* are digitoxin, gitoxin, and gitaloxin.

Alkaloids

Pain is a seemingly horrible curse that humans have to endure (Fig. 12-15). The sensation of pain may be caused by physical or physiological factors, or a combination of the two. Often, the same drugs will be effective against physical pain and psychological anguish because they numb the senses and produce a feeling of disassociation with the real, conscious world. One of the oldest, and still predominant, sources of pain-relieving drugs is the opium poppy (*Papaver somniferum,* Papaveraceae, Fig. 12-16, see also Chapter 13). Fossilized poppy capsules have been found from prehistoric settlements around the Mediterranean. Pictorial representation of opium poppies can be found in Egyptian, Greek, and Roman art. As in the case today, early use was probably in large part narcotic (see Chapter 13), but the use of opium as a calming agent and to help quell dysentery dates from at least the first century B.C.

FIGURE 12-14
The leaves of the garden ornamental foxglove which yields digitoxin, a steroidal glycoside effective in stabilizing the heart's action.

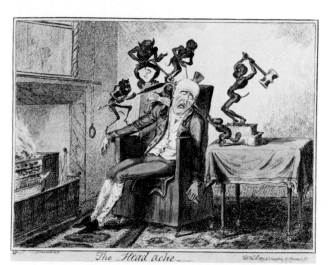

FIGURE 12-15
The Headache by George Cruikshank, published in London in 1819. (Reproduced from "Medicine and the Artist" by permission of the Philadelphia Museum of Art.)

The opium poppy is native to the region that includes eastern Europe and western Asia. Most of the modern production is in Nepal, India, Laos, and Cambodia. Because only about 4 percent of the opium harvested is used for legitimate medical purposes, cultivation of the crop is prohibited in many countries, including the United States. The depiction of the poppy capsule in ancient cultures (see Fig. 3-10) shows that the same method of obtaining opium latex from the fruits has been employed for centuries. The fertilized, swollen capsules are slashed by a worker backing through the fields to avoid brushing against capsules that he has already cut. The incision across the capsule must be deep enough to cut the latex vessels running through the outer fruit wall, but shallow enough to prevent severe injury. With care, a capsule can be tapped 3 to 10 different times. After the incision is made, the capsule is left for a day while the latex oozes from the cut and dries slightly in the air. The next day, the tacky mass is scraped from the capsule, stored, and eventually dried (see Fig. 13-8).

In ancient times, the dried latex was simply powdered and dissolved in some liquid, usually wine. Today, the latex is also dried, powdered, and dissolved, but the solution is then subjected to chromatography, a process which separates the various chemical components of a mixture. For medicinal purposes, the important chemicals present in the latex are alkaloids. Over 26 different alkaloids have been isolated from opium, but only 3 of these, morphine, codeine, and papaverine (see Fig. 13-10), are extensively used. Morphine, isolated from opium in 1803 by a German pharmacist, F. W. Serturner, is the most abundant of the three in the latex, and the most potent pain killer. It is also the most dangerous of the three because it is exceedingly habit forming. Administration is always by means of a hypodermic needle and in carefully regulated amounts. Even in situations in which it is administered medicinally, the patient can become addicted. Nevertheless, one should appreciate how important morphine was before the advent of modern anesthetics when all surgery was performed while the patient was awake (or drunk) and strapped to a table. Morphine provided practically the only relief from the pain of these operations as well as from wounds incurred in wars. It was also used, mixed with scopolamine during the 1930s and 1940s to produce "twilight sleep," a common anesthetic

FIGURE 12-16
The opium poppy is the source of morphine, papaverine, and codeine.

used in childbirth during labor and delivery. Today, morphine is administered only in cases of extreme pain.

In contrast to morphine, codeine is a milder, and usually non-habit-forming, pain killer. It is used in some "over-the-counter" pain relief medicines and in many prescription drugs. The alkaloid can be isolated from opium or synthesized from morphine.

The last of the three major alkaloids obtained from opium is papaverine, used primarily in drugs for the treatment of internal spasms, particularly those of the intestinal tract. Paragoric, a simple tincture (solution) of opium used historically for diarrhea and cramps, is effective in large part because of the papaverine it contains. While paragoric is now difficult to obtain because of strict laws governing the dispensing of opiates, it forms a component of several commercial products used for simple diarrhea.

Another group of alkaloids used for controlling a variety of muscle spasms and in psychiatry are obtained from members of the potato family (Solanaceae). Like the chemicals isolated from opium, these are often classed as *analgesics* because they relieve pain by soothing the muscle spasms. The solanaceous alkaloids most commonly used are hyoscyamine (and l-hyoscyamine), atropine, and scopolamine (or hyosine). Technically, these are known as *tropane alkaloids* because they contain a dicyclic portion formed by combining ornithine and three acetate-derived carbon atoms (Fig. 12-17). Although they occur in several species, the tropane alkaloids are most commonly extracted from *Atropa belladonna* (Fig. 12-18, 12-19), native to central and southern Europe, *Hyoscyamus niger* and *H. muticus* from Eurasia, and *Duboisia myoporoides* and *D. leichtardtii*, native to Australia and neighboring islands (see Chapter 13 for a different view of these plants).

The first of these, commonly called belladonna or deadly nightshade, has been used medicinally since the time of Dioscorides. Its scientific name comes from the name of the Greek god Atropos, one of the three fates responsible for cutting the thread of life. Needless to say, the plant is very poisonous. The common name belladonna comes from the medieval use of eye drops prepared from dried leaves of the plant. Women would use the drops to make their pupils expand. The wide-eyed, innocent look produced was considered charming (hence, *bella* meaning beautiful and *donna*, lady). Species of *Hyoscyamus*, or henbane (Fig. 12-19), also figured prominently in ancient European medicine and sorcery, although it was apparently never used to enhance beauty. *Duboisia* has been a source of alkaloids used in western medicine only since about 1877, but the Australian bush natives seem to have recognized the potency of these species long before this time.

Although all parts of these solanaceous species contain alkaloids, usually only the leaves are dried and placed in solvents. Today, extracts are fractionated, and the different alkaloids purified. The action of the alkaloids is to soothe the smooth muscle system and thus control painful cramping. Among the muscles which are responsive to alkaloid treatment is the heart. Consequently, tropane alkaloids are used in many medicines prescribed for cardiac problems. Scopolamine and atropine are also used to dilate pupils during eye examinations. Atropine is used mixed with methylene blue, phenyl salicylate, and benzoic acid to combat infections and pain associated with the urinary tract.

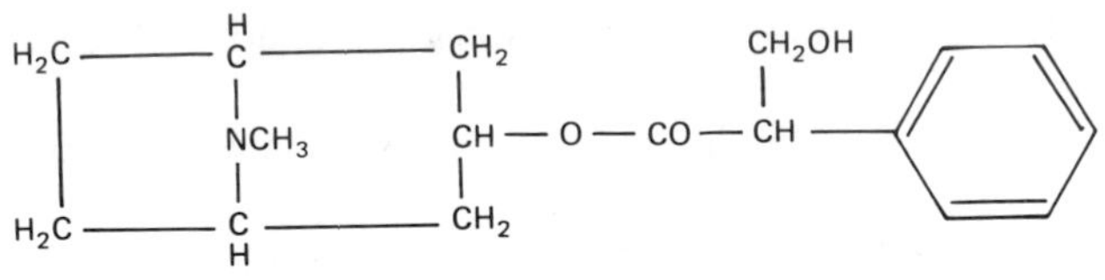

Hyoscamine

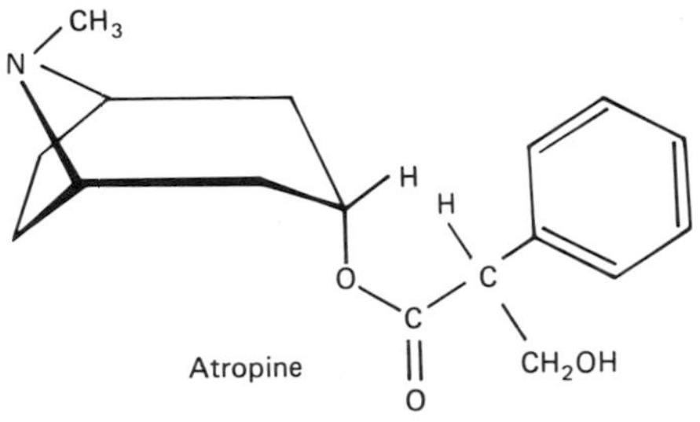

Atropine

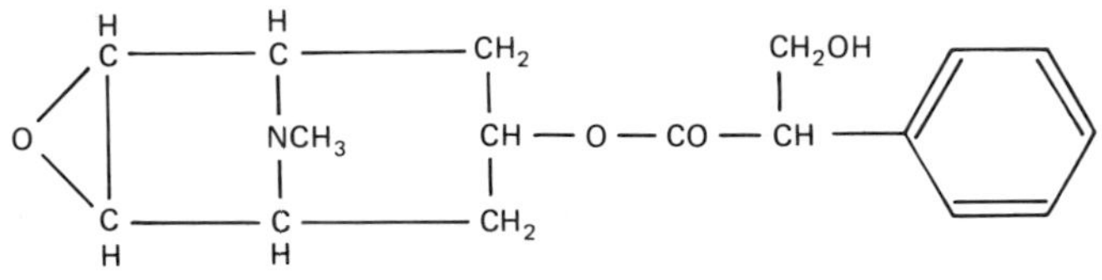

Scopolamine

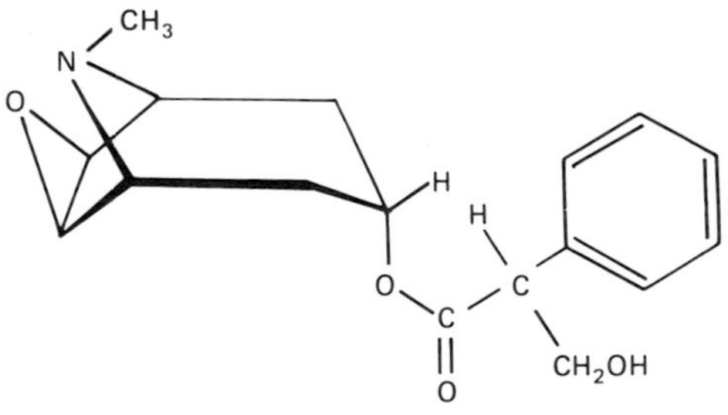

Scopolamine

FIGURE 12-17
Tropane alkaloids, found predominantly in members of the Solanaceae, are dicyclic nitrogen compounds formed by combinations of orthinine and molecules of acetic acid. Hyoscamine and atropine are isomers of the same compound. The two bottom figures show different ways of representing the scopolamine molecule.

The belladonna alkaloids are prescribed for stomach and bladder cramps and to prevent nausea and vomiting caused by motion sickness. They are prescribed for victims of Parkinson's disease to decrease stiffness and tremors and are often given to patients before surgery as a relaxant and to reduce salivation. Finally, these alkaloids are helpful in cases of nerve gas or mushroom poisoning.

Not all alkaloids used to treat heart disease are obtained from members of the potato family. Another important source of different compounds is a lily, hellebore, *Veratrum viride* (Liliaceae), a species indigenous to North America. The most important chemicals obtained from the plants are veratrosine and veratramine (both glycosides) and germidine and germitrine (alkaloid esters). The ester alkaloids reduce the pressure of the heart beats, slow down the heart, and stimulate blood flow. American hellebore, the common name for the species, was used by North American Indians and adopted by settlers from Europe. Undoubtedly, the adoption of the plant for medicinal purposes was

FIGURE 12-18
Stamps commemorating bioactive plants. From top left: Czechoslovakian stamp showing a poppy; Polish stamp of *Datura* and another with belladonna; a pyrethrum flower and plantation from the Republic of Rwanda (lower right).

facilitated by the fact that a related species, *V. album*, had been used in Europe for centuries to treat, among other things, cholera. Alkaloids are extracted from the bulbous roots of both of these species after they have been cleaned and dried.

In addition to quinine, *Cinchona* yields among its 30 other alkaloids, quinidine, a compound useful in treating heart disease. Quinidine inhibits fibrillation of the auricle and corrects improper heart rhythms. The alkaloid is usually administered as a sulfate salt.

A common illness in the United States today that is often associated with, or a result of, heart disease is hypertension, caused by difficulty in pumping blood through the arteries. One of the most effective agents known in treating hypertension comes from a plant that was first mentioned in the Vedas of India dating several centuries before Christ. These scientific poems noted that snakeroot (*Rauvolfia serpentina*, Apocynaceae, Fig. 12-20) was useful for treating snakebites. Hindis applied the name *chandra* (moon) to this plant since it was also used to treat "moon disease," or lunacy. Indian sages have chewed on snakeroots for centuries because of its calming effects. Mahatma Ghandi was known to have been extremely fond of snakeroot tea. But despite the abundant folk wisdom about *Rauvolfia*, it was not until 1949 that Indian scientists realized its potential for regulating hypertension. Soon after it was verified that rauvolfia alkaloids were also effective in the treatment of mental illness.

The most important chemical obtained from snakeroot is reserpine (Fig. 12-7), isolated in Switzerland in 1952. While the alkaloid occurs throughout

FIGURE 12-19
Henbane (left) and belladonna (right) are the sources of alkaloids historically used for both medicine and murder. Hamlet's father was poisoned with a solution of henbane poured in his ear, and Juliet drank a potion of deadly nightshade to feign death.

FIGURE 12-20
The snakelike root of *Rauvolfia* contains an alkaloid used in the treatment of hypertension and schizophrenia.

the plant, it is more concentrated in, and thus primarily extracted from, the root. The dramatic effects of this drug completely altered practices in mental institutions once it was released for general use. Before reserpine began to be administered, schizophrenics were treated with electrical shocks or by injections of insulin, both of which cause violent reactions (Fig. 12-21). The use of relatively large intravenous amounts of reserpine produces a pronounced calming effect with relatively little trauma. The drug does, however, have side effects when used in these quantities. Smaller doses are used to treat hypertension and menstrual and menopause problems. A related New World species, *R. tetraphylla*, is also used as a source of alkaloids, but most of the world's supply still comes from India, Pakistan, and Java.

Cancer is one of the most dreaded diseases of our time. Drug companies continually search for anticancer agents and people afflicted with cancer and their relatives grasp at any possible panacea. The continued public pressure to allow the use of laetril (extracted from the seeds of apricots) as a treatment, despite constant documentation that it is ineffective in controlling cancer, illustrates the desperation with which cures are sought. The search for effective anticancer agents has not been without success, however. In a few cases, substances that were able to arrest cancer turned out to be plant alkaloids.

The common periwinkle (*Catharanthus roseus*, Apocynaceae, Fig. 12-22) has been used in its native range in Europe for hundreds of years as a folk treatment for diabetes. Because of its use in lay medicine, the plant was tested for effectiveness against various illnesses. In 1957, it was tried for the treatment of leukemia and found to be effective in curing some forms of the disease (especially those that commonly afflict children). The active chemicals, identified as vinblastine and leurocristine, were marketed shortly after 1957 under several

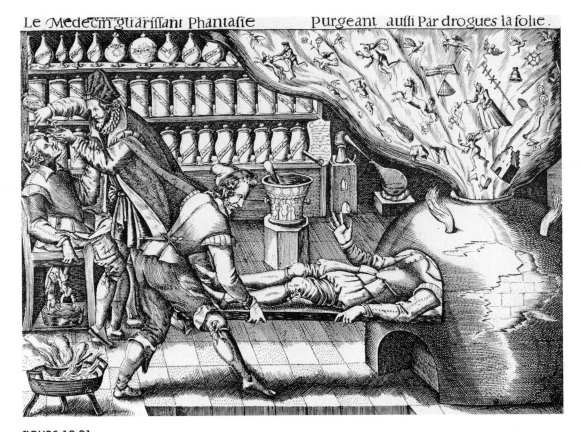

FIGURE 12-21
Before reserpine, treatments for mental illness were barbaric and only moderately successful. This anonymous illustration from seventeenth century France shows a physician curing "fantasy" using drastic measures to purge the demons causing the madness. The drawing also shows another physician administering wisdom *(sagesse)* from a bottle. On the shelf reside bottles labeled reason *(raison)*, truth *(verité)*, humility, etc., implying that personality qualities could be dispensed in liquid form. Although meant as a parody, drugs have since been shown to have profound effects on mental illnesses. (Reproduced from "Medicine and the Artist" by permission of the Philadelphia Museum of Art.)

trade names. Rates of successful cures using drugs manufactured from these alkaloids ranged from 50 to 70 percent in the case of some lymphonias. In cases of lymphocytic leukemia, remission was found in 99 percent of the cases where vinblastine derivatives were used. Patients subsequently exhibited a 50 percent survival rate after three years. Treatment has also been effective for Hodgkin's disease. Continued experimentation has indicated that mixtures of the alkaloids with other chemicals may produce even better results.

A second plant found to contain antitumor alkaloids is the mayapple (*Podophyllum peltatum*, Berberidaceae). Plants of this species are herbaceous perennials that flower early in the spring in the deciduous forests of Canada and eastern United States. The active compounds podophyllin and alpha and beta peltatin are particularly abundant in the rhizomes. Extracts of the roots were used by native Indians as purgatives and for skin disorders and tumorous

FIGURE 12-22
The periwinkle alkaloids have been found effective in the treatment of some leukemias.

FIGURE 12-23
The autumn crocus yields colchicine, used to treat gout and cell malignancies. (After Baillon.)

growths. Today, mayapple alkaloids are used as the basis of VM-26, a drug used to treat lymphocytic leukemia. Although newer than the drugs based on *Catharanthus* alkaloids, VM-26 appears to be as effective as vincristine in treating childhood leukemias.

A final plant used with some success in the treatment of cancer is the autumn crocus (*Colchicum autumnale*, Liliaceae, Fig. 12-23). The alkaloid colchicine extracted from the corms of this species interferes with mitotic cell division. As a chemotherapeutic agent for the treatment of cancer, it makes use of this property by, hopefully, preventing the proliferation of cancer cells. However, since colchicine also affects normal cells, its use must be carefully monitored, and it has produced only limited successful results.

OTHER PLANTS OF MEDICINAL USE

Many other plants are employed in the production of the enormous array of medicines available to us today. Several mucilaginous compounds of plant origin (see Chapter 11) are used in soothing ointments and as carriers for other medicines. Aloe and plantain constitute two of the most important sources of plant mucilages *Plantago ovata* and *P. psyllium* (Plantaginaceae) are sources of psyllium, a colloid mucilage used for intestinal problems. These plantains are annual, rosette herbs native to southern Europe, northern Africa, the Canary Islands, and Asia as far west as Pakistan and India. The seed husks of these species, which can contain up to 30 percent psyllium, are removed, ground, sieved, and screened. The resultant powder is administered and absorbs water in the intestinal tract, providing a smooth bulky mass that is unaffected by bacteria. As the mass moves through the intestine, it provides relief for irritations caused by both constipation and chronic diarrhea.

Species of *Aloe* (Liliaceae), primarily *A. barbadensis,* have also long been used for their soothing gels. In the Mediterranean region, *A. barbadensis* has been used since Greek times for a long list of disorders, but primarily for skin problems such as burns, ringworm, or venereal sores. The soothing agent derived from aloes is the gelatinous pulp scraped from under the epidermis or the juice of the perennial succulents. The fresh gel is applied topically to inflamed areas. In addition to the gelatinous pulp, aloes produce a latex that contains anthraquinone glycosides. A mixture of these glycosides isolated from the water-insoluble parts of the latex is known as *aloin*. The latex has been used as a powerful purgative because it irritates the intestine. Today, aloe gel is used in shampoos, conditioners, and hand lotions as well as mixed with lanolin into salves that provide relief for burns originating from fires, overexposure to sunlight, or radiation. Aloin is often used in combinations with the belladonna alkaloids in medications prescribed for intestinal ailments.

Another plant-derived drug that should be found in every home is *ipecac syrup*. Ipecac is obtained from *Cephaelis ipecacuanha* (Fig. 12-24), a member of the coffee family (Rubiaceae). It contains several alkaloids that cause vomiting. While this may appear to be more of an illness inducer than a cure, ipecac is a standard item in poison control kits. When poison ingestion occurs, vomiting can cause expulsion of the poison before it is assimilated. While ipecac is effective for only some kinds of poisonings, it can often safely and quickly avert a tragic accident.

A final plant-derived drug that has recently received considerable attention

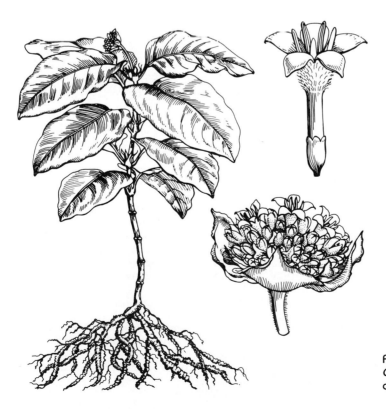

FIGURE 12-24
Cephaelis, from which ipecac is obtained. (After Baillon.)

is *chymopapain*, an enzyme obtained from papaya (*Carica papaya*, Caricaceae (see Fig. 5-27)). This enzyme is a sister to the more familiar papain (Chapter 5) used for tenderizing meats. In medicine, chymopapain is used because it exhibits specificity in its dissolution of proteins. Formerly, surgery was the recommended treatment of severe back pain due to the slippage of a disk in the spinal column. The pain caused by pressure of the displaced cartilagenous disk necessitated drastic action, and the excision of the misplaced piece of cartilage seemed to be the only solution. Now, doctors have found that injection of chymopapain into the area of the spinal column where the disk is causing problems relieves the pressure by dissolving the cartilage. Such a procedure is much simpler and less costly than major surgery. Drawbacks include the sensitivity of some individuals to the enzyme.

It is possible that more and more useful botanical drugs will be found as different plants continue to be tested. In China, herbal medicine has been practiced for thousands of years, and today, Chinese scientists are actively investigating substances in plants that have traditionally been used in folk cures. Several of those being investigated show promise and are being clinically tested. In the United States private and government agencies also carry out large-scale screening programs in attempts to locate potential medicinal plants. Hopefully, this search will be able to continue despite the destruction of natural vegetation and the extinction of species that is being caused by population increases.

ADDITIONAL READING

Balandrin, M. F., J. A. Klocke, E. S. Wurtele, Wm. H. Bollinger. 1985. Natural plant chemicals: Sources of industrial and medicinal materials. Science 228:1154–1160.

Duke, J. A., C. R. Gunn, E. E. Leppik, C. F. Reed, M. L. Solt, and E. E. Terrell. 1973. *Annotated Bibliography on Opium and Oriental Poppies and Related Species.* ARS-NE-28. Beltsville, Maryland, USDA.

Farnsworth, N. R. 1984. How can the well be dry when it is filled with water? *Economic Botany* 38:4–13. A discussion of the decline of pharmacognosy in the United States and an appeal for the continued investigation of plants as potential sources of medicines.

Gilbert, L. E. 1980. Ecological consequences of a coevolved mutualism between butterflies and plants. pp. 210–240 in L. E. Gilbert and P. H. Raven. (eds.) *Coevolution of Animals and Plants,* 2nd ed. Austin, University of Texas Press.

Harborne, J. B. 1982. *Introduction to Ecological Biochemistry,* 2nd ed. New York, Academic.

Klayman, D. L. 1985. *Quinghoasu* (artemisinin) an antimalarial drug from China. Science 228:1049–1055.

Kreig, M. B. 1964. *Green Medicine. The Search for Plants that Heal.* Chicago, Rand McNally.

Lawrence, G. H. M. 1965. Herbals, their history and significance. pp. 1–35 in *History of Botany.* Pittsburgh, Hunt Botanical Library.

Lewis, W. H., and M. P. F. Elvin-Lewis. 1977. *Medical Botany.* New York, Wiley.

Morton, J. F. 1977. *Major Medicinal Plants.* Springfield, Illinois, Charles C Thomas.

Rock, J. F. 1922. Hunting the chaulmoogra tree. *National Geographic* 41:242–276.

Stern, W. L. 1974. The bond between botany and medicine. *The Bulletin (Pacific Tropical Botanical Garden)* 4:41–60.

Swain, T. (ed.). 1972. *Plants in the Development of Modern Medicine.* Cambridge, Massachusetts, Harvard University Press.

Taylor, N. 1965. *Plant Drugs that Changed the World.* New York, Dodd, Mead.

Tippo, O., and W. L. Stern. 1977. Medicinal Plants. Chapter 9 in *Humanistic Botany.* New York, W. W. Norton.

Tyler, V. E., L. R. Brady, and J. E. Robbers. 1976. *Pharmacognosy,* 7th ed. Philadelphia, Lea and Febiger.

Chapter 13

Psychoactive Drugs
and Poisons
from Plants

*I*n the previous chapter, we discussed drugs that are used medicinally to relieve pain or control disease. Many of the same drugs can also alter an individual's sense of perception by producing feelings of tranquility, invigoration, or otherworldliness. Sometimes people want to escape from, or "alter," reality and thus seek out these drugs, but in excessive quantities, most are toxic. This chapter focuses on plant substances used because of their psychoactive properties, and on the overlapping group of poisons obtained from plants. Table 13-1 lists the plants discussed in this chapter.

There are many ways to group psychoactive drugs. They can be classified in terms of the chemical nature of the compounds involved, the effects they produce, or the source from which they are obtained. None of these classification schemes is perfect. The psychological effects of the compounds involved sometimes differ between chemically very similar substances, and a classification based on chemical structure is rather difficult for a nonchemist to remember. Attempts to place psychoactive drugs into categories of stimulants or depressants,

TABLE 13-1

Plants Discussed in Chapter 13

COMMON NAME	SCIENTIFIC NAME	FAMILY	CHROMOSOME NUMBER
Ayahuasca	*Banisteriopsis caapi*	Malpighiaceae	$2n = 20$, diploid
Belladonna	*Atropa belladonna*	Solanaceae	$2n = 50, 72$
Coca	*Erythroxylum coca*	Erythroxylaceae	$2n = 24$
Cohoba	*Anadenanthera peregrina*	Fabaceae	$2n = 26$, diploid
Curare	*Chondodendron tomentosum*	Menispermaceae	Unreported
	Strychnos nux-vomica	Loganiaceae	$2n = 24, 44$, di & polyploid
Datura (jimsonweed)	*Datura* spp.	Solanaceae	$2n = 24, 48$
Hemlock, poison	*Conium maculatum*	Apiaceae	Unreported
Henbane	*Hyoscyamus niger*	Solanaceae	$2n = 34$
Iboga	*Tabernathe iboga*	Apocynaceae	$2n = 22$
Kat	*Catha edulis*	Celastraceae	Unreported
Mandrake	*Mandragora officinarum*	Solanaceae	Unreported
Marijuana	*Cannabis sativa*	Cannabaceae	$2n = 20$ diploid
Mescal bean	*Sophora secundiflora*	Fabaceae	$2n = 18$
Ololiuqui	*Rivea corymbosa*	Convolvulaceae	$2n = 30$ diploid
Opium poppy	*Papaver somniferum*	Papaveraceae	$2n = 22, 44$ di, tetraploid
Peyote	*Lophophora williamsii*	Cactaceae	$2n = 22$ diploid
Pyrethrum	*Chrysanthemum cinerariifolium*	Asteraceae	$2n = 18, 27, 36$
Rotenone	*Derris elliptica* D. *uliginosa*	Fabaceae	$2n$ 20, 22, 24, 36
	Lonchocarpus nicou L. *urucu*		$2n = 44$ tetraploid Unreported
San Pedro	*Neoraimondia macrostibas*	Cactaceae	Unreported
	Trichocereus pachanoi	Cactaceae	Unreported
Tobacco	*Nicotiana tabacum*	Solanaceae	$2n = 48$ tetraploid
	N. rustica		$2n = 48$ tetraploid

TABLE 13-2

Psychological Classification of the Major Plant-Derived Psychoactive Drugs

KIND OF PSYCHOLOGICAL EFFECT	PLANT-DERIVED COMPOUNDS	PHYSIOLOGICAL ACTION
Behavior stimulants and convulsants	Cocaine	Blocks reuptake of norepinephrine
	Caffeine	Activates intracellular metabolism
	Nicotine	Stimulates acetylcholine receptors
	Convulsants (i.e., strychnine)	Blocks inhibitory synapses
Narcotic analgesics (opiates)	Opium, morphine, codeine	Probable mimics of endogenous protein neurotransmitters that relieve pain
Psychedelics	Tetra hydrocannabinol, mescaline, myristicine, elemicin	Effect unknown; norepinephrine mimics
	Atropine, scopolamine, ololiuqui, harmine	Block acetylcholine action; serotonin mimics
Antipsychotic agents	Reserpine	Depletes norepinephrine
Sedative-hypnotic compounds	Alcohol	Central nervous system depressants

Source: Adapted from various tables in R. M. Julien. 1981. *A Primer of Drug Action*, 3d ed. San Francisco, Freeman.

hallucinogens or narcotics sometimes leads to confusion because many drugs can act in a combination of ways. For example, many of the so-called depressants, like the narcotic analgesics (opiates) and the solanaceous alkaloids, can also act as hallucinogens. Stimulants such as nicotine and cocaine will likewise cause visions if taken in sufficient quantities. The use of the word "narcotic" itself is frought with problems because some people use it to refer strictly to those drugs obtained from the opium poppy (as here), while others employ it for any habit-forming drug or even for any drug used illicitly. Table 13-2 provides one way of grouping psychoactive drugs by their primary effects, but since this is a botany text, we will follow yet another system and discuss drugs taxonomically in terms of the species from which they are obtained. The two most widely used psychoactive compounds, caffeine and alcohol, are each treated separately in the next two chapters. These merit individual attention because of their widespread economic and cultural importance and because each is obtained from a variety of plant sources.

THE CHEMISTRY AND PHARMACOLOGY OF PSYCHOACTIVE DRUGS

Almost all chemicals that have psychoactive properties contain nitrogen, and most belong to one of the classes of alkaloids. The most notable chemical exception is the active compound of marijuana, Δ-*trans*-tetrahydrocannabinol

(THC) (see Fig. 13-7). Some other alcohols (including ethanol) and terpenes have psychoactive effects and have been designated as hallucinogenic agents.

Before they can act, psychoactive drugs must be absorbed into the bloodstream and transported to sites where they can exert their effects. Like medicinal drugs, psychoactive drugs can be taken orally, injected into the body, or absorbed through membranes such as those lining the nose, rectum, vagina, or lungs.

Once the active compounds enter the bloodstream, they are transported to all parts of the body. A drug may exert its influence in only one place in the

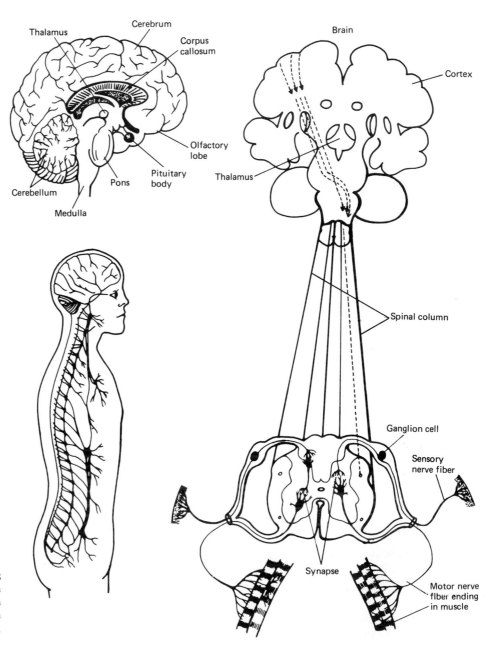

FIGURE 13-1
The central nervous system is composed of the brain and the spinal cord. A cross section of the brain and spinal cord shows the path of a simple reflex arc.

body, but it is still present (at least initially) throughout the bloodstream because blood and the products it carries continuously circulate around the body. As the blood passes through the liver on its cycle, some drug substances are removed, chemically degraded, and excreted. Other drugs are degraded in the places where they act, and the breakdown products are picked up by the blood and transported to the kidneys for excretion. It is this process of circulation, ending in degradation and excretion, that accounts for the initial "rush" followed by an eventual "wearing off" of a drug's effect.

Most psychoactive drugs act on cells of the central nervous system (the brain and spinal column, Fig. 13-1). Since the blood flow to the brain is 10 times that to other tissues, drugs reach the brain more quickly than other tissues. However, blood capillaries around the brain are more tightly packed that those elsewhere and are covered by sheaths that make passage of compounds from them into the brain difficult. Thus the actual amounts of drugs that enter the brain are less than the blood flow to it might suggest.

Once they have reached the central nervous system, psychoactive drugs generally act by altering the natural interactions between *neurons,* or sensory cells. These cells transmit information by chemical signals. Neurons (Fig. 13-2) release chemical substances called *neurotransmitters* in response to stimulation. Once released, these transmitters flow across the space, or *synapse,* between a transmitting neuron and a receiving neuron. Specific sites on the receptor recognize the transmitted compound and bind briefly with it. Once the compound has been accepted, it triggers a response in the receptor neuron. Among the receptor neurons involved in psychological reactions are those responsible for our perception of pain and emotion, as well as interpretations of audio and

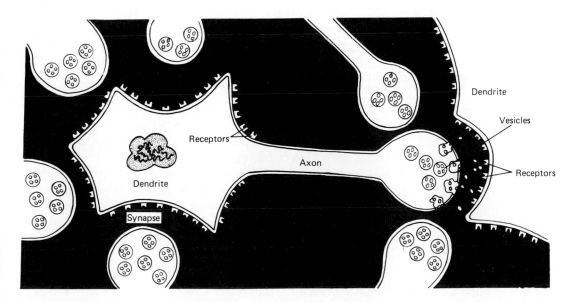

FIGURE 13-2
This diagram shows a neuron that has been electrically stimulated. The small vesicles at the end of the axon respond by releasing their contents which travel across the synapse to the receptors on the adjacent neuron's surface. (After a National Institute of Health Clinical Center photo.)

visual stimuli. Psychoactive drugs, for the most part, alter or mimic the behavior of four kinds of natural neurotransmitters: acetylcholine, norepinephrine, serotonin, and neuropeptides. The way in which THC works is still unclear. Caffeine (see Chapter 14) appears to act as a stimulant by activating intracellular metabolism, and alcohol (see Chapter 15) is a general central nervous system depressant.

Acetylcholine is a neural transmitting chemical that is released by neurons of the brain and the peripheral nervous system. It causes muscle contractions and speeds up the heart beat. The compound is synthesized within the transmitting neuron and stored until the neuron is triggered. Once stimulated, the neuron releases acetylcholine into the synaptic region. After a neighboring neuron has received the chemical at specific places on its dendrite and reacted, the acetylcholine is broken down by an enzyme so that it can no longer produce an effect. Drugs can effect this sequence by blocking the transmission of acetylcholine (atropine and scopolamine), preventing the breakdown of the chemical, or by mimicking its action (nicotine). Drugs that block its action, prevent rapid or jerky muscle reactions and thus produce a relaxed sensation. Compounds that prevent acetylcholine breakdown, or mimic its effects, act as stimulants because they cause the receptor neurons to fire at an increased or continuous rate.

Norepinephrine is another of the brain's endogenous (made within the body) neurotransmitters. It is synthesized in transmitting neurons and released when these cells are stimulated. However, after it exerts its effect, norepinephrine is reabsorbed, not broken down. Apparently, the same molecules are reused over and over. Some drugs (e.g., reserpine) deplete norepinephrine. Others such as cocaine prevent reabsorption of norepinephrine or mimic its action (mescaline, myristicine, and elemicin).

Serotonin is a brain neurotransmitter that stimulates cells regulating body temperature, sensory perception, and sleep. Some psychoactive drugs such as the LSD-type compounds appear to alter the functioning of the neurons that transmit serotonin, thereby producing illusions of strange images.

Finally, a recently discovered group of neurotransmitters are peptides, polymers of amino acids. These include enkephalin, beta endorphin, and oxytocin. These chemicals are produced in minute quantities and are received by very specific receptors. Some of these appear to act like the body's own pain killers. It is believed that opiates effect the same receptor sites as these endogenous pain killers and thus produce a dull, relaxed sensation. The evidence for this hypothesis is that opiates are very specific in their effects, effective in low concentrations, and blocked by antagonistic agents. Opiates influence receptor cells in the brainstem (see Fig. 13-1), the medial thalamus, and at several places in the spinal cord.

HISTORY OF DRUG USE

We tend to think of our society as "the drug culture," but the use of mind-altering drugs is very ancient. What perhaps distinguishes our culture from previous ones is the modern dissociation of drugs from formal cultural or religious customs. It is easy to understand why humans would appreciate a substance that alleviated pain or reduced hunger and fatigue. In such cases, psychoactive properties of drugs merge into medicinal usage. Nevertheless, many drugs have always been taken in excessive quantities when there was no physical need for them. These doses led to audio and visual hallucinations. The original appeal of these kinds of psychological effects was quite different from the modern enjoyment of the sensations as recreational experiences.

Primitive peoples without rational explanations for the phenomena they observed everywhere around them ascribed events to the actions of supernatural agents. Under normal conditions, people cannot see or speak with such powers, but under the influence of psychoactive agents, they could be transported to another world where communication with gods, demons, or even the dead was possible. In most cultures, only specific individuals were allowed to ingest psychedelic substances. Such people, usually men, were variously called healers, medicine men, witches, or shamans. Shaman, originally a Siberian word, has been adopted by anthropologists for this group of diviners because it lacks the connotations attached to most of the other designations.

The initiation of an individual into the role of shaman was often painful and dangerous. Initiation rites sometimes involved starvation, some form of self-mutilation, and consumption of psychoactive drugs. The combined effect produced a physical condition that enhanced the illusion of visiting the supernatural world and acquiring the insights necessary to serve as the inter- mediary between the earthly and spiritual worlds. The shaman repeated the ingestion of drugs when he needed the advice of ancestors or divine guidance to solve a problem.

Anthropologists and botanists who have studied psychoactive drug use have found that New World peoples employed many more species of plants for their psychoactive properties than their counterparts in the Old World (40 versus 6 species). Assuming that the distribution of psychoactive plants is the same in the two hemispheres, some authors have suggested that the explanation for the use of few plants in the Old World lies in the cultural superiority of Eurasian civilizations. This explanation assumes that shamanism is a primitive trait and that the cultural environments of the ancient European and Asian civilizations provided a milieu that replaced the need for spiritual intermediaries. Such authors forget, however, the localized distributions of many of the plants used by New World peoples and the overwhelming dominance of two psy- choactive drug plants, the opium poppy and marijuana, in the Old World.

Cannabaceae

According to a Neolithic Chinese legend, the gods gave humans one plant to fulfill all needs. The plant was *Cannabis sativa*, ma, marijuana, or hemp. This assertion is not so far-fetched. *Cannabis* assuredly ranks among the world's most remarkable plants. Its durable fibres can be turned into ropes, fish nets, and clothing (Chapter 16), its seeds are highly nutritious; and the oil expressed from them can be used for lamps or in paints and varnishes. Ten-thousand-year-old pot shards imprinted with twisted hempen fibers are part of the evidence that *Cannabis* was among the oldest cultivated plants. Today, the species is best known for the psychoactive chemicals it produces. Marijuana rivals alcohol, caffeine, and nicotine as the most widely used nonmedical drug. It is said to be the most profitable of California's agricultural commodities and perhaps the second most lucrative crop in the United States. How it came to occupy this position is a fascinating story, involving cultures spanning almost every continent.

Cannabis, native to central Asia, has seeds that can be dispersed by water, wind, birds, or animals. Still, it is humans who have spread the species worldwide, often without knowledge of its psychoactive properties. The first to use *Cannabis* were the Chinese, who had such high respect for the plant that they referred to their country as the "land of mulberry and hemp." The ingenious Chinese found many uses for ma. It was hemp fiber that T'sai Lun used to make the first true paper in 105 A.D. The ancient Chinese *Book of Rites* instructed that mourners wear hempen clothes to show respect for the dead. The Chinese apparently knew of hemp's psychoactive properties, but they were primarily interested in its medicinal virtues. The legendary Shen Nung (ca. 3000 B.C.) observed that the female plants contained a greater proportion of the creative (yin) principles than the male (yang), and recommended cultivation of the former and its use in correcting spiritual imbalances.

The ancient Hindu texts, the Vedas, describe Siva, the Lord of Bhang, bringing *Cannabis* down from the Himalayas for the enjoyment of the Indian people (Fig. 13-3). India was the country in which the hallucinogenic properties of *Cannabis* were first exploited. Whereas the Chinese grew their plants tightly spaced to discourage branching and therefore to improve their fiber production, the Bengalese (from "Bhangland") developed a cultivation strategy that maximized production of the psychoactive compounds. The Indians also realized that marijuana was dioecious (Fig. 13-4) and that female plants were more potent than males. The resins exuded on the upper leaves and bracts of the female inflorescences are particularly rich in intoxicating substances. Since the resins serve to protect vulnerable plant parts from desiccation, they are produced in greatest abundance when the plants are exposed to heat and sun. By growing the plants widely spaced and removing the male plants (to prevent fertilization and prolong female flowering) resin production is enhanced. The stimulation of resin production by dry conditions is the reason why plants grown in semiarid climates are more potent than those from cool temperate regions. Likewise, sinsemilla (unfertilized, or without seeds) marijuana is one of the strongest forms.

The most common way of ingesting *Cannabis* in India is in the form of bhang, a milk-based beverage concocted with ground *Cannabis* leaves, sugar, and an assortment of spices. Bhang is widely drunk and is commonly offered as a gesture of hospitality. Indians classify *Cannabis* products into ganja, consisting of the potent female flowers and upper leaves, and hashish (charas), which is relatively pure resin.

There are many colorful stories about traditional hashish collection. In Nepal, naked men are said to have run through the fields of flowering female plants and then to have scraped off the clinging resin globules. In Persia, the plants were beaten on rugs which were subsequently washed to dislodge the resins.

Marijuana may have also had a role in Western civilizations, but it is generally believed that the species had not spread farther west than Turkey in ancient times. The history of Arab use of hashish is not documented, but many

FIGURE 13-3
Because the Vedas credited the god Siva with giving humans *Cannabis*, he became known as the Lord of Bhang. As the plant itself acquired sacred meaning, the devout started calling it "indracanna," literally, "food of the gods."

legends exist to fill this gap. One of them credits an ascetic monk with finding *Cannabis* growing wild on the hillsides and sharing it with the poor (Sufis). The monk Haydar was so delighted with the plant that he ate nothing else until his death 10 years later in 1221 A.D. The Sufis are still associated with the use of hashish to ease the pain of their bleak existence.

When Marco Polo returned from his travels to the east in 1297 A.D., he brought back a story that was destined to become a classic. He recounted that there was a terrorist, Hasan-ibn-Sabah, the "Old Man of the Mountain," who had a large band of followers willing to do anything, even commit murder or

FIGURE 13-4
Cannabis sativa. Male shoot (A)
and flowers (B, C); female shoot
(D) and flowers (E, F).

suicide, for their leader. This blind loyalty was attributed to a very clever recruitment strategy. Potential candidates were drugged and brought unconscious into an exquisite mountain garden. There, among the exotic flowers and water works, beautiful women were ready to minister to every need. After experiencing the pleasures of this paradise, the recruit was drugged once again and, upon awakening, told that he could return to the garden in this life or after death only if he swore complete allegiance to the Old Man. The technique was apparently very successful, and the band of loyal followers became known as assassins, a name purportedly derived from hashish, the drug they were supposed to have been given. There is no evidence to substantiate this story, and much to refute it, but it was compelling enough to be retold countless times and has been misinterpreted even in recent history by antimarijuana zealots who imply that the hashish was given to the assassins to produce violence rather than to ensure loyalty.

In the fifteenth and sixteenth centuries, Arab traders introduced marijuana to Africa where its use quickly spread. The drug was commonly given to women in childbirth and fed to babies when they were weaned. The Kafirs called the plant dagga, a name later applied to a culture that made widespread use of *Cannabis*. Africans originally mixed the dried plant parts into beverages or chewed them. The technique of smoking became popular only after the Dutch colonized the continent (Fig. 13-5).

By the time of the Crusades, when European contact with the Arabs was reestablished, marijuana use was common throughout Africa and Asia. The *Tales of the Arabian Nights* (written sometime between 1000 and 1700 A.D.) was widely read in Europe and provided a romantic introduction for many to hashish. "The Tale of the Hashish Eater," in particular, described the effects of the drug on a degenerate who used it to delude himself that he was wealthy and handsome. When Napoleon's army was stationed in Egypt in 1798, recruits experimented with the local intoxicant and returned to France with stories and samples. French students suddenly felt that a visit to northern Africa was a necessary part of their education. They returned home with glowing youthful reports of marijuana use.

A French psychiatrist, Dr. Jacques Joseph Moreau de Tours, read of the medicinal uses of hashish in combating plague and dysentery and decided to test the drug himself. Because he had read that the drug produced a profound change in perception, he thought that he might be able to administer it to sane individuals in order to reproduce the symptoms of psychosis and thereby learn about the causes of mental illness. As part of his studies, he asked an artist friend, Theophile Gautier, to try the drug and report his reaction. Gautier was impressed with the effect of hashish on his artistic senses, and he recommended it to his friend, the painter Boissard. Eventually the group expanded, meeting monthly in Boissard's hotel suite where they consumed hashish while being observed by Dr. Morcau. The "Club des Hachichins," as they called themselves, got its name from the story of the Old Man of the Mountain. The club eventually became a meeting place of the great French artists and writers of the day: Alexander Dumas, Victor Hugo, Eugene Delacroix, and Charles Baudelaire. Not surprisingly, much of the art and literature of this period reflects the preoccupation with the newly discovered drug. Despite Moreau's eventual conclusions published in "On Hashish and Mental Alienation" (1845) that hashish use had

FIGURE 13-5
The Dutch taught the world how to smoke, but the North Africans contributed the waterpipe, or hookah, a device which cools smoke by drawing it through water. Lewis Carroll drew attention to the hookah in his *Alice's Adventures in Wonderland* (1865) by his depiction of the Caterpillar sitting on a mushroom smoking a hookah and giving advice to Alice about parts of the mushroom to eat to make her grow tall again: "In a minute or two the Caterpillar took the hookah out of its mouth . . . and crawled away remarking . . . 'One side will make you taller and the other side will make you shorter.' " (Chapter 5). Some authorities believe that Lewis purposefully used a mushroom because he was familiar with reports that certain mushrooms caused hallucinations in which size and distance were greatly altered.

overall deleterious effects on mental health, the use of the drug continued to spread. The fascination with hashish traveled across the English Channel, from the French to the British intellectual community, where W. B. Yeats, Oscar Wilde, and Ernest Dawson experimented with it to see if it enhanced creativity.

Marijuana's introduction into the Americas was sporadic. The Spaniards were apparently the first to bring *Cannabis* to the New World when they started hemp cultivation in Chile in 1545. The British, hoping to produce a lucrative fiber crop, brought hemp to Jamaica around 1800. Although their commercial attempts failed, *Cannabis* managed to escape cultivation and spread around the island. When African slaves were brought to Jamaica in the middle of the century to harvest sugar cane, they found a local source of their familiar drug already established.

Before 1800, the British had sent formal orders for the colonists in North America to produce hemp. Even Thomas Jefferson and George Washington realized the military need for rope-making fibers and took up hemp cultivation. An appreciation of the other uses of *Cannabis* did not develop in the United States until the 1800s. The introduction of hemp's psychedelic effects followed the same pattern as in France and England. First there was tremendous curiosity, primarily by intellectuals and artists, a rash of drug-inspired art, and then a general spread of marijuana use. In fact, the 1876 Centennial Exposition had a Turkish bazaar which featured hashish smoking as a special attraction (Fig. 13-6). Despite the sensationalized stories that soon began to appear about decadent hashish dens and rampant consumption by the poor, the use of the drug spread.

During prohibition, marijuana smoke floated up and down the Mississippi to the tunes of Dixieland. Jazz musicians found "moota" a pleasurable way to enhance their music without experiencing the stupefying effects of alcohol. They expressed their appreciation of the drug with such songs as Benny Goodman's "Sweet Marijuana Brown," Cab Calloway's "That Funny Reefer Man," "Texas Tea Party," the "Mary Jane Polka," and others.

It has been suggested that the desire to suppress the increasing popularity of black and Latin-inspired music began the movement that eventually led to the criminalization of marijuana. However, racism would have been only one of many factors that countered marijuana use. Perhaps the most effective antimarijuana campaign was carried out by Harry Anslinger who was struggling to justify the existence of the newly created Narcotics Bureau of the Federal Government. He publicized stories of the horrors of marijuana "addiction" and depicted the pot smoker as a savage fiend whose aroused sexual desires and violent tendencies led to criminal activities and the use of stronger drugs. Anslinger revived the story of the "Old Man of the Mountain," but twisted it so that the drug itself was the cause of irrational violence. Marijuana was tried by the press and condemned by an emotional, ill-informed public. In the late 1920s and 1930s, marijuana use was banned in Louisiana, Texas, and Illinois. Other states soon followed, and in 1937 the use of *Cannabis* came under federal jurisdiction with the passage of the 1937 Marijuana Tax Law which actually heavily taxed, but did not prohibit, the drug.

Other countries had previously tried to reduce or prevent *Cannabis* use. South Africa passed the first anti-*Cannabis* law in 1870, and by 1925, the League of Nations agreed to an international statute against its use. All of these legal sanctions against *Cannabis* had two things in common, they were not based on

FIGURE 13-6
Two gentlemen smoking chiboques in the Turkish bazaar at the U.S. Centennial Exposition in 1876. It is thought that these pipes were filled with hashish since the custom of smoking hashish had recently become popular among the cultured and well-traveled. (Photo courtesy of the Fitz Hugh Ludlow Memorial Library.)

any substantial medical evidence of severe effects of marijuana use, and they universally failed to halt its spread. Most medical studies beginning in 1893 with the Indian Commission's thorough report failed to find any detrimental effects of moderate use. Sanctions against marijuana only seem to increase its popularity. The explosive spread of pot smoking in the 1960s was a drastic demonstration of this kind of effect. Young people throughout the world, disillusioned with the ways in which the older generation was handling affairs, made marijuana smoking a symbol of their rebellion.

While attention was focused on marijuana in the 1960s, the suggested medical virtues as well as the possible detrimental effects were investigated. Once THC (Fig. 13-7) was finally isolated in 1965 and measured quantities could be used for testing, it was discovered that the active principal is effective in reducing the pressure exerted within the eyes of glaucoma patients. Use of the compound also reduces the nausea experienced by cancer patients undergoing radiation or chemotherapy treatment. Since THC dilates the bronchial vessels, it provides relief for asthma sufferers. Because of the useful medicinal qualities, marijuana cigarettes or concentrated THC in the form of pills can now be prescribed by physicians when appropriate.

In 1972, a federal government study on the effects of marijuana was released. The report suggested decriminalization of marijuana because there was no evidence of medical or mental damage from use of the drug and because the drug was so widely used that laws against it were unenforceable.

In 1981, the results of another study made by the National Academy of Sciences were released. The academy panel concluded that the use of marijuana should be viewed with concern since low or moderate use impaired the sense, sensibility, and sensitivity of users. They could find no evidence of addiction or permanent deleterious medical effects with low or moderate use, but heavy use was correlated with several psychological and physiological problems. In males, heavy use was linked to low sperm counts. Since marijuana is usually smoked, heavy users can develop lung disorders similar to those incurred by cigarette smoking. Finally, the Academy found that there was a high probability that heavy users would turn to "hard drugs." The panel did conclude, however, that marijuana was safer than its legal, more widespread counterparts, alcohol and tobacco, and suggested that the present legal sanctions against possession and trade be removed.

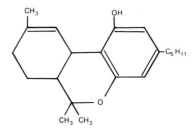

FIGURE 13-7
The action of Δ-trans tetrahydrocannabinol (THC), a complex alcohol responsible for marijuana's intoxicating effects, is still not understood.

Papaveraceae

Poppies are probably best known in the United States as garden ornamentals or for their culinary seeds, but they rose to fame because of opium. As we discussed in Chapter 12, the capsules of the fruits from which the seeds are obtained are rich in an alkaloid-containing latex (Fig. 13-8). The latex has been extremely important in medicine, but opium itself, and morphine, the most abundant of the opium alkaloids, are most often used as mind-altering drugs.

The collection of opium from fruits of the opium poppy (*Papaver somniferum*) extends back to at least 3000 B.C. Sumerian tablets from 2500 B.C. refer to opium as the "joy plant" and describe the ingestion of small balls of the latex to induce sleep and relieve pain. Opium is mentioned in every medical treatise written during ancient Greek and Roman times. The name opium comes from the Greek "opion" for poppy "juice." The early Greeks realized that the drinking of wine in which latex had been dissolved led to a trancelike state, and they therefore associated poppies with several divinities such as Hypnos, the God of Sleep, Morpheus, the God of Dreams, and Thanatos, the God of Death.

It is commonly thought that the Chinese introduced opium to the rest of the world, but China was ignorant of the drug until the seventh century A.D. when Arab traders first brought samples to the Orient. The original use in the

FIGURE 13-8
The pericarp of the developing
opium poppy capsule is slashed so
that the latex in the vesicles under
the epidermis will exude. (Photo
courtesy of Wayne Ellisens.)

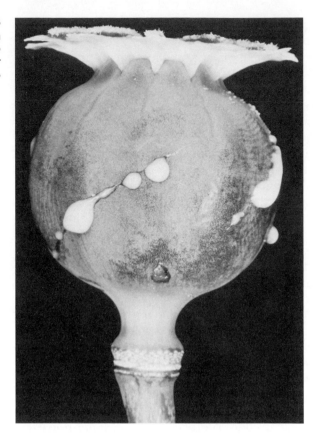

East was to cure dysentery in much the same way that paragoric (tincture of opium) is used for diarrhea today. In the seventeenth century, the Dutch introduced tobacco smoking to Formosa and began to mix tobacco with opium in their pipes as a treatment for malaria. The practice of smoking opium spread to the mainland where it was rapidly adopted throughout the country. Soon, tobacco disappeared from the mixture, and smokers inhaled vapors from heated balls of latex that were dropped into the bowl of a pipe (Fig. 13-9).

Opium smoking became so popular that Chinese officials sought to ban the sale of opium. Their antiopium edicts, issued as early as 1729, were ignored by both the Chinese populace and the Portuguese suppliers. In 1800, the British gained a monopoly over trade rights with China. Since China had little use for Western goods, the English felt compelled to continue the Portuguese practice of opium dealing because it was one of the few commodities for which the Chinese would exchange coveted products such as silk and spices. Eventually, England even established opium plantations in her Indian colonies to provide opium for trade. Importation of opium into England was, however, illegal at this time, but distance and the need to maintain international trade seems to have exonerated the practice elsewhere. The Chinese lords tried again and again to halt the drug trafficking because they could see the debilitating effects on their people, but without the support of the British, their efforts failed.

The situation came to a head in 1839 when the Chinese confiscated and destroyed British opium supplies in Canton. England retaliated by invading

FIGURE 13-9
A Chinese opium den.

China, touching off the first Opium War. The war ended with a treaty that ceded Hong Kong to the British. A second Opium War broke out in 1865 and the Chinese were again defeated. This time they were forced to agree to British opium importation. Naturally, opium consumption among the Chinese continued to escalate. Trafficking in the drug was finally reduced in 1906 when the last emperor of China managed to get the British to agree to a reduction in importation. Still, use did not drastically drop until the revolution and the establishment of the People's Republic of China in 1949. Opium trading and use are now strictly forbidden.

Opium consumption was not common in Europe until 1525 when Paracelsus (see Chapter 12) discovered, or rediscovered, a way to dissolve it in alcohol. The tincture produced by dissolving opium in alcohol was known as *laudanum*. The ease with which laudanum could be consumed and its purported medical benefits made it the most popular drug of its time.

In 1803, morphine was isolated from opium (Fig. 13-10), producing a purified alkaloid that could be given in measured doses. On a per weight basis, morphine is 10 times as strong as opium (which contains at least 24 other alkaloids and numerous other compounds). Purified morphine can also be administered intravenously. When the hypodermic syringe was developed in 1853, it presented doctors with a rapid method of introducing this potent pain killer into the bloodstream and morphine use rose.

Morphine's effectiveness was both a blessing and a curse. In many cases, the speed with which a drug enters the bloodstream (and thus the intensity of the "rush") is correlated with the level of addiction it can produce. Only after 45,000 soldiers returned home from the Civil War addicted to the pain killer were the dangers of morphine use recognized.

In an attempt to develop a nonaddicting pain killer, scientists discovered that morphine could be chemically altered by the addition of two acetyl groups. The end product is a semisynthetic compound known as *heroin*. Heroin is an even more powerful analgesic than morphine, and it, like morphine, was originally described as lacking the addicting qualities of its parent drug. While vestiges of this idea persisted until 1905, it soon became apparent that heroin was even more dangerous than morphine because it crosses cell membranes

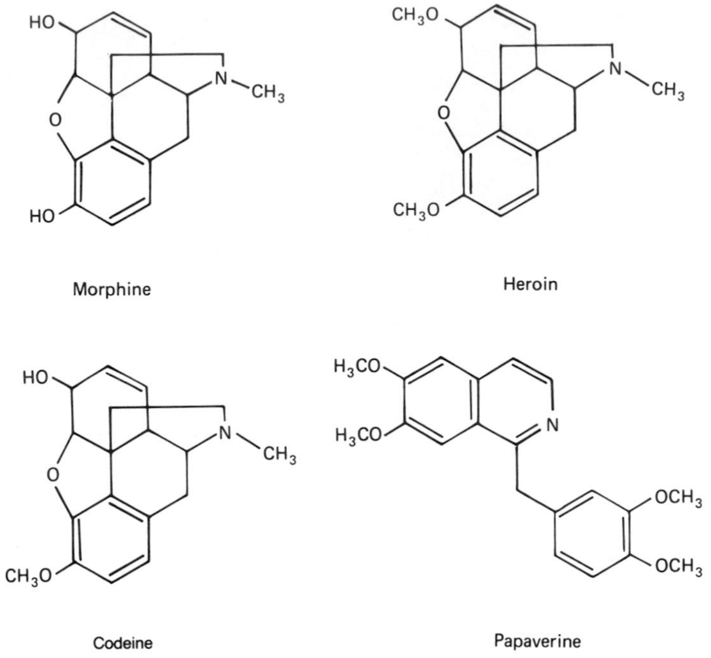

FIGURE 13-10
The major opium alkaloids.

Morphine

Heroin

Codeine

Papaverine

more rapidly than its natural counterpart. Because of this fast absorption (and hence cycling around the bloodstream), heroin must be more frequently administered than other opiates. In addition to the effects produced by heroin itself, users of the drug suffer from disruption of the blood flow, infections, diseases such as hepatitis, and collapsed blood veins. Heroin is very physically addicting and produces pronounced withdrawal symptoms once the habit has become established. Two to three weeks of injecting 30 mg per day is considered enough to create a tenacious habit. The main cause of death for heroin addicts is by overdose. Because of the common usage of adulterants in heroin and the use of heroin in combination with other drugs, "overdose" levels for an individual are not constant and therefore cannot be adequately gauged. The lifestyle of the heroin addict who no longer seeks the drug for pleasure, but because he or she desperately needs it to avoid withdrawal, tends eventually to lead to overall physical problems. The death rate for heroin users is more than twice the normal rate.

In 1914, a law was passed prohibiting the possession of opiates for nonmedical purposes. In 1924, the manufacture of heroin was declared illegal in the United States, and in 1956, existing legal supplies of the drug were destroyed.

Because of its addictive properties, even morphine is seldom used in its pure form as a modern-day medicine. In contrast, the use of heroin, all illegal, has dramatically increased during the last 30 years. Since heroin is made from morphine, the total production of opium has correspondingly increased. Until 1972, most of the opium that eventually ended up in the United States was grown in Turkey. When Turkey and the U.S. government reached an agreement to limit opium poppy cultivation, the center of production shifted to southeast Asia. Economically, opium is an extremely lucrative crop. In 1972, 10 kg of crude opium which sold for $250 where it was grown, would eventually gross $200,000 when sold on the street. With profits such as these, people are willing to risk the penalties if caught.

Fabaceae

Many species of the bean family (Fabaceae) are rich in alkaloids, and some of them have been used as hallucinogenic agents by various New World Indian tribes. Snuff made from powdered pods of cohoba (*Anadenanthera peregrina*) was first reported from Hispaniola in 1496 and was much later found to be used by the Indians of the Orinoco region of Venezuela. In South America, the snuff is called "yopo," and it is often mixed with lime (calcium carbonate) when taken. The snuffs can be inhaled through a number of devices, one of which is a long, hollow pole. One end of the pole is placed in a nostril and the other into a pile of snuff on the ground.

Mescal, or red bean (*Sophora secundiflora*; Fig. 13-11), a common ornamental shrub or small tree of the American southwest, was used by native peoples of this region to induce trances. The plant has received some recent publicity because mescal beans have been confused with mescaline and unknowing youths have been tempted to try them. However, *Sophora* contains cytisine, a much more dangerous alkaloid than mescaline, and one that can easily cause death due to respiratory failure if slight overdoses are taken.

Celastraceae

As Americans, we are accustomed to reaching for a cup of coffee, tea, or other caffeine-containing beverage when we need a "lift," but in other parts of the world, similar effects are produced by consuming plants with different but equally effective stimulatory alkaloids. Among these is kat (*Catha edulis*), a native of Arabia but now most widely grown along the eastern edge of Africa. The alkaloids are ingested by chewing wads of freshly cut leaves usually mixed with small amounts of lime. Each lump, or quid, is masticated for about 10 minutes until all of the juice has been expressed. The remaining cellulose mass is then swallowed.

Erythroxylaceae

Like kat, coca (varieties of either *Erythroxylum coca* or *E. novogranatense;* see Fig. 12-11) has traditionally been chewed to extract the stimulatory alkaloids. Coca is native to the north-central Andes. Discoveries of bags of coca leaves and utensils in 3,000-year-old Andean burial sites indicate that it was used by natives of this region long before Europeans discovered South America. By the time Pizarro conquered Peru, coca was an integral part of Inca life and was considered to be a sacred plant.

Preparation of coca involves simply collecting the leaves, drying them, and sometimes allowing them to "sweat" (lightly ferment) in order to render them pliable rather than crisp. Indians normally dip the leaves into lime that they carry in a small bag before chewing them (Fig. 13-12). As in the case of its use with kat, lime aids in the extraction and absorption of alkaloids. The masticated residue is either expelled or swallowed.

FIGURE 13-11
Texas mountain laurel is valued as an ornamental plant because of its fragrant purple flowers and glossy green foliage. The red and black seeds, sometimes called mescal beans, contain cytisine, an alkaloid which has been used to produce hallucinogenic trances, but which is extremely toxic and can produce respiratory failure and death.

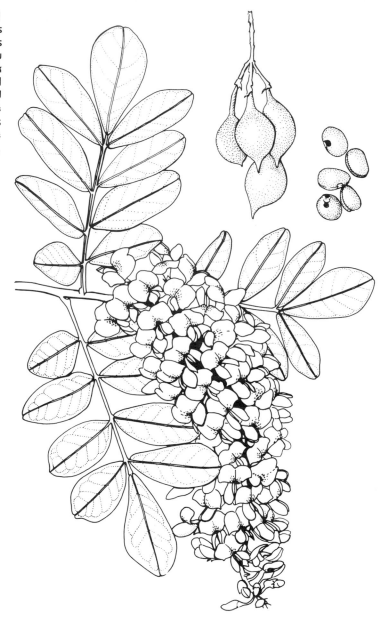

Cocaine, the active principle in coca, acts directly on the central nervous system by blocking readsorption of norepinephrine (Table 13-2) in the brain. This blockage has the result of making the coca chewer feel invigorated and relatively immune to fatigue and hunger. Peasants of the Inca Empire depended on coca, as do an estimated 90 percent of their modern descendants, to mitigate the harsh environment of the high elevations of the Peruvian and Bolivian Andes.

The Spanish conquerors tried to prohibit coca use until they realized that the Indians they enslaved would work harder if allowed to chew it. Coca was taken back to Europe by the Spanish, but in leaf form it never received much

FIGURE 13-12
This Peruvian woman sells both coca leaves and oca, the lime in which the Indians dip the leaves before masticating them.

attention. In 1860, cocaine was isolated and, unlike the homely leaves, it soon became extraordinarily popular. Sigmund Freud publicized the drug in his treatise *Uber Coca* (1884). Among other things, Freud recommended coca as a treatment for alcoholism and morphine addiction. He also lauded it as a local anesthetic and praised its use in psychotherapy to relieve depression. In both Europe and the United States, doctors began to prescribe its use (Fig. 13-13).

An enterprising Italian, Angelo Mariani, developed a coca wine beverage in 1860 which was soon the rage of Europe. Mariani's wine earned commendations from scores of celebrities including Jules Verne, Ulysses S. Grant, President McKinley, Emile Zola, Henrik Ibsen, and Thomas Edison. Coca cola, which originally included both caffeine-rich extracts from *Cola nitida* (see Chapter 14) and coca extracts, was first marketed in 1886 as a headache remedy. Since 1904, however, federal law has prohibited the inclusion of cocaine in any beverage. Ironically enough, after the Coca Cola Company complied with the federal law, it was sued for misleading advertising on the grounds that the name implied that the beverage contained coca products. As a result, coca leaves, with the cocaine removed, are now used to flavor the syrup from which the soda is made.

As more and more reports of cocaine-induced violence and other unsubstantiated effects of the drug circulated after the turn of the century, the government took steps to curb its use. In 1914, the drug was formally declared illegal by the Harrison Narcotic Act. As a result, cocaine costs escalated, but despite illegality and high costs, its use continues to increase. Status seekers appear delighted to pay exorbitant prices for a brief experience. Indulgers usually chop or grind crystals of cocaine hydrochloride before sniffing ("snorting") a line of the fine powder. Cocaine may also be smoked or injected. "Free basing," the conversion of cocaine from the salt form to the base form which alters its solubility (increasing the drug's potency), has proved to be particularly harmful. Both these latter methods of ingestion produce stronger sensations but are much more dangerous to the user than sniffing.

Just how dangerous is cocaine? While researchers have not been able to

FIGURE 13-13
"What do you do, my dear, to look so beautiful?" "I drink Coca of the Incas every day," exclaims a French poster printed in 1876. Coca des Incas was a French tonic wine made from coca leaves—a forerunner of Coca Cola. (Courtesy of Tim Plowman.)

find any organic damage from prolonged coca use per se among Andean Indians, their findings do not mean that cocaine is a harmless drug. It should be remembered that the Indians who chew coca leaves absorb only limited quantities of the drug because the alkaloids are present in the leaves in relatively small amounts (0.65 to 1.25 percent on a dry weight basis). An average native user might consume 2 oz of leaves per day or about 0.7 grains of cocaine. A person who "snorts" or injects cocaine could be consuming as much as 6–8 grains per day of the drug plus adulterants. A lethal dose of cocaine is considered to be 1.2 grams. Cocaine is not physically addicting, but its use can become a pronounced habit as indicated by the burgeoning number of treatment centers for cocaine abusers. After becoming accustomed to the elevated feelings it produces, habitual users suffer from depression when denied the drug. While occasional use produces feelings of mood elevation, vitality, mental clarity, sexual stimulation, and reduction in appetite in many users, chronic users can suffer from anxiety, confusion, insomnia, and impotence. Paranoid psychosis can develop from long-term heavy use. Recent experiments with pregnant animals have suggested that cocaine can be detrimental to fetal development. As more research is conducted, other harmful effects may be discovered.

Malpighiaceae

One of the most important groups of plants used by South American Amazonian tribes as a source of hallucinogenic compounds are members of the genus *Banisteriopsis,* primarily *B. caapi* (Fig. 13-14). Depending on the tribe, this

FIGURE 13-14
Banisteriopsis caapi is a tropical vine with stems that contain the alkaloid harmine, a psychoactive substance known as yaje or ayahuasca by the Amazonian Indians who use it in rituals. (After E. W. Smith in P. Furst, 1974, *Flesh of the Gods*)

species is called caapi, ayahuasca, yaje, or cipo. Infusions of mashed bark or stems are usually prepared, but stems can also simply be chewed in order to release the active alkaloid. People using ayahuasca describe the perceptions produced as feelings of having experienced death or of experiencing a separation of spirit and body. Sometimes the hallucinations are pleasant, other times apparently terrifying with vivid apparitions of jaguars or snakes (Fig. 13-15). Like many other alkaloid-rich preparations, mixtures of ayahuasca often cause vomiting and diarrhea. Unlike the use of many other psychoactive drugs by native peoples, ayahuasca is often taken in communal rituals. It is thought that this practice of communal ingestion enhances feelings of clairvoyance or telepathy. The group, under the supervision of a yaquero, is led through chants to experience shared hallucinations. This effect was reflected in the name ''telepathine'' given to the mixture of active chemicals first isolated from *Banisteriopsis caapi*. Since this initial chemical work, the active compounds have been further purified and named harmine and harmaline (see Fig. 13-22). In some tribes, the rituals commemorate an act of incest by a deity. In almost all cases, the ceremonies are filled with sexual symbolism, but there is no sexual gratification during the experience.

Cactaceae

Several species of the cactus family are used as sources of hallucinogenic compounds. The best known is peyote, *Lophophora williamsii,* a small, globose, gray-green cactus native to the Rio Grande Valley bordering the United States and Mexico (Fig. 13-16). Native American Indians appear to have extended its distribution to Arizona and New Mexico centuries ago. No one is sure when

FIGURE 13-15
Many drugs produce hallucinations featuring animals. People under the influence of ayahuasca often report seeing jaguars or reptiles such as snakes. The solanaceous (tropane) alkaloids are associated with visions of wolves. (Collage of animals from J. Harter, 1979, *Animals*, New York, Dover Publications)

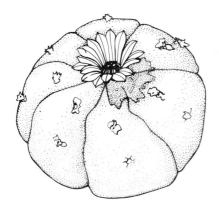

FIGURE 13-16

Mescaline is the psychoactive compound in the gray-green peyote buttons that grow natively in southwestern U.S. and adjacent Mexico. Several other species of cacti also contain the alkaloid and cause hallucinations.

peyote was first used, but sixteenth century reports by European explorers describe its use by the Aztecs as a divinatory plant. These accounts refer to peyote as the "diabolic root," because the Spanish observed the Aztecs using the plants ritualistically. After the collapse of the Aztec Empire, use of peyote survived among a few Mexican Indian tribes such as the Huichol, Cora, and Taramara. In the United States, the Plains Indians began to use it as late as the 1880s.

Indians harvest the plants by cutting off the top of the stems and leaving the sturdy taproot for regeneration. The stem tips, or buttons, were eaten fresh, or dried for later consumption. Dried buttons require a period of softening and mastication before they can be swallowed. The initial reaction after swallowing is nausea. One to two hours later, queasiness disappears and is replaced by kaleidoscopic illusions of vivid colors, sensual hallucinations, and distorted perceptions of time. During this last period, which can last from 5 to 12 hours, the faithful report hearing the voices of their ancestors who help them to diagnose and cure any problems they may have.

Although 30 to 40 different alkaloids and numerous chemicals are ingested when peyote is consumed, the most active compound present is the alkaloid mescaline (Fig. 13-17). Isolated in 1896, mescaline was the basis for research into psychedelic compounds for the next 50 years.

While there is no evidence that peyote or pure mescaline is addicting, repeated use may permanently affect the mind. Because of their potentially

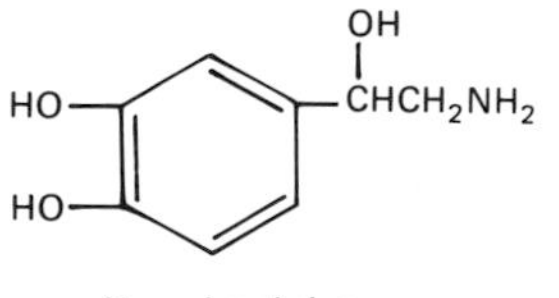

Norepinephrine

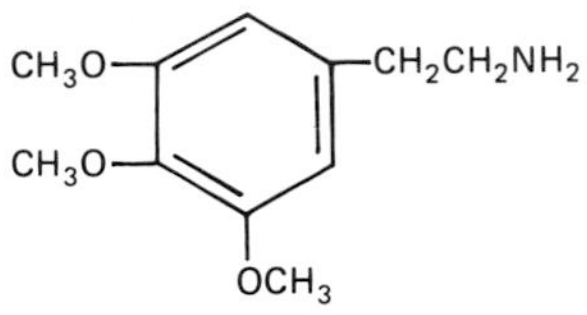

Mescaline

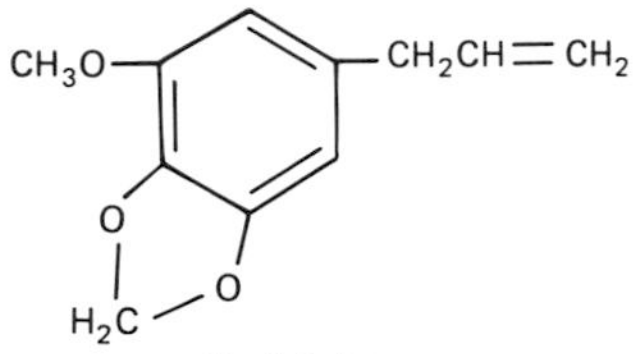

Myristicine

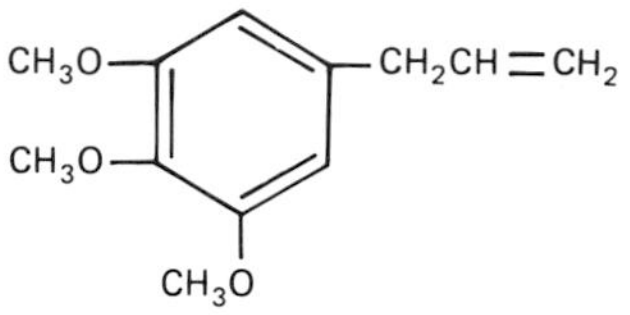

Elemicin

FIGURE 13-17
The chemical structure of mescaline, the dominant psychoactive alkaloid in peyote and related alkaloids which mimic the action of norepinephrine.

harmful effects, both the plant and the compound are illegal to possess or sell in the United States. However, one religious sect, the Native American Church, uses peyote as an integral part of its services. In a case that went before the Supreme Court, the church received an exemption from the prohibition against peyote use. Originally founded by natives in northern Mexico, the sect holds Christian services modified by the use of peyote and espouses as doctrine, brotherly love, abstinence from alcohol, and high standards of moral conduct. Today, there are over a quarter million members of the church within the United States (Fig. 13-18).

Other cacti such as the San Pedro cacti (columnar species of *Neoraimondia* and *Trichocereus*) have also been used by indigenous peoples in their native regions of Peru and Chile. All of the species used appear to contain mescaline and are used to produce hallucinations.

FIGURE 13-18
"A white man uses prayers out of a book, these are the first words on his lips. But with us, peyote teaches us to tell from the heart" says a Cheyenne peyote leader. Because of the importance in their worship services, peyote use is legal in the Native American Church. A mescal tepee on the Kiowa Reservation. (Photo taken in 1892; courtesy of the Smithsonian Institution National Anthropological Archives.)

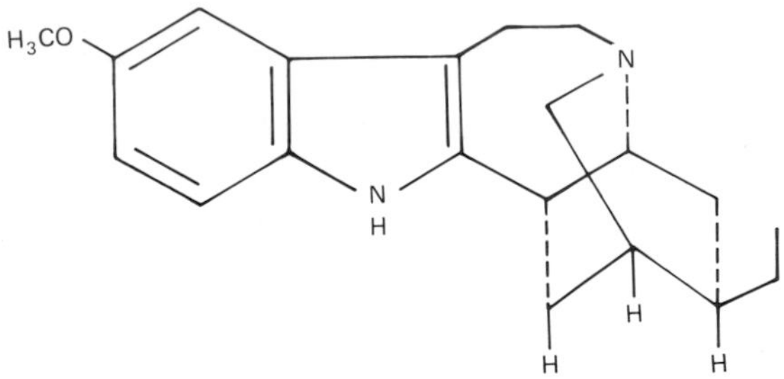

FIGURE 13-19
Ibogamine, the principal alkaloid responsible for the psychoactive effects of iboga.

Apocynaceae

While fewer drug plants were used by native Africans than by South and Central American natives, one, iboga (*Tabernathe iboga*) was important as an hallucinogen. Iboga is, in fact, one of the very few psychoactive drugs in the entire periwinkle family. Although the species of the family are rich in alkaloids, most of them are toxic (Fig. 13-19). A few species such as *Rauvolfia serpentina* and *Catharanthus roseus* are used medicinally (see Chapter 12), but there is no indication that they were used in rituals. In preparing iboga, the bark or roots of the shrubs are powdered or chewed, often with other plant species. Ingestion of the alkaloids, one of which is ibogamine, is said to lead to frightening visions and in large doses can produce severe convulsions, respiratory arrest, and death.

Convolvulaceae

Before 1937, no species of the morning glory family was known with certainty to be used for psychoactive purposes. In 1955, Humphrey Osmond described the use of *Rivea corymbosa* (Fig. 13-20) by Mexican natives, and, in 1960, T. MacDougall detailed the use of another member of the family, *Ipomoea tricolor*

FIGURE 13-20
Rivea corymbosa, or ololiuqui, a white-flowered vine of the morning glory family contains the psychoactive compound, D-lysergic acid amide.

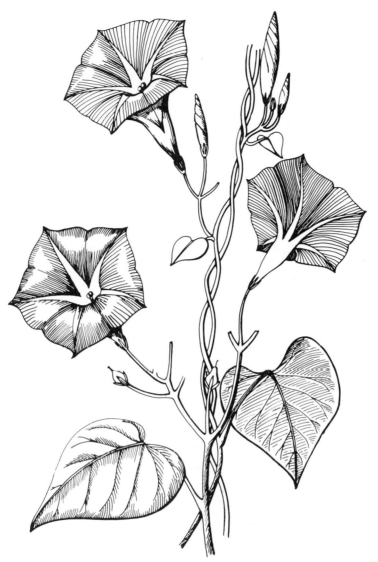

FIGURE 13-21
Because of the extremely fine line between effective and lethal doses, the seeds of hallucinogenic morning glories have rarely been used even by primitive peoples.

(Fig. 13-21) for altering perceptions. Still more startling was the discovery in 1960 that the seeds of these plants contained D-lysergic acid amide (LSD) (Fig. 13-22). Prior to this time, lysergic acid alkaloids were thought to be restricted in nature to ergot (*Claviceps purpurea*), a rust that infests grains, and to a few other fungi (Fig. 13-23). Synthetically produced LSD had become a part of the American counterculture in the 1960s and was regarded as something very modern and exciting. Osmond's and MacDougall's discoveries showed that the effects of these kinds of compounds had been recognized and utilized by native peoples, perhaps for thousands of years.

Both of the convolvulaceous species now known to be used are called *ololiuqui* (pronounced o-low-lee-oo-key) (Fig. 13-21) by the peoples that use them. Very carefully measured quantities of seeds are ingested and usually only by experienced individuals during divinatory rituals. Because there is little

FIGURE 13-22
Serotonin, a naturally produced neurotransmitter, and its mimic alkaloids, harmine and harmaline (from yage), D-lysergic acid amide (from ololuiqui), and semisynthetic LSD-25 (D-lysergic acid diethylamide). Although D-lysergic acid amide structurally resembles its synthetic counterpart LSD-25, it is only one-tenth as powerful.

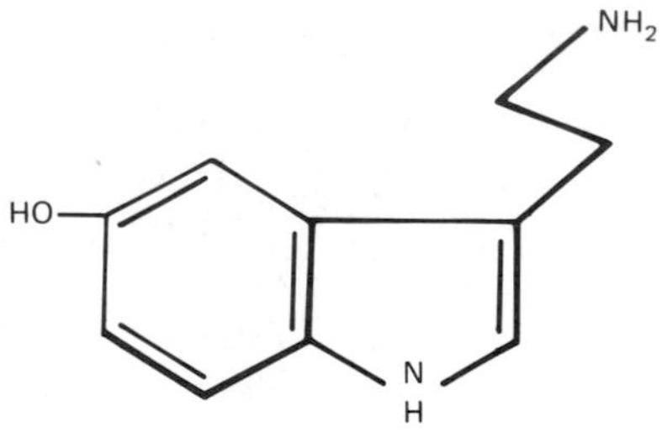

Serotonin

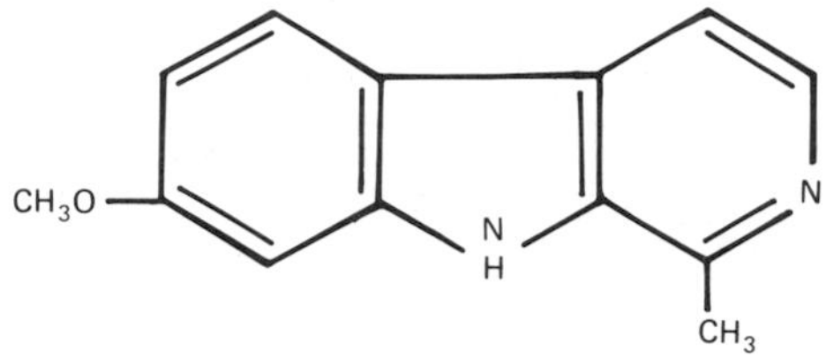

Harmine

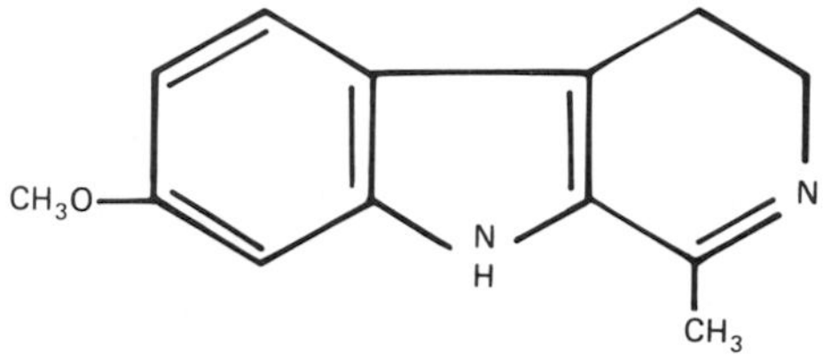

Harmaline

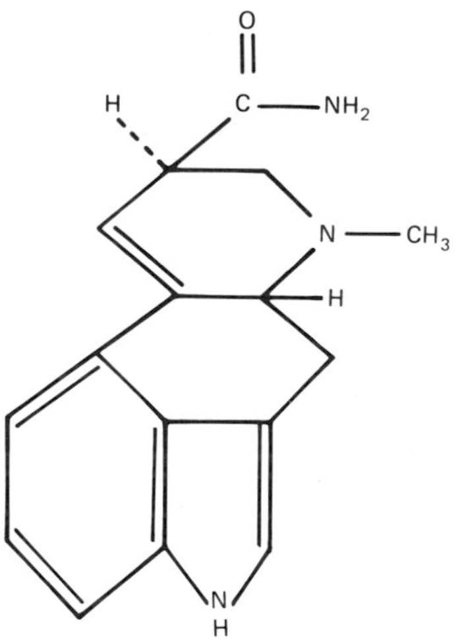

d-Lysergic acid amide
ergine
(from ololiuqui)

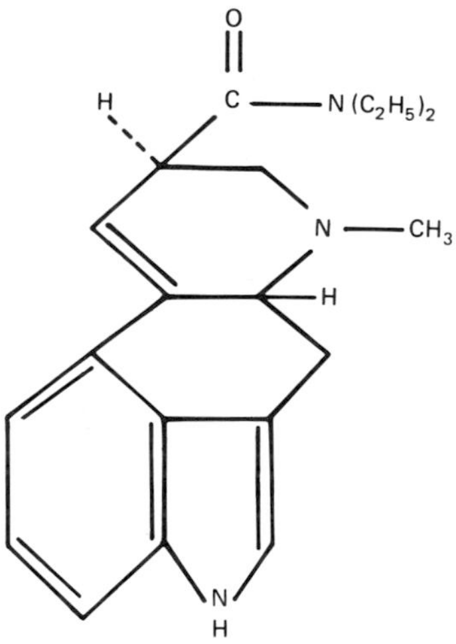

d-Lysergic acid diethylamide
LSD-25
(semisynthetic)

difference between the quantities of seeds that produce vivid hallucinations and the quantities that produce death, there has been little lay experimentation with the seeds, despite the fact that one of the species is a commonly cultivated ornamental plant.

Solanaceae

Although used by some Central and South American peoples, plants of the Solanaceae are rarely used today as sources of mind-altering drugs. Undoubtedly,

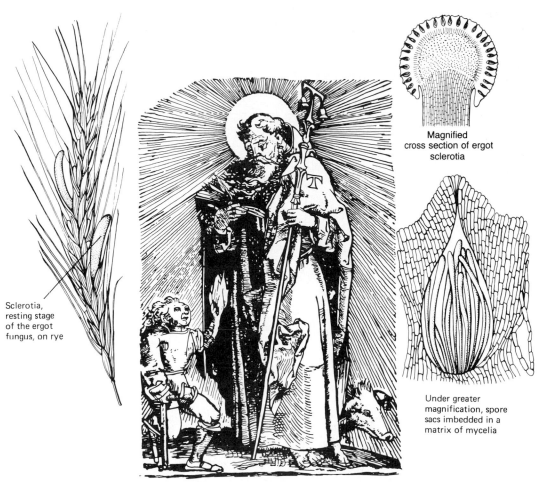

Sclerotia,
resting stage
of the ergot
fungus, on rye

Magnified
cross section of ergot
sclerotia

Under greater
magnification, spore
sacs imbedded in a
matrix of mycelia

St. Anthony and a victim of ergotism

FIGURE 13-23

Ergotism is caused by the ingestion of sclerotia (the resting stage of the ergot fungus which parasitizes rye grain). Epidemics of ergotism were especially common among European peasants who unwittingly ate infested rye. The symptoms of intense burning pains with the victim's limbs eventually becoming gangrenous are depicted by the burning hand in this redrawn woodcut produced in Strassbourg in 1540. St. Anthony, also shown in the drawing, reflects the common name of ergotism, "St. Anthony's fire." In the late sixteenth century, ergot was commonly employed by midwives to quicken labor and reduce the incidence of postpartum bleeding. The administration of ergot became a routine procedure following childbirth until modern medicine realized its dangers and found other drugs that had the same beneficial effects.

the major reason for their restricted use is their toxicity. At one time, however, various species were used in both Eurasia and the Americas as sources of hallucinogenic compounds. Historically, the most important of these were species of *Datura, Hyoscyamus, Atropa belladonna,* and *Mandragora officinarum* (Fig. 13-24). Many of these were also used medicinally, and a few have become important modern sources of pharmaceutical compounds (see Chapter 12). The

FIGURE 13-24
Mandrake, or *Mandragora* (from man, and dragon), was long thought to be an aphrodisiac because of the human shape of its root. This belief is reflected in the biblical story of childless Rachel (Genesis 30) who exchanged an evening with her husband (Jacob) in return for her sister Leah's son's mandrakes. (From Matthias de L'Obel, 1581, *Plantarum sev Stirpium Icones*, Antwerp.)

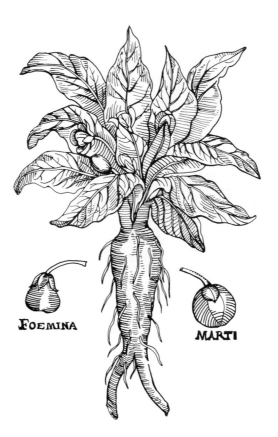

tropane alkaloids (see Fig. 12-17) (atropine, scopolamine, and hyoscyamine) which these species contain are effective in controlling smooth muscle action. Carefully measured amounts are thus useful in treating heart irregularities. In larger, but sublethal, doses, the same alkaloids produce hallucinations.

Of the plants in this group, only datura is native to the New World (Fig. 13-25). Datura is particularly rich in scopolamine, which is considered to be the most hallucinogenic of the solanaceous alkaloids. In Central and South America, seeds or roots of various species are eaten to produce visions, but even among people that use datura, it is considered dangerous and too strong for general use. Usually datura was ingested only by boys during puberty rites or by trained shamans. In South America, alcoholic beverages made from datura fruits were given to women and slaves of dead warriors to stupefy them before they were buried alive with their deceased masters. Infusions of bark and leaves were also used by various tribes.

Species of datura native to Europe and Asia were less widely used than congeneric species in the New World, but belladonna, mandrake, and henbane assumed great importance in Europe during the Middle Ages. The ancient use of solanaceous alkaloids in witchcraft and folk medicine led European scientists after the Renaissance to investigate their action and eventually to the discovery

FIGURE 13-25
Datura acquired its common name jimsonweed in 1705 when British soldiers in Jamestown, Virginia, inadvertently ate the plant leaves in a salad and went mad for 11 days.

of their medicinal properties. Likewise, their uses as hallucinogens and poisons prompted writers and artists to include them in the writings and paintings of the time.

Atropine, obtained from either belladonna or henbane, can be absorbed directly through the skin, and during the Middle Ages, witches rubbed ointments containing it on their bodies. The psychological sensations produced give one the feeling of flying. While under the influence of atropine, witches were supposed to be transported to rendezvous with spirits or demons. These meetings, called sabats, are often figured in European art of the period. Our modern portrayal of a witch riding a broom may stem from these ancient rites. The depiction of the flying comes from the hallucinogenic sensation of being transported, and the broom purportedly represents the stick used to apply atropine-impregnated ointments to vaginal membranes for absorption (Fig. 13-26).

Another figure commonly attributed to visions seen while under the influence of solanaceous alkaloids is the werewolf. Similar to the visions of jaguars and snakes produced when South American natives use caapi, wolves were associated with trances induced by henbane, belladonna, and related species.

A quite different member of the Solanaceae, however, has had the greatest worldwide impact. Common tobacco, *Nicotiana tabacum*, and, to a lesser extent its relative *N. rustica*, have become important commercial items in almost every country of the world (Fig. 13-27). The genus *Nicotiana* is thought to be native to the New World, but at least one species, *N. suavolens*, which occurs in Australia, was independently domesticated. At least 1,000 years before Columbus landed in the West Indies, tobacco was smoked, eaten, and snuffed by native peoples throughout the New World (Fig. 13-28). Among American Indians, tobacco was used medicinally to ease childbirth pains and stave off hunger on

FIGURE 13-26
Atropine produces the sensation of
flying, which gave rise to the idea
of witches riding on their
broomsticks.

long hunts. Dried leaves were so valuable that they were used as money and
incorporated into ritualistic and symbolistic practices such as the smoking of a
peacepipe.

Tobacco was considered sacred by many American tribes. Mayan priests
thought that smoke rising from their pipes carried messages to the gods (Fig.
13-29). Recent studies of Amazonian tribes have shown that the use of tobacco
in divination rites continues today. Shamans of the Warao tribe of Venezuela,
for example, fast almost to starvation during their initiation rites and then eat
and smoke large quantities of tobacco. The resultant hallucinations produce the
sensation of being transported to another world for meetings with the spirits
that govern the lives of the people. Later in their careers, shamans frequently
turn to tobacco to help them solve tribal problems.

Although Columbus brought "cigars" to Queen Isabella after his first
voyage, it was not until Andre Thevet brought seeds of *Nicotiana tabacum* from
Brazil to France in 1556 that tobacco cultivation started in Europe. Linnaeus
named the genus after Jean Nicot, the French ambassador to Portugal who
made a fortune importing and popularizing the use of the plant in Paris. Claims
of tobacco's medicinal virtues as a remedy for female problems, a snakebite
antidote, lung strengthener, ulcer remedy, cure for the plague, and potent
aphrodiasiac, accelerated its sweep across Europe and Asia.

Tobacco use became so widespread that King James I was alarmed to find
that purchases of tobacco were depleting England's silver supply. Neither the
high cost of the leaves nor King James's blistering attack in his *Counterblaste to
Tobacco* (1604) checked the spread of smoking or sniffing. In other countries as

FIGURE 13-27
Two species of tobacco, *Nicotiana tabacum* (right) and *N. rusticum* (left) were both
domesticated by American Indians and used for smoking.

FIGURE 13-28
Indians smoking a tabac (pipe).
(Redrawn from Thevet, 1573.)

FIGURE 13-29
Mayan priests blew smoke to the four winds during religious ceremonies. (Redrawn from a frieze at Palenque, Mexico.)

well, leaders tried to discourage tobacco use (Fig. 13-30). One Chinese emperor ordered smokers to be decapitated, and a Russian Tsar ordered the nostrils of snuffers to be split so that they could no longer practice the habit. Still, tobacco continued to attract followers.

Eventually, the British promoted tobacco cultivation in their colonies to ensure a national supply. Virginians started growing tobacco in 1612. Since an acre planted in tobacco yielded four times the revenue of an acre planted to corn, colonists turned more and more exclusively to tobacco farming. The Virginia monopoly on tobacco production granted by the Queen was broken when settlers in Maryland began to cultivate the crop. Soon, much of the eastern seaboard of the United States was engaged in tobacco production. For many years, tobacco was the most important item of trade between England and North America.

Primitive cultivation of tobacco consisted simply of saving seeds of the previous year's crop and sowing them in a cleared area. Plants were harvested at maturity and the leaves dried in the sun. In contrast, modern tobacco farming is quite elaborate and employs various cultivation techniques for the many different cultivars and the types of tobacco being produced. On modern U.S. farms, tobacco is usually first sown in seed beds and the seedlings transplanted into fields after 2 months of growth. The spacing of plants and fertilizer regime varies with the final type of tobacco to be produced. Nitrogen-rich fertilizers are applied to encourage the development of large, pliable leaves suitable for cigar wrappers (Fig. 13-31), while the small, more brittle leaves favored for cigarettes are produced by limiting nitrogen and supplying phosphorus and potassium at critical stages.

While it is growing, the flower stalks and the side shoots are usually removed to prevent the plants from diverting resources to seeds and low branch development. Plants producing leaves to be used for cigar wrappers are generally grown under cheesecloth where the high humidity and protection from sunscald also promote formation of large, thin, blemish-free leaves. Two to 4 months after planting, the crop is considered mature. Good grades can be picked a leaf at a time, or whole stalks can be harvested. In either case, the leaves or stalks are tied into bunches called "hogsheads" for curing (Fig. 13-32).

During the curing process, the moisture content in the leaves is reduced from about 80 to 20 percent. Starches are converted to sugars, and some proteins are broken down enzymatically. Slow drying permits aerobic fermentation to take place but prevents the growth of molds or fungi. Curing can be done by circulating air or by smoking. Occasionally, bundles of leaves are sun-dried. Following curing, the leaves are aged for periods varying between a few months and a year. For cigarette and cigar fillings, the tobacco is remoistened before marketing and the veins and petioles removed. The softened leaves (Fig. 13-33) are then cut by machines into strips. Depending on the ultimate tobacco product, various materials can be added. Moisture-retaining substances such as glycerin, cider concentrate, and diethyl glycol are common additives, but honey or sugar, oil of hops, licorice, coumarin, rum, or menthol can be added for flavoring.

Additives have been increasingly used in tobacco mixtures since the middle 1970s to compensate for the flavor and aroma lost when manufacturers produce cigarettes with low tar and nicotine. While most of these are drawn from the FDA list of "Additives Generally Recognized as Safe," there is concern that, when burned, many may constitute health hazards. Coumarin, a compound removed from the FDA list of safe drugs because it has been shown to be carcinogenic, is still being purchased in large quantities by cigarette manufacturers. Licorice, likewise, produces carcinogenic compounds when burned. Even the sugars added produce tars when burned. Ironically, the federal government exempts tobacco manufacturers from the labeling requirements that other food and drug producers must observe.

While tobacco smoking was widely enjoyed before 1880, its popularity greatly increased after this year because of a change in the curing process. Before 1880, most tobacco was cured by directly placing it over the hot smoke of a charcoal fire. After 1880, farmers began to use hot air brought to the drying rooms by flues. This indirect drying produced a milder form of tobacco than that produced by smoke drying. Mild tobaccos were perfect for cigarettes, a relatively "genteel" way of smoking. However, these light tobaccos produce an acid smoke when they are burned. Tobacco cured directly tends to produce an alkaline smoke. Acid tobacco smoke has little physiological effect on the body if it is simply puffed. Consequently, it must be inhaled in order to produce an exhilarating effect. The inhaled smoke is neutralized on the surface of the lungs, and the nicotine carried in the smoke is readily absorbed through the lung membranes. Smokers who inhale tend to become addicted to tobacco because they become physiologically dependent on the strong sensory reaction produced when nicotine is absorbed. Pipe and cigar smokers can become accustomed to smoking and dearly miss their habit if they quit smoking, but, in general, there is little physical reaction to the absence of nicotine in their blood since they normally do not inhale.

Nicotine, the major alkaloid in tobacco, is a stimulant of the central nervous system (Table 13-2) and can cause nausea, dizziness, and hallucinations in large doses. It is physiologically addicting, and withdrawal symptoms appear when

FIGURE 13-30
Despite James I's characterizing tobacco smoking as "A custome lothsome to the eye, hatefull to the nose, harmefull to the braine, dangerous to the lungs, and in the blacke stinking fume thereof, neerest resembling the horrible Stigian smoke of the pit that is bottomless" (*Counterblaste to Tobacco*), the British continued to smoke and went on to develop the genteel art of blowing smoke rings.

FIGURE 13-31
Examining tobacco leaves on a Georgia farm. (Photo by Steve Wade; courtesy of the USDA.)

FIGURE 13-32
Stacks of cut tobacco called
hogsheads are allowed to stand
until they wilt. (Photo by E. O.
Brewer; courtesy of the USDA.)

it is kept from habituates. In pure form, nicotine is a poison and is often used, after treatment with sulfuric acid, as an insecticide. In addition to nicotine, smoking leads to the inhalation of tars. These tars are known carcinogens, and cigarette companies have therefore tried to manufacture cigarettes that reduce the intake of tars. However, none of the various filtering or extraction devices is perfect, and some tars are drawn in by smokers who inhale.

FIGURE 13-33
A stretched leaf of properly cured
tobacco. (Photo by Murray Lemmon;
courtesy of the USDA.)

In light of the toxicity of nicotine and tars, it is not surprising that there are now well-established correlations between smoking and lung cancer and smoking and heart disease. Smokers age prematurely, die at a younger age than nonsmokers, and tend to be susceptible to emphysema, bronchitis, and other respiratory ailments. Smoking during pregnancy increases the risk of miscarriage and problems in early infancy. Infants born to mothers who smoke can emerge into the world underweight and addicted to nicotine.

Despite the warning of the Surgeon General that smoking is bad for your health, tobacco smoking is a legal form of drug ingestion. In direct contrast to its stance concerning many less harmful drugs, the government continues to subsidize nicotine drug addiction by providing price supports for tobacco farmers. Cigarette companies also blissfully ignore the effects of tobacco smoking and continue to lure people into smoking with advertisements that portray smokers as healthy, particularly masculine, independent, and/or beautiful.

PLANT POISONS

So many plants are poisonous to varying degrees that an enumeration would be futile. Regional agricultural stations usually publish lists of local plants that are dangerous to humans or livestock, and many horticulture books indicate which ornamental plants can be toxic if eaten (see Kingsbury in the list of Additional Readings). Some of these plants have been used by humans for their poisonous compounds. These are the ones in which we are interested here.

As we indicated earlier, many plants used for their psychological effects are poisonous if used in large quantities. It is difficult to know exactly how humans discovered which plants were or were not toxic and the quantities that could be ingested relatively safely. Ancient Greek and Roman medical records attest to an early interest in documenting the occurrence and usefulness of plant toxins. Administration of plant poisons as a form of capital punishment is well known from the story of the death of Socrates, after being sentenced to drink the juice of poison "hemlock" (*Conium maculatum*, Apiaceae). In the Middle Ages, "succession powders" were deviously employed to eliminate potential heirs to the throne. The employment of these powders became so common that the wealthy and powerful hired tasters as protection against sudden death from "food poisoning." Today, scientists often seek out plants known by natives to be poisonous. Since many poisons are either alkaloids or steroids, such plants are potential sources of drugs or commercial poisons for use as insecticides, herbicides, or fungicides.

Some modern primitive tribes in the tropics use arrow or fish poisons such as curare or barbasco. Curare is made from extracts of the bark and stems of *Chondodendron tomentosum* (Menispermaceae) and from seeds of species of *Strychnos*, particularly *S. nux-vomica* (Loganiaceae). *Chondodendron*, a vine native to South America, contains tubocurarine. Species of *Strychnos* occur natively in South America, southeast Asia, and Indonesia and many yield the potent alkaloids curarine, strychnine, and toucine. Curare alkaloids, primarily tubocurarine chloride, have had limited applications in medicine because they block neuromuscular activity and thus help relax muscles that remain contracted under anesthesia. Strychnine was formerly used medicinally as a central nervous system stimulant but has now been replaced by safer compounds.

FIGURE 13-34
In Kenya, women pick pyrethrum flowers that will be used as a natural source of a powerful insecticide. (FAO photo.)

Barbasco and tuba are fish poisons used by Amazonian peoples. The first is prepared from various species of *Derris*, such as *Derris elliptica*, and the second from *Lonchocarpus nicou* (both Fabaceae). Recently, species of both these genera have been shown to be effective insecticides because they contain rotenone. This isoflavanoid compound is fatal to insects but comparatively harmless to humans because it leaves no residue.

Other "natural" insecticides have been extracted from pyrethrum (*Chrysanthemum coccineum*, *C. cinerariifolium*, and *C. marschallii*, Fig. 13-34) and from tobacco. Insecticidal compounds from species of *Chrysanthemum* are sold as Persian insect powder or Dalmation. Pyrethrins, the active compounds, are esters that break down into acids and alcohols. Since they stun insects, but do not instantly kill them, pyrethrins are usually mixed with other substances that ensure death of the insect. In some regions, notably Kenya, Tanzania, and parts of northeastern South America, pyrethrum is grown commercially for natural compound extraction.

It is estimated that there are at least 1,200 plants of potential use as sources of insecticides. Yet, the fact that these would be natural insecticides does not mean that they are, or would be, less harmful than synthetic compounds. In fact, nicotine sulfate, a commercially available, natural insecticide is one of the most toxic pesticides on the market today.

Undoubtedly, people will continue to search for additional plant sources of medicines, psychoactive drugs, and poisons. In such a search, Shakespeare's advice on plant poisons still applies:

> Virtue itself turns to vice, being misapplied,
> And vice sometimes by action dignified.
> Within the infant rind of this small flower,
> Poison hath residence, and medicine power.

(Friar Laurence in *Romeo and Juliet*)

ADDITIONAL READING

Austin, G. A. 1979. *Perspectives on the History of Psychoactive Substance Use*. National Institute on Drug Abuse. Washington, D.C., U.S. Government Printing Office. A succinct compendium of important dates in the history of drug use in selected areas.

Anderson, E. F. 1980. *Peyote, the Divine Cactus*. Tucson, University of Arizona Press.

Becker, C. E., R. L. Roe, and R. A. Scott. 1979. *Alcohol as a Drug*. Huntington, New York, R. E. Krieger.

Brecher, E. M., and the Editors of *Consumer Reports*. 1972. *Licit and Illicit Drugs*. Mount Vernon, New York, Consumers' Union.

Castaneda, C. 1973. *The Teachings of Don Juan*. New York, Ballantine.

Emboden, W. 1979. *Narcotic Plants*. New York, Macmillan.

Furst, P. T. (ed.). 1972. *Flesh of the Gods*. New York, Praeger.

Heiser, C. B. 1969. *Nightshades, the Paradoxical Plants*. San Francisco, Freeman.

High Times Encyclopedia of Recreational Drugs. 1978. New York, Stonehill Publishing Co. A very readable and interestingly illustrated assessment of the so-called recreational psychoactive drugs.

Julien, R. M. 1981. *A Primer of Drug Action*, 3rd ed. San Francisco, Freeman. A first book on how drugs act, and the effects of many of the drugs commonly used in the United States.

Kingsbury, J. M. 1972. *Deadly Harvest: A Guide to Common Poisonous Plants*. New York, Holt, Rinehart and Winston. A usable text, written for the lay person about poisonous plants that are likely to be encountered in the United States. Pictures of toxic plants and terms used are included.

Lingeman, R. R. 1974. *Drugs from A to Z: A Dictionary*, 2nd ed. New York, McGraw-Hill.

Schleiffer, H. (ed.). 1973. *Sacred Narcotic Plants of the New World Indians*. New York, Hafner. Selected excerpts from a variety of authors that recount experiences or stories of native drug plants.

Schleiffer, H. (ed.). 1979. *Narcotic Plants of the Old World*. Monticello, New York, Lubrecht and Cramer. The companion to the book on new world psychoactive plants.

Schultes, R. E. 1969. Hallucinogens of plant origins. *Science* 163:245–254.

Schultes, R. E., and A. Hofmann. 1973. *The Botany and Chemistry of Hallucinogens*. Springfield, Illinois, Charles C Thomas. A very thorough treatment of plant-derived hallucinogens that deals with the plants themselves, the methods of preparation, the chemistry of the active principles, and the effects of the drugs.

Taylor, N. 1966. *Narcotics, Nature's Dangerous Gifts*. New York, Dell.

Chapter 14
Stimulating Beverages

The drinking of chocolate milk, coffee, and tea have become such integral parts of American life that we take them for granted. Many Americans profess that they could not face the day without a morning cup of coffee, and the 4:00 coffee break or tea time is indulged in by most of the Western world who need a lift at the end of the day. We drink these beverages not only for the boost they give us, but for their flavor as well. Other stimulating beverages such as maté, kola, and guarana have become locally important for the same reasons, but coffee, tea, and cocoa reign supreme on a world wide basis (Fig. 14-1).

FIGURE 14-1
This drawing represents the world's three most important nonalcoholic beverages and the civilizations associated with them: coffee in an Arab's cup, tea being drunk by a Chinese, and cocoa in an American Indian's goblet. (From Dufour, Philippe, *Traitez nouveaux et curieux du café, du thé, et du chocolat,* Lyons, 1671.)

TABLE 14-1

Active Compounds in the World's Major Stimulating Beverages*

PLANT PART	CAFFEINE	THEOBROMINE	POLYPHENOLS
Coffee, unroasted, dried	1–1.5	—	—
Teas, dried leaves	2.5–4.5	—	25.0
Cacao:			
Dried nibs	0.6	1.7	3.6
Fresh cotyledons	0.8	2.4	5.2
Kola, fresh seeds	2.0	—	—
Guarana, dried fruit	3.0–4.5	—	—

*Figures given in percent weight. Amounts of the compounds in a particular beverage depend on how the beverage is made.

TABLE 14-2

Amounts of Caffeine in Commonly Consumed Beverages and Medicines

ITEM	CAFFEINE, MG
Coffee:	
5-oz cup, drip method	146
5-oz cup, percolator method	110
5-oz cup, instant	53
5-oz cup, decaffeinated	2
Tea:	
5-oz cup, brewed 1 min	9–33
5-oz cup, brewed 3–5 min	20–50
12 oz, canned	22–36
Cocoa and chocolate:	
6 oz, made with canned powder	10
1 oz milk chocolate	6
1 oz (1 square) baking chocolate	35
Soft drinks:	
12 oz Mountain Dew	52
12 oz Dr. Pepper (regular or sugar free)	37–38
12 oz Pepsi, regular	37
12 oz Coca Cola	34
Nonprescription drugs:	
Stimulants:	
NoDoz (standard dose):	200
Pain relievers (standard dose):	
Excedrin	132
Midol	65
Anacin	64
Cold remedies:	
Dristan	32
Diuretics (standard dose):	
Aqua-Ban	200
Weight-control aids (daily dose):	
Prolamine	280
Dietac	200

Source: Adapted from *Consumer Report,* vol. 46. 1981, pp. 598 and 599. All soft drinks and medications containing caffeine are not included. Consumers should read the labels of such products to determine whether or not they contain caffeine. Several manufacturers have recently removed the caffeine from one form of their medications labeled "caffeine free."

All these beverages are stimulants because they contain chemicals (Table 14-1), primarily caffeine and its relatives (Fig. 14-2), that cause particular physiological reactions in humans. As pointed out in Chapter 13, caffeine is a general cellular stimulant, activating intracellular metabolism. Five minutes after drinking a cup of coffee, the caffeine in it reaches the bloodstream. As it circulates throughout the body, it stimulates the heart, increases stomach acidity and urine output, and causes a 10 percent rise in the metabolic rate. If a person is tired, caffeine makes her or him feel more alert because it mimics the feelings produced when the body releases adrenaline. In excessive doses however, caffeine can produce unpleasant symptoms. One gram of the compound (about 10 cups of coffee) causes anxiety, headache, dizziness, insomnia, heart palpitations, and even mild delirium. Heavy tea or coffee drinkers can develop a tolerance to caffeine and can even suffer withdrawal symptoms if they quit their habit. Still, coffee and tea drinking are so ubiquitous that caffeine has become the world's most popular drug. We now add it to numerous soft drinks and medications (Table 14-2). Unfortunately, children who have neither the body mass nor the tolerance for large amounts of caffeine consume appreciable quantities of soft drinks. There is some debate about the permanent effects of moderate caffeine intake on adults, but doctors now advise pregnant women to reduce or eliminate caffeine consumption because laboratory studies have shown that the alkaloid causes chromosome damage in laboratory animals.

In view of their world wide importance, we discuss coffee, tea, and cacao in detail, looking at the places where humans first used them for beverages, their discoveries by Europeans, and the subsequent rise of the three to important positions in international trade. We also discuss their methods of cultivation and the processing necessary before each reaches a cup or a glass. Maté, kola, and guarana are discussed only briefly because of their limited use. Table 14-3 lists the plants yielding stimulating beverages that we discuss in this chapter.

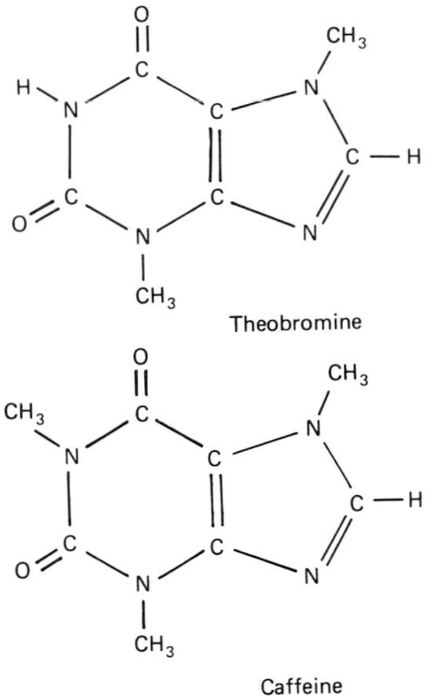

FIGURE 14-2
Caffeine and theobromine both have stimulating effects.

COFFEE

Coffee is second in importance only to petroleum and its by-products in terms of the value traded annually on the international market. Although the genus *Coffea* (a member of the madder family, Rubiaceae) is native to eastern Africa, comparatively few Africans drink coffee. The earliest records of coffee use are from Ethiopia where natives chewed leaves and fruits gathered from wild trees growing in the understories of montane forests. A mixture of ground-roasted, or green, coffee fruits and fat was taken along on hunts as a survival staple similar to the pemmican used by American Indians. Caffeine that dispelled fatigue and relieved hunger was leached out of the leaves or fruit during chewing. Beverage making seemed to have been a later development, and some authors have suggested that a fermented beverage was made from the fruits before a steeped one.

Coffee was taken to Arabia in the sixth century, and, as alluded to by its name *Coffea arabica,* the Arabs were the first to brew coffee (Fig. 14-3). Coffee drinking spread from Arabia to Egypt about 1510, and the beverage reached Europe in 1616. By 1650, coffee had arrived in England where it became an important part of the social and political environment. England alone had over

TABLE 14-3

Plants Discussed in Chapter 14

COMMON NAME	SCIENTIFIC NAME	FAMILY	CHROMOSOME NUMBER
Achiote	*Bixa orellana*	Bixaceae	$2n = 14, 16$ diploid
Cacao	*Theobroma cacao*	Sterculiaceae	$2n = 20$, diploid
Chicory	*Cichorium intybus*	Asteraceae	$2n = 18$ diploid
Coffee:			
Arabian	*Coffea arabica*	Rubiaceae	$2n = 44$ tetraploid
Liberian	*C. liberica*		$2n = 22$ diploid
Robusta	*C. canephora*		$2n = 22$ diploid
Guarana	*Paullinia cupana*	Sapindaceae	Unreported
Kola	*Cola nitida*	Sterculiaceae	$2n = 40$ tetraploid
Maté	*Ilex paraguariensis*	Aquifoliaceae	$2n = 40$ tetraploid
Tea	*Camellia sinensis*	Camelliaceae	$2n = 30$ diploid

3,000 coffee houses by 1675, many of which functioned as forums for political and religious debates. The popularity of the houses and the spread of the ideas discussed within them so alarmed King Charles II that he labeled them "seminaries of sedition" and tried to have them closed. The uproar that resulted from Charles's edict forced him to rescind his order, and soon the demand for coffee sent all of Europe looking for new sources of the seeds.

For years, the Arabs monopolized the coffee trade and tried to prevent the cultivation of coffee by other countries. They shrewdly dipped the seeds in boiling water before marketing to kill the embryos and prevent germination. Eventually, the Dutch managed to secure live seeds from Mocha, the traditional Arabian source of coffee. With viable seeds, the Dutch started extensive plantations in Sri Lanka (Ceylon) and the East Indies thus breaking the Arabian monopoly. Trees from these plantations were sent to the Botanical Garden in Amsterdam in 1706, but only one survived the journey. Seeds from this single tree were later given to other botanical gardens in Europe, including the Jardin de Plantes in Paris.

FIGURE 14-3
A lithograph of an Arabian coffee service, redrawn from *A Thousand and One Arabian Nights.*

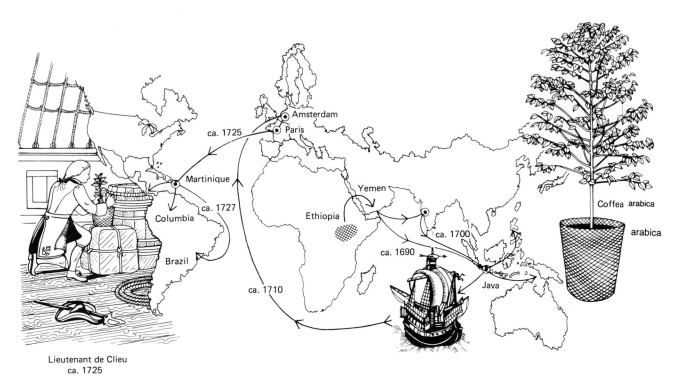

FIGURE 14-4
From its origins in Africa, virtually one clone of *Coffea arabica* spread around the world and established successful coffee production in the West Indies and South America. On the left is Lt. de Clieu tending the coffee plant he brought from France to Martinique. (Adapted from N. W. Simmonds. 1976. *Evolution of Crop Plants.* New York, Longman.)

The tree in the conservatory of the Paris garden was considered a curiosity by most people, but one man realized its potential value. This young Frenchman, Gabriel de Clieu, argued that, if coffee grew well for the Dutch in the East Indies, it should grow equally well for the French in the West Indies. Stories vary as to whether or not de Clieu obtained seeds from the tree legally, but they all agree that, about 1723, he arrived in Martinique with a single surviving offspring of the Parisian tree. By 1777 there were thousands of coffee trees on Martinique. From this small island, seeds were dispersed throughout the Antilles, the northern coast of South America, and, eventually, Brazil (Fig. 14-4) which now leads the world in coffee production (Table 14-4).

Although *Coffea arabica* has figured most prominently in history and now accounts for more than 90 percent of the world's coffee production, it is not the only species of the genus that yields coffee-producing seeds. Two other species, *C. canephora* (called robusta coffee) and *C. liberica* (Liberian coffee) are used to produce about 9 and 1 percent, respectively, of the world's crop.

Plants of both robusta and Liberian coffees are more hardy than those of Arabian coffee and both are easier to grow. In addition, robusta trees produce more fruits per tree than *C. arabica.* Nevertheless, Arabian coffee is by far the most important crop species. The predominant use of *Coffee arabica* can be attributed to several characteristics. First, it is a self-compatible polyploid

TABLE 14-4

World Production of Coffee, Tea, and Cacao

CROP	TOP 5 COUNTRIES, 1,000 t		TOP 5 CONTINENTS, 1,000 t		WORLD, 1,000 t
	COUNTRY	t	CONTINENT	t	
Coffee	Brazil	1,680	South America	2,731	5,537
	Colombia	798*	Africa	1,188	
	Mexico	240*	North & Central America	962	
	Indonesia	233	Asia	599	
	Ivory Coast	255*	Oceania	57	
Tea	India	595*	Asia	1,580	2,020
	China	429	Africa	218	
	Sri Lanka	175	South America	58	
	U.S.S.R.	150	Oceania	9	
	Kenya	112	—	—	
Cacao	Ivory Coast	400*	Africa	865	1,557
	Brazil	346	South America	469	
	Ghana	160*	North & Central America	111	
	Nigeria	150*	Asia	78	
	Cameroon	90*	Oceania	34	

*Indicates an estimated or an unofficial figure.
Note: The U.S.S.R. is considered a country only. Its figures are not added into those of either Asia or Europe. Figures are in metric tons.
Source: Data from FAO Production Yearbook for 1983. Volume 37. Published in 1984. FAO, Rome.

($2n = 44$). The successful introduction of coffee into the New World was possible because the plant sent to Amsterdam was able to produce viable seed by self-pollinating and self-fertilization. The other two species are self-incompatible diploids ($2n = 22$). More importantly however, coffee made from *C. arabica* has a better flavor than that made from the other two species. One of the few places where the bitter flavor of *C. canephora* is often preferred is in parts of Africa. *Coffea canephora* is grown primarily for use in blended coffees or to make decaffeinated or instant coffee where the taste is disguised or altered. Likewise, *C. liberica*, the most bitter of the three, is used primarily as a filler in mixtures with other coffees.

All cultivated species of *Coffee* are small trees with glossy leaves and fragrant, jasminelike white flowers produced in the leaf axils (Fig. 14-5). Once fertilized, the ovule takes 7 to 9 months to mature into what is called a "cherry." Actually, the cherry is an accessory "berry" with a tough outer layer composed of the floral cup and exocarp, fleshy mesocarp, and thin, fibrous endocarp, or "parchment" (Fig. 14-5). Within the endocarp are two seeds pressed together in such a way that the inner side of each is flattened. Each seed, commonly called a "bean," is surrounded by a thin, silvery seed coat. The seed itself is composed mostly of endosperm surrounding a small, curved embryo.

Coffee grows best under conditions similar to those of its native habitat. Because coffee plants cannot tolerate sustained freezing, cultivation is restricted to tropical and subtropical latitudes. Coffee prefers fertile soils and an annual rainfall of at least 190 cm (75 in) with a dry season during which flowers develop. An average coffee plant produces its first crop after it is 3 years old,

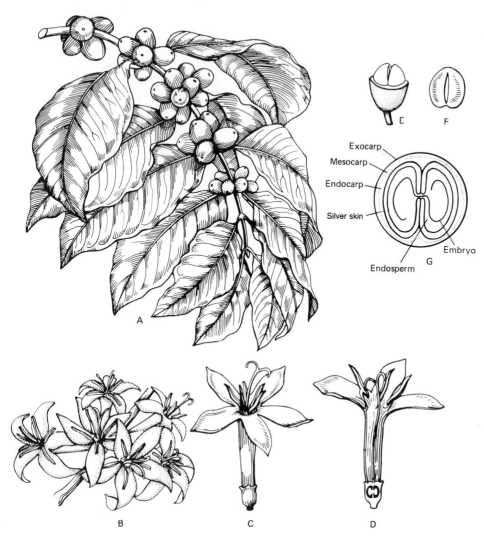

FIGURE 14-5
(A) A branch of coffee with clusters of berries in the leaf axils redrawn from Baillon; (B) a cluster of jasminelike flowers; (C) an individual flower in (D) cross section; (E) the position of the two seeds in the fleshy fruit; (F) the ventral (flattened) side of the seed; (G) a cross section of the whole berry showing the component parts.

and it has a productive life of about 40 years. Most coffee is grown today in open orchards, but some cultivators still claim that superior coffee is produced by shaded trees. Coffee is generally grown on hillsides at altitudes between 1,500 to 2,000 m (4,920 to 6,560 ft). Because of the size of the trees and the slope of the terrain on which it grows, coffee is seldom harvested mechanically. Moreover, the best coffee comes from berries picked just at the point of ripening (Fig. 14-6).

To produce the aromatic ground coffee that we use for beverage making, the seeds must be separated from the rest of the fruit and roasted. Either a wet or a dry process can be used to remove the outer fruit parts (Fig. 14-7). In the former, the fresh fruits are dried in the sun and the pericarp subsequently rasped away. In the wet process (which produces a superior final flavor), the fresh fruits are depulped by a machine and the seeds washed. The wet seeds are then allowed to ferment for 12 to 24 h. This fermentation is not the same process as that involved in the production of alcohol (see Chapter 15). In the cases of

FIGURE 14-6
A coffee picker such as this Colombian can pick 91 kg (200 lb) per day. About 2,000 beans (1,000 cherries) are needed to produce a pound (453 g) of roasted, ground coffee. (United Nations photo by Jerry Frank.)

FIGURE 14-7
Steps in the processing of coffee. (A) selective hand picking of only the red, ripe berries produces coffee with the finest flavor; (B) washing the berries; (C) fermenting the fruits; (D) drying the beans once the pericarp has been removed; (E) transporting the green beans to market; (F) an older method of roasting the beans. Modern roasters use automatic tumblers.

coffee, cocoa, tea, and kola, fermentation refers to an enzymatic, chemical alteration of several compounds. Coffee fermentation produces substances that will eventually develop into the characteristic coffee aroma and taste. It also causes the seeds to take on a gray-green hue. After fermentation, the seeds are dried by turning them in the sun for at least 1 week. Any remaining endocarp and the seed coats are removed mechanically before shipping.

Roasting is done in cylinders that simultaneously tumble and heat the seeds, causing water loss and partial sugar caramelization. During the process, water-soluble aromas develop and begin to diffuse. The kind of roast depends on the final temperature that the beans are allowed to reach. Light roasts are roasted until the temperature reaches 204°C (400°F). Roasting is stopped at 218°C (425°F) for dark roasts. The beans are often sprayed with water or cold forced air when the proper temperature is reached in order to stop the process quickly and slow down the loss of aromatic oils.

Despite the statement that dark-roast coffees are "stronger" than light roasts, they are not stronger in the sense of containing more caffeine. Dark roasts seem stronger because they have more flavor than light roasts. Dark-roast beans also feel oily because their higher roasting temperatures cause some of the aromatic oils to come to the surface of the beans. Light- or medium-roast beans are dry on the outside. Almost all of the roasting for American palates is of either the light or medium type. Europeans and Latin Americans, on the other hand, prefer dark roasts, one of which we commonly call *espresso*. It should be noted that espresso is a method of making coffee and could be performed with a light-roast coffee. It just happens that espresso coffee is generally made with dark-roast beans. Once roasted, the beans can be shipped as such, or ground and packaged (often under a vacuum to preserve the flavor) before being shipped to market.

Methods of brewing coffee are almost limitless. Until recently, the percolator method was the most popular in the United States for home use, despite the fact that most connoisseurs consider it to produce a relatively poor cup of coffee. Perhaps the popularity of the percolator method can be ascribed to its convenience and the pleasant sounds and aromas emitted during percolation. Automatic drip machines have recently replaced the percolator in many homes because they are more convenient than percolators and make good coffee quickly. Many other conveniences in coffee making, some not necessarily leading to a good cup of coffee, have also become popular. Among these are premeasured packets for coffee makers, throw-away filters for drip coffee pots, and most importantly, instant coffee.

The first marketable instant coffee was produced in 1909 by an American chemist who produced Washington's "soluble" Red E coffee. Today numerous brands and types are available. To make instant coffee, coarsely ground beans are put into huge sealed stainless steel percolators and brewed under pressure for hours. Sealing is supposed to keep the aroma and flavor from escaping, but the damage done to the coffee by the high temperatures and constant recycling of the brew is irreparable. Aromas are often later added to make up for some of the lost ingredients. The coffee from the percolators is then sprayed through nozzles under high pressure into the top of a room several stories high. As the spray falls, it dries to a powder by the time it reaches the floor. Often, to make the instant powder more closely resemble real ground coffee, it is tumbled with

FIGURE 14-8
The same wild cornflower we know as an introduced roadside weed provides the roots that are roasted and ground to make chicory.

water or steam, causing it to adhere together in small chunks. Freeze drying is another method of producing instant coffee. During this process, the brewed coffee is poured onto trays which are subjected to very low temperatures under a vacuum. Sharp-edged particles are produced when the sheet of freeze-dried coffee is broken into small fragments.

Another recent innovation is decaffeinated coffee. In 1981, it was estimated that 17 percent of the coffee drunk in the United States was decaffeinated. Because caffeine is removed from green coffee beans, not from ground coffee, it is possible to buy unroasted decaffeinated beans, as well as decaffeinated roasted beans, ground coffee, and instant coffees. It was a German, Ludwig Roselius, who developed the first successful process for removing caffeine from coffee in 1906. The young man was motivated to perfect the process because he believed that his father's death had been caused by his drinking large amounts of coffee.

Today, there are three basic processes used to remove caffeine from coffee beans: solvent extraction, water extraction, and steam extraction. The last of these has only recently been perfected in Switzerland, and the procedure is a well-guarded secret. Most commercial decaffeinated coffees are produced by a solvent method in which beans, presoftened by steam, are extracted with an organic solvent such as methylene chloride. The solvent is drawn off the beans and any remaining traces left in the coffee are evaporated by steam or heat during the roasting process. The caffeine is removed from the solvent with water and then purified by crystallization. About 20 kg (44 lb) of caffeine is recovered from each ton of processed coffee.

In the water extraction process, green beans are percolated with water that is saturated with all of the water-soluble compounds in coffee except caffeine. Consequently, only caffeine is supposedly dissolved out of the coffee. The caffeine is removed from the extraction water with organic solvents and purified. Many people prefer coffee decaffeinated in this way because no toxic organic solvents actually come into contact with the coffee. Nevertheless, it is more costly than the direct solvent extraction process, and thus less commonly used.

After decaffeination, the solvents are recycled, the coffee beans set aside for roasting, and the caffeine sold. Over 900,000 kg (2 million lb) of caffeine are produced from the decaffeination of tea and coffee each year in the United States. Another 2 million lb are imported each year. About three-fourths of this caffeine is used in soft drinks. The rest is added to headache and cold medicines (Table 14-2).

A final treatment of coffee consists of the addition of other substances. In some cases, such additions constitute adulterants, in others, they serve as flavor enhancers. Adulterants include ground, roasted peas, beans, orris root, or grains. Flavor modifiers include chocolate, liqueurs, orange or almond extract, and vanilla. Perhaps the most common coffee additive is chicory (*Cichorium intybus*, Asteraceae, Fig. 14-8), which can be considered an adulterant or a flavor enhancer, depending on the reason for its addition to ground coffee. Chicory superimposes on the normal coffee flavor a distinctive, pronounced taste that many individuals enjoy (Fig. 14-9). Dried, powdered roots of this southern European native have been used alone to produce a beverage, as well as mixed with coffee to "stretch" the beverage, or substituted for coffee when the beans were scarce. In the United States, coffee with chicory is a regional specialty of Louisiana. Unlike coffee, chicory contains no caffeine.

FIGURE 14-9
"La chicoreé, la plante qui fait du bien" (Chicory, the plant that makes you feel good). Steps in the processing of chicory root shown in the inset on the left include breaking and drying the roots, roasting, and grinding. (Photo courtesy Le Roux et Companie.)

CACAO

Cocoa is America's contribution to the list of important nonalcoholic beverages. But today, eating chocolate surpasses cocoa in popularity. Cocoa, it should be noted, is a beverage. Cacao refers to the plant *Theobroma cacao* (Sterculiaceae, Fig. 14-10) from which chocolate products are obtained. Despite the long history of cacao use, the procedures used to manufacture modern cocoa powder and chocolate have all been developed within the last 150 years. Cacao was collected by native peoples and the seeds roasted for use in a stimulating beverage long before Europeans came into contact with America, but the beverage made by Central American natives was quite different from our modern sweet cocoa. The Mayans thought that cacao had a divine origin, a belief reflected in the name *Theobroma*, literally, "food of the gods."

Europeans first encountered cacao when Columbus and his men landed in Nicaragua. Columbus reported the consumption of a strange beverage, but it was not until Hernando Cortez visited the royal Aztec court that its importance in the New World was appreciated. Like the Mayans, the Aztecs roasted cacao beans and used them in religious ceremonics. Quetzalcoatl (see Fig. 3-7) was credited with giving humans not only corn, but also cacao. Montezuma was reported to have drunk 50 golden goblets of the bitter native beverage each day. This brew, offered Cortez as a sign of respect, was concocted by mixing ground cacao seeds with ground achiote (*Bixa orellana,* Bixaceae; see Figs. 16-30, 16-31). Red pepper and other spices were added to the chocolate mixture and the entire mass heated until it turned into a homogeneous paste that could be molded into tablet-shaped pieces. A drink was made by dropping one of these tablets into hot water. A special stick, a molinet, was used to stir the drink

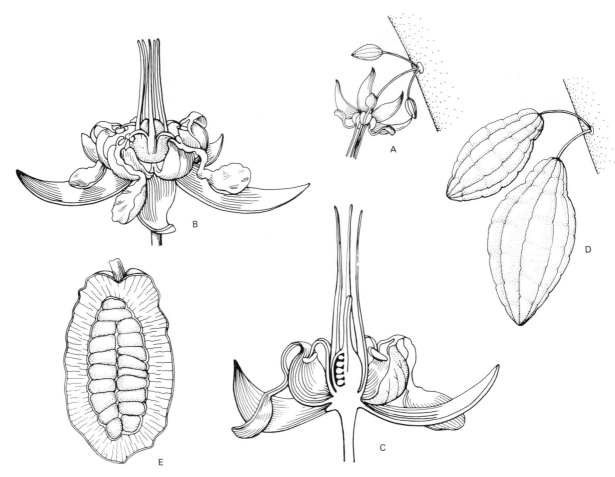

FIGURE 14-10
Flowers and fruits of a chocolate plant. (A) A cauliflorous flower, (B) flower greatly magnified, (C) cross section of a flower, (D) fruits borne directly on the trunk, (E) cross section of a fruit.

until the tablet dissolved. The Indians often thickened the drink by adding atolle (ground cornmeal).

When Cortez brought cacao back to Europe, it was not well received. Later, when Europeans deleted the spices and added sugar, the popularity of the beverage soared. Sweet hot chocolate became the rage of Europe by the middle of the seventeenth century. Spain established cacao plantations on Trinidad and Hispaniola, thereby commencing a long monopoly of the cacao trade. The Dutch were the first to break the Spanish monopoly by establishing plantations in southeast Asia in 1670. Germany later brought plants to Samoa and New Guinea. Africa began to produce crops after 1879 when cacao was introduced into Ghana. As in the case of coffee, production of cacao is now highest in areas far removed from its place of origin, but within the same latitudes (between 20° north and south of the equator). As shown in Table 14-4, Africa, rather than Central America, is today the world's largest producer.

Because cacao was first brought back to Europe from Mexico, it was initially assumed that the genus was native to Central America. However, taxonomic studies have shown that cacao probably arose in the eastern Andes and subsequently migrated northward. By the seventh century, native peoples in Mexico were cultivating the species. The Mayans introduced *Theobroma* to the Yucatan Peninsula, and the Aztecs later continued to spread its cultivation. Sometime before 1492, cultivation caused a divergence of Central American cacaos from their South American ancestors. The domesticated Central American cacaos were called *criollos* and the South American forms, *foresteros*.

Cacao plants are small trees propagated either by seed or by cuttings. Because they grow natively as understory trees, they have traditionally been planted in shaded orchards. Cacao belongs to a group of plants that bear their flowers cauliflorously, or directly on the trunks and branches (Fig. 14-10) rather than in leaf axils or on branch tips. Flowers are produced in flushes twice a year. After pollination by midges that live in the orchard debris, the flowers produce fruits that mature in about 3 months. Individual pods are 10 to 32 cm (4 to 13 in) long and contain 20 to 60 seeds (Fig. 14-11). The pods are picked by hand, cracked open, and the seeds and pulp scraped from the husks. Scraping is often done with a stick because many people are allergic to the pulp. The moist seeds are allowed to ferment chemically in piles for 4 to 7 days (Fig. 14-12). During fermentation, polyphenols are converted into precursors of what we perceive as a chocolate flavor, and the beige seeds take on a purplish color. At the same time, water loss causes the seeds to shrink from the seed coats.

FIGURE 14-12
The steps in the processing of cacao. (A) Pods are picked and broken open (B) to obtain the seeds. Once at the factory, the beans are placed in fermentation boxes (C) and turned into the next bin each day. The beans are then washed in large basins (D) and, at this point shipped to the manufacturing plants in other countries (E). To produce cocoa nibs, the beans are roasted and their seed coats removed. At this point, the nibs contain about 57 percent fat, 30 percent of which is later pressed out during the making of modern cocoa. In the making of milk chocolate, cocoa nibs, sugar, cocoa butter, and condensed milk are stirred in revolving tubs to produce a smooth, homogeneous paste (F).

Fermentation also kills the embryos, preventing unwanted germination. While these chemical processes are occurring inside the seeds, microbes liquefy the pulp surrounding them. Throughout the process, the piles of seeds are often turned, but they can be left simply to "sweat" for several days.

Once fermentation is complete, the seeds are dried, usually by spreading them in the sun, until the moisture content reaches about 60 percent. The seeds are then "polished" by tumbling in a mechanical polisher (or by stamping) to remove any bits of pulp left on the seeds. At this point, the cacao seeds are sorted and graded. Batches of seeds are inspected and tested for flavor, aroma, size, and potential mold or fungal infestation. The sorted seeds are then shipped to processing plants. The seeds now smell good and have a slightly chocolate flavor, but they are still very bitter and oily. A series of procedures still has to follow before a good grade of cocoa or chocolate is produced.

When the seeds arrive at the factory, they are cleaned by fans and rotating brushes that blow and scrape off bits of foreign matter still adhering to them. Once cleaned, they are ready for roasting. Roasting is done in batches or by moving the seeds on a conveyor belt through ovens. The oven temperature is set at 121°C (250°F) for cacao to be used for chocolate, and slightly higher for seeds to be processed into cocoa powder. Roasting drives off water and acids and develops the final chocolate flavor. Once roasted, the seeds are cracked to release the large cotyledons. Fans blow away the debris consisting of the seed coats (commonly called "shells" or "husks"). The shells can be sacked for mulch, used for the extraction of theobromine, or pressed to yield cocoa butter. Extracted theobromine is converted to caffeine that is subsequently added to beverages and medicines.

The cotyledons left after roasting and separation from the seed coats are called *nibs*. It is from this part of the fruit that chocolate is actually obtained. The nibs are ground to a fine paste by a series of rollers. The action of grinding generates enough heat to melt the fat present in the nibs and to produce a thick, dark liquid called *chocolate liquor*. The fate of the liquor is determined by the eventual use of the chocolate. If the liquor is simply molded into small squares, baking chocolate results. Until the middle of the nineteenth century, this was always the final product. Chocolate beverages were made by dissolving blocks of this kind of chocolate in hot water or milk (an English innovation) to which sugar had been added. The result was rich and very heavy because of the large amounts of fat present in the chocolate. In 1828 in Holland, C. J. van Houten developed the first process for producing cocoa as we know it. The process included subjecting the nibs to pressure to press out much of the fat. Cocoa butter was thus separated from relatively dry cocoa powder, yielding a product with 30 percent less fat than that of the nibs. At about the same time, the process of alkalinization, or the adding of alkali to neutralize many of the organic acids, was also developed in Holland. *Dutching,* as this process is often called, darkens the cocoa, makes its flavor mild, and increases its solubility. About 90 percent of all cocoa is now Dutched.

The final way in which the chocolate liquor can be used is to make eating chocolate. Two additional European innovations led to the smooth, creamy milk chocolate that has surpassed cocoa in popularity. In 1875, the Swiss started to add condensed milk to chocolate and produced the first solid milk chocolate. A short time later, another Swiss, Lindt, found that continual stirring of the chocolate liquor (with or without condensed milk added), led to an extremely smooth, creamy texture (Fig. 14-12). Stirring causes the cacao particles to be ground into finer and finer pieces. By adding extra cocoa butter (obtained from the production of cocoa powder), even creamier chocolate is produced. While the procedures of fermenting and roasting cacao beans were known to the Aztecs, innovations such as Dutching, continual stirring, and adding milk have extended the uses of cacao far beyond those envisioned 200 years ago.

TEA

Tea is not as important an international commodity as coffee, but it is probably drunk by a greater number of people. Most of the tea which is produced is consumed locally with the result that comparatively small quantities enter international trade.

Although the precise origin of tea, *Camellia sinensis* (Camelliaceae), is unknown, evidence indicates that the species is native to China. Asians have many legends describing the first uses of tea. According to one story, an ancient emperor discovered tea when a leaf from a tea plant accidentally fell into his cup of boiling water. According to another legend, the first tea plant sprung from the eyelids of a pious monk who was a follower of Bodhidharma. The monk vowed he would stay awake until Buddha's work on earth was completed, but although he valiantly tried, he fell asleep after many days. When he awoke, the monk was so mortified by his weakness that he tore off his eyelids. Buddha, pleased by the intentions of his follower, caused the eyelids to sprout into tea plants when they touched the ground. Tea was thus thought to have been a gift from Buddha to aid humans in their vigils by keeping them awake and, at the same time, serving to enhance the peaceful, harmonic state necessary for contemplation.

Whatever the origin, elaborate methods of brewing tea were devised early in its history. One of these involved pouring boiling water over pieces of a crumbled cake of pressed tea leaves and rice. Sliced onions, ginger, and orange were added to the brew. By 780 A.D., the *First Tea Classic*, a Chinese compendium of information on the cultivation and preparation of tea was published by Lu Yu.

Brought to Japan in 593 A.D., the beverage assumed a central role in Buddhist ritual there. Accordingly, each aspect of the Japanese tea ceremony (Fig. 14-13) reflects principles inherent in Zen Buddhist belief. The management of the tea garden, with a portal so low that one must stoop to enter, a floral display of a precise form in the alcove, and a ritualized method of brewing tea leaves are all parts of the elaborate tea service. The tea itself is prepared by whipping powdered tea leaves into steaming water with a bamboo whisk. The green, foamy beverage is served in freshly rinsed raku bowls.

The mongols also learned to brew tea from the Chinese and began to use tea as a trade commodity as they wandered across Asia. Dried tea was packed into bricks (hence the term brick tea) that were easily transported. Russia eventually acquired tea by means of this northern trade link with China, and the Russians developed their own method of brewing tea in a samovar. The practice of sweetening tea with a spoonful of jam and the addition of a slice of lemon were both Russian innovations.

Europeans were first exposed to tea when the Portuguese began to explore

the coast of China in the sixteenth century. By the eighteenth century, tea had become an important trade item for Europeans. Both the British and the Dutch East Indian Trading Companies were intimately involved with the purchase of tea in the Orient and its sale in Europe and the New World.

In colonial times, Americans of European descent were primarily tea drinkers. While tea was expensive, it was cheaper than coffee, and only small quantities were needed to produce a drinkable cup of liquid. The colonists were so accustomed to their tea that they became incensed when the British declared that they had to pay a tax on tea which was brought to the colonies. To show their resentment, the colonists staged the famous Boston Tea Party (December 16, 1773) during which they dumped large quantities of the British East India Tea Company's cargo into Boston Harbor.

The British initiated commercial plantings of tea in India between 1818 and 1834. India is now the world's largest producer of tea (Table 14-4). Sri Lanka, which is the second largest modern producer, began to grow tea only after 1880 when coffee rust finally destroyed all of the British Empire's coffee industry there. In fact, it was only after the English coffee plantations on this island fell victim to the coffee rust and were replaced with tea orchards that tea supplanted coffee as the favorite beverage of the British. The custom of British tea time is purported to have been initiated by the Dutchess of Bedford sometime early in the twentieth century. The lady, feeling a lack of energy in the late afternoon, used tea and cakes to bolster her spirits. Her practice was subsequently adopted by the wealthy, and eventually by all of Britain.

Plantations in the New World have met with little success. Argentina is the only country in the Americas with any appreciable acreage devoted to tea cultivation.

Tea plants are small, evergreen trees. As indicated by the generic name, tea belongs to the same genus as the beautiful horticultural camellias (Fig. 14-14). Plantations of tea are generally started from seed which has been carefully

FIGURE 14-14
A tea branch in flower.

germinated in protected enclosures. When sufficiently robust, the young plants are set into fields. Less commonly, cuttings are used to start plants. Plantations are located primarily in areas with a good rainfall and a constant, cool temperature throughout the year. After about 3 years, the trees are pruned and subsequently repruned every 12 or so years. The type of pruning varies with the country in which the tea is grown, but the purpose is the same in all cases, to force the trees into a bushlike growth form that facilitates picking and encourages rapid shoot growth (Fig. 14-15).

Picking can begin when trees are 4 years old. For fine teas, only the two or three youngest leaves and the terminal bud of each branch are picked (Fig. 14-16). It has been shown that these parts contain the highest quantities of caffeine and other constituents which give tea its flavor. A single shrub can be picked about once a week.

The type of processing of the leaves depends on the final kind of tea desired. For green tea, the leaves are shredded (or rolled) and heated to inactivate leaf enzymes. The leaves are then dried and packaged. Green tea is usually consumed locally. For the international market, most tea is processed into black tea. To produce black tea, the leaves are first spread out on a flat surface or placed in drums to wither (Fig. 14-17). The withered, flaccid leaves are then rolled and twisted. Historically, twisting was manual work, done by simply rolling the leaves on a hard surface with the palm of the hand. Now, rolling and twisting are done with mechanical rollers. The purpose of rolling is to crush the cells of the leaves releasing the enzymes in the cytoplasm. The rolled leaves are then allowed to sit and chemically ferment for a few hours.

As in the cases of coffee and cacao, tea fermentation refers to the alteration of various chemical constituents of the leaf. Tea leaves contain, in addition to caffeine, tannins, and polyphenols. The caffeine is responsible for the stimulating

FIGURE 14-15
Because the ends of the branches are constantly being picked, tea shrubs assume rounded shapes. Tea is ideally grown on hillsides where cool, moist conditions produce leaves with the finest flavor. (Photo courtesy of the Japan National Tourist Organization.)

FIGURE 14-16
Only the tips of the shoots of tea plants are picked as this Indonesian worker demonstrates. (Photo courtesy of UNESCO.)

effects of tea, and the tannins impart the characteristic bite and brown color. The flavor and quality of black tea is determined by the ratios of products which result from the chemical fermentation of the polyphenols. Once fermentation is judged to be complete, the tea is fired, or heated, to stop further chemical activity and to reduce the water content to about 3 percent. The dried tea is tasted, graded, and packaged.

Oolong tea combines the taste properties of green and black teas because it is semifermented, or allowed to undergo an enzymatic fermentation for only a very brief period. It is also made from a particular variety of tea (*Camellia sinensis* var. *chemisa*) and is produced primarily in southern China and on Taiwan.

The drinking of iced tea is a relatively recent innovation. This might be expected from the fact that ice was a scarce commodity until the advent of modern refrigerators. The introduction of iced tea to the American public (the only people who drink appreciable quantities of it) is credited to an enterprising Britisher trying to sell tea at the 1904 St. Louis World's Fair (the same fair at which ice cream cones made their debut). The gentleman, finding that no one wanted to drink hot tea in the middle of a St. Louis summer, added ice to his wares and immediately increased his sales.

Tea bags, now so commonly used in America to prepare tea, were initially made by a New York wholesaler who sent small samples of tea tied in silk bags to potential customers as a promotional scheme. The idea obviously went far beyond an advertising idea. Tea bags were made for many years from gauze sewn into bags. By 1934, 8 million yards of gauze were used annually in the manufacture of tea bags. Now, as we shall see in Chapter 16, tea bags are made from a kind of paper derived from plant fibers.

FIGURE 14-17
The steps in the processing of tea differ in different parts of the world and have often changed from hand to mechanical methods. Various methods for harvesting and withering and some of the early machinery employed are shown by (A) harvesting of tea by plucking the upper two or three leaves and the apical bud; (B) withering and fermenting the tea by hand; (C) drying leaves in bamboo baskets over charcoal in China; (D) mechanical cutting of tea was done in 1870 by Reid's Original Teacutter; (E) withering and fermenting tea on wicker trays as is the custom in some parts of the Orient; (F) withering drums; (G) hand rolling in black tea production; (H) an early machine for the rolling of tea was this Empire roller; (I) drying in a power tea dryer. (Redrawn from photographs in Ukers. *All about Tea.* 1935.)

LESS IMPORTANT STIMULATING BEVERAGES

Maté

We associate maté today with Argentina and, to some extent, Uruguay and Paraguay. The species from which maté comes, *Ilex paraguariensis* (a member of the holly family, Aquifoliaceae) is native to the mountains of northern Argentina, southeastern Brazil, and Paraguay. However, the name maté is derived from a Quechuan word, and historical and anthropological remains show that maté was widely used in pre-Columbian times by Andean Indians as well as by Indians of the Paraguayan lowlands.

As in the case of tea, leaves are used to produce maté. Unlike tea, all of the leaves of the tall evergreen maté trees are used. Branches are cut from the trees and the leaves stripped and carried to a processing area where they are separated from twigs and debris. The leaves are then dried on a frame over a fire. The type of wood used in the drying fire influences the final maté flavor. After being dried, the leaves are crushed into rather coarse pieces, sieved, and shipped. Maté produced by this process is the most commonly encountered green type. Some maté is also toasted by exposing the leaves to more intense heat. In this case, the final product is brown, and the maté is sold as *maté cocido*.

Maté is traditionally drunk from a dry, hollowed gourd called a maté (Fig. 14-18). Maté powder is placed in the gourd (with or without sugar) and the gourd filled with boiling water. After the mixture has steeped and settled, a metal straw (*bombilla*) with an expanded, flattened end is inserted into the gourd. The flat end has holes large enough to allow movement of the hot tea, but small enough to prevent pieces of the leaves from entering the straw. Maté can also be prepared in a pot by pouring boiling water over the powdered leaves and allowing the leaves to settle. The decanted liquid can be drunk plain or with sugar.

Guaraná

While the Andean Indians and cultures of southeastern South America used maté as a source of a stimulating beverage, the Amazonian forest Indians used

FIGURE 14-18
These maté gourds are used for brewing the caffeine-rich maté leaves *(foreground)*, and the infusion is then sipped up through the bombilla, which strains out the leaf particles.

FIGURE 14-19
Guaraná is a popular stimulating beverage in Brazil where it is available in many forms. In the center are the native utensils used to prepare guarana. The tongue of a large amazonian fish, the *pirarucu*, is used to rasp a small amount of powder from the rods of ground guarana seed paste. The powder is dissolved in water. On the *right* in the foreground is a figurine of the fish made of guarana paste. Next to the figurine are scales of the *pirarucu* (which are used as fingernail files). In the back are two tin jars of guarana powder and a bottle of concentrated syrup.

guaraná (*Paullinia cupana,* Sapindaceae). The species, a monoecious climbing vine (but a sprawling shrub under cultivation) native to central Brazil, produces caffeine-rich seeds that are removed from the fruits when they are mature. One to three arillate seeds are borne in each capsule. To prepare guaraná for beverage making, the seeds are dried after their arils are removed. The seeds are then roasted for 2 to 3 h and shaken to remove the seed coats. The cooled, toasted seeds are then crushed into a paste that is formed into rods (*bastoes*) about 2.5 cm (1 in) wide and 20 cm (8 in) long (Fig. 14-19). The final product looks rather like a segment of a rusted metal rod. To make a glass of guaraná, the cylinder is scraped and the resultant powder mixed with water.

Guaraná is probably second only to coffee as the most popular drink in Brazil. However, the term guaraná is now used much like the term cola in the United States to refer to a kind of soft drink. Originally the bottled drink was made with guaraná, but it is now usually artificially flavored, colored, and impregnated with caffeine. Guaraná pills which are supposed to provide energy are sold in health food stores. The stimulating substance in the pills is, of course, simply caffeine.

Kola

A relative of cacao, *Cola nitida* (Sterculiaceae) is best known to Americans because of cola soda pops. In the original formulation of Coca Cola, kola, as well as coca (see Chapters 12 and 13), was used to make the beverage. Now, coca leaves from which the cocaine has been removed, artificial flavorings, and caffeine are added. The seeds of the kola plant, like those of its relative *Theobroma,* are used in beverage making, but they are treated differently from those of cacao. Another difference is that, unlike cacao, kola is native to West Africa where it has long been used to produce a caffeine-rich beverage.

The flowers of the kola plant are borne in axillary inflorescences. The fruits each contain about eight seeds which are scraped from the harvested pods. Kola fruits lack the pulp of cacao pods, and the fleshy seed coats are removed before the seeds are allowed to ferment or "sweat" for a few days to develop their flavor. The seeds (or "kola nuts") are simply dried and then pulverized into a powder that is mixed with boiling water to make a beverage. In addition to caffeine, kola seeds contain a glycoside, kolanin, which acts as a heart stimulant. Several other species of *Cola* are chewed as a source of caffeine in the Far East.

ADDITIONAL READING

Cutrecasas, J. 1964. Cacao and its allies: A taxonomic revision of the genus *Theobroma. Contributions to the United States National Herbarium* 35:379–614. The classical treatment of the systematics of the species in the genus which yields cocoa.

Eden, T. 1976. *Tea,* 3rd ed. London, Longman.

Erickson, H. T., M. P. F. Correa, and J. R. Escobar. 1984. Guarana (*Paullinia cupana*) as a commercial crop in Brazilian Amazonia. *Economic Botany* 38:273–286.

Harler, C. R. 1964. *The Culture and Marketing of Tea,* 3rd ed. New York, Oxford University Press.

Kolpas, N. 1977. *The Chocolate Lovers' Companion.* New York, Quick Fox. A very readable, but accurate account of the process of producing cacao and chocolate.

Porter, R. H. 1950. Maté—South American or Paraguay tea. *Economic Botany* 4:37–51. Somewhat dated, but one of the few detailed articles in English on the production of maté.

Purseglove, J. W. 1974. Theaceae (= Ternstroemiaceae). pp. 599–612 in J. W. Purseglove. *Tropical Crops. Dicotyledons.* New York, Wiley.

Purseglove, J. W. 1974. *Theobroma* L. pp. 571–598 in J. W. Purseglove. *Tropical Crops. Dicotyledons.* New York, Wiley.

Schafer, C., and V. Schafer. 1976. *Coffee.* Yerba Buena Press. San Francisco. A somewhat disorganized, but interesting account of various aspects of the history, production, and consumption of coffee.

Ukers, W. H. 1935. *All About Tea,* 2 vols. New York, The Tea and Coffee Trade Journal Co. The "Bible" of the history and cultivation of tea.

Ukers, W. H. 1935. *All About Coffee,* 2nd ed. New York, The Tea and Coffee Trade Journal Co. The authoritative companion to the tea volume.

Chapter 15

Alcoholic Beverages

*A*lcoholic beverages are as important commercially as stimulating beverages but in many ways are their opposites. Alcohol is produced by a true microbial fermentation rather than by endogenous chemical changes such as we discussed in the cases of coffee, cacao, and tea. More importantly, alcohol acts as a depressant rather than as a stimulant on the human body. While one might think that humans would have shunned products that dulled the senses, the production of alcoholic beverages has been independently "discovered," or readily adopted, by almost every society in the world. Humans are also the only animals that will drink alcohol to excess under natural conditions and are therefore the only animals that can become truly addicted to it. We concentrate in this chapter on the methods used to produce alcoholic beverages, the effects of alcohol as a drug, the various plants from which these beverages are made, and the history of the use of fermented beverages. Table 15-1 lists the plants discussed in this chapter.

The production of all alcoholic beverages depends initially on the process of fermentation by yeasts. This process yields two basic types of alcoholic beverages, wines and beer. Distillation or fortification of either of these produces the wide array of beverages with high alcoholic contents that we now consume.

FERMENTATION

Virtually all beverage alcohol is produced by the action of fungi, in all cases species of the genus *Saccharomyces*. This genus belongs to a group of fungi known as *yeasts*. Like all fungi, yeasts lack chlorophyll and cannot manufacture their own food. Yeasts differ from other fungi in that each cell lives independently and reproduces primarily by budding, or by simple mitotic divisions (Fig. 15-1). Most fungi are multicellular and form chains of cells called

TABLE 15-1

Plants Discussed in Chapter 15

COMMON NAME	SCIENTIFIC NAME	FAMILY	CHROMOSOME NUMBER
Agave	*Agave angustifolia*	Agavaceae	$2n = 60, 120$
	A. palmeri		unreported
	A. tequilana		unreported
Apple	*Malus pumila*	Rosaceae	$2n = 34, 51$ di and hexaploid
Barley	*Hordeum vulgare*	Poaceae	$2n = 14$ diploid
Corn	*Zea mays*	Poaceae	$2n = 20$ diploid
Dates	*Phoenix dactylifera*	Arecaceae	$2n = 36$ diploid
Grape	*Vitis vinifera*	Vitaceae	$2n = 38$ diploid
Hops	*Humulus lupulus*	Cannabaceae	$2n = 20$ diploid
Juniper	*Juniperus communis*	Cupressaceae	$2n = 22$ diploid
Marijuana	*Cannabis sativa*	Cannabaceae	$2n = 20$ diploid
Maguey	*Agave pacifica*	Agavaceae	unreported
Pear	*Pyrus communis*	Rosaceae	$2n = 34, 51$ di and triploids
Rice	*Oryza sativa*	Poaceae	$2n = 24$ diploid
Rye	*Secale cereale*	Poaceae	$2n = 14$ diploid

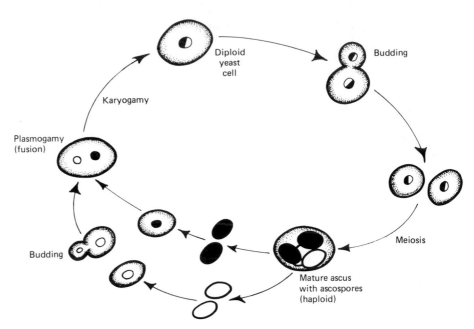

FIGURE 15-1
This life cycle of *Saccharomyces cerevisiae*, the yeast most commonly used in brewing processes, shows that yeasts are single-celled organisms which do not form mycelia or specialized asci bearing structures. While budding (mitotic divisions) of diploid cells is the most common form of reproduction, sexual reproduction involving meiosis can also occur.

hyphae that mat together into the vegetative body or mycelia. These fungi typically form reproductive structures (asci) that contain haploid or diploid spores. Although many fungi can carry out fermentation, species of *Saccharomyces* are generally used because they are very efficient at alcohol production, and they can tolerate higher levels of ethanol than most other fungi. During fermentation they manufacture compounds other than alcohol which are believed to impart subtle, desirable characteristics to the fermented liquid. Some strains of *Saccharomyces* also have the ability to clump into masses during the later stages of fermentation (a phenomenon known as *flocculation*), which facilitates their removal from the final beverage. The species of *Saccharomyces* that are used for alcohol production (primarily *S. cerevisiae* and *S. uuvarum*) live and reproduce in anaerobic (oxygen-free) conditions, usually in a solution.

Species of *Saccharomyces* live by ingesting sugar and nitrogenous compounds produced by other organisms. Sugars vary from simple sugars composed of a single saccharide unit, to disaccharides or polysaccharides, formed by the linkage of two or more such units (Fig. 15-2). The only sugars that can be used by *Saccharomyces* are simple six-carbon monosaccharides. These monosaccharides, also called *hexoses*, are the most common type of simple sugar found in nature. In addition to sugars, yeasts need amino acids in order to live and reproduce. The need for these two kinds of compounds is absolute: *Saccharomyces* will die in a medium which does not contain hexoses and will grow poorly, if at all, in one that is lacking nitrogen. Yeast cannot effectively metabolize starch even though starch is composed of units of sugar molecules. Consequently, in the production of all fermented beverages, the material that is to be fermented must contain hexoses that are naturally present or that have been produced by enzymatic degradation of starches. Essentially this means that enzymes in some form must be added to almost everything that is fermented except honey and some fruit juices.

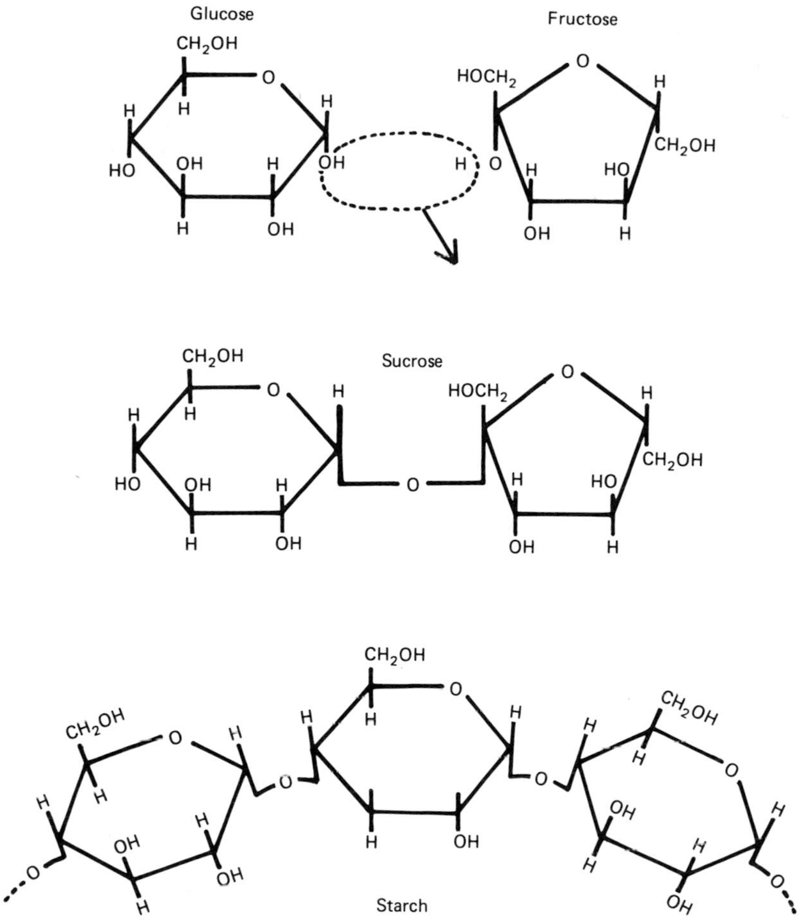

FIGURE 15-2
The monosaccharides glucose and fructose, both six-carbon sugars, can join to form sucrose, a disaccharide. Sucrose is the most common sugar transported in plants. Several monosaccharides can also link together to form polysaccharides or starches. Disaccharides and polysaccharides including starch must all be broken down to monosaccharides before they can be consumed by yeasts and used to produce alcohol.

During the process of fermentation, simple sugars are broken down (via a number of intermediate steps) into ethanol and carbon dioxide. From the yeast's point of view, these two are waste products. During fermentation the yeasts also produce small quantities of alcohols of long-chain lengths (called *fusel oils*), acetaldehyde, acetic acid, and traces of sulfur-containing compounds. The chemical steps involved in fermentation are shown in Fig. 15-3. Theoretically, taking into account only the energy required for the chemical reaction, 51.1 percent of the sugar could be converted into alcohol. In practice, because the yeast is using a portion of the sugar for growth and maintenance, about 47 percent is converted.

As the yeast culture grows and metabolizes more and more of the sugar in the liquid, carbon dioxide builds up and causes the solution to become bubbly. This gas is often simply allowed to escape from the solution. With unlimited sugar, the alcohol level increases during fermentation until it reaches a concentration of between 14 and 18 percent. Levels of alcohol above 18 or 19 percent are usually toxic to the yeast and lead to death of the cells. This tolerance limit places an upper value on the percentage of alcohol produced solely by fermentation. To achieve a higher concentration of alcohol, the solution must be fortified or distilled.

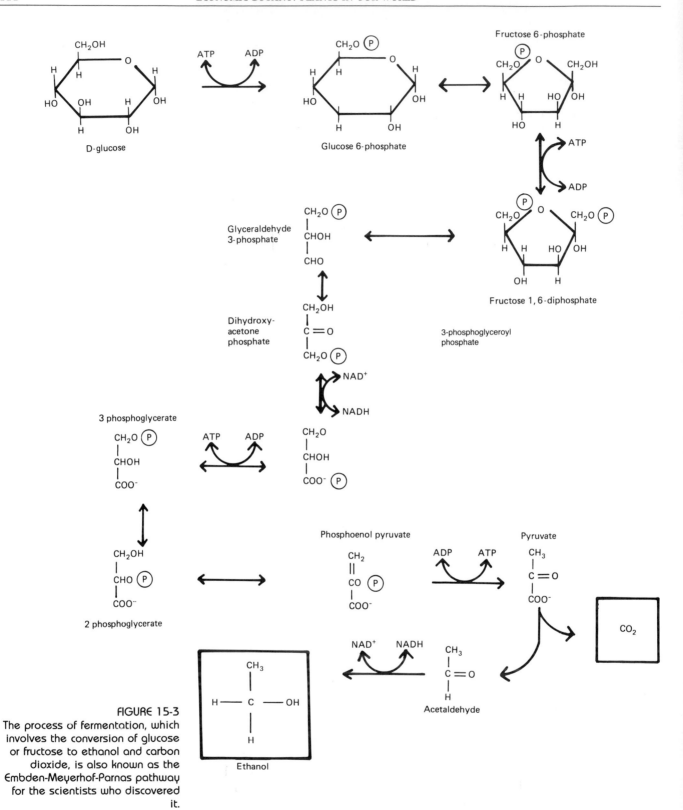

FIGURE 15-3
The process of fermentation, which involves the conversion of glucose or fructose to ethanol and carbon dioxide, is also known as the Embden-Meyerhof-Parnas pathway for the scientists who discovered it.

ALCOHOL AS A DRUG

Few people would deny that alcohol is a drug. Yet it is used legally by most adult Americans. Deaths related to alcohol consumption (including automobile accidents) rank fifth in the list of major killers of Americans. Perhaps it is our long association with alcoholic beverages or the fact that, until fairly recently, most fermented beverages contained low levels of alcohol that explains the worldwide acceptance of alcoholic beverages. It is true that in low to moderate doses, no evidence has been found of persistent, harmful effects on the bodies of individuals consuming alcohol. However, because alcohol is absorbed, metabolized, and excreted differently from most other drugs, it rapidly affects one's sense of reason and physical coordination. As a result, the social drinker who drives a car or takes tranquilizers while drinking, often inadvertently harms him- or herself or others.

Beverage alcohol is *ethanol*, a compound that is soluble in both water and fats. Consequently, it moves rapidly across membranes, and almost all alcohol consumed is completely absorbed in the stomach and upper portion of the intestine. The only way for alcohol to leave the system is by metabolic breakdown, a process that is linear over time. The breakdown products are carbon dioxide and water. However, ingestion rates usually greatly exceed the rate of metabolism. Once quantities ingested exceed the amounts that can be metabolized, alcohol levels in the body simply rise. Alcohol is a nonselective central nervous system depressant affecting all of the neurons in the brain and brainstem. Elevated concentrations in the blood lead initially to a general numbness perceived as a relaxed feeling, and then to disorientation, reduced judgmental ability, and loss of the power to reason clearly.

In addition to these temporary effects, alcohol can lead to permanent physical damage. Drinking during pregnancy appears to be correlated with fetal abnormalities. This effect, known as *fetal alcohol syndrome*, seems to be most pronounced if drinking is in "binges" or during critical times of fetal development. Epidemiological studies show that individuals who both smoke and drink stand higher risks of dying from cancer than those who indulge in only one, or neither, of these activities. Alcohol can also be an addictive drug, with some individuals apparently more prone than others to addiction. Prolonged, excessive use often leads to severe malnutrition (due to lack of interest in eating properly), permanent brain damage, proliferation of fat in the liver (cirrhosis), and heart problems.

Attempts to make alcohol use illegal in 1920 with the passage of the Eighteenth Amendment (prohibition) failed with its repeal in 1933 (Twenty-first Amendment). The belief that alcohol was a dangerous drug did not change during the 13 years of prohibition, the country simply realized that the numbers of deaths, proliferation of crime, and amounts spent for legal fees were higher with prohibition than without it. For the past 50 years, we have been able to buy and consume alcohol freely in the United States. Nevertheless, as alcohol becomes a proportionately more important mortality factor, there is pressure to institute legal sanctions against alcohol abuse, while at the same time providing rational treatment for those who have become victims of alcoholism. Educational programs about the use and abuse of alcohol have been initiated in many states

FIGURE 15-4
The blood level concentration of alcohol in a person's body is related to body weight and the number of drinks consumed per unit of time (assuming that each drink is 1 oz of 80 proof whiskey).

Number of drinks	Body weight (pounds)							
		100	125	150	175	200	225	250
1		.03	.03	.02	.02	.01	.01	.01	Sober
2		.06	.05	.04	.04	.03	.03	.03	Sober
3		.10	.08	.06	.06	.05	.04	.04	
4		.13	.10	.09	.07	.06	.06	.05	
5		.16	.13	.11	.09	.08	.07	.06	Driving impaired
6		.19	.16	.13	.11	.10	.09	.08	Driving impaired
7		.22	.18	.15	.13	.11	.10	.09	
8		.26	.21	.17	.15	.13	.11	.10	
9		.29	.24	.19	.17	.14	.13	.12	
10		.33	.26	.22	.18	.16	.14	.13	Illegal
11		.36	.29	.24	.20	.18	.16	.14	
12		.39	.31	.26	.22	.19	.17	.18	

along with stiffer penalties for offenses such as driving while intoxicated (Fig. 15-4). Perhaps we will eventually find a balanced policy toward alcohol and other drugs that consistently takes into account their harmfulness to individuals and society.

STILL WINES AND MEAD

By definition, wine is fermented fruit juice. In practice, however, the term is overwhelmingly used for the fermented juice of grapes (*Vitis vinifera*, Vitaceae). Other wines are generally specified by using the name of the fruit from which they are made, as, for example, peach wine, or blackberry wine. Wine may be the oldest fermented beverage made by humans, although some authors contend that mead (or even beer) was produced before it. *Mead* is a fermented solution of honey and water. The sugar in pure honey is so concentrated that *Saccharomyces* or other fungi or bacteria cannot live on it (hence its long shelf life). However, if honey is diluted with water, it provides an excellent medium for yeasts because it consists of simple sugars. Consequently, before humans could have consciously made mead, they would have had to gather wild honey and dilute it (either by washing combs or sweetening water). Today, mead is made by boiling a dilute solution of honey to which nitrogen-containing compounds and, in some cases, aromatic herbs are added. Fermentation is completed in 6 to 8 weeks at temperatures between 15 and 25°C (59 to 77°F). Mead is classified according to the kind and dilution of the honey from which it was made.

Wine, in contrast, is made in nature. The yeasts responsible for the fermentation of fruit sugars are usually present on fruit skins, and fermentation can occur naturally if the skin of a ripe, sweet fruit is punctured. Human production of wine would merely involve collecting fruits, bruising or crushing them, and letting them ferment. Still, no one knows when wine making began, and estimates varying from 8000 to 3000 B.C. have been proposed. The older date is probably the more realistic.

Since grapes are the dominant fruit used for wine, we will concentrate on grape wine production. Botanical evidence suggests that the species of grape most widely used for wine was originally domesticated in western Asia about 4000 B.C. Any wine making before this time would have used wild grapes (Fig. 15-5). Grapes, like most fleshy fruits, are primarily water, but they also contain an appreciable amount of fructose, a simple monosaccharide. Egyptians used wine primarily for religious ceremonies (Fig. 15-6), and it was only between 2000 and 1000 B.C. under the Grecian Empire that wine became a popular beverage (Fig. 15-7). The classical Greeks made a heavy wine that they mixed

FIGURE 15-5
This branch and flower of a mustang grape show the tendrils and the three- to five-lobed leaves characteristic of the grape family. In this species, the greenish flowers have reduced sepals and five nectariferous glands alternating with the stamens at the base of the two-celled ovary. The fruit is a two- to four-seeded berry.

FIGURE 15-6
Harvesting grapes and producing wine. (Redrawn from an Egyptian wall frieze in the Tomb of Nakht.)

FIGURE 15-7
Bacchus as the God of Wine. [Engraved by Gabriel Muller (1771). Courtesy of Christian Brothers Collection.]

FIGURE 15-8
According to Hebrew folklore, when Adam planted the first grapevine, the devil buried a pig, a lion, and a lamb at its roots so that those who drank too much of the juice of its fruits would become fat, ferocious, and feeble. Joshua and Caleb are shown in this woodcut carrying the "big grape." [Woodcut by Jost Amman (1531–1591 A.D.). Courtesy of the Christian Brothers Collection.]

with water for drinking. They stored their wine in vessels smeared with pine pitch to prevent leakage, a practice that accounts for the Greeks' fondness for a resinous flavor in their wine. Now, this flavor is added to produce the Greek retsina wines. The Romans did not use pitch on their wine vessels, which is one reason why Italian wines surpassed Greek wines in popularity under the Roman Empire.

Wine grape cultivation spread from the eastern Mediterranean region (Fig. 15-8) about 600 B.C. to France and later to Spain, Portugal, and Algeria. Until recently, Europe and north Africa were the undisputed world leaders in quantity and quality of wine production. Now, the United States, Argentina, and Russia rank among the top 10 wine-producing countries of the world (Table 15-2).

Wine making in the United States got off to a slow start, probably because the portions of the country originally most densely settled were too cold for growing European wine grapes. Although it could not be successfully grown in New England, the vinifera grape was hardy in many other areas of the New World. Columbus introduced plants into the West Indies on his second voyage (1493), and the Spanish began its cultivation in California around 1769. By

TABLE 15-2

Leading Wine Countries and Regions of the World

TOP 5 COUNTRIES, 1,000 t		TOP 5 CONTINENTS, 1,000 t		TOTAL WORLD, 1,000 t
COUNTRY	t	CONTINENT	t	
Italy	8,000	Europe	23,366	33,388
France	6,000	South America	3,440	
U.S.S.R.	3,300*	North & Central America	1,548	
Spain	3,157	Africa	1,100	
Argentina	2,500*	Oceania	390	
United States	1,450			
Germany, Federal Republic of	1,356*			
Romania	1,000*			
Portugal	795*			
Yugoslavia	720*			

*Indicates an estimated or an unofficial figure.
Note: The U.S.S.R. is considered as a country only. Its values are not included in the figures for either Europe or Asia. Figures are in metric tons.
Source: Data from *FAO Production Yearbook for 1983,* vol. 37. 1984. Rome, FAO.

the middle of the nineteenth century, California had a small but respectable wine industry (Fig. 15-9).

It is lucky that Americans became involved in viticulture because in 1860, European vines began to die from an infestation of *Phylloxera,* an insect commonly called a root aphid (although it is not a true aphid). This insect is a native of North America and had been inadvertently carried to Europe between 1850 and 1860. Populations of the insect expanded rapidly in the fields of the susceptible *V. vinifera* and decimated the vines. The French sent a commission to the United States, which determined that many North American species and hybrids were naturally resistant to the insect. As a result of the commission's findings, literally thousands of American cuttings and seeds were sent to Europe

FIGURE 15-9
The vintage in California, a wood engraving which appeared in *Harper's Weekly* in 1878. (Courtesy of the Christian Brothers Collection.)

to replace the dying vines (Fig. 15-10). The plants produced from the cuttings and seeds, mostly from Texas, were used as rootstocks onto which the traditional varieties of the vinifera grape were grafted. Almost all European grapes are still grafted, even though resistant strains of *V. vinifera* have now been developed. In California, both grafted and nongrafted vines are commonly used.

Vines begin to produce several years after planting and can bear well for over 50 years. The deciduous plants are pruned between February and April (in the northern hemisphere) to remove the long branches produced during the previous year and to ensure symmetrical branch growth. The sprouts that emerge later in the spring are trained onto trellises that vary in structure depending on the areas in which the grapes are grown. For example, in France, Italy, and California, wires are strung just above the stumps and along the rows. In Chile and Argentina, tall arbors train the vines upward. In many vineyards, several varieties of grapes are grown together so that when the grapes are picked, they will automatically produce the desired blend of grape types. Blending can, however, be done at any number of points in the wine making process, up to the final bottling.

Making Wine

Fine vintage wine is made today in much the same way as it was centuries ago, although the bulk wines that are now commonly produced are often treated slightly differently. Large wine making companies in the United States monitor each step of the process with sophisticated chemical and computer analyses. Additional chemicals ranging from fructose to minute amounts of mineral salts are added whenever the analyses indicate a deficiency. In our discussion, we follow the traditional steps in wine making, pointing out some of the recent innovations.

Since grapes are naturally equipped with everything necessary for fermentation, the process of wine making is basically very simple. The grapes are crushed and the juice is allowed to ferment. Juice can be expressed by stomping on the grapes (a method still used in some parts of Europe, or with hand-operated, electric, or fuel-powered presses, Fig. 15-11). In areas where grapes

FIGURE 15-10
This allegorical statue commemorates the rescue of the French wine industry after the devastation by *Phylloxera* blight in the middle of the nineteenth century. The young woman represents the resistant American rootstocks which replaced the more susceptible European rootstocks. The statue is in Montpelier, France, the home of Gustave Foex who was head of the Montpelier School of Viticulture at the time of the blight. (Courtesy of the École Viticulture de Montpelier.)

FIGURE 15-11
Grape crushing in the Salinas Valley, California. (Courtesy of Paul Masson Vineyards.)

FIGURE 15-12
The processing of vinifera and labrusca grapes is contrasted here. As the flow chart for
red vinifera wines shows, natural yeasts are traditionally used and pasteurization, if

are grown as part of a cooperative venture, they are often pressed at some central point and the liquid mass transferred to trucks for transport to the winery. Sulfur dioxide is introduced into the closed container to kill bacteria. Grapes can even be shaken loose by mechanical pickers and pressed in the field. Again, the liquid is placed in an atmosphere of sulfur dioxide.

If the expressed juice is to be made into white wine, the free juice is run into fermentation tanks and the peels and stems re-pressed (Fig. 15-12). The juice from the second pressing can be added to the original juice or used for lower grade wines. For red wine, the skins go into the fermentation vat with the juice. The red color of red or rosé wine (from which the skins are removed after a short time) is due to epidermal pigments that dissolve in the juice. Consequently, white wine can be made from red grapes if the skins are removed right after pressing. Red wine cannot be made naturally from white grapes. Bulk red wines made from white Thompson seedless grapes have the color added to the wine.

Once the juice is in the fermentation tank, preferred strains of yeast are often added (although the process could proceed without additional yeast), and fermentation is allowed to continue for about 8 to 10 days, after which the incipient wine is drawn off of the skins if they are still present. Any additional liquid obtained from pressing the skins that remained throughout fermentation is considered inferior in quality to the initial juice and is either used in the blending of poorer grade wines, or for vinegar.

After the initial fermentation, the liquid is allowed to ferment for 20 days to about 1 month. During this second fermentation period, particles and dead yeast cells settle to the bottom of the tanks. When the process is complete, the wine is drawn off from this sediment and placed in aging tanks. The concentrations of acids in the grapes, principally malic and tartaric acids, change during the fermentation and throughout the aging process. The tartaric acid, which precipitates from the solution as "cream of tartar," was once one of the important commercial sources of this baking aid. Because particle sedimentation continues over time, the wine is often transferred across a series of tanks during aging (a process known as *racking*), each time leaving behind a sedimentary layer. Aging tanks are used over and over again but are cleaned and doused with sulfur (to kill bacteria) after each use. The large wooden tanks traditionally used in wine making can acquire individual characters and some become known as the producers of "superior" wine. The modern use of stainless steel tanks has removed some of the charm provided by these fragrant wooden vessels.

Time of aging in tanks varies. White wines are usually aged from 1 year to 18 months, but red wines can be aged for as long as 5 years. Bulk wines can

carried out, occurs at the end of the process. Sugar is rarely added. White vinifera wines undergo the same procedure but the skins are removed before the juice is fermented. For labrusca grapes, the must is sterilized before selected yeasts are introduced. Sugar is often added, resulting in sweeter wines. In both cases, sulfur dioxide (a gas) is used to kill unwanted bacteria. To clear the wine of particles which are small enough to remain in suspension, "fining agents," such as diatomaceous earth, are added. These attract and flocculate the particles so that they can settle out. This final step helps prevent the wines from developing off flavors during aging.

FIGURE 15-13
During the process of aging, a wine's acidity is reduced and the aroma and final color develop. Workers periodically check a wine's development by removing samples for taste tests. (Photo courtesy of Almaden Vineyards.)

be filtered to remove particles suspended in the liquid and then aged for only 1 year. At stages during aging, the wine is sampled and judged by a wine master (Fig. 15-13). Following the judgment of this authority, the wine is bottled after the tank aging is complete, or used only for blending. Again, depending on the wine master's decision about the potential of the wine, it can be sold soon after bottling or aged further in the bottle. White wines can benefit by aging in the bottle for up to 5 years, after which they tend to deteriorate. Red wines, in contrast, can continue to improve for 30 or perhaps 40 years. The quality of the wine that ultimately results depends on the kinds of grapes from which it was made, the year in which the grapes were harvested, and the procedures followed during fermentation and subsequent aging. If correctly judged as to potential and properly aged, a truly fine wine can result.

The naming and labeling of wines varies considerably between countries (Fig. 15-14). Within a given country, names and labeling practices can follow various schemes that give indications about the kind of grape used, the place of the vineyard, the slope of the area on which the vines were growing, the place where the wine was bottled, the level of sweetness, and, of course, the year in which the grapes were picked. The proper reading and assessment of a wine label is an art and is described in great detail in several of the references given at the end of this chapter.

Wines produced in the way described above are called *still wines* because they contain no gaseous carbon dioxide. The fermentation tanks are open or have outlets that allow escape of gases as they are produced. If the fermentation ceases before all of the sugars have been metabolized by the yeast, the wine will be sweet. If the yeast has converted all, or almost all, of the sugar before fermentation stops, the wine will be dry.

Champagne and Other Sparkling Wines

Sparkling wines, among them champagne, can be made by adding a little extra sugar to a wine which still has some active yeast cells, and tightly capping the bottle. The carbon dioxide produced from the final spate of fermentation is trapped inside the bottle. This late fermentation presents a problem, because the dead yeast cells and various particulate matter that settle out in the aging tanks of a still wine are sealed in the bottles. In order to remove the sediment in the champagne without losing all of the bubbles, ingenious methods have been devised.

Removal of the sediment is facilitated by storing the champagne bottles at an angle during the last part of fermentation and while aging. The bottoms of the bottles are kept higher than the necks (Fig. 15-15), and they are regularly turned a little so as to aid in the accumulation of the sediment in the bottle neck. Traditionally, the sediment was removed by opening the bottle and letting the champagne shoot outward. A little replacement wine was added and the bottle closed as rapidly as possible. Naturally, a great deal of champagne as well as part of the effervescence was lost. Today, the bottle neck is plunged into liquid nitrogen which freezes the liquid in the neck containing the sediment (Fig. 15-16). The cork is removed and the pressure from inside the bottle forces the frozen cylinder of champagne plus debris outward. The part of the frozen, extruded cylinder of champagne that contains the sediment is cut off, leaving the lower frozen portion as a temporary cork. The bottle is then immediately recorked. In another modern method, known as the transfer method, the second fermentation occurs in a closed tank, and the sparkling wine produced is drawn off the sediment and bottled under pressure so that the carbon dioxide is not lost. For less expensive sparkling wines, carbon dioxide is simply pumped into a still wine at the time of bottling in the same way that it is pumped into carbonated beverages.

Other Fruit Wines

Fruit wines can be made from many different kinds of fruits. Two such wines, cider and perry, made from apples (*Malus pumila*, Rosaceae, Fig. 15-17) and pears (*Pyrus communis*, Rosaceae), respectively, are often separated from other fruit wines because they have a much lower final alcohol content than wines made from grapes, or dates (*Phoenix dactylifera*, Arecaceae). Since apples and pears contain relatively little natural sugar, cider and perry are traditionally made with the addition of extra sugar, but even so, the beverages produced by fermentation have only 5 to 7 percent alcohol. For other fruit wines, the juice is expressed and sugar is added (usually about 150 to 300 g per liter of fruit juice). The final alcoholic content of these wines can consequently reach as high as 18 percent. In some cases, ammonia salts are added to provide the yeasts with nitrogen for growth. These fruit wines usually take about 5 to 6 weeks for fermentation to reach completion. The wine can be mellowed (real aging is not practiced), blended, possibly further flavored or colored with caramel color, and then bottled. If a sparkling "pop" wine is to be produced, carbon dioxide is forced into the wine when it is bottled.

Fortified Wines

Fortified wines are fermented wines to which distilled spirits are added. These *spirits* are concentrated ethanol or a highly distilled beverage like cognac. There are basically two kinds of fortified wines. The first group, which includes sherry, port, and Madeira, is made exclusively from grapes and spirits. The second

FIGURE 15-14
Labeling formats vary according to the country of origin of a particular wine and are sources of truth or chicanery as to what one can expect to find in the bottle. This label of a French Bordeaux wine serves as an example. (A) The distinction "Grand vin" is no assurance of a great wine. (B) "Grand Cru Classe," however, is a designation created by the bureaucracy of Bordeaux winemakers which implies that the wine has been graded and ranks among the vintages just below the top four growths which can be labeled as "Premier Grand Cru." (C) 1976 is the year of the vintage (harvest), which happens to have been a fine year in this region. (D) "Mis en bouteille au chateau" indicated that the wine was estate-bottled and should not be confused with the similar phrase "Mis en bouteille dans nos caves," which means, essentially, factory-bottled. (E) "Appelation Graves controllee" is an assurance that the wine was produced in the small region of Bordeaux known as Graves.

FIGURE 15-15
Champagne bottles are stored on tilted racks and turned periodically in order to collect the sediment in the necks of the bottles. (Courtesy of the Champagne News and Information Bureau.)

group includes aperitif wines such as Dubonnet and vermouth which have, in addition to wine and spirits, flavorings from a variety of plants.

The practice of fortifying wines began in Europe in the seventeenth century. Specific regions, Jerez in Spain, the Douro Valley in Portugal, and Madeira Island soon became the centers for the production of these types of wines. While fortified wines given the same names are now produced elsewhere, these three regions still produce the majority of, and to most connoisseurs the finest, fortified wines. Differences in grape varieties and production procedures give each of them a characteristic flavor.

Sherry is traditionally made from palomino grapes fermented into a dry wine that is fortified after aging. The spirits used for fortification are carefully blended using 95 percent alcohol and some of the fermented wine mixed in equal proportions and then clarified. Some sherries, notably the "fino" types, are fermented by a variety of *Saccharomyces* that forms a cap on top of the fermenting liquid. Other sherries are fermented by strains that behave as those in the fermentation of most wines. Sherry is matured dry. If a sweet sherry is desired, special sweet wines are blended with the dry, mature sherry. Before marketing, sherry is usually filtered using egg whites and/or diatomaceous earth.

Port is fortified as soon as fermentation has produced a wine judged to be suitably sweet. The wine itself is made from a combination of particular varieties of red and white grapes. The spirits added to fortify the wine are strictly regulated by law and must come from the Casa do Douro or Junta Nacional dos Vinhos. Maturation and the initial aging occur after blending and fortification. Tawny ports are traditionally aged for shorter periods than ruby ports. As in the case of sherry, the fortified wine is decanted and filtered before bottling.

The production of Madeira is unique. The initial process is like that leading to port. Red or white grapes can be used, and fermentation is allowed to proceed until a wine of the desired sweetness has been produced. At this point, however, neutral spirits, essentially 95 percent alcohol, are added. The newly fortified wine is then sealed by a customs officer in large tanks which are subsequently gradually heated over a period of 2 weeks to a temperature of 58°C (136°F). This temperature is maintained for 3 months, after which the wine is cooled. Once cool, the seals are removed, the wine is checked by government officials, and if it is judged acceptable, it is filtered and bottled. Madeira is often aged for about 2 years before marketing.

FIGURE 15-16
During "le degorgement," the cork of the champagne bottle is momentarily removed so that the sediment and frozen liquid in the neck of the bottle can extrude and be cut away. A fresh cork is expertly fastened to the bottle before more of the champagne can escape. (Courtesy of the Champagne News and Information Service.)

FIGURE 15-17
An apple press used to make juice from which cider is made.

BEER, ALE, STOUT, AND SAKE

As in the case of wine, no one knows when humans first began to brew beer. Educated guesses usually converge on a date of about 6,000 years ago. Certainly, by the time of recorded history, the practice was well established. Written records indicate that much of the grain of the Sumerian civilization was used for making beer. One Greek myth claims that Demeter fled from Mesopotamia disgusted at the beer drinking of its inhabitants. Early brewing is usually linked with bread making. As described in Chapter 6, one of the early ways of making grains digestible was to sprout them. Barley breads were initially made from sprouted grain that had been dried and ground into flour. A soft dough of sprouted barley flour would be, as we shall see, a good place for yeasts to live. Egyptian beers, in fact, were made from a solution of water and pieces of dough made from sprouted barley flour that was subsequently allowed to ferment. After fermentation, the liquid was poured off the sludge that had settled on the bottom of the container. Although wine was never much used by Egyptians except for special occasions, beer was widely drunk and many different kinds were made.

Early beers were relatively simple to make, but since the liquid mixture to be fermented was left exposed to the air so that wild yeast would fall into it, there was little certainty of the final beer quality. In addition to species of *Saccharomyces*, other fungi and bacteria would fall into the brew and multiply, often producing a foul-tasting or rotten-smelling batch of beer. Some quality

FIGURE 15-18
Modern beer making is an exacting science as the elaborate equipment and computerized control board of the Lone Star Brewing Company in San Antonio, Texas shows.

control was assured by using a small amount of the yeasty liquid from a previous good batch of beer. Uniform production of palatable beer as we now know it, is a relatively recent affair. In fact, the entire process of beer brewing, much more than that of wine making, has changed greatly in the last 200 years. Today, brewing is a complicated process involving several ingredients never used in brewing in earlier times. Since most of the beer we drink is made in large American breweries, we will follow the process as it now occurs in one of these operations (Fig. 15-18).

The three basic ingredients used in modern beer making are barley malt, hops, and water. In addition, most U.S. breweries also use adjuncts derived from plants other than barley. Before we go into the process of brewing itself, we will look into the origin and production of each of the important plant-derived ingredients.

Malt

Strictly speaking, malt is any sprouted grain that has been subsequently dried, but in practice the term refers to germinated barley grain. Barley (*Hordeum vulgare*, Poaceae) is preferred over the other grains used in the past for several reasons. First, during malting, barley husks stay on the kernels. Other grains usually shed their husks once they begin to germinate. During the brewing process, the husks add some flavor to the brew and later collect at the bottom

FIGURE 15-19
Barley in the malting room is turned continuously to ensure even germination. (Courtesy of the Rahr Malting Company.)

of the mashing tank where they form a bed through which the beer can be filtered prior to fermentation. Most importantly, however, is the fact that of all of the possible malts, barley malt contains the largest amount of enzymes capable of converting starches to sugars.

The first step in malting is to steep the grain in huge tanks where it is washed by a flow of water for 8 to 10 h. The washing process causes the grains to absorb water and initiates the germination process. The barley then sits in still water for another 40 h, after which the water is drained and the barley conveyed to large, warm germination rooms where it is periodically turned (Fig. 15-19).

The chemical processes set in motion by this artificially induced germination are the same as those which would occur in nature. We pointed out in Chapter 6 that the stored energy in a grain is starch. At the beginning of germination, the embryo produces enzymes that break down the starch into sugars that can be readily absorbed by the developing seedling and used as a source of energy. From the point of view of the brewer, several important things happen during this brief period. The grain synthesizes an abundant supply of hydrolytic enzymes that converts some starch to fermentable sugar, the cell walls of the endosperm break down, and many of the endosperm constituents other than starch become reduced to compounds with relatively low molecular weights. Among this last group of chemicals are many proteins that are degraded to polypeptides and amino acids. In malting, the germination process is stopped when the emergent seedling is about one-third as long as the grain itself (Fig. 15-20). Kilning, or

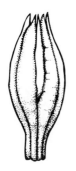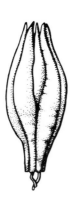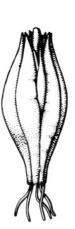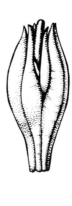

FIGURE 15-20
In the process of malting, barley is permitted to germinate until the emerging embryonic seedling is one-third the length of the fruit. At this point, the fruits are kiln-dried, (which shrivels the emerging rootlet), and shipped to breweries.

heating, the sprouted grain to temperatures between 130 and 200°C (266 to 392°F) kills the emerging seedling and any microorganisms that may be present on the grain. Some large breweries malt their own barley, but others buy malt from the huge malting corporations in Wisconsin where most of the varieties of barley grown for malt are produced.

Hops

The second important plant ingredient in beer is hops. Despite our modern assumption that hops have always been an integral part of beer, they were originally added to the beverage by the Dutch and Germans about 746 A.D. The English began to use hops in brewing only after 1524. Before the use of hops became popular, other plants, such as bog myrtle, were often added to beer as flavoring agents. Up until 1860, plant-derived flavoring agents other than hops were still being added to beer in England. Hops probably rose to ascendency because they not only imparted a pleasant taste and aroma to beer, but they also added enzymes that helped to coagulate unwanted proteins. If too many

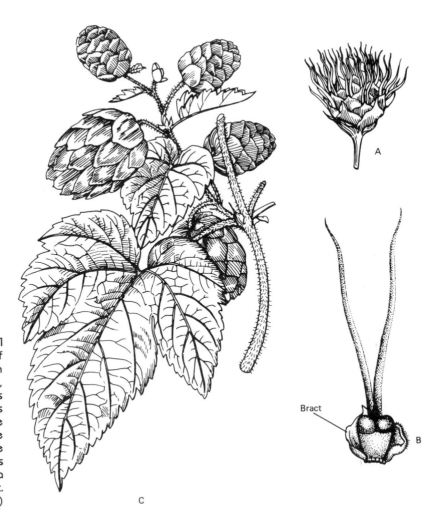

FIGURE 15-21
(A) The cone-like inflorescence of the female hop plant is similar in structure to that of its relative, marijuana. Both (B) the flowers and the bracts of the female hops have glands that produce the characteristic hop flavors, but since most hops that are grown are parthenocarpic, only the bracts provide the flavoring. (C) shows a branch with parthenocarpic fruit. (After Baillon.)

FIGURE 15-22
In harvesting hops, the vines are cut individually and laid on the truck bed for hauling to the processing plant. (Photo courtesy of the USDA.)

proteins are present in beer, the brew tends to become cloudy. Hops, therefore, contribute to the production of a sparkling, clear beer.

Hops belong to the genus *Humulus,* one of the two genera of the Cannabaceae (the other is *Cannabis sativa,* marijuana, or hemp; see Chapters 13 and 16). Only one species of the genus, *H. lupulus,* is used in brewing. Hops are dioecious vines which produce clusters of flowers (Fig. 15-21). Each female flower is subtended by a leafy bract. When mature, the female inflorescences resemble soft pine cones because all of the bracts overlap one another. The bracts of the female flowers possess numerous glands that contain volatile oils. In southeastern Washington state, one of the largest hop-growing regions of the world, hops are grown in orchards consisting of rows of female vines trained onto trellises over 6 m (ca. 18 ft) tall (Fig. 15-22). During harvesting, entire shoots are stripped from the trellises and the hops removed. Hop producers carefully label each crop of hops and send brewers samples of the different lots. Once a brewer has chosen the batch with the qualities desired, he places an order. The shipment of hops which finally arrives has often been dried and pressed into pellets. Pelletizing reduces shipping costs, prevents spoilage, and appears to have no effect on the properties of the hops.

Adjuncts

The last major ingredient, or actually set of ingredients, now used in brewing are the adjuncts. These are unmalted grains (barley, rice, and wheat), corn grits, corn syrup, or, more rarely, potatoes. These carbohydrates can constitute today up to 30 percent, by weight, of the plant material used in brewing. Adjuncts are a recent brewing innovation, and they are much more commonly used in the United States than in Europe. They are used because they are less expensive than barley malt. They also allow a brewer to use malt made from a lower grade of barley than if malt alone were used, because the flavor of the malt becomes relatively less important in the resultant light-flavored beers than in full-bodied, European beers. A light flavor seems to be preferred by American beer drinkers. Alternative starch sources usually have less protein than barley and thus reduce the number of clouding proteins that must be removed from the beer. Corn and rice are both precooked before use in order to convert their starches into a form that the malt enzymes can degrade into sugars.

Brewing

When all of the ingredients have been assembled, brewing can begin (Fig. 15-23). In large breweries, the malt and the adjuncts are mechanically conveyed to scales and the proper amounts automatically weighed and added to a mashing tun. Slightly acidified water is poured into the malt-adjunct mixture in the mashing tun. The water can be heated to a temperature between 68 and 73°C (154 to 163°F) before it is added, or it can be added cold and the entire mixture heated. In either case, the mash is allowed to stand for 2 to 6 h. During mashing, the enzymes in the barley malt diffuse into the solution and break down some of the starch of both the barley endosperm and the adjuncts. Proteins are also degraded into amino acids.

Once mashing is complete, the liquid portion of the mixture contains simple sugars, some starches and other carbohydrates, proteins, amino acids, and various other compounds. This liquid, called the *wort,* is filtered, usually by draining it through the bed of barley husks that has formed on the floor of the mashing tun. The filtered wort is boiled to inactivate the enzymes and to sterilize and concentrate the solution.

Hops are added a little at a time during the brewing process. The quantity of hops added varies from beer to beer and brewery to brewery, but it is always quite small compared to the other ingredients. Once brewed, the beer is cooled and the particles of hops removed by filtering or centrifuging. Once these operations are complete, the wort is pumped to fermentation tanks and selected strains of *Saccharomyces cerevisiae* or *S. uuvarum* (considered here to include *S. carlsbergensis*) are added. It is essential that the cultures of yeast be continually checked in the laboratories maintained by the breweries to ensure that contamination has not occurred.

Fermentation is carried out today in large cylinders with conical bases and at cool or cold temperatures. If *Saccharomyces cerevisiae* is used, and fermentation allowed to proceed at room temperature, "top fermentation" occurs. The yeast rises to the top of the tank as the process continues, forming a frothy mass. At low temperatures, the mat of cells drifts to the bottom of the tanks where it collects in the narrow part of the conical base. *Saccharomyces uuvarum* is by nature a bottom fermenter and is used to produce ale and lager beers. Much of the carbon dioxide produced during the fermentation is generally allowed to escape from the tops of the tanks.

Fermentation continues for 7 to 12 days at cool temperatures. The resultant brew, called *green beer,* is run off from the fermentation tanks and allowed to age for about 2 to 3 weeks. During this time, insoluble proteins settle out of the liquid and some undesirable chemicals, such as polyphenols, are broken down. Most breweries in the United States pasteurize and then filter or centrifuge the beer. Since, by law, there is a maximum number of microorganisms (live or dead) per milliliter permitted in commercial beers, these processes assure that there are no, or very few, yeast cells remaining in the beer. However, pasteurizing and filtering remove the natural carbonation. Consequently, carbon dioxide is later readded before bottling. The few beers that are not pasteurized generally use millipore or some other filtering system to remove the microorganisms.

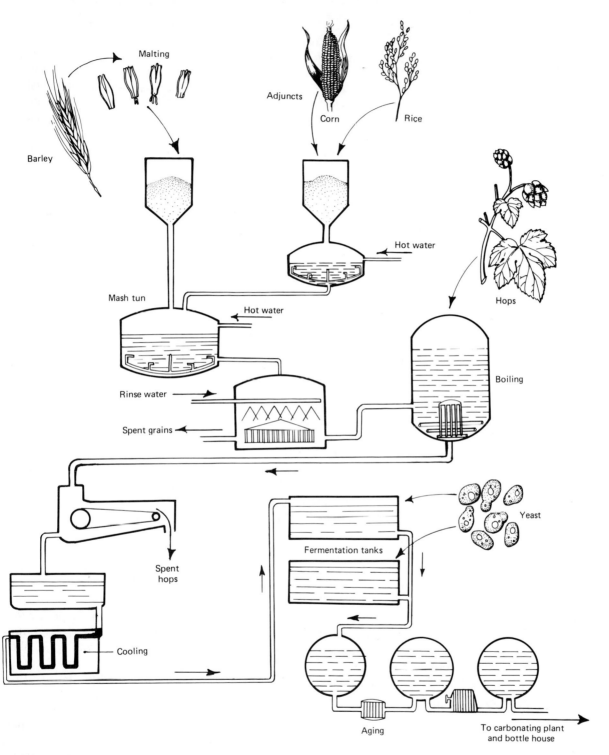

Malting

Adjuncts

Corn

Rice

Barley

Hot water

Mash tun

Hops

Hot water

Rinse water

Spent grains

Boiling

Spent
hops

Yeast

Fermentation tanks

Cooling

Aging

To carbonating plant
and bottle house

FIGURE 15-23
The brewing process.

As we mentioned earlier, ale and lager beers differ from regular beers in the kinds of *Saccharomyces* that are used. Lager beer can be classified into two types. One has a strong hop component and is known as a Pilsner beer, while the other, brewed with comparatively few hops, yields a less bitter, Munich beer. *Stout* is made using roasted barley as well as malt and has 6 percent or more caramel added for color and body.

A recent innovation in the production of beer is the brewing of "light" beer, or beer with fewer carbohydrates (and consequently fewer calories) than regular beer. Beer companies were capable of making light beers in the 1960s and made attempts to introduce them to the American public. However, the first commercially successful light beer was not produced until 1975 when Miller began to market "Lite". Within 5 years there were more than 30 kinds of light beer being sold in the United States. The success of light beers after 1975 can be attributed for the most part to the fact that beer companies had improved their taste. Moreover, at the same time, there began to be a nationwide emphasis on good nutrition and the consumption of fewer calories. All of these factors undoubtedly played a role in finally making the new beers successful.

Light beer can be made in two ways. One way is simply to add fewer starches to the same amount of water used in normal brewing. In regular beer making, the yeast converts all of the sugars in the wort into alcohol, but there is a large amount of unconverted (nonhexose) carbohydrates left in the liquid. In brewing light beer, the same amount of starch is converted to sugar by the malt enzymes and fermented, but fewer carboydrates are left in the brew than in conventional brews. In the second method, additional enzymes are added to the mashing tun. The result is that more of the starches are converted to sugar than in the usual mashing process. When this wort is fermented, a beer is produced that has a higher alcoholic content than regular beer, but fewer carbohydrates. In order to make a final product that matches regular beer in alcoholic content, water is added. When marketed, light beers tend to have a slightly lower final alcoholic content than regular beers.

The success of modern beer brewing is due to the understanding of the complex series of chemical reactions that take place, beginning with the malting of the barley. The production of these barley enzymes, their action on starches and proteins, and the process of fermentation itself are all chemical processes. As in any chemical reaction, each one proceeds best at a certain temperature and in an environment with a specific acidity or alkalinity. Today, these factors are carefully regulated during the brewing process. Consequently, modern breweries, much more than modern wineries, are like industrial plants with automated, precise weighings and additions of various chemicals. Computers often monitor ionic concentrations and temperatures throughout the various steps (see Fig. 15-18).

Sake

Sake, known to many of us as rice wine, is the traditional beverage of Japan. Since it is made from fermented grain, sake is actually more properly considered a beer than a wine, but a beer made quite differently from those we discussed before. The fermentation itself, like that of all the others, is carried out by species of *Saccharomyces*, particularly *S. cerevisiae*. The distinction lies in the production of the fermentable sugars. Sake is made from rice (*Oryza sativa*, Poaceae). If the rice were simply malted, or if enzymes were added to it, a form of regular beer would result. Instead, the conversion of rice starch to sugar is performed by another fungus, *Aspergillus*. Raw rice used to make sake is first polished by removing the pericarp-seed coat complex of the grains. The polished rice is washed, steeped, and then steamed for 30 to 60 min. Steaming changes the form of the starches present (the same reason for which it is cooked if it is used as an adjunct in beer) and sterilizes the mass. The rice is then cooled, spread in special rooms, and inoculated with cultures of *Aspergillus oryzae*. The rice is mounded into piles and the fungus allowed to grow. By the end of 40 h, a mycelium, or layer of fungal hyphae, forms and the rice begins to smell distinctly moldy. During this period, the *Aspergillus* produces enzymes that break down some of the starch. The rice is then spread out, mixed with water, and heated. Lactic acid is added to prevent growth of other organisms. *Saccharomyces* cultures are added to the water-rice mixture over the next 25 days. During this period, there is simultaneous breakdown of starch by enzymes released by the *Aspergillus*, and fermentation of the sugars by *Saccharomyces*. By the end of the growth period of the two fungi, the final alcoholic content of sake can reach 19 percent. The completely fermented thick slurry is placed in cloth filters and squeezed to force out all of the liquid. Pure ethanol is usually added to raise the final concentration to between 20 and 22 percent. In a sense, therefore, sake is a fortified beer. It is not aged as such, but allowed to mature for about 40 days. Sake is filtered and pasteurized before marketing. Traditionally, sake is consumed within 1 year after its production.

Chicha and Pulque

In Central and South America, native Indians commonly used corn to make a beer called *chicha*. Since barley malt was not available, they used a different source for amylases to break down the starch. Kernels of corn were chewed for a short period to mix them with salivary amylases. The chewed mass was spat into a container of ground corn and water and allowed to ferment. Chicha is still produced in this way in remote regions of Peru and Bolivia.

In arid and semiarid regions of Central America, natives tap the stems of species of *Agave* and collect the exuded sweet sap which is fermented into a beverage called *pulque*. Pulque can also be made by severing *Agave* plants at the base and cutting off the leaves. The stem portions are baked to render them sweet. The cooked plant material is mashed, mixed with water, and allowed to ferment into a beverage of low alcoholic content. After the Spaniards taught the practice of distillation, natives began to distill the beverage into tequila and mescal. Pulque is still locally available in parts of Mexico, but the primary use of pulque today is as a precursor of tequila.

DISTILLATION

The process of distillation involves the separation of chemicals on the basis of their boiling points. The Arabs are usually credited with the discovery of the procedure. During the Middle Ages the Arabs distilled scents (Chapter 9) and

FIGURE 15-24
The top of the distillation columns
used in the production of Jack
Daniel's bourbon. (Courtesy of Jack
Daniel's Distillery.)

FIGURE 15-24
The top of the distillation columns used in the production of Jack Daniel's bourbon. (Courtesy of Jack Daniel's Distillery.)

antimony salts called *kuhl* which were used by the women to darken their eyelids. When the Arabs began to distill wine, they called it *al'kuhul* from which our word alcohol is derived. Distillation later reached Italy where the distillate of wine was called *aqua vitae*, or water of life. In general, early uses were medicinal, or the liquid obtained was used as a solvent. By the fifteenth century, however, the British and Scotch were distilling barley beer, and by 1688, brandy was being produced in Cognac, France. These early uses are examples of the two main classes of distilled beverages. The first includes the whiskeys, which are all distilled from solutions that would be classified as beers. The second group includes cognacs and brandies, which come from the distillation of wine.

Whiskeys are all made in essentially the same way, by fermenting malted barley alone or malted barley mixed with other grains, and distilling the product. The differences lie in the kinds of grains used, the places and length of time of aging, and the presence or absence of blending. Distillation can take place in a *pot still* or in a *patent still*. The first uses batches of liquid and distills each separately. The second, which allows continuous distillation, is the most widely used kind of still in modern whiskey operations. Actually, distillation of whiskey is not a simple operation; it involves a series of stills, each with several plates that allow the liquid to condense and revaporize several times (Fig. 15-24). This type of multiple system produces a very pure solution. The concentration of ethanol that results can reach as high as 99.99 percent, but most distillations for beverages to be consumed are at percentages between 80 and 95 percent. The distillate is then diluted with water to the desired strength. The "proof" of a whiskey that we buy is equal to twice the concentration of ethanol. Thus a 90 proof bourbon contains 45 percent ethanol. Once distilled, and diluted, the

FIGURE 15-25
Cooperages have been kept in business by laws requiring that bourbon must be stored in new oak barrels during aging. This photo shows a cooperage in 1911. (Courtesy of the Guinness Museum.)

raw whiskey is usually aged in some sort of barrel (always wood for good whiskeys) for at least 2 years (Fig. 15-25).

For connoisseurs of whiskeys, the king is *scotch*. Scotch is supposed to be made from only barley malt. The characteristic taste is caused by the kilning of the malt over fires fueled by peat moss. Most scotch is not, however, made from pure malt, but from 40 percent malt and 60 percent grain. The aging can be in new or used casks which have been charred. The names of scotches are often derived from the hamlets or glens in which they are made.

Bourbon is an American invention, originally developed by Scotch and Irish immigrants in Pennsylvania (Fig. 15-26). The English settlers in New England produced and drank mostly cider and beer. By law, bourbon must today be made from at least 51 percent corn (*Zea mays*, Poaceae) and distilled to produce a solution of less than 95 percent alcohol (Fig. 15-27). It must also be aged for at least 2 years in new, charred, oak barrels. Since the barrels cannot be reused, they constitute a major part of the expense of producing bourbon.

Rye is made from at least 51 percent rye grain and distilled to no higher than 80 percent alcohol. It must also be aged for at least 2 years in new barrels. Irish whiskey is made from a mash of primarily barley malt, but wheat and rye are often added. Its distinction lies in the fact that the malt from which it is made is not dried over peat. Distillation occurs in pot stills, and the top and the tail of the distillate are readded to the next batch and redistilled.

Straight whiskey can be any of the distilled beverages discussed above if it has been distilled to less than 80 percent ethanol and aged for at least 2 years in new charred barrels. Blended whiskeys can be blended straight with other spirits or with neutral spirits (flavorless ethanol).

FIGURE 15-26
Checking the mash at the Jack
Daniel's plant. (Courtesy of Jack
Daniel's Distillery.)

Gin and vodka differ from whiskeys in that they are distilled to a very high percentage of alcohol and therefore lack many of the flavoring agents that are carried across with the ethanol in other whiskeys. In addition, neither is aged. Gin has traditionally been made from a distillate of a fermented mash made from malt and other grains. Vodka can be made from malt and grains, or potatoes, whichever is cheaper. Gin is flavored with juniper (*Juniperus communis*, Cupressaceae, a native of Europe) "berries" (actually fleshy cones) but can also have a number of other flavoring agents.

Rum is the distillate of fermented molasses or sugarcane juice. Light and dark rums differ in the degree to which they were distilled and the amount of aging they receive. Tequila and mescal, both native beverages of Central America, have been produced only since the Spanish introduced the practice of distillation. Both are distillates of particular kinds of pulque. In the first case, only pulque from *Agave tequilana* is used, in the later, pulque from either *A. angustifolia* (maguey) or *A. palmeri*. Modern operations crush the cooked stems by machine and often ship the fermented "pulque" to Mexico city for distillation and aging (Fig. 15-28). Most tequila production, however, still centers around the small town of Tequila in west-central Mexico.

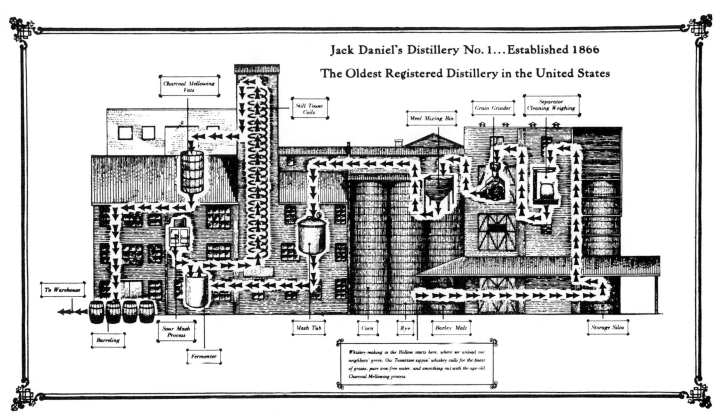

Jack Daniel's Distillery No. 1...Established 1866

The Oldest Registered Distillery in the United States

Charcoal Mellowing Vats

Still Tower Coils

Grain Grinder

Separator Cleaning Weighing

Meal Mixing Bin

To Warehouse

Barreling

Sour Mash Process

Fermenter

Mash Tub

Corn   *Rye*   *Barley Malt*

Storage Silos

Whiskey-making in the Hollow starts here, where we unload our neighbors' grain. Our Tennessee sippin' whiskey calls for the finest of grains, pure iron-free water, and smoothing out with the age-old Charcoal Mellowing process.

FIGURE 15-27
Diagram of the process used to produce Tennessee sour mash whiskey at the Jack
Daniel's Distillery. (Courtesy of Jack Daniel's Distillery.)

FIGURE 15-28
Trimmed stems of agave await
processing at the Sausa Tequila
factory in Mexico. (Photo courtesy
of Jack Neff.)

FIGURE 15-29
In 1605, an order of Carthusian monks living in the monastery of La Grande Chartreuse were given a recipe for an "elixir of long life," which included 130 medicinal and aromatic herbs. The brothers worked with the recipe to produce a therapeutic beverage and, later, Chartreuse liqueur which they dispensed to the poor as medicine. Chartreuse is now sold worldwide as an aperitif, but it is still made by monks who have taken vows of silence. Thus the recipe remains a guarded secret. (Photo courtesy of the Chartreuse Distillery.)

Brandies are distilled wines. The most famous brandies are those from Cognac and Armagnac in France. The exquisite flavors of these particular brandies are due to the wines from which they are distilled and the fact that they are carefully matured for many years after distillation. Many distilled wines are labeled as brandies with a qualifying adjective indicating the kind of wine from which they were distilled (e.g., peach brandy, raspberry brandy), or have special names such as calvados or applejack (apple brandies) or kirsch (cherry brandy). Fine brandies are always made by distilling a wine or flavored wine. Inexpensive brandies can be made by flavoring distilled wine or even relatively pure spirits.

Liqueurs differ from brandies in that they have sugar and/or syrup added to the distilled liquid. Liqueurs also have characteristic flavors. These can be imparted to the liqueur before or after distillation, but fine liqueurs are all made by adding the flavoring agents before distillation and distilling the volatile oils with the ethanol. Flavoring agents used include a broad spectrum of plants and plant products, herbs, fruits, and barks. The kinds of plants used and their proportions are often a carefully guarded secret. It is rumored that Chartreuse, a fine liqueur that has been made since 1605 by Carthusian monks in France and Spain, contains over 130 different flavoring agents (Fig. 15-29).

ADDITIONAL READING

Ayers, J. C., J. O. Mundt, and W. E. Sandine. 1980. *Microbiology of Foods*. San Francisco, Freeman.

Beech, F. W., and J. G. Carr. 1977. Cider and perry. pp. 139–313 in A. H. Rose (ed.). *Alcoholic Beverages*. New York, Academic.

Braidwood, R. J., J. D. Sauer, H. Helbaek, P. C. Manglesdorf, H. C. Cutler, C. Coon, R. Linton, J. Stewart, and J. Oppenheim. 1953. Did man once live by bread alone? *American Anthropologist* 55:515–526.

Corran, H. S. 1975. *A History of Brewing*. David and Charles, Newton MA, Abbot.

Einset, J., and C. Pratt. 1975. Grapes. pp. 130–153 in J. Janick and J. N. Moore. (eds.). *Advances in Fruit Breeding*. West Lafayette, Indiana, Purdue University Press.

Gillette, P. A., and P. Gillette. 1974. *Playboy Book of Wine*. Chicago, Illinois, Playboy Press.

Goswell, R. W., and R. E. Kunkee. 1977. Fortified wines. pp. 477–535 in A. H. Rose (ed.). *Alcoholic Beverages*. New York, Academic.

Johnson, H. 1978. *The World Atlas of Wine*: *A Complete Guide to Wines and Spirits of the World*. New York, Simon and Schuster.

Kodama, K., and K. Yoshizawa. 1977. Sake. pp. 423–475 in A. H. Rose (ed.). *Alcoholic Beverages*. New York, Academic.

Lichine, A. 1982. *Alexis Lichine's New Encyclopedia of Wines and Spriits*, 3rd ed. New York, Knopf.

Jarczyk, A., and W. Wzorek. 1977. Fruit and honey wines. pp. 387–421 in A. H. Rose (ed.) *Alcoholic Beverages*. New York, Academic.

Olmo, H. P. 1976. Grapes. pp. 294–298 in N. W. Simmonds. (ed.). *Evolution of Crop Plants*. New York, Longman.

Rose, A. H. (ed.). 1977. *Alcoholic Beverages. Economic Microbiology I*. New York, Academic.

Rose, A. H. 1977. History and scientific basis of alcoholic beverage production. pp. 1–41 in A. H. Rose (ed.). *Alcoholic Beverages*. New York, Academic.

Simpson, A. C. 1977. Gin and vodka. Pp. 537–593 in A. H. Rose (ed.). *Alcoholic Beverages*. New York, Academic.

Wile, J. 1978. *Frank Schoonmaker's Encyclopedia of Wine*. New York, Hastings House.

Winkler, A. J. 1949. Grapes and wine. *Economic Botany* 3:46–70.

Chapter 16

Fibers, Dyes, and Tannins

*F*ibers, in the broad sense, have at various times played a part in providing humans with shelter, vessels in which to carry water and cook food, and thread for making fabrics. While most of the world has abandoned mud and wattle construction, thatched roofs, or baskets smeared with clay as water and cooking utensils, the importance of plant fibers as the source of weaving materials remains.

Initially, humans would have obtained flexible plant fibers by simply pulling off strips of bark or by cutting stems and leaves into strips. While these materials could be used for lashing and interlacing into mats and baskets, they can only produce rather coarse, stiff items. A major innovation was the discovery that individual fibers could be separated from the other cells with which they occurred and used to weave textiles. Undoubtedly, animal skins predated any woven material, but plant fibers were used long before animal fibers for weaving. By 10,000 B.C., coastal Peruvian hunter-gatherers were twining and interlocking wild cotton. Flax was being woven over 8,000 years ago, at least 1,000 years before the domestication of sheep and about 3,000 years before silk extraction was practiced. Still, these dates are relatively recent considering the fact that the association of humans with plants and animals dates back millions of years. This recency is understandable because only a small number of animals (sheep, camels, vicunas, guanacos, some goats and rabbits, and the silk moth) and relatively few plants, produce fibers that can be twined or spun. Thus, humans had to appreciate the nature of fibers, learn which plants contained them, and how to extract them before they could learn to spin and weave them.

In general terms, a fiber is a strand that is very long relative to its width. To a plant anatomist, a fiber is a specific kind of elongate cell with thick walls and tapering ends. In commerce, a plant fiber is a strand that can be made of one cell or hundreds of cells. Such fibers range from a fraction of a millimeter to over 2 m in length.

Vegetable fibers share many features that distinguish them from animal or synthetic fibers. The primary difference between plant fibers and those obtained from other sources is that plant fibers are composed of cellulose (Fig. 16-1). Animal fibers are made of protein (Fig. 16-1). This fundamental chemical difference determines how the two react to heat, various other chemicals, water, and predacious organisms. For example, cellulose molecules are not subject to denaturation by high temperatures, but heat will crack the protein backbone of animal fibers and make them brittle. Because of this differential reaction to high heat, cotton sheets can be boiled when they are washed, but woolen garments are ruined by washing in hot water. Plant and animal fibers behave differently in dye baths as well. The complexity of the protein molecules of animal fibers promotes their acceptance of dyes, while plant fibers require relatively elaborate treatments to ensure successful color adherence. Animal fibers are also particularly susceptible to attack by animal pests such as moths and silverfish. Plant-derived fabrics are essentially immune to these pests, but they, and even paper, are readily attacked by fungi, mold, and even termites. Finally, plant fibers tend to be less elastic than animal fibers, yet they have a higher affinity for water.

It is more difficult to contrast plant fibers with synthetic fibers because the properties of synthetically produced fibers vary with the polymer chemistry of different synthetic materials. Some synthetic fibers (discussed in Chapter 17) are made from cellulose, but most are manufactured from chemicals derived

FIGURE 16-1
Cellulose is a polysaccharide consisting of long chains of glucose molecules which are cross-linked to neighboring molecules forming a rigid framework. (A) In contrast, the ... basic molecules that make up proteins are amino acids. (B) Each amino acid has an amine (NH) and a carboxyl (COOH) group attached to different carbon atoms that form the carbon backbone of the molecule. Amino acids differ from one another by having different side chains, indicated here by the letter R. When amino acids combine with one another, the amine end of one bonds with the carboxyl end of another with the loss of a molecule of water. (C) Protein molecules are usually composed of thousands of amino acids. The long, unbranched chains often twist or coil into complicated configurations. (D)

A. CELLULOSE MOLECULE

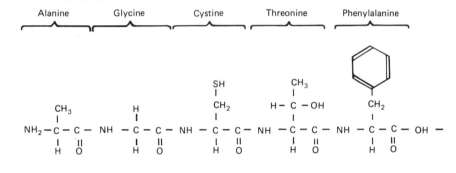

B. PROTEIN MOLECULE

from petroleum. There has also been some experimentation with fatty acids cleaved from seed oils (see Chapter 10) as precursors of textile polymers. Synthetically produced fibers have challenged natural fibers for supremacy in terms of world production in recent years, but their previous advantage of being inexpensive has dwindled as petroleum costs have risen. Even though they have at times been more expensive than their synthetic counterparts, natural fibers have never been totally replaced because people value their special properties. No synthetic material can match the comfort of cotton or the crispness of linen.

Commercially important plant fibers are extracted from many unrelated species, but they can be functionally grouped into those which can be plaited,

TABLE 16-1

Classification of Vegetable Fibers by Use

USE	EXAMPLES
Textile fibers:	
Soft, or bast, fibers	Hemp, jute, ramie, linen
Seed and fruit fibers	Cotton, coir
Hard, or leaf, fibers	Sisal, henequen, abaca, pineapple
Plaiting and weaving	Palms, stems of several grains, cyperus, pandanus, bamboos
Brush fibers	Palms, sorghum, broomroot
Filling fibers	Kapok, milkweed, cattails
Felting fibers	Paper mulberry, lace bark
Papermaking fibers	See Chapter 17

twined, woven, and spun into threads, and those which cannot. This latter group includes fibers which are short, brittle, and/or slippery. Important plant fibers of this type include wood fibers which are used for items such as paper, fiberboard, cellophane, and rayon. These fibers are discussed in Chapter 17. Here we concentrate on fibers that can be interlaced, twined, or spun to form yarns and threads.

Spinning can occur only if the fibers have structural properties that cause the individual strands to clasp one another when twisted. Although the forces that keep spun fibers together in a strand are simply mechanical, they prevent the strands from slipping free from one another and thus allow production of a continuous yarn or thread. Most animal hairs (including human hair), and some plant fibers such as those on milkweed seeds, cannot be spun because they are too slippery to stay together when twisted around one another.

Plant fibers used for basketry, caning, and weaving into mats are not separated from the other cells with which they occur, but they provide flexibility and strength to the strips cut from the plant material.

Vegetable fibers can also be further classified by their use (Table 16-1) or by the part of the plant from which they are obtained. A classification based on usage divides fibers into those used for textiles (woven goods), brushes, plaiting or coarse weaving, stuffing material, paper, and specialty products. Here we deal primarily with textile fibers which are commonly grouped into seed and fruit fibers; soft, or bast, fibers; and hard, or leaf, fibers. Bast fibers come from the phloem tissues of dicotyledons. Hard fibers come from the leaves of certain monocotyledonous plants. In the following discussions, we point out the probable natural ecological roles of fibers produced in different plant organs as well as the ways in which humans have artificially selected for specific characteristics. Table 16-2 lists the plants discussed in this chapter.

HOW PLANT FIBERS DIFFER FROM ONE ANOTHER

A fiber to be spun and woven into soft fabrics that will withstand repeated washings needs very different characteristics from one to be used in ropes that will be continually wetted with seawater. The value of a fiber and the uses to which it lends itself are determined by a number of factors. The look and feel

TABLE 16-2

Plants Discussed in Chapter 16

COMMON NAME	SCIENTIFIC NAME	FAMILY	CHROMOSOME NUMBER
		FIBERS	
Abaca	*Musa textilis*	Musaceae	$2n = 20$ diploid
Coir	*Cocos nucifera*	Arecaceae	$2n = 32$ diploid
Cotton	*Gossypium hirsutum*	Malvaceae	$2n = 52$ tetraploid
	G. barbadense		$2n = 52$ tetraploid
	G. arboreum		$2n = 26$ diploid
	G. herbaceum		$2n = 26$ diploid
Flax	*Linum usitatissimum*	Linaceae	$2n = 30, 32$ diploid
Hemp	*Cannabis sativa*	Cannabaceae	$2n = 20$ diploid
Henequen	*Agave fourcroydes*	Agavaceae	$2n = 140$ pentaploid
Jute	*Corchorus capsularis*	Tiliaceae	$2n = 14$ diploid
Kapok	*Ceiba pentandra*	Bombacaceae	$2n = 72-88$ polyploid
Milkweed	*Asclepias syriaca*	Asclepiadaceae	$2n = 22$ diploid
Ramie	*Boehmeria nivea*	Urticaceae	$2n = 14$ diploid
Sisal	*Agave sisalana*	Agavaceae	$2n = 138-149$ pentaploid
		DYES	
Annatto	*Bixa orellana*	Bixaceae	$2n = 14$ diploid
Bloodroot	*Sanguinaria isabellinus*	Papaveraceae	unreported
	S. canadensis		$2n = 18$ diploid
Butternut	*Juglans cinerea*	Juglandaceae	$2n = 32$ diploid
Cutch (catechu)	*Acacia catechu*	Fabaceae	$2n = 26$ diploid
Fustic	*Chlorophora tinctoria*	Moraceae	unreported
Henna	*Lawsonia inermis*	Lythraceae	$2n = 24, 32$ diploid
Indigo	*Indigofera tinctoria*	Fabaceae	$2n = 16$ diploid
Logwood	*Haematoxylon campechianum*	Fabaceae	$2n = 24$ diploid
Madder	*Rubia tinctoria*	Rubiaceae	$2n = 22$ diploid
Safflower	*Carthamus tinctorius*	Asteraceae	$2n = 24$ diploid
Weld	*Reseda luteola*	Resedaceae	$2n = 24, 26, 28$ variable
Woad	*Isatis tinctoria*	Brassicaceae	$2n = 28$ diploid

Note: Species included in Table 16-3 are not included.

of a fiber are affected by its structure (Fig. 16-2). Round fibers tend to feel silkier and have a smoother appearance than flat ones. The breaking, or tensile, strength of a fiber varies with its cross-sectional area, the length of the individual cells of which it is composed, and the way in which the cells are held together. Elastic properties are a measure of the amount to which a fiber can be stretched before breaking and the facility with which it can regain its original shape once

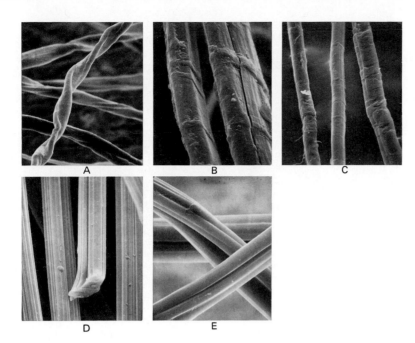

FIGURE 16-2
Scanning electron micrographs of (A) cotton (×284), (B) flax (×474), (C) ramie (×234), (D) rayon (×474), and (E) acetate (×474), show that the semisynthetic rayon and acetate fibers are smoother and more uniform than the natural plant fibers. The gradual twist of the cotton fiber causes the fibers to interlock and form threads when spun. (Photos courtesy Burlington Mills)

released. The amount a fiber is twisted, the way in which the cells are held together, and the numbers of cells per fiber all contribute to its elasticity. Fibers also have different densities, or weights relative to an equal volume of water, which affect how fabrics made from them will drape. Finally, the particular chemistry of various fibers causes them to react differently with water, sunlight, heat, acids, alkalis, solvents, electricity, and microorganisms. All the various properties are taken into account in choosing a fiber for a particular purpose.

FIBER EXTRACTION

Despite the fact that various fibers come from different plant parts and from an array of plant species, the same basic procedures are used to separate them from the masses of other cells in which they are embedded. The primary processes are termed retting, scutching, decorticating, and ginning. *Retting,* a process used mainly for bast fibers, takes advantage of the fact that fibers have thicker cell walls than most other plant cells and are therefore comparatively resistant to breakdown by bacteria. Consequently, retting, or bacterial rotting, is often used to decompose soft plant tissues and dissolve the gums and pectins holding plant cells together. Plant material can be retted by dumping it into stagnant water (Fig. 16-3) or by simply allowing it to remain on the ground where it will be repeatedly covered with dew. The process takes from a few to several weeks, during which the mass of rotting material must be continually tested to ascertain the point at which the soft tissues, but not the fibers, have disintegrated. If the plant matter is allowed to ret too long, even the fibers will fall apart.

Retting does not remove all of the nonfibrous material, however. At the end of the process, thick-walled woody xylem cells also persist. To remove

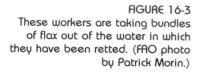

FIGURE 16-3
These workers are taking bundles of flax out of the water in which they have been retted. (FAO photo by Patrick Morin.)

FIGURE 16-4
After retting and scutching, flax stems are drawn over the hasps on a hackling board (*foreground*) in order to separate and align the fibers. In the background, an FAO consultant is shown testing the strength of the fibers. (Photo courtesy of FAO.)

these cells from the fibers, the material remaining after retting is washed, dried, and "broken" by feeding it under fluted rollers. The breaking process crumbles the brittle woody material but not the more flexible bast or leaf fibers. The broken pieces of woody matter are then removed from the fibers by beating and scraping, a process known as *scutching*. Finally, the fibers can be hackled to separate and align them. *Hackling* is accomplished by drawing a mass of fibers across a set of vertical pins resembling a comb (Fig. 16-4).

In some cases, fibers can be most easily and inexpensively separated from the other tissues by *decorticating*. This mechanical procedure entails crushing the plant material and scraping the nonfibrous material from the fibers. In general, decorticating is used only for leaf fibers.

Ginning is a process unique to seed fibers. During the process, seeds are pulled free from the fibers covering them. In the case of cotton, the fibers subsequently undergo extensive cleaning and combing.

Once extracted and cleaned, fibers can be further processed by bleaching and/or heating in alkali. Fibers are usually bleached before dyeing so that their natural tan or brown pigments will not affect their final color. In our discussions of various plant fibers, we will indicate those which are commonly dyed, and the processes used to prepare them for dyeing. Later in the chapter, we explain how dyes adhere to fibers and which plant dyes have historically been used to color weavings and textiles.

SEED AND FRUIT FIBERS

Although seed and fruit fibers come from different parts of the fruit, both serve to aid in seed dispersal. Yet the mechanisms of dispersal are different. Long fibers on the surfaces of seeds promote dispersal by wind. Anyone familiar with the silky tufts of hairs on the tops of milkweed (*Asclepias* spp., Asclepiadaceae) seeds knows how effectively they help the seeds float in the air.

Fibers located within the fruit wall generally serve to protect the seed and/ or to provide buoyancy for water dispersal. Nevertheless, few plants have seed or fruit fibers long enough for any practical purpose such as spinning.

Cotton is a notable exception because it produces seed fibers that can be spun into thread. The seed fibers of milkweed and kapok (*Ceiba pentandra*, Bombacaceae; Fig. 16-5) are too fine and slippery to spin. They are therefore used commercially primarily as stuffing material. Kapok was also formerly used for life preservers because it is so light and water-resistant.

Coconuts are also unique because they produce the only fruits from which fibers are extracted for commercial use. The entire coconut mesocarp consists of a fibrous mass that provides buoyancy for the water-dispersed fruits. Coconuts can apparently stay afloat without damage in seawater for months. The natural distribution of coconuts on coastal areas throughout tropical areas of the world attests to the success of this method of dispersal.

Cotton Is King

Cotton (Fig. 16-6) is by far the most important fiber in the world today. It is, moreover, the world's most important nonfood plant commodity. The popularity

FIGURE 16-5
Kapok pods split at maturity,
releasing their seeds covered with
long, silky hairs. (After Baillon.)

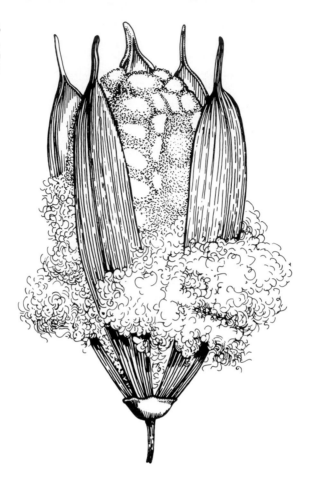

of cotton can be ascribed to the large amount of fiber produced per plant combined with the fact that picking, processing, and manufacturing of textiles from cotton costs less than the processing of other plant fibers. The extensive use of modern machinery brought the cost of cotton down to its present level. The processing of many other fibers is not as easily done by machine. Cotton is also a very versatile fiber which produces fine textiles that dye well and withstand vigorous washings.

Each cotton fiber is a single long epidermal seed coat cell (Fig. 16-7). These cells are so long that they resemble hairs. In many cultivated cottons there is a second layer of short fuzzy hairs underneath the long fibrous hairs. When cottonseed is used as an oilseed crop (Chapter 10), these short hairs, or *linters,* are removed and used for papermaking.

Both the history of the use of cotton and the taxonomy of the cotton genus *Gossypium* (Malvaceae) are complicated. There is still not complete agreement among botanists working with cottons about the ancestors of the cultivated species. All agree, however, that species of cotton were independently domesticated in the Old and New Worlds. Some authors have suggested a preagricultural use of cotton around the eastern Mediterranean Sea, but all the archeological evidence indicates that the first use in the Old World was in

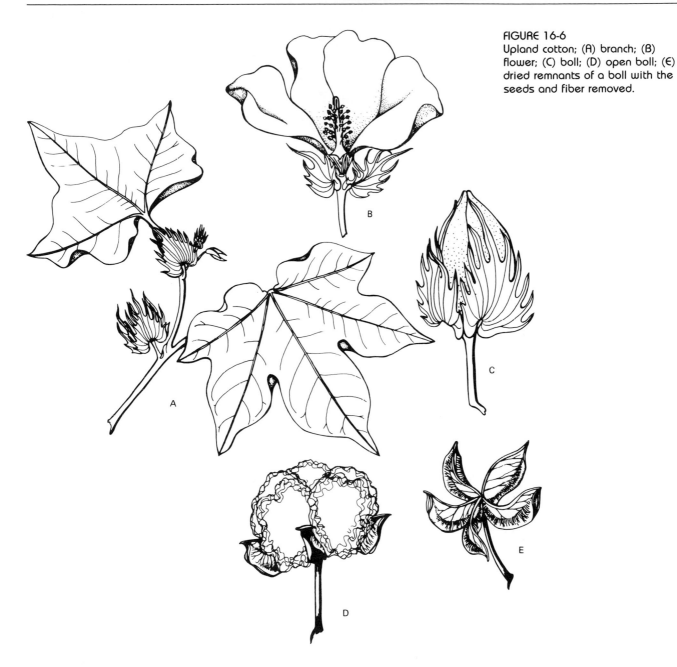

FIGURE 16-6
Upland cotton; (A) branch; (B) flower; (C) boll; (D) open boll; (E) dried remnants of a boll with the seeds and fiber removed.

south-central Asia. Cotton fabrics dated as having been made about 3000 B.C. have been recovered in Pakistan. These would have been made from either *Gossypium arboreum,* known only as a cultigen, or *G. herbaceum,* another diploid which was also originally domesticated in Asia.

By the middle of the fifteenth century, cultivation of these two species had spread throughout the Near East and into Europe via Spain by the Arabs. Both of these cottons have short fibers, or *staples,* and both were eventually supplanted by New World cottons with longer staples. They are grown today primarily in India and Pakistan.

FIGURE 16-7
Each cotton fiber develops from the longitudinal growth of a single epidermal cell. (A) Cross section of a boll; (B) seed; (C) individual fibers developing on the seed surface.

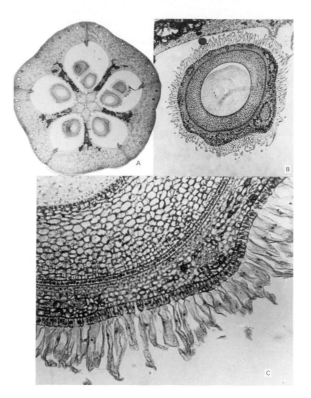

FIGURE 16-8
Cotton weaving became a highly developed craft in the pre-Hispanic societies of the north coast of Peru, and cotton textiles assumed an important place in mortuary practices from at least 900 to 200 B.C. up to the time of the Spanish conquest in 1532. This mummy of an 18- to 20-year-old woman, interred about 1200 A.D., was wrapped in 24 cotton cloths that had a combined area of 286 m² (3078 ft²) and a weight of over 150 kg (331 lb). (Photo courtesy of James Vreeland.)

Two species of cotton were also domesticated in the New World. Both are tetraploids that have a diploid New World and a diploid Old World parent. The most likely parents appear to be *Gossypium raimondii,* a diploid species endemic to northern Peru, and wild *G. herbaceum* from Africa. One of the New World cultigens, *G. hirsutum,* has been found in archaeological excavations that date back to 3400 B.C. in Mexico, its presumed area of domestication. Columbus recorded being offered cotton thread by West Indian natives on October 12, 1492, upon his first landing in America. *Gossypium hirsutum,* known as cotton belt, upland, or West Indian cotton, accounts for about 95 percent of the world's crop. Early settlers chose this species for cultivation because it is more resistant to boll weevil attack than its cultivated American relative. Once it became cultivated on a large scale, it was naturally the species with which most of the research and development was conducted.

The other important American cotton is *Gossypium barbadense,* commonly known as Sea Island, Egyptian, or pima cotton. None of these common names is particularly appropriate, but all reflect some aspect of its recent history. Despite its modern secondary status, *G. barbadense* appears to have been used for fiber before *G. hirsutum.* The earliest dates now available from Peru indicate that this species was used about 8000 B.C. by nonagricultural people, and domesticated by 2500 B.C. Looming of cotton textiles has been traced back to 1800 B.C. By the height of the Inca Empire in the thirteenth and fourteenth centuries, weaving had become an integral part of the culture of western South America. The enormous amounts of cloth used in Inca mortuary practices (Fig. 16-8) suggest that there was a large group of professional weavers who either fulfilled their obligations to the state by, or were forced into, weaving.

By the time of the arrival of Europeans, *Gossypium barbadense* had been spread by humans from its native region in western South America across the continent and into the West Indies. It is believed that it was given the name Sea Island cotton because it was introduced to plantations in the southern United States from the West Indies. It is, however, no longer grown in the eastern part of this country. Plants of the species were also successfully introduced to Egypt. The cultivars that we now include under the designation of Egyptian cotton are all descended from a selection made sometime between 1817 and 1819 by a Frenchman working in Cairo. The staple of these fine cottons can reach lengths up to 6 cm (2.4 in) and produce fine-quality threads. Pima cotton, another cultivar of this species, was also selected from Egyptian plantings of *G. barbadense*. The name is derived from its early cultivation in Pima County, Arizona.

All species of *Gossypium* are perennials in nature, but humans have selected for an annual habit and the ability to bloom and fruit in temperate latitudes. By growing cotton as an annual, farmers ensure short stature, uniformity in plant size (semicultivated perennial plants of *G. barbadense* can be shrubs 4 m tall), and synchronous fruiting. On modern farms, plants are often sprayed with defoliants when the cotton is mature so that the foliage will not get in the way of the harvesting machines which pluck the mature bolls from the plants (Fig. 16-9). Once picked, the seeds with their fibers have to be removed from the fruit and then the fibers have to be separated from the seeds. Most Americans know that Eli Whitney invented the cotton gin, a machine that pulls cottonseeds from the fibers, in 1794 (Fig. 16-10), but it is often not appreciated how important this labor-saving device was in the development of the cotton industry. In 1791, 2 years before Whitney's invention, the United States exported 400

FIGURE 16-9
Cotton plants are sometimes defoliated at the end of the season to facilitate harvesting by machines. (Courtesy of the USDA.)

FIGURE 16-10
The workings of a cotton gin.
(Adapted from Klein, R. 1979, The
Green World.)

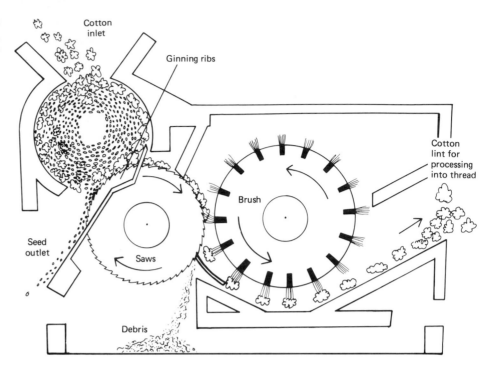

FIGURE 16-10
The workings of a cotton gin. (Adapted from Klein, R. 1979, The Green World.)

bales of cotton. By 1800, 7 years later, exports had jumped to 30,000 bales (Fig. 16-11).

Once ginned, cotton fibers undergo extensive processing. Different lots of cotton with staples of varying length are usually blended together by feeding pieces of different bales into bins. The cotton is then processed by picking machines which pluck and beat the fibers with steel spikes attached to rollers. During this procedure, unwanted matter is removed and the process of combing is begun. In large factories, cotton bales are opened, blended, and picked in a single operation. After being picked, the fibers are *carded,* or combed parallel to one another. In earlier times, cotton, like wool, was hand-carded using two instruments resembling brushes that were repeatedly pulled across one another. Today, masses of fiber pass under huge rollers covered with wires of different lengths which tease the fibers apart. After being fed through a series of such rollers, the fibers are rapidly pulled downward and outward across a long comb, producing a thin web of relatively parallel fibers. The web is then gathered and twisted into a loose rope called a *sliver* about 2 cm in diameter. Cotton fiber is stored as slivers until it is ready to be spun (Fig. 16-12).

Prior to spinning, slivers are mixed and drawn by pulling them with different amounts of pressure through a series of pairs of rollers. *Drawing* completes the alignment of the fibers parallel to one another and considerably reduces the diameter of the original sliver. The drawn cards are drafted, or stretched across a moving frame and then spun by twisting them tighter and tighter as they are pulled onto the drafting rollers. The result is yarn or thread that is made up of many overlapping, parallel fibers held together by simple mechanical forces.

Once made, either the thread, or a textile woven from it, is cleaned and

FIGURE 16-11
40,000 bales of cotton stand
waiting for shipment in one area of
Texas. (Courtesy of the USDA.)

mercerized. *Cleaning* consists of boiling the thread or cloth with caustic soda for 8 h and then bleaching it with a chemical such as hydrogen peroxide. These treatments remove pectins and waxes and make the fibers light in color so that they will readily accept dyes. *Mercerization* was invented in 1844 by John Mercer, an English textile maker. Mercerization increases the luster of cotton, promotes its uptake of dyes, and increases its durability. The process consists of placing a thread or woven textile, stretched under pressure, into a cold bath of caustic soda for several hours. The treatment causes the fibers to swell and the cellulose molecules to deform. The material being mercerized is kept under pressure for some time and not allowed to shrink back to its former shape. Consequently, it retains its swollen, circular form caused by the alkali treatment.

Before weaving, threads, particularly those that will be used for the warp (the yarns placed lengthwise in the loom), are sized to increase their strength. *Sizing* is the addition of a thick substance such as starch or a gel to fabric (or paper) in order to stiffen it and to fill surface irregularities. Sizing is generally washed out after weaving is completed. Woven fabrics can also be singed or passed rapidly over flames to burn off any irregular, protruding pieces of fiber.

Cotton fabrics are often *sanforized* to reduce shrinkage when they are washed. Sanforization dates back only to 1970 when the Sanforizer Company introduced an ammonia process that swells cotton fibers and prevents their shrinking after they have been washed. Sanforizing eliminated the old tradition

FIGURE 16-12
This Ethiopian family spins its own cotton and weaves it on a homemade loom. (FAO photo by Patrick Morin.)

of buying a cotton garment "one size too large" because of the invariable shrinkage after the initial washing.

In addition to shrinkage, a problem with cotton has historically been its tendency to wrinkle when washed. *Permanent press* cotton fabrics have consequently been developed that greatly reduced their need for ironing after laundering. Although there are several processes used to produce a permanent press fabric, all involve the use of chemicals that crosslink the cellulose polymers of the cotton cloth and thus cause the fabric to retain the shape it had when the chemicals were applied. Not only does permanent press keep flat surfaces smooth, but it also allows pleated or ruffled garments to retain their shapes after washing.

Coir

In American grocery stores, we usually find coconuts (*Cocos nucifera*, Arecaceae) that are about 15 cm (5.9 in) in diameter and rough brown on the outside. What we buy is only part of the fruit, the endocarp with the mature seed inside. All of the exocarp and the mesocarp were stripped from these fruits before marketing. A completely mature coconut is about twice the size of our supermarket items and smooth green or greenish brown on the outside (Fig. 16-13). Occasionally, in southern parts of the United States, the West Indies, or Hawaii, street vendors sell green coconuts for their juice and one can watch as they hack off thick pieces of the husk with machetes before inserting a straw so that their customers can drink the sweet coconut water.

The bulk of a mature coconut fruit is the thick, fibrous mesocarp that constitutes the source of a fiber known as coir. Coir fibers are made up of bundles of cells that are longer than cotton fibers, but shorter than most bast or leaf fibers. To produce high-grade coir, 10-month-old (immature) coconuts are harvested and husked. The husks are retted for 8 to 10 months, usually in brackish water. When soft, the husks are thoroughly washed, beaten to remove

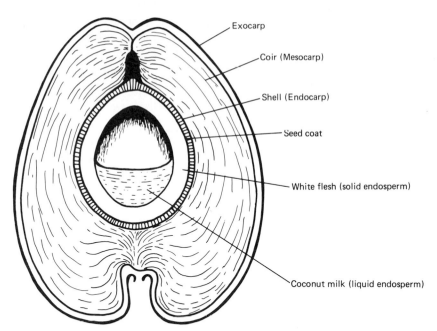

Exocarp

Coir (Mesocarp)

Shell (Endocarp)

Seed coat

White flesh (solid endosperm)

Coconut milk (liquid endosperm)

FIGURE 16-13
Coconut fiber, or coir, comes from the fibrous fruit coat that surrounds the seed valued for its nutritious endosperm.

all of the pulpy remains, shaken, and washed again. The clean, pure fibers are spun into yarns that are used primarily for ropes and matting.

The fact that the most valuable coir fibers come from unripe coconuts presents something of an economic problem since the most valuable commodity obtained from coconuts is copra (dried coconut endosperm) which is eaten directly or used for oil (see Chapter 10). In immature coconuts, the endosperm is liquid, thus copra can be obtained only from mature fruits. Consequently, although the mature fruits yield fibers that are tougher than those of young coconuts, most coir is extracted from husks pulled from ripe fruits used for copra production.

Fibers of mature husks are extracted in much the same way as they are from young fruits, but after retting, the mass of fibers of each husk is usually decorticated. Decortication is done by a combination of hand and machine labor. Workers take each husk individually and shove half of it (holding onto to other half) between two bars. On the other side of the bars, the fibers are roughly combed as nails projecting from huge rotating drums tear through them. After decorticating half of a shell, the worker pulls the husk out, turns it around and pushes in the other half. During this operation, two types of fibers are produced. The first, called *mattress fibers,* are relatively short fibers that are pulled out of the husk by the nails on the drum. They fall to the ground under the drum and are later collected and used primarily as stuffing material. The fibers remaining in the mass are longer than mattress fibers. They are called *bristle fibers* and generally end up in brushes or brooms. In some areas, this labor-intensive decorticating procedure has been replaced by completely mechanized operations. However, mechanical decortication produces only one type of fiber similar to the mattress fibers produced by older methods.

Because of its natural resistance to salt water, coir is often used to produce netting for shellfish or seaweed harvesting. In general, however, coir fibers are inferior in most respects to those extracted from other plants, and its continued production stems in large part from the availability of enormous quantities of husks as a by-product of the copra–coconut oil industry.

BAST FIBERS

Bast, or "soft," fibers are thick-walled fibers and associated phloem cells of many dicotyledonous species. While the phloem itself serves as part of the conduction system, the surrounding cells appear to function mainly to support the stems. Individual bast fiber elements can be made up of hundreds of cells and can be over 2 m (6.6 ft) long. The fibers are usually separated from the stems by retting, followed, in some cases, by pounding. Depending on the quality of the fiber,

FIGURE 16-14
Jute. (A) Fruiting branch, (B) flower, (C) plant.

TABLE 16-3

World Production of Major Textile Crops

CROP	TOP 5 COUNTRIES, 1,000 t		TOP 5 CONTINENTS, 1,000 t		WORLD, 1,000 t
	COUNTRY	t	CONTINENT	t	
Cotton (lint)	China	4,637*	Asia	7,440	14,692
	U.S.S.R.	2,760*	North & Central America	2,084	
	United States	1,682	Africa	1,230	
	India	1,260*	South America	914	
	Brazil	552	Europe	190	
Flax fiber and tow	U.S.S.R.	420*	Europe	159	669
	China	66*	Asia	69	
	France	54	Africa	25	
	Poland	30*	South America	3	
	Egypt	25	Oceania	2	
Hemp fiber and tow	China	60*	Asia	126	230
	India	55*	Europe	45	
	U.S.S.R.	57*	South America	4	
	Romania	20*	—		
	Turkey	14*			
Jute and substitutes	China	1,590	Asia	3,897	4,057
	India	1,019*	South America	74	
	Bangladesh	908*	Africa	21	
	Thailand	234	North & Central America	11	
	Brazil	61	—		
Sisal	Brazil	184	South America	190	384
	Tanzania	82*	Africa	179	
	Kenya	51*	North & Central America	12	
	Angola	20*	Asia	4	
	Madagascar	13	—		

*Indicates an estimated or an unofficial figure.
Note: The U.S.S.R. is treated as a country only. Its values are not added to those of either Asia or Europe.
Source: Data from *FAO Production Yearbook for 1983,* vol. 37. 1984. Rome, FAO.

most bast fibers can be bleached and dyed. Bast fibers are not mercerized or sanforized.

Jute

Jute (*Corchorus capsularis,* Tiliaceae; Fig. 16-14) is the world's foremost bast fiber and is second only to cotton in terms of production (Table 16-3). The tall reedlike plants have been used since prehistoric times and probably yielded the fibers for some of the sackcloth referred to in the Bible. Today's primary use is still for sacking and similar material, but it is seldom worn, even in penance.

The species appears to be a native of the Mediterranean region from which it was spread throughout the Middle and Far East. Jute plants are herbaceous annuals that can reach 5 m (16.4 ft) in height. Individual fibers can be 2.3 m (7.5 ft) long but are generally somewhat shorter. Jute fibers are relatively inelastic and tend to disintegrate rather rapidly in water. They are separated from the stems by retting, which sometimes causes individual cells of a fiber strand to loosen, giving the fibers a rough feel. This roughness, the brittleness of the fibers, and their inability to hold dyes promotes their use for coarse goods such as carpet backing, canvas, twine, and "gunny" sacks. The popularity of jute comes from the facts that it grows very rapidly and that the separation of the fibers can be carried out quickly and inexpensively.

Flax

Flax (*Linum usitatissimum*, Linaceae) was considered the oldest textile fiber used by humans until recent findings of ancient cotton fabrics showed that cotton was used at an equivalent time. Archaeological digs have uncovered remains of a flax species in ancient settlements occupied by the Swiss Lake Dwellers about 10,000 years ago, and Egyptian mummies dated to be over 5,000 years old were generally wrapped in linen cloth. While it is difficult to reconstruct with certainty the extent to which the Lake Dwellers used flax, carvings on Egyptian tombs beautifully document its cultivation along with wheat, figs, and olives (see Fig. 10-10). The classical Greeks used linen, and the Romans spread the cultivation of flax all across Europe. Our word "line" (as in a straight line) and the generic name *Linum* are derived from the Latin word for a flax fiber.

Flax is presumed to be native to the eastern and southeastern Mediterranean region, but its geography has been so altered by humans that its natural distribution is difficult to determine. Today, there are no populations of *Linum usitatissimum* considered to be truly wild. The plants from which fibers are obtained are tall, little-branched annuals that yield linseed (see Chapter 10) as well as fiber (Fig. 16-15). Although plants in ancient times were used for both products, cultivars selected for either long fibers or high seed yield are planted today.

Flax fibers are naturally smooth, straight, and two to three times as strong as cotton fibers. The beauty and luster of linen, the textile made from flax, has long been appreciated. The growing of flax and the production of linen was an important local industry throughout the Low Countries of Europe, England, and Ireland between the seventeenth and nineteenth centuries. While wool was the principal textile fiber used in temperate areas during this time, flax was the leading plant textile fiber. Often, the two kinds of fibers were used together. The linsey-woolsey of the Pilgrims who settled North America was made by weaving wool across a warp of linen thread. Cotton supplanted linen as the favorite plant fiber for clothing textiles only in the nineteenth century.

The modern ascendency of cotton is based largely on economics. Cotton can be readily and inexpensively processed by machines. Machine-processed cotton is much superior to that processed by hand. Flax is generally obtained by dew retting the cut stalks of mature plants, but since this process can take weeks, plants are now sometimes uprooted and chemically retted. Once the fibers have been lifted from the disintegrated tissues, they are dried, scutched, and hackled (Figs. 16-4, 16-16). In contrast to cotton, machine-processed flax is not superior to that produced by hand. In fact, linen woven from hand-processed flax is generally superior to that produced mechanically. The flax (linen) industry therefore never really became industrialized. As a consequence, linen eventually became too expensive for most of the world's population.

FIGURE 16-15
Flax; (A) Flowering branch; (B) Seed capsule; (C) Seed in section.

FIGURE 16-16
Farmers in Addis Ababa, Ethiopia, bring in bundles of flax to be scutched. (Courtesy of the FAO.)

Hemp

True hemp comes from the same species, *Cannabis sativa* (Cannabaceae), as marijuana (see Fig. 13-4). Despite the current popularity of the resinous constituents of hemp plants (see Chapter 13), *Cannabis* was initially spread around the world because of its fiber, not its chemicals. Like jute and flax, hemp has been cultivated since prehistoric times. A native of western Asia, hemp was used in the Chinese Neolithic Yang Shao culture which flourished about 4000 B.C. The seeds were also probably consumed along with millet, rice, barley, and soybeans in ancient China. The first indications of its use in the Mediterranean region are from the first century A.D.

Hemp plants are dioecious (or polygamodioecious) annual herbs (Fig. 16-17) that produce the best fibers when grown under temperate conditions with over 100 cm (39.4 in) of rain per year. The fibers (Fig. 16-18) are like those of flax in appearance but tend to be stiffer, and contain more lignin. They are extracted from the stems by retting, often followed by scutching and pounding. Well-processed hemp is creamy white, soft, and has a silky sheen. Most of the fiber is extracted as quickly and inexpensively as possible, however, and tends to be darkly colored and rough. As a result, hemp is typically used for cordage, rope, canvas (traditionally made from *Cannabis*), and sailcloth. Hemp was also used to make the original set of Levis. Levi Strauss, an enterprising young businessman, was able to pick up a large amount of hempen sailcloth at a

FIGURE 16-17
A rendering of a hemp plant in Leonard Fuch's *De Historia Stirpium*, Basel, 1542.

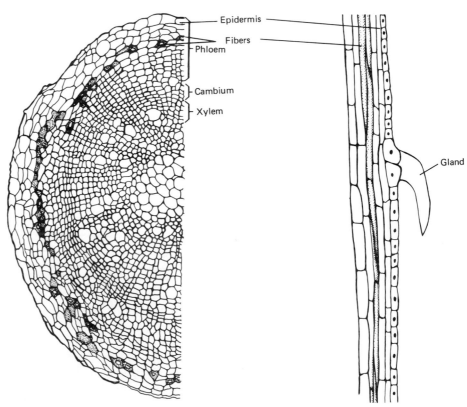

Epidermis
Fibers
Phloem
Cambium
Xylem
Gland

FIGURE 16-18
A cross section of a *Cannabis* stem showing the location of the fibers (*stippled*) that are extracted by retting.

FIGURE 16-19
Levi Strauss used hemp cloth, imported from Nimes, France, to manufacture the first of his now famous trousers. The French name for the cloth, "serge de Nimes," became "denims" when it crossed the Atlantic. A shipment of hempen cloth from Genoa, Italy (a city called "Genes" by the French) eventually gave rise to the word "jeans," now used for the denim pants. This lithograph shows an early jeans factory. (Courtesy of Levi Strauss Company.)

very reasonable price. It was with this lot of material that he executed his now famous design for working pants with copper rivets at the seam joints. For subsequent production of his surprisingly successful trousers, he used cotton cloth (Fig. 16-19).

Recently there has been a renewed interest in the production of hemp, particularly in France, where new machines have been developed for breaking the fibers. The woody fragments separated from the fibers are collected and used in the manufacture of composition board.

Ramie

Ramie (*Boehmeria nivea,* Urticaceae; Fig. 16-20), or China grass, has never been one of the world's most important fibers because of the agronomic problems

FIGURE 16-20
Flowers of ramie. (Photo by Miloslaw Petruska, courtesy of the National Academy of Sciences.)

involved with growing the crop and the pronounced chemical impregnation of the fibers. Stands of ramie tend to be uneven, making harvesting difficult, and the fibers contain considerable amounts of gum and pectin that must be removed. However, once harvested and properly processed, ramie produces the longest and silkiest of all the plant fibers. Ramie is a perennial species native to tropical Asia where it has been gathered for fiber extraction for over 7,000 years. Some of the Eyptian mummies (ca. 5000 to 3300 B.C.) were bound with ramie strips rather than in linen. A related species, *B. cylindrica*, was used by the New World Indians as a source of twine used to attach spear and arrow heads to shafts.

The stalks of ramie are not retted to free the fibers. Instead, the bark and phloem tissues are peeled from the stems and the fibers decorticated by beating and scraping. After the fibers are finally separated from the woody matter and soft tissues, they remain in ribbonlike strips because they are held together by gums. The gums can be dissolved away by repeated washing and scraping or, in modern operations, by treating with caustic soda. During the last decade, ramie has been increasingly planted in this country as a crop rotation species that is harvested, decorticated, and degummed by machine. If these mechanized operations can reduce the price and truly eliminate the need for hand labor, ramie could become an important fiber for textiles and materials requiring exceptional strength.

LEAF, OR HARD, FIBERS

The widespread use of leaf fibers is a comparatively recent phenomenon. The use of these fibers has escalated dramatically during the last 50 years as improved machinery and transportation systems have allowed greater exploitation of tropical crops. Most leaf fibers are obtained from rapidly growing tropical monocotyledons with fibers that are easily and inexpensively extracted by decorticating machines. The fibrous strands can occur throughout the leaves and around the circumferences of the leaf bases. Within the leaves, the thick-walled fiber cells provide support. Once extracted, the fibers can be spun, but they are too stiff to be made into textiles suitable for modern clothing. The name "hard" applied to this type of fiber reflects this stiffness. Nevertheless, leaf fibers make better ropes than most bast fibers (Fig. 16-21). While other plants are occasionally used, virtually all leaf fibers commercially extracted come from two genera, *Agave* and *Musa*.

Sisal and Henequen

Both sisal and henequen come from the leaves of species of *Agave* (Agavaceae), the first from *A. sisalana* and the second from *A. fourcroydes*. Both species are native to Central America where the Mayans and Aztecs are known to have extracted and woven sisal fibers into rough garments. *Agave sisalana* has, in addition to fibers, sharp spines on the ends of the leaves that have been used by native peoples as needles. The provision of both the fiber and a sewing utensil gave rise to the species' common name of "needle and thread plant."

FIGURE 16-21
Although most rope is now made using modern machinery, some is still produced using the same processes employed by Central American natives before the arrival of Columbus. (Photo courtesy of the United Nations.)

The sap tapped from this agave can also be fermented into a beer known as maguey (see Chapter 15).

Fibers are removed from both *Agave* species in the same way. The outer, mature leaves are cut at the base (Fig. 16-22), carted to the factory, and fed between rollers (Fig. 16-23) that squeeze out most of the water and turn the soft tissues into an amorphous mush that is scraped away from the fibers. The

FIGURE 16-22
Harvesting sisal leaves in El Salvador. The leaves are cut at their bases. Each plant yields about 40 to 80 leaves per cutting and has a productive life of 10 to 20 years. (Photo courtesy of the United Nations.)

FIGURE 16-23
Workers feed agave leaves under rollers that crush the leaves, squeezing out water and freeing the fibers. (Photo courtesy of J. L. Neff.)

fibers are then washed and hung in the sun to dry (Fig. 16-24). They can be dyed or used directly since they are naturally a creamy white if properly washed and dried. Henequen is still grown primarily in its native area of Mexico, but sisal has been planted in Brazil and Africa where huge plantations produce a large proportion of the world's supply. One advantage sisal and henequen have over other fiber crops is that the plants are native to, and grow best in, arid regions. Consequently they are excellent crops for regions little suited for other types of cultivation.

Abaca

Abaca, or Manila hemp, comes from *Musa textilis* (Musaceae), a relative of the banana. The fibers are extracted primarily from the outer peripheries of the leaf bases that make up the "stem" of the giant herbaceous plants. Within the last few decades, abaca has been converted from a relatively obscure fiber plant

FIGURE 16-24
Henequen, like sisal, is extracted from a species of *Agave* native to the New World. However, most henequen is now produced in Africa. (Courtesy of B. L. Turner.)

known only in its native home in the Far East to one of the most widely used sources of plant fibers. Perhaps you cannot recall ever having seen the fiber mentioned, but you have undoubtedly come into contact with tea bags, dollar bills, "Manila" envelopes, German or Italian salamis wrapped in clothlike casings, and filter-tipped cigarettes. All of these products are made, or have been made, from abaca. These uses might seem to place abaca in a category similar to wood fibers, but abaca has also been used for textiles. Long before Europeans arrived in the Philippines, abaca textiles had replaced bark cloth as the chief source of clothing materials.

Like sisal and henequen, abaca fibers are often extracted by feeding the leaves into a decorticating machine. Entire plants are harvested when they are about 4 years old. The expanded leaf blades are cut away and the petioles hauled to the processing plant. The fibers can also be obtained by peeling strips (called *tuxies*) from the outer edges of the petioles. The strips are placed on wooden blocks and all extraneous material scraped from the fibers.

DYEING AND DYE PLANTS

The ability to perceive color is a wonderful aspect of being human. While many animals are able to see various colors and some, such as butterflies, may even see a wider range of colors than we do, they cannot make use of color simply for their own enjoyment. Humans, in contrast, can pick, choose, and manipulate colors strictly for their own pleasure.

It is easy to see how humans would have learned about dyes. Children quickly see that colors are produced by mashing fruits or flowers, and parents can certainly attest to the fastness of some of these extracts. But, while many plant parts are pigmented, most plants do not yield good dyes because the chemicals responsible for the colors fade or turn muddy over time, or will not adhere to objects to be dyed. Historically, dyes were water-soluble plant extracts, while pigments were insoluble substances derived from inorganic sources.

Despite the antiquity of dyeing, the chemistry of the processes by which dyes are bonded to fibers is still not completely understood. For a dye to color a thread, fabric, or other substance, it must become completely bound to the object it is to dye. Otherwise, the color will not be "fast," and it will fade quickly or simply wash off when the object is immersed in water. It is believed that the direct bonding of charged parts of the fiber and dye molecules, hydrogen bonds, and hydrophobic interactions are all involved in the adsorption of classic dyes. It is also thought that in traditional plant dyeing processes, the structural configuration of the cellulose polymers is altered in such a way that the number of binding sites is increased, thereby enhancing the affinity of the fiber for dye molecules.

By Egyptian times, it was known that specific substances, called *mordants*, increased the adherence of various dyes to fabrics. The term mordant comes from the Latin *mordere*, meaning "to bite," because it was believed that the mordant literally bit into the fabric to make holdfasts for the dye. We now know that mordanting agents, usually salts of metals (such as tin, aluminum, iron or chromium) form a chemical bridge between the dye and the fiber

FIGURE 16-25
Roots of madder yield a red dye that has been used since ancient times. This illustration from Pomet, *Historie des Drogues,* was based on that originally published in Gerald's herbal of 1597. (Courtesy of Rita Adrosko, Division of Textiles, Smithsonian Institution.)

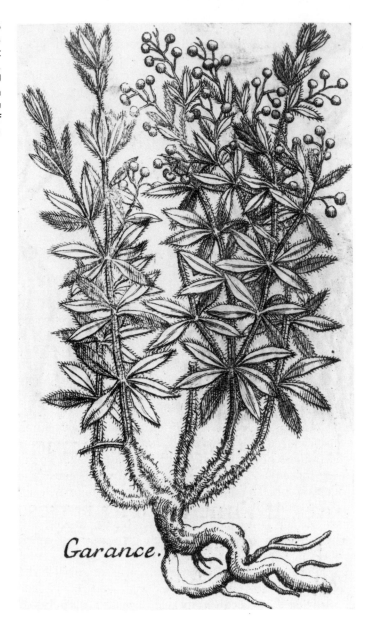

Garance.

molecules. In ancient times, these agents would have come from the metal vessels in which the dyeing was done or from additives such as dung or urine. In addition to producing strong dye-fabric bonding, mordants can also affect the final color of a dye. Iron is said to "sadden" colors, tin and chrome "brighten" them, and copper enhances bluish hues. A few dye plants such as madder (*Rubia tinctoria,* Rubiaceae; Fig. 16-25) produce particularly good dyes because they naturally contain substances that act as mordanting agents.

Every culture seems to have discovered its own repertoire of dyes, and we know that by 3000 B.C., Eurasians had already identified numerous plants as good dye sources. Henna (*Lawsonia inermis,* Lythraceae) was used by Grecian

FIGURE 16-26
Henna was used by women in ancient Greece to tint their hair a reddish brown. Today it is used as a hair rinse and leather dye. Classical hair care from a nineteenth century lithograph.

and Egyptian women as a hair dye (Fig. 16-26). The leaves of this shrubby species can be ground with water into a paste that normally produces an orange dye with a great affinity for protein. Today, chemically altered henna is used as the base for a wide array of hair colorants. The excavations of Pompeii revealed elaborate dye shops with evidences of the ancient dyes fustic (*Chlorophora tinctoris,* Moraceae) and cutch (*Acacia catechu,* Fabaceae). The use of safflower (*Carthamus tinctorius,* Asteraceae; see Fig. 10-12) in India and other regions of the Old World dates from at least several centuries before Christ.

Historical associations with colors reflect the importance of former dye plants. In the Old World, indigo (*Indigofera tinctoria,* Fabaceae; Fig. 16-27) has been considered an opulent color since its first recorded use in China over 6,000 years ago. The red stigmas of saffron (*Crocus sativus,* Iridaceae) dyed the royal robes of Irish kings. The green that we associate with Robin Hood's men (ca. 1265 A.D.) was produced by dipping their garments first in a dye bath of woad blue (*Isatis tinctoria,* Brassicaceae; Fig. 16-28) and then in a bath of deep yellow weld (*Reseda luteola,* Resedaceae). The color of the jackets of the Tory redcoats

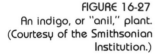

FIGURE 16-27
An indigo, or "anil," plant.
(Courtesy of the Smithsonian
Institution.)

of the American Revolution was produced by madder, and the gray of the southern forces in the Civil War with butternuts (*Juglans cinerea*, Juglandaceae).

While some dyeing processes are relatively straightforward, others attest to human ingenuity. A case in point is indigo, which comes from a legume with tiny reddish flowers (Fig. 16-29). The leaves and branches of the plant are harvested, placed in a vat, covered with water, and permitted to ferment. The sludge of partially rotted plant matter which settles to the bottom of the vat is collected and pressed into cakes. Once dry, these cakes can be ground into a powder that produces a colorless solution when dissolved in water. The characteristic strong blue color of indigo develops only after an item which has been steeped in the solution is removed and exposed to the air.

FIGURE 16-28
Woad, a source of a blue dye, was one of the dyes used to make the green outfits worn by Robin Hood's merry men. The band of generous robbers would have known a lot about dyeing since it was actively practiced in the depths of Sherwood forest. The dye process produced odors so foul that dyers were forced to carry out their trade away from towns. In fact, Queen Elisabeth I (1533–1603), decreed that no woad processing could take place within 5 miles of her residence so that she would be spared smelling the fumes.
(Redrawn from Fuch's *De Historia Stirpium*, Basel, 1545.)

The plant that yields indigo and the process of indigo dye production originated in India. Because of the scarceness of good blue colors and the exceptional quality and color-fastness of the dye, indigo was an exceedingly important item of trade between India and other parts of the world from at least three centuries B.C. Although, as mentioned earlier, Europeans had their own source of blue from woad, it never truly rivaled indigo for richness of color. In fact, the only reason woad continued to be used after indigo became readily available was because an international group known as the "woadites" lobbied for laws in Britain to prohibit importation of indigo, which they labeled "the devil's weed."

When the New World was discovered, numerous new dye plants such as annatto (*Bixa orellana*, Bixaceae; Figs. 16-30, 16-31), bloodroot (*Sanguinaria canadensis*, Papaveraceae), and logwood (*Haematoxylon campechianum*, Fabaceae) were discovered. Logwood produced a dye that made a good black possible for the first time in Europe. It is used today as a histological stain. Annatto, a red dye produced in the pulp surrounding the seeds of the achiote, is still used in Latin American cooking and as a coloring agent for margarine, cheese, and cosmetics. Annatto is added seasonally to butter when cows are not getting fresh grass, and therefore enough vitamin A, to produce cream that yields yellow-colored butter.

FIGURE 16-29

An old illustration (1694) of indigo production in the West Indies shows: the harvest of indigo (negres coupant l'anil), placing the stalks of the plants in vats of water to ret (negres jetant l'anil dans l'eau), agitating the stalks to enhance fermentation (negres remnant l'anil dans l'eau), and collecting the precipitate and hanging the dyed cloth to dry (negres portant l'indigo dans les caisses pour faire secher). (Courtesy of Rita Adrosko, Division of Textiles, Smithsonian Institution.)

While many of the classic dyes of the Old and New World are of historical interest (Fig. 16-32), the use of natural dyes is today almost completely restricted to culinary purposes and to histological work. The reason for the decline in the use of natural dyes was the discovery in 1850's and 1860's of synthetic dyes made from derivatives of coal tar. These synthetic dyes, called *aniline dyes,* produce a wide array of vibrant colors that are much more color fast than even the best natural dyes like indigo.

TANNINS

Like dyeing, tanning, the process of turning raw skins into leather, has traditionally been carried out using plant products. The first records of the tanner's art come from the Mediterranean region and date from 1500 B.C., but

Gousses de Roucou

Negres fabriquant le Roucou

Roucou

FIGURE 16-30
In this lithograph, annatto is being extracted from plants in the West Indies. The plant itself is shown to the *right* of the picture and an enlargement of the spiny capsule to the *left*. In the background is the vat in which the seeds are soaked and fermented. The seeds with their orange-red pulp were then ground to a paste in a mortar. (Courtesy of Rita Adrosko, Division of Textiles, Smithsonian Institution.)

FIGURE 16-31
A branch of an annatto plant with flowers and fruit. (After Baillon.)

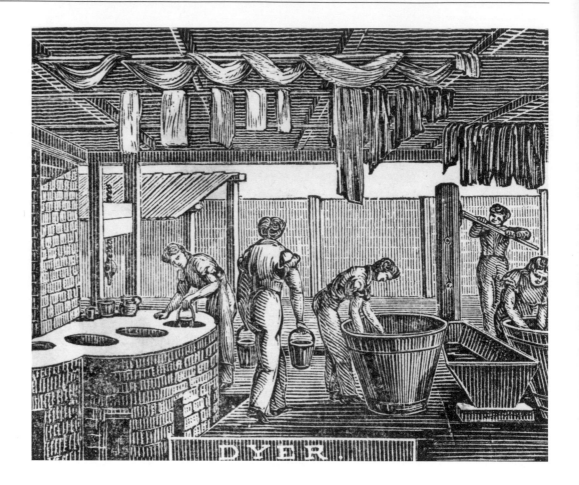

FIGURE 16-32
This lithograph of an American dye factory (ca. 1832) shows a worker lifting a bucket of dye from a furnace heater and other workers engaged in dyeing or hanging up dyed materials. Eighteen years later, the discovery of aniline dyes revolutionized the dyeing industry. (Courtesy of the Smithsonian Institution, Division of Textiles.)

it is probable that tanning was actually practiced before this date. During the process of tanning, animal skins become resistant to the effects of water, heat, and various microorganisms. The chemicals responsible for this change are collectively called *tannins,* but actually consist of a wide array of compounds that fall into two main groups, *hydrolizable* and *condensed* tannins. The first are complex esters of phenolic acids. They are called hydrolizable tannins because they are easily split by water (hydrolized) into alcohols (usually sugars) and acids. The second major group, condensed tannins, are not easily broken down and are polymers of cyclic compounds. Table 16-4 lists the important plant sources for each of these kinds of tannins.

Animal skin is made of fibers of the protein collagen. The molecular fibrils of this protein are usually highly organized and resistant to degradation. However, certain regions are disorganized and highly susceptible to attack. It appears that tanning acts by impregnating these disorganized regions and chemically bonding

TABLE 16-4

Important Vegetable Sources of Tannins

COMMON NAME	PLANT PART	SPECIES	FAMILY
	HYDROLIZABLE TANNINS		
Algarobilla	Pods	Caesalpinia brevifolia	Fabaceae
Chestnut	Wood	Castanea sativa	Fagaceae
Chinese tannin	Galls on	Rhus semialata	Anacardiaceae
Divi-divi	Pods	Caesalpinia coriaria	Fabaceae
Knoppern nuts	Fruits	Quercus robur	Fagaceae
Myrobalans	Fruits	Terminalia chebula	Combretaceae
Pomegranate	Fruits, twigs	Punica granatum	Punicaceae
Sumac, tanner's	Leaves	Rhus coriaria	Anacardiaceae
		R. typhina	Anacardiaceae
Tara	Pods	Caesalpinia spinosa	Fabaceae
Turkish tannin	Galls on	Quercus infectoria	Fagaceae
	CONDENSED TANNINS		
Catechu	Heartwood	Acacia catechu	Fabaceae
Chestnut	Bark	Castanea sativa	Fagaceae
Eucalyptus	Bark, wood	Eucalyptus wandoo	Myrtaceae
Larch, European	Bark	Larix decidua	Pinaceae
Locust, black	Bark	Robinia pseudoacacia	Fabaceae
Mallee, brown	Bark	Eucalyptus astringens	Myrtaceae
Mangrove	Bark	Rhizophora candelaria	Rhizophoraceae
		R. mangle	
Oak, cork	Bark	Quercus suber	Fagaceae
Pine	Bark	Pinus sylvestris	Pinaceae
Quebracho	Heartwood	Schinopsis balansae	Anacardiaceae
		S. lorentzii	
Spruce	Bark	Picea abies	Pinaceae
Wattle, black	Bark	Acacia mearnsii	Fabaceae

Data from Haslam, E. 1966. Chemistry of Vegetable Tannins. Academic Press, N.Y.

with the collagen molecules. In addition to tanning leathers, tannins have been used as mordants in regular dyeing operations. Because they have astringent properties, they have been occasionally used medicinally. In recent decades, tannins have been added to mud in oil drilling operations in order to increase their viscosity.

Although tannins are present in almost all plants, their ecological roles are still not fully understood. It is known that they are often concentrated in the heartwood and bark of trees and in special structures such as galls produced in reaction to infestation by insects. Perennial plants often have more tannins in their leaves than annuals, and evergreen trees have leaves with higher tannin contents than most deciduous trees. Consequently, many workers think that tannins serve to deter feeding by various insects, especially since laboratory experiments have shown that some tannic acids interfere with the digestive processes of insects. It is possible that tannins also inhibit microbial growth.

Early humans may have discovered which plants contained tannins by soaking hides in water into which pieces of particular plants had fallen. No one knows for sure, but leather sandals 3,300 years old found in Egyptian excavations clearly show that by this date both the processes of tanning and dyeing of leather were quite advanced. In the Mediterranean region, tannins were usually extracted from sumac (*Rhus* spp., Anacardiaceae) or various oak (*Quercus*, Fagaceae) species. Later European sources of tannins included spruce (*Picea*, Pinaceae) and pomegranate (*Punica granatum*, Punicaceae).

While there have been subtle improvements in tanning processes over the last several thousand years, the basic procedures are remarkably similar to those employed by primitive humans. The initial step in the tanning of leathers is to remove the hair on the outer surfaces. Originally, the hides were singed, but by Greek and Roman times, the hair was removed by soaking the skins in lime water. Today, enzymes are often used as depilatories. Originally, tanniferous plant parts were simply included between skins to be tanned or thrown into water. It was soon discovered that concentrated solutions could be made by soaking the plant parts and then evaporating some of the water. Yet, these potent solutions posed a problem with thick hides because the tannins quickly impregnated the outer skin layers and sealed off the inner layers. The inner layers would consequently be improperly tanned. To avoid this problem, thick hides are often pretreated with a "spent" (previously used) tannin solution. Another method of tanning used by American frontiersmen was to tie skins into a sack which was filled with a solution of tannins.

After tanning for an appropriate period, leathers are washed and often treated with oil or grease for softness. The leather can then be finished by coating it with a layer of gum, resin, or wax.

In North America, Indians treated hides with native plants. The colonists learned to use these indigenous sources of tannins, and they subsequently girdled hemlocks (*Tsuga canadensis*, Pinaceae) and stripped them of their bark. Originally, hemlocks grew from Maine to Pennsylvania with Boston the center of the tanning industry. As the hemlocks were depleted, hardwoods such as the chestnut (*Castanea dentata*, Fagaceae) began to be used. In the 1930s, as thousands of chestnut trees succumbed to the chestnut blight caused by the fungus *Endothia parasitica*, over 100,000 tons of tannin extract from dead trees alone became available. Pennsylvania, in the heart of the natural oak-chestnut forest region, then became the center of the American tanning industry and remains so today.

During the last 100 years, the heartwood of the quebracho tree (*Schinopsis lorentzii*, Anacardiaceae) of South America and species of wattle (*Acacia*, Fabaceae) have become important tannin sources. Mangroves (*Rhizophora* spp., Rhizophoraceae) are also rich in tannins and would seem to provide a limitless tropical source of tannins because of their ubiquity in coastal tropical regions. Generally, tannins have been extracted from wild trees, but some efforts have been made to cultivate species of wattle (*Acacia* spp., Fabaceae) as tannin sources in Australia, South Africa, and Ceylon. There is also some interest in cultivating New World species of sumac.

Today, all of the vegetable tannins used in the United States are imported. Quebracho and wattle account for 90 percent of the imported supply. Tannins are prepared by soaking pieces of wood and bark in water. The tanniferous water extract is then spray-dried or concentrated. Powdered or solid tannins are redissolved in water during tanning operations.

Unlike vegetable dyes, which have practically disappeared since the advent of aniline dyes, natural tannins have remained important in the leather industry. Large commercial operations do use synthetic tanning agents. In the United States, only 15 percent of the processed leathers are tanned using vegetable tannins, but many of the other 85 percent originally tanned with chrome are secondarily retanned with vegetable tannins. In the underdeveloped and semideveloped countries, most of the tanning is still done with vegetable tannins.

ADDITIONAL READING

Ash, A. L. 1948. Hemp production and utilization. *Economic Botany* 2:158–169.

American Fabrics and Fashion Magazine (eds.). 1980. *Encyclopedia of Textiles,* 3rd ed. Englewood Cliffs, New Jersey, Prentice-Hall. A thorough work that treats natural and synthetic fibers from their origin in a plant, animal, or test tube to the finished product.

Brooklyn Botanical Garden. 1973. *Natural Plant Dyeing. A Handbook. Plants and Gardens* 29:1–64. An easy to use and informative guide to plant dyeing.

Cook, J. G. 1968. *Handbook of Textile Fibres,* 4th ed. *I. Natural Fibres.* Watford, England, Merrow Publishing Company.

Haslam, E. 1979. Vegetable tannins. pp. 475–523 in T. Swain, J. B. Harborne, and C. F. van Sumere (eds.). *Recent Advances in Phytochemistry,* vol. 12. New York, Plenum.

Phillips, L. L. 1963. The cytogenetics of *Gossypium* and the origin of New World cottons. *Evolution* 17:460–469. Evidence for the allopolyploid origin of the New World cultivated cottons from *G. herbaceum* and *G. raimondii.*

Schetky, E. M., C. H. Woodward, and E. Scholtz (eds.). 1980. *Dye Plants and Dyeing—A Handbook.* Brooklyn, New York, Brooklyn Botanic Garden. A good companion to the 1973 book published by the Brooklyn Botanical Garden.

Smith, C. E., Jr. 1965. Plant fibers and civilization—Cotton, a case in point. *Economic Botany* 19:71–82.

Spencer, J. E. 1953. The abaca plant and its fiber, Manila hemp. *Economic Botany* 7:195–213.

Stephens, S. G., and M. E. Moseley. 1973. Cotton remains from archeological sites in central coastal Peru. *Science* 180:186–188.

Vreeland, J. 1977. Ancient Andean textiles. Clothes for the dead. *Archaeology* 30:167–178. A fascinating discussion of ancient cotton use in Peru.

Chapter 17

Wood, Cork, and Bamboo

We might feel we live in an age of petroleum, gas, concrete, and steel, but wood is still used to heat more homes and for the construction of more family dwellings than any other material. The forests that yield this wood now cover about one-fifth of the earth's land surface and constitute an estimated 90 percent of the total terrestrial biomass (Table 17-1). Yet projections are that by the year 2000, the land covered by forest will have shrunk drastically, creating hardships in both the lesser developed countries, which depend on wood for energy and construction, and the developed countries, which have become accustomed to consuming vast quantities of wood products such as paper and fiberboard.

In Chapter 1, we contrasted woody plants with herbs, stating that herbs do not contain wood, but we did not define wood. We use the word wood for the same substance present in a wide variety of plants, but woods of different species have individual characteristics and appearances that make them useful

TABLE 17-1

Productivity and Biomass for the Major Habitats of the Earth

HABITAT TYPE	AREA, 10^6 KM^2	WORLD NET PRIMARY PRODUCTION, 10^9 T/YEAR	World Biomass, 10^9 T
Tropical rain forest	17.0	37.4	765
Tropical seasonal forest	7.5	12.0	260
Temperate evergreen forest	5.0	6.5	175
Temperate deciduous forest	7.0	8.4	210
Boreal forest	12.0	9.6	240
TOTAL FOREST	48.5	73.9	1,650
Woodland and shrubland	8.5	6.0	50
Savanna	15.0	13.5	60
Temperate grassland	9.0	5.4	14
Tundra and alpine	8.0	1.1	5
Desert and semidesert scrub	18.0	1.6	13
Extreme desert, rock, sand, ice	24.0	0.07	0.5
Swamp and marsh	2.0	6.0	30
Lake and stream	2.0	0.8	0.08
TOTAL NATURAL NONFOREST	86.5	34.47	172.55
Cultivated land	14.0	9.1	14.0
TOTAL TERRESTRIAL	149	117.5	1,837.0
TOTAL MARINE	361	55.0	3.9
GLOBAL TOTAL	510	172.5	1,840.9

Source: Modified from Table 15-1 in R. T. Whittaker and B. E. Likens. 1975. The biosphere and man. pp. 305–328 in H. Lieth and R. H. Whittaker. Primary Production of the Biosphere. New York, Springer Verlag.

TABLE 17-2

Plants Discussed in Chapter 17

COMMON NAME	SCIENTIFIC NAME	FAMILY	CHROMOSOME NUMBER*
Cork	*Quercus suber*	Fagaceae	$2n = 24$
Cucumber tree	*Magnolia acuminata*	Magnoliaceae	$2n = 76$
Douglas fir	*Pseudotsuga menziesii*	Pinaceae	$2n = 26, 27$
Eastern white pine	*Pinus strobus*	Pinaceae	$2n = 24$
Eucalyptus	*Eucalyptus spp.*	Myrtaceae	$2n = 22, 24$
Hemlock	*Tsuga spp.*	Pinaceae	$2n = 24$
Kenaf	*Hibiscus cannabinus*	Malvaceae	$2n = 36, 72$
Leucaena	*Leucaena spp.*	Fabaceae	$2n = 52, 56, 104$
Mahogany	*Swietenia macrophylla*	Meliaceae	$2n = 46, 48, 54$
Mulberry, paper	*Broussonetia papyrifera*	Moraceae	$2n = 26$
Papyrus	*Cyperus papyrus*	Cyperaceae	$2n = $ ca. 102
Red maple	*Acer rubrum*	Aceraceae	$2n = 68–104$
Rice paper plant	*Fatsia papyrifera*	Araliaceae	$2n = 48$

*The ploidy levels of these species have not been given since many of the species and their relatives have not been biosystematically studied.

for a wide array of purposes. Wood has also become the starting material for the manufacture of numerous other products. In this chapter we discuss what wood is, how it is formed, the differences between hard- and softwoods, and how individual characters of woods influence their utilization. Table 17-2 lists the plants discussed in this chapter.

WHAT IS WOOD?

As outlined in Chapter 8, plants grow in length because of elongation of cells produced by their terminal meristems (Fig. 17-1). Most cells formed by these meristems mature to form the epidermis, general matrix tissue (parenchyma), and primary xylem and phloem. In truly herbaceous dicotyledons and the monocotyledons, all, or almost all, of the vascular tissue is produced by terminal meristems. In plants which subsequently become woody, a cambium layer forms between the primary xylem and phloem cells. The *cambium* is a lateral meristem capable of dividing to produce additional xylem toward the inside of the stem and phloem toward the outside. The conductive tissues formed by the cambium are called the "secondary" xylem and phloem to differentiate them from the primary tissues derived from apical meristems. In botanical terminology, secondary xylem is wood. Secondary xylem cells (Fig. 17-2) are shorter than primary xylem cells and as they mature, their walls become impregnated with lignin. By the time a xylem cell is fully functional, it is dead and its cytoplasm has disintegrated.

As we saw when we examined stems in Chapter 8, isolated bundles of xylem and phloem cells with cambial cells separating them occur in rings in some herbaceous dicotyledons. Strictly speaking, therefore, these herbs contain wood as we defined it. However, in everyday usage, wood is applied only to

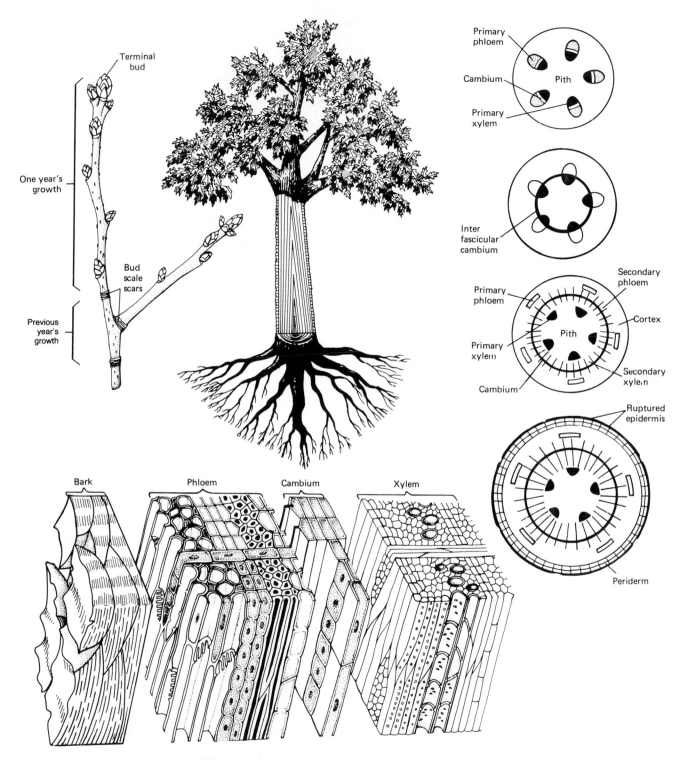

FIGURE 17-1 (Continued top of p. 512)
All increases in a tree's height occur at the very end of its branches where new cells are
produced by terminal meristems (upper left). Increases in girth result from cell divisions by

the cambium, a lateral meristem. As the cambium cells divide, phloem cells are created to the outside and xylem cells to the inside. As new phloem cells are produced, the old cells are pushed outward, crushed, and eventually sloughed off with the bark. Only the outermost cells of the xylem which form the sapwood actively conduct water and solutes, even though they, like the older xylem cells, are dead. The inner, nonfunctional xylem cells can eventually become impregnated with tannins, giving rise to a dense, dark-colored wood known as the heartwood. On the extreme right of the drawing are serial cross sections showing the development of the stem from the youngest portions of the plant (top). At the bottom of the drawing is a magnification showing the different components of a woody stem.

the large accumulations of secondary xylem that make up the bulk of a tree's or shrub's trunk and branches. In plants with persistent woody stems, the cambial cells grow together forming a complete ring. Continued production over the years by the circle of cambium cells leads to the production of wood in the vernacular sense.

An understanding of how cells are added through lateral growth of woody stems shows why wood is composed of secondary xylem and not secondary phloem, or a mixture of xylem and phloem cells. Think of a stem as a cylinder with the cambium forming a circle near the circumference. During the year, it adds xylem cells to the inside. As it does so, the core of xylem increases in girth. The cambium layer adds some cells to increase its own circumference which compensates for the expansion in diameter. The old phloem on the outside of the cambium cannot divide or stretch to accommodate the increase in the radius of the stem. Consequently, it is torn and crushed as the stem grows.

The only part of the phloem that is composed of normal, functional cells is that closest to the cambium. The xylem cells closest to the cambium are also the only functional (although not living) conductive cells. The region of the xylem that is actively conducting water is known as the *sapwood*. The older, no

FIGURE 17-2
Xylem elements from a dicotyledan such as this cucumber tree (*Magnolia acuminata,* Magnoliaceae) *(left)* have scalariform pits along their walls through which water moves from cell to cell. The xylem of softwoods *(right)* is composed of tracheids similar to those of the eastern pine *(Pinus strobus,* Pinaceae). (Courtesy of W. A. Cote of the N. C. Brown Center for Ultrastructure Studies.)

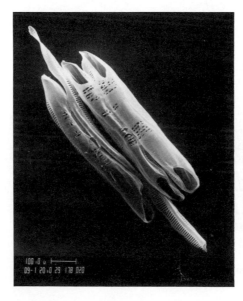
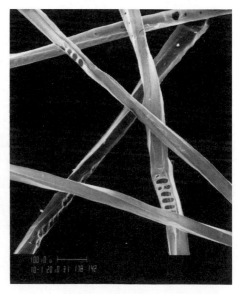

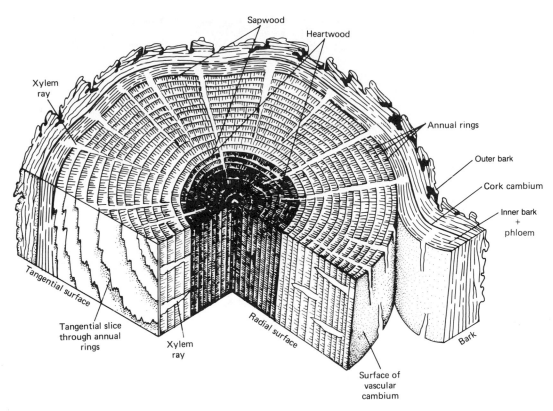

Sapwood

Heartwood

Xylem ray

Annual rings

Outer bark

Cork cambium

Inner bark + phloem

Tangential surface

Tangential slice through annual rings

Xylem ray

Radial surface

Surface of vascular cambium

Bark

FIGURE 17-3
A cross section through an oak stem shows the heartwood (the dark central area of the xylem), the sapwood (the lighter, still functional outer xylem), the vascular cambium, and the layers of the bark.

longer functional, xylem cells remain intact and form the *heartwood*. Heartwood is often darker in color than sapwood because it becomes impregnated with chemicals as it ages. Most of what we commonly think of as wood is, therefore, the heartwood of large, single-stemmed species that have accumulated appreciable amounts of secondary xylem.

In addition to xylem and phloem, the cambium makes a number of other kinds of cells. Among these are living *ray cells* in the functional part of the xylem which serve as a horizontal conduction system (Fig. 17-3). The number of cells in the rays and their arrangement differ in different species.

Surrounding the vascular tissues of woody perennial plants is the *bark* (Fig. 17-3). The bark includes all of the tissues outside of the functional vascular cylinder. Part of the bark is therefore composed of dead phloem cells. Most of the bark, however, is *periderm*, a tissue that replaces the epidermis as the protective layer. One part of the periderm is the *phellogen*, a secondary, lateral, meristem that produces *phellem*, or cork. The periderm of most species also contains *phelloderm*, a kind of parenchyma tissue. Because the function of the periderm is to protect the vascular tissues and the cambium, it must continually change to accommodate the growth of the trunk and branches. It does this by repeatedly making new phellogen and sloughing off the old, cracked layers of

FIGURE 17-4
With age, a tree's bark is shed in a way that is often characteristic of the species. Shown is a sample of bark types. Those indicated with an asterisk are turned 180° from the way in which they are normally borne. From the upper left across the top row: sycamore*, honey locust, pin oak*, Chinese elm, middle row: black gum, white ash, juniper*, white pine*, bottom row: shagbark hickory, tulip tree, white birch, and sugar maple*. (Photo courtesy of Garry Fox.)

cork. Dead, crushed phloem cells become incorporated in the periderm and are sloughed off with the outer bark. The barks of different species of trees and shrubs are distinctive because of the variations in the ways in which the bark meristem produces new cells and sheds the old (Fig. 17-4).

Hard and Softwoods

Wood is divided into two major categories, hardwoods and softwoods. These terms refer to the kinds of plants that produce the wood, not to the hardness or softness of the wood itself. All wood produced by gymnosperms (conifers) is considered softwood. Hardwood refers to any wood that comes from a dicotyledon. Anatomically these kinds of wood differ because the xylem of gymnosperms is composed primarily of *tracheids,* long cells that conduct water only through openings in their side walls (see Fig. 17-2). Almost all dicotyledons have xylem composed of *vessels,* relatively short cells that conduct water primarily through openings on their end walls. Mixed in with the vessels are tracheids and other kinds of cells. Because gymnosperm wood contains fewer

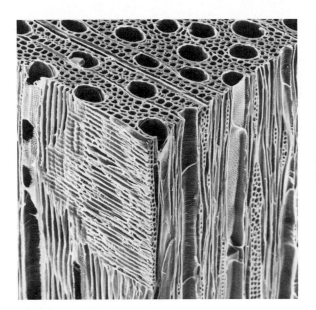

FIGURE 17-5

The scanning electron micrograph on the *left* (×110) is a piece of red maple, a dicotyledan (*Acer rubrum,* Aceraceae) sliced to show the rays running perpendicular to the stem axis and an assortment of xylem elements with different patterns of secondary wall thickenings. The douglas fir (*Pseudotsuga menziesii,* Pinaceae) on the *right* (×60) is a gymnosperm which has relatively uniform wood in contrast to the maple. The differentiation between the small cells in the background and the large cells in front show the appearance of annual rings. (Courtesy of the N. C. Brown Center for Ultrastructure Studies.)

kinds of cells than angiosperm wood, it tends to be comparatively uniform (Fig. 17-5). Rays form a horizontal system in both types of wood. Resin canals can occur in the xylem, phloem, or even bark of both soft- and hardwoods but are more common in softwoods. Laticifers (see Chapter 11) are found only in angiosperms, where they can occur in either the xylem or the phloem.

Tree Rings

Seasonal variations in climate can be reflected in the activity of the cambium and, ultimately, in the appearance of the xylem. Annual temperature changes and seasonal aridity are the climatic fluctuations that have the most pronounced effects on xylem production. During the spring in temperate regions, or the wet season in tropical semiarid areas, the cambium produces numerous, large xylem cells. In the summer or dry season, only a few, small cells are added. During the freezing months of the winter, all cell division ceases. The visible results of these changes in growth are rings in the xylem created by the concentric layers of large and small cells. In cold temperate regions, the rings can be counted to give an accurate idea of the age of a tree (Fig. 17-6). *Dendrochronology,* or the study of tree ages, has helped date archaeological ruins and provided records of recent climatic events. Unfortunately, the rings produced by seasonal aridity in tropical habitats are poor age indicators because tropical arid seasons occur less predictably than temperate winters.

FIGURE 17-6
The age of trees growing in temperate regions, and something of the history of their growth, can often be read by examining their annual rings. In the case of this hemlock (*Tsuga* sp., Pinaceae), the number and width of the rings indicate that the tree spent its first 75 years growing slowly in dense shade. In the seventy-sixth year, the trees surrounding this tree were cut, allowing it to receive more sunlight. The effect of the increased sunlight is reflected in the increased wood production after age 76.(From Brodatz, P. 1971. Wood and Wood Grains, Dover, NY.)

Characteristics of Woods

Woods differ in color, porosity, grain, and figure, all of which contribute to their appearance (Fig. 17-7). The differing colors of woods, ranging from black to green, yellow, red, and white, are produced by the impregnation of the xylem cells by differently colored compounds. *Porosity,* a feature of dicot wood, refers to the way in which large vessels are dispersed within a given part of the year's growth. Some hardwoods such as maple and poplar have a few large vessels scattered about within a ring. Others such as oak and ash have the large vessels arranged parallel to the annual rings. *Grain* is a technical term referring to the alignment of the xylem cells. All the cells can lie parallel to the vertical axis of the tree, or they can be variously tipped to produce an irregular pattern. In some cases, the cells are spirally arranged, or they occur in bands oriented in alternate directions. The *figure* of a wood is determined by the number of rays, the porosity (in hardwoods), the grain, and the arrangement of the annual rings. The presence or absence of knots, formed by the inclusion of branches into the xylem, also contributes to the figure of a wood. Even within a given

FIGURE 17-7
Figures of woods of different species. From the top left: live oak, Texas ebony, cedar, zebrawood, middle row: yellow pine, wenge, pecky cypress, Douglas fir, bottom row: mesquite, ash, crosscut oak and plainsawn oak.

species of wood, the appearance can differ depending on the way in which it was cut (Fig. 17-8).

Different species of trees produce wood that differs in density and mechanical properties. These structural qualities are important considerations for engineers, architects, carpenters, cabinet makers, and even the do-it-yourselfer. *Density* is defined as the weight of a piece of oven-dried wood of a standard size, usually 1 cm³. Since the weight of 1 cm³ water is 1 g, the density of 1 cm³ of a wood is equal to its weight in grams. Measured in this way, if the density of wood is less than one, it will float. One of the lightest woods is balsa with a density of 0.13. Lignum vitae has a density of about 1.23. Balsa is much too light to provide structural support, and lignum vitae is so dense that working with it is extremely difficult. Pine, the most common wood used in home construction, has a density of about 0.35 to 0.50, and a dense wood for fine furniture (such as oak), is usually around 0.60. The mechanical properties of woods are determined using standardized pieces to ascertain characteristic features of tension, static bending, shear, and compression strength (Fig. 17-9).

When we buy wood, we balance its visual and structural properties against its prospective use (Fig. 17-10). If a wood is to be used in construction where it will not be seen, the structural properties and the cost will largely determine the choice. For furniture and visible surfaces, the color and figure are among the primary considerations.

The structure and chemical composition of woods also determine their suitabilities as fuels (Fig. 17-11, 17-12). Since softwoods and light hardwoods burn quickly with a flash of heat, they are best used as kindling to start a fire. Many conifer woods should be avoided because they are impregnated with resins that cause deposits of flammable material to build up in chimneys.

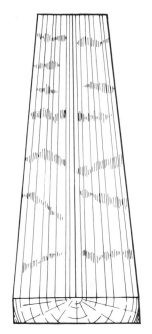

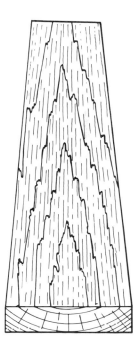

FIGURE 17-8
Quartersawn (*above*) and plainsawn (*below*) boards. By examining the figure of a board, one can tell how the board was cut from the log.

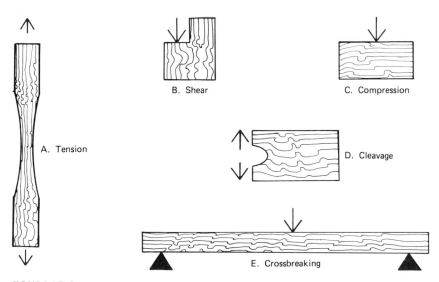

B. Shear

C. Compression

A. Tension

D. Cleavage

E. Crossbreaking

FIGURE 17-9
Tension, cross-breaking, sheer, and compression strength are mechanical properties of wood that are tested by using standardized pieces of wood. Architects, engineers, and contractors use these measurements to help them determine what kind, what size, and how many boards are needed for a particular job.

FIGURE 17-10
Because of differences in their characteristics, woods of different species lend themselves to different purposes.

Which kind of wood, for what?

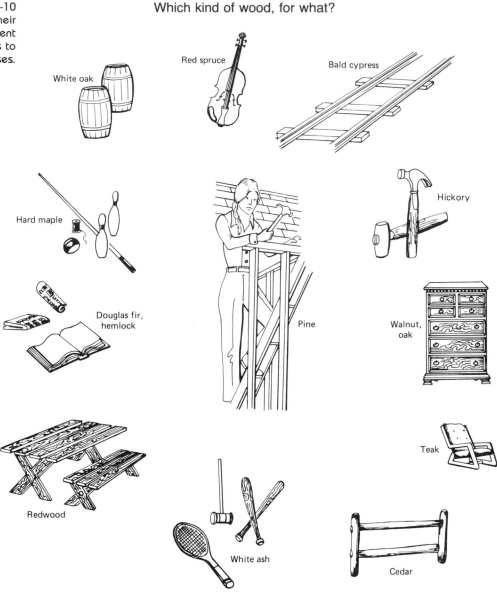

White oak

Red spruce

Bald cypress

Hard maple

Douglas fir, hemlock

Pine

Hickory

Walnut, oak

Redwood

White ash

Teak

Cedar

FIGURE 17-11
The standard unit of measurement for firewood is the cord, defined as a pile of wood 8 ft long, 4 ft high, and 4 ft deep. A "face cord" has the same 8 ft by 4 ft front dimension but is only as deep as a single log. Thus a face cord of 24-in logs is half of a full cord, but a face cord of 16-in logs has only one-third the volume of a full cord.

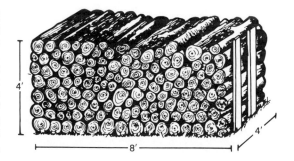

4'

8'

4'

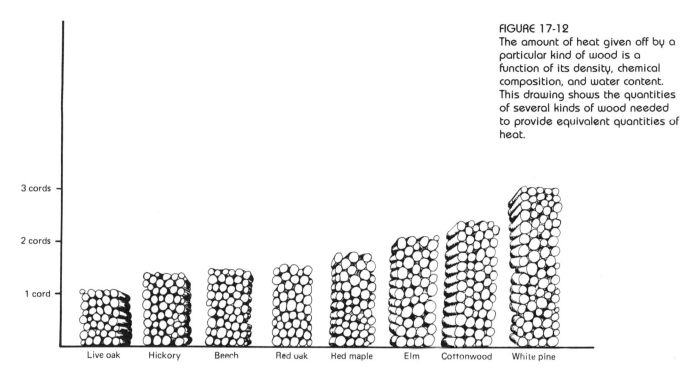

FIGURE 17-12
The amount of heat given off by a particular kind of wood is a function of its density, chemical composition, and water content. This drawing shows the quantities of several kinds of wood needed to provide equivalent quantities of heat.

Hardwoods with a medium to high density usually make the best firewood and charcoal. Charcoal is produced by slowly burning wood in an atmosphere of reduced oxygen (Fig. 17-13). During this initial burning, volatile hydrocarbons are driven off, leaving a material that burns at a much higher temperature than wood. The heat of a charcoal fire can reach temperatures high enough to smelt metals.

With the increasing costs of fossil fuels, there has been a renewed interest in the United States in using wood-burning stoves as sources of home heat. However, the extensive use of wood-burning stoves poses pollution problems

FIGURE 17-13
Charcoal is produced by stacking piles of wood in such a way that they burn slowly in an oxygen-limited environment. (Courtesy of Jack Daniels.)

greater than that of the old coal furnaces. Wood is thus impractical in very high density areas, but in areas of low density, or for isolated factories with a plentiful nearby supply of wood, it can profitably be substituted for fossil fuels.

LUMBERING AND MILLING OF WOOD

In the building and furniture industry, wood has to be harvested and milled. The milling of lumber varies from one kind of wood to the next and from one commercial operation to another (Fig. 17-14). There are, however, several standard ways in which a large round trunk is reduced to boards of various dimensions (Fig. 17-15). Fresh (green) wood contains a large amount of water and cannot be used directly for construction because later drying results in warping. Consequently, lumber is allowed to season, or dry. Uncut or milled lumber can be air dried in sheds. Such a passive system takes from 20 to 300 days for every 2.5 cm (in) thickness in the wood. The great range in time reflects differences between woods and local humidity conditions. Air drying can be hastened by using fans. A much more rapid method of drying wood involves the use of kilns, closed chambers with circulating hot air systems. By using a kiln, drying time can be reduced to 2 to 50 days per 2.5 cm of thickness. Lumber can also be treated with preservatives (Fig. 17-16) and/or various insecticides.

Wood that is not milled, but stripped of its bark, dried, and impregnated

FIGURE 17-14
A board being cut with a large rotary blade. (Courtesy of the USDA.)

with chemicals (if necessary) is called *round wood*. It is used for a number of purposes such as posts, poles, and mine timbers. Logs can also be cut into relatively small, slightly wedge-shaped pieces called shingles that can be used for siding or roofing. Shakes, used for similar purposes, differ from shingles in that they are the same thickness at both ends. Both shingles and shakes are usually impregnated with chemicals that help to resist rotting.

WOOD PRODUCTS

While the traditional uses of wood rely on its strength and workability, many modern uses depend on the natural properties of wood that allow it to be used for semisynthetic products. Such manufactured wood products include veneers, composition boards, paper, and textiles.

Veneers are thin sheets of wood of uniform thickness that are often affixed to another surface, usually a different wood or composition board. The thin sheets can be made by shaving across a flat wood surface or by "peeling" a continuously revolving log (Fig. 17-17). Most veneers are now produced by the rotary (peeling) method because it yields larger, more uniform sheets. The thickness of a veneer sheet can vary from 0.25 mm (0.01 in) to almost 1 cm (0.40 in), but most are about 0.25 cm (0.10 in) thick. After cutting, sheets of veneer are dried, either in the air or in kilns. Rough pieces of veneer are used to produce numerous crude objects such as orange crates or small berry boxes. Finer pieces are used for furniture and plywood surfaces. The cost of making furniture can be reduced by laminating a fine wood veneer onto a cheaper structural material. As tropical hardwoods have become increasingly scarce, this method of producing attractive surfaces has become increasingly common.

The most economically important use of veneer is in the manufacture of *plywood*. To make plywood, sheets of veneer are glued together with the grains of each sheet at right angles to those of the sheets above and below it. An odd number of sheets is always used, giving rise to the terms 3-ply, 5-ply, etc. Some plywoods have a solid core consisting of small boards glued together. For solid-core plywoods, even numbers of sheets of veneer are used. The outer layers of veneer on plywood are often of a better quality and/or appearance than those of the inner layers.

In making a piece of plywood, care must be taken to balance correctly the sheets on either side of the center sheet or the core. An improper balance of sheets will cause the plywood to warp and can lead to the separation of the layers. Depending on the kind of wood, the number and thicknesses of the layers of veneer, and the kinds of glue used, plywoods can range from slightly stronger than a board of the same thickness to one that, pound for pound, is stronger than steel.

While we probably think of plywood as a relatively modern product, pieces of boards with analogous structure have been found in Egyptian tombs dated to be over 3,500 years old. It is still not clear how artisans of ancient civilizations were able to cut veneer sheets. The use of veneers as decorative objects dates from Roman times. During the reign of King Louis XIV of France, the use of inlaid pieces of veneer in furniture, a process known as *marquetry*, was very popular and led to the production of beautiful works of functional art.

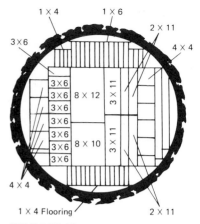

FIGURE 17-15
The mill operator decides what kind of cutting procedure will yield the most lumber from each log. Structural pieces are usually cut from near the center of a log where the wood is dense and not likely to crack. The outer portions of logs are most suitable for smaller boards such as 1 by 4 in flooring boards.

FIGURE 17-16
Timbers or boards are placed in these cylinders, where they will be treated with wood preservatives. Creosote (a coal tar product) and Wohlman salts (a poisonous mixture of fluorine, phenol, copper naphthenate, and other compounds) are often used to protect the wood from fungal decay and attack by insects. Since many of these are carcinogenic, their use is now regulated. (Photo courtesy of the Koppers Company.)

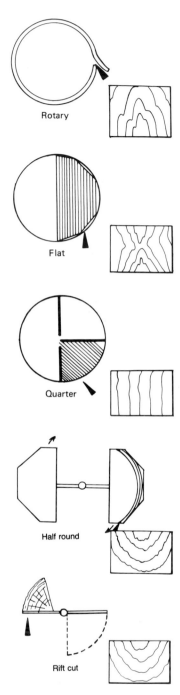

FIGURE 17-17
Methods of cutting veneer and the figures of the resulting products. (Redrawn after P. Villiard. 1975. A Manual of Veneering. New York, Dover.)

Plywood has limitations because its manufacture is dependent upon the availability of large sheets of veneer. Much of the available wood is unattractive, irregularly shaped, or too small for making veneer, but these otherwise useless pieces of wood can all be processed and made into particle, chip, or fiberboard. Consequently, the production of processed boards has gradually supplanted that of plywood.

Manufacture of particle board started in the 1940s after the development of synthetic resins and has continued uninterrupted ever since. For both particle- and fiberboard, wood is first reduced to small chips by shaving or splintering. The particles are then sorted and graded. Once particles of the proper size have been produced, they can be mixed with resins, pesticides and/or fire retardants, and pressed into the desired shapes and sizes. There is less waste in the manufacture of particle board than in the production of milled lumber. Often, the two operations are used to complement one another because the sawdust, bark chips, and small pieces of wood left from the milling process can be used for particle- or fiberboard.

Fiberboard differs from particleboard in that wood fibers, not small pieces of wood, are used. The fibers are xylem elements that have been separated from one another by placing small wood particles in chemical solutions that dissolve out the pectins holding the fibers together. Fibers can also be produced mechanically by grinding wood chips. The processes used to reduce wood to fibers for fiberboard are basically the same as those used to make pulp for paper. Once the fibers have been separated, they can be mixed with various additives, such as synthetic resins for different degrees of strength, and protective compounds for resistance to fire and pests. The pulp can also simply be pressed into sheets without the addition of resins to produce wet-felted fiberboard. The amount of pressure applied while the boards are drying, and the amount of resin added determine the final classification of the fiberboard.

PAPER

It is often said that the Egyptians invented paper because they wrote on sheets of papyrus. However, these sheets were formed by pressing together strips of the leaves of papyrus (*Cyperus papyrus*, Cyperaceae). Likewise, other paper-like substances have been made by various cultures. Among these are the famous rice "paper" of the Orient, which is made by pounding sheets spirally cut from the pith of the rice paper plant *Fatsia papyrifera* (Araliaceae). Both the Mayans of Mexico and the Polynesians independently developed a method of producing paperlike sheets by pounding the bark of the paper mulberry, *Broussonetia papyrifera* (Moraceae). The South Pacific natives, but apparently not the Mayans, used the bark cloth for clothing. None of these substances were true paper.

Paper, as we now know it, is made from plant fibers which have been separated from one another and then matted together in a thin sheet. The Chinese apparently first produced a paper made by this process about 100 A.D. In its production, they also used the paper mulberry, but separated the fibers and mixed them with other plant fibers such as flax, hemp, and even fibers extracted from rags. The fibers, floating in liquid, were allowed to settle in a thin film on a screen. The screen was shaken as it was lifted to interlock the fibers. Upon drying, the resultant sheet of paper was peeled from the screen.

It took 1,000 years for papermaking to spread from the Orient to Europe, reaching Spain when the Moors conquered it in the twelfth century. While some leaf and wood fibers were used to make early European papers, most were made from fibers extracted from linen, cotton, or hempen rags. For the next hundred years, paper was essentially made by hand using variations of the process developed in China. Because of the hand labor involved, paper was expensive and not widely available. Hand production was finally supplanted after the invention of a papermaking machine in France just after the French Revolution (1789). After the British made some improvements on the machine which sped up the process, the increased production created a shortage of rags. Rags were in such demand that even mummy wrappings were irreverently removed from Egyptian tombs and shipped to Britain. It soon became obvious, however, that new sources of fiber would have to be found. Wood was a practical source, but the initial methods of separating out the wood fibers by rubbing logs on stones left much to be desired. The resultant fibers were short, weak, and retained their pectins and lignins, which produced a paper that was short-lived and rapidly acquired a yellow tinge.

Chemical methods of separating wood fibers were developed after 1851. The first processes involved boiling wood chips in caustic alkali until the wood was reduced to a mass of fibers. Once washed, the fibers could be bleached. Various improvements on this process involving different dissolving agents led to the sulfite process developed in Paris in 1857 and the sulfate process in 1884. Today almost all papers are made from wood using one of these two processes, but sheets labeled 100 percent rag bond are still made from the fibers extracted from rags. In addition to wood pulp, pulp from abaca (see Chapter 16), sugarcane pressings (bagasse), and bamboo are also used. Conifer woods are generally preferred over hardwoods for papermaking because their xylem tracheids are longer than the xylem vessels of hardwoods (about 2.0 to 4.0 mm vs. 0.5 to 1.5 mm).

As in the case of fiberboard, the first step in making paper is the production of pulp. Either mechanical or chemical methods can be used. For either purpose, the logs are stripped of bark, and the wood reduced to chips. Enormous machines are now sometimes driven into the forest to receive pulp trees as they are felled. The machines take entire logs, strip them, and reduce them to chips that are trucked to the paper plant. In other cases, logs are floated to the mills where the bark is stripped by raspers and often subsequently used to fuel the mill itself. The logs are then chipped at the mill.

Mechanical pulping used in the ground wood process consists of grinding the chips to free the fibers which can be bleached or used directly to make paper (Fig. 17-18). Chemicals can be added to the pulp before the fibers are floated onto draining screens. Several different kinds of chemicals, including resins, gums, starches, or rosin, can act as sizing agents which fill in surface irregularities and improve a paper's ability to accept ink. Most paper produced via a mechanical process has little sizing because the paper is of such poor quality that it does not justify the expense.

Once prepared, the slurry of fibers is flooded under a roller to produce uniformly thick sheets. The sheets drain as they move across a screen and onto rollers that press and mat the fibers before the sheets are rolled into wet pulp rolls. Paper produced from fibers that were mechanically extracted, yellows rapidly and eventually crumbles because the pectins and lignins left in the pulp degrade. However, because of the demand for cheap newsprint, a large amount of the world's paper is produced by this process.

The sulfite and sulfate chemical processes (Fig. 17-19) differ from one another principally in the agents used to reduce the wood to pulp and to

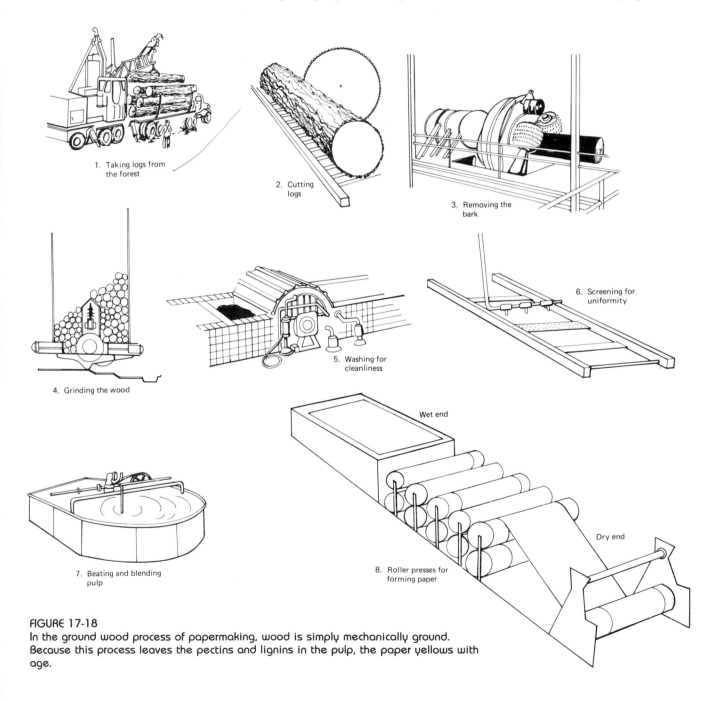

1. Taking logs from the forest

2. Cutting logs

3. Removing the bark

4. Grinding the wood

5. Washing for cleanliness

6. Screening for uniformity

7. Beating and blending pulp

8. Roller presses for forming paper

Wet end

Dry end

FIGURE 17-18
In the ground wood process of papermaking, wood is simply mechanically ground. Because this process leaves the pectins and lignins in the pulp, the paper yellows with age.

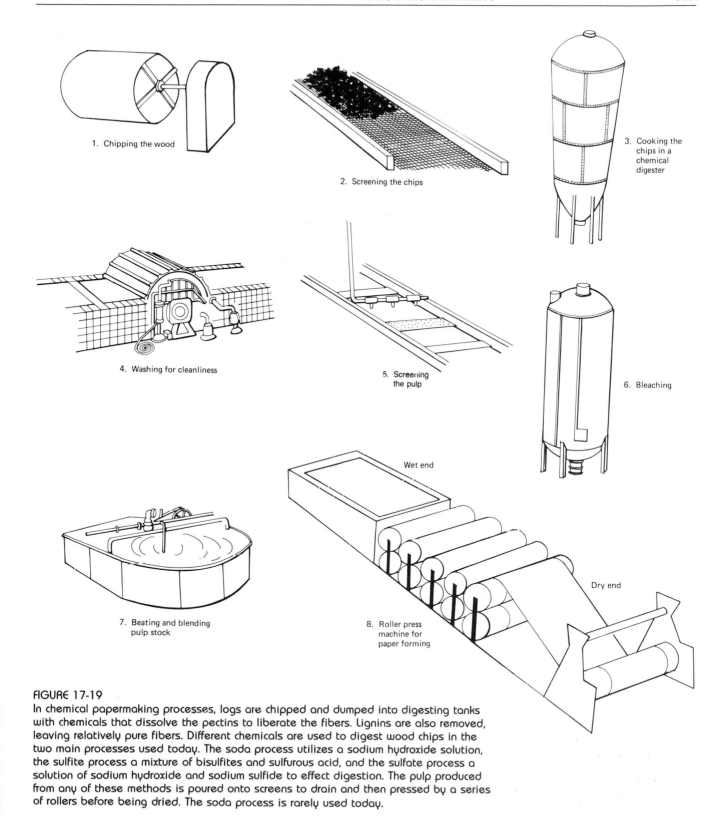

1. Chipping the wood

2. Screening the chips

3. Cooking the chips in a chemical digester

4. Washing for cleanliness

5. Screening the pulp

6. Bleaching

Wet end

Dry end

7. Beating and blending pulp stock

8. Roller press machine for paper forming

FIGURE 17-19

In chemical papermaking processes, logs are chipped and dumped into digesting tanks with chemicals that dissolve the pectins to liberate the fibers. Lignins are also removed, leaving relatively pure fibers. Different chemicals are used to digest wood chips in the two main processes used today. The soda process utilizes a sodium hydroxide solution, the sulfite process a mixture of bisulfites and sulfurous acid, and the sulfate process a solution of sodium hydroxide and sodium sulfide to effect digestion. The pulp produced from any of these methods is poured onto screens to drain and then pressed by a series of rollers before being dried. The soda process is rarely used today.

dissolve out the lignin. In the *sulfite process,* the chips are cooked in a digester with bisulfites. Hot acid is then pumped in and the cooking completed. The softened fibers are then forcefully blown into a chamber to separate them.

The *sulfate process* is an alkaline rather than an acid process and uses as digestive agents, sodium sulfate, sodium sulfide, and caustic soda (sodium hydroxide). This process is now the most widely used because, unlike the sulfite process, it dissolves the resins out of the pulp and can therefore be used for gymnosperm woods such as Douglas fir and pine. After digestion, the tenuously held together fibers are beaten to separate them.

Once separated from one another, the fibers produced by either process are washed and sometimes centrifuged to separate them from bits of bark and debris. The clean fibers are treated with a chlorine bleach that is flushed out with sodium hydroxide. The fibers are then rinsed and mixed with enough water to make a slurry. Sizings, most commonly rosin soaps, are generally added to chemically produced pulp. Since the sizing substances tend to bind poorly with the pulp, a binding agent such as alum (aluminum sulfate) is added. This is the same type of agent that is added to dyeing solutions in order to bind dyes to various fibers (Chapter 16), but in the textile industry, they are called mordants. Binding agents also serve to ensure the adherence of pigments to the fibers if the paper is to be colored.

Finally, the pulp is ready to be formed into paper. It is flooded into a large box with a slit in the bottom that allows an even, smooth flow of pulp across the screen (Fig. 17-20). The sheet drains as it moves with the screen until it reaches a first set of pressing rollers. It continues around a series of rollers through a dryer. If necessary, the paper is coated after it is produced so that it will be more suitable for the printing of halftones (photographs).

Paper produced by the sulfite process has a high acid content that causes brittleness and disintegration of the paper after about 100 years. Unfortunately, most books printed since 1850 were printed on sulfite papers. We are now

FIGURE 17-20
A headbox of a papermaking operation is used to regulate the flow of pulp onto the draining screen. (Courtesy of the Potlatch Information Service.)

facing the possible loss of millions of valuable books. The Library of Congress has consequently embarked on an enormous project to treat their books with diethyl zinc, a compound that bonds with cellulose and neutralizes the acid. Most smaller libraries will not be able to carry out such expensive programs.

Because of the large quantities of energy and forest productivity required for papermaking, it would seem that the recycling of paper would be highly desirable, but a problem with the recycling of paper is the energy that must be expended to remove inks and dyes. Most paper plants are not equipped with machinery for recycling operations, or they are able to process only one kind of pulp and cannot handle the mixture of pulps that invariably comes with a collection of paper for recycling. So far, the cost of buying wood and producing new pulp is still cheaper than the costs of recycling paper. Yet, as population sizes grow, and the demand for paper increases, the balance may eventually turn.

RAYON, CELLOPHANE, ACETATE, AND ARNEL

Rayon and cellophane are the same product in a different form. Both are made from pure cellulose derived from wood. Initial steps toward the production of rayon were made in Europe in the nineteenth century (1855), but suitable technology was really developed only after 1900. The first steps in the production of the two are the same as those employed for making paper. Chips of wood are reduced to pulp by a chemical treatment, usually with hot caustic soda. The resultant mass, once washed, is ground and allowed to oxidize in the air. Carbon disulfide is then added to produce cellulose xanthate. Caustic soda is once again added, causing the mass to become viscous. The mixture is allowed to sit until it reaches the proper consistency, filtered, and either extruded in fine sheets that are thoroughly washed to produce cellophane or forced through small openings to produce rayon threads. These threads are drawn through an acid bath that neutralizes the alkali and makes the fibers flexible. The strands are then washed, dried, and wound onto spools. High-performance rayons can be sanforized and mercerized (see Chapter 16), accept dyes well, and match cotton in general performance. The newest rayons (marketed as Avril) are like natural fibers in terms of the crimping or lobing of the individual strands.

Acetate and acetate fibers were once grouped with rayon and cellophane, but now they have come into their own. They differ primarily from rayon and cellophane in that they are made from purified cellulose to which one or more acetyl groups have been added. Despite this partially synthetic production, fibers made from acetates still have many properties of completely natural plant fibers. Processes for the spinning of acetate fibers were developed in the 1920s, but it was only after the development of triacetates (about 1950) which do not shrink, that they began to be extensively used.

The ingredients used to make acetates are wood pulp and, sometimes, straggling fibers (linters) left on cottonseeds after ginning. The purified pulp, prepared by one of the chemical methods, is mixed with acetic acid, acetic anhydride, and a catalytic agent. The chemical reaction that results leads to the linkage of acetyl groups with the cellulose molecules of the fibers. The acetate is then chilled in water, forming flakes that are dissolved in acetone. Dyes can be added to the solution, which results in deep, vibrant colors that do not fade when the fibers age. Triacetates are made by dissolving triacetate in alcohol and methylene chloride and then spinning it. These fibers, marketed under the name of Arnel, are now extensively used in double knits, tricot, and various permanent press items. Acetate fibers dry rapidly, are soft, resist wrinkling, and are not attacked by moths or molds. Masses of acetate fibers that have not been spun are used to make the filters of some modern cigarettes.

CORK

Strictly speaking, cork is not wood since it is the bark of a tree, not secondary xylem. Still, it is usually included in a discussion of wood since it is produced by trees and often used for woodlike products. As shown in Fig. 17-21, cork (phellem) is part of normal bark, but most trees produce only small quantities of cork cells. One species, the cork oak (*Quercus suber*, Fagaceae), produces layers of cork several inches thick which can be stripped from the trees without damaging them. The areas around the Mediterranean Sea (Fig. 17-22) where this species grows naturally are covered by an evergreen scrub vegetation similar in aspect to the scrub along the coast of southern California. As in California, this region is prone to frequent, large-scale fires. The bark of the cork oak is presumably the product of natural selection because it provides protection against fire damage.

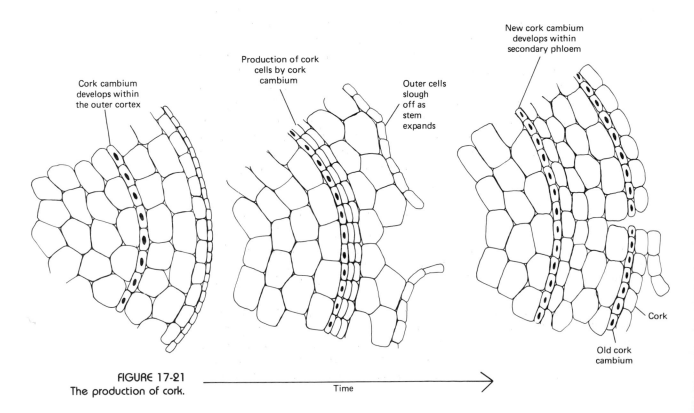

FIGURE 17-21
The production of cork.

Time

FIGURE 17-22
The major cork-producing areas of the world (*stippled*).

Cork acts as an insulating material because it is composed of numerous, air-filled cells. These cells can withstand a compression of up to 10,000 lb/in² without rupturing. When the pressure is removed, the cells return to their former size. The air-filled cells also make cork lightweight and cause it to float. The Greeks and Romans appreciated the qualities of cork and used it for the soles of shoes and for fishing floats. We still use cork for shoe soles and fishing floats, but its modern uses have expanded because we have not been able to produce a synthetic with all of its qualities. One recent use has been for insulating spaceships.

Cork trees are large, evergreen trees that build up enough bark to be stripped when they are about 25 years old. This first stripping yields "virgin" bark that is considered inferior to the bark that replaces it. The virgin cork is mostly ground and used to make composition cork, flooring, and wall coverings. Successive crops of cork can be harvested about every 9 to 10 years after the first stripping (Fig. 17-23). The superior cork produced by these later layers is used for stoppers, barrel bungs, cork veneers, and specialty purposes.

FIGURE 17-23
Stripping bark from cork trees. (Courtesy of the Armstrong Cork Company.)

BAMBOO

Like cork, bamboo differs from wood because it does not contain any secondary xylem, but in the Orient, bamboo is used as a structural material that fulfills many of the functions of wood. The name bamboo can be applied to a number of genera and species that belong to the Bambusoideae, a subfamily of the grass family, Poaceae. New growth in bamboo is produced only by apical meristems (Fig. 17-24). Consequently, there is no secondary, or radial, growth. All growth is vertical and caused by the production of new apical cells and cell elongation. Stems of mature bamboo consist of segmented, hollow tubes which give bamboo a unique combination of flexibility, light weight, and strength. In some tropical areas, entire towns are built only from bamboo. In Japan, bamboo scaffolds are used in the construction of skyscrapers. The canes of bamboo can be split and the strips used for fishing poles, caning for chairs, and wall or window coverings.

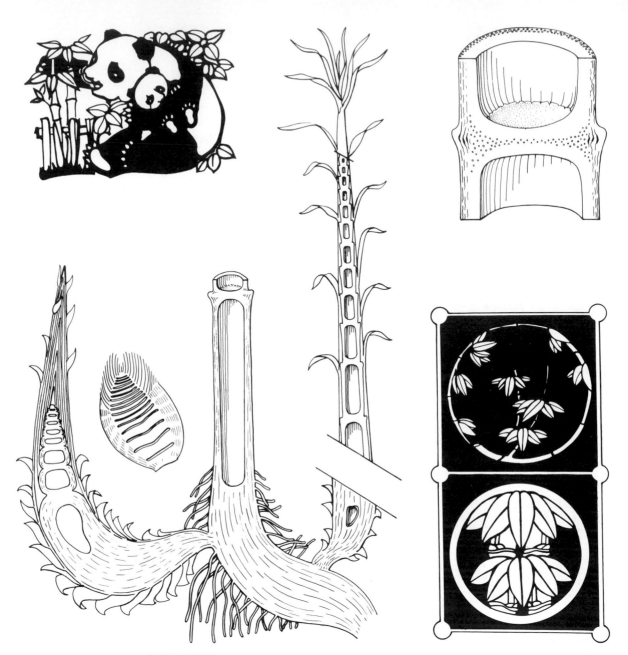

FIGURE 17-24

Bamboo is among the world's more curious and versatile plants. Like other monocots, bamboo has no vascular cambium, and its stems cannot increase in girth by adding lateral layers of cells. The diameter of a stem is essentially determined at the time of sprouting (*bottom left*). Bamboo spreads by sending up shoots from rhizomes. Bamboo's structural strength combined with flexibility is due to the morphology of the stems which are made up of a series of hollow segments separated by solid internodes (*center & upper right*). Vascular bundles and associated fibers are scattered in the outer walls of the segments. In the orient where many species of bamboo are native, the plants are the sole food of the panda (*upper left*), and bamboos have come to symbolize resistance to hardship because they stay green all year and bend (without breaking) under the weight of the snow. The long internodes are thought to represent virtue, while the hollow stems are reminders of the value of humility. These associations and the importance of bamboo in construction and horticulture have led to the adoption of bamboo by many oriental families as a symbol for the family crest (mons), two of which are shown on the lower right.

SUPPLIES OF WOOD IN THE FUTURE

With the increased demands for wood and wood pulp that will come as world populations increase, there is growing concern about the ability of the earth's forests to meet the projected needs. Some predictions are that the consumption of forestry products will rise by 75 percent during the next 16 years. As the demand increases, forested areas will be drastically reduced. Most of the losses of natural forests are, and will continue to be, in the tropics (Fig. 17-25). Population growth rates are highest in tropical areas where the need for agricultural land, wood for fuel, and external demands for hardwoods have escalated the cutting of forests in the last 30 years.

Unfortunately, large-scale deforestation has drastic results. Tropical logging operations tend to denude large areas from which the small amounts of natural soil are soon lost by erosion. It is estimated that 10 tons of soil per person are lost each year from the eastern slopes of the Himalayas where one-quarter of the forests have been cut in the last decade. Inhabitants of the overpopulated tropical regions where logging is rampant are already turning to grass, dung, and crop residues as sources of fuel. The burning of the last potential sources of organic material that could otherwise be used for soil improvement only aggravates the problems caused by the clearing of the forests. Another severe problem is the inevitable extinction of species that destruction of tropical rain forests will cause. It has been estimated that hundreds of thousands of species will be lost in tropical environments because of the destruction of natural forest habitats.

While trees are supposedly a renewable resource, the use of billions of

FIGURE 17-25
Mahogany is one of the valuable hardwoods that is fast disappearing from tropical forests. Here, mahogany logs are floated down river to a harbor on the Ivory Coast. (FAO photo by A. Defever.)

FIGURE 17-26
A 20- to 25-year-old red pine plantation (Photo courtesy of the USDA.)

FIGURE 17-27
Leucaena species are among trees being looked at as potential sources of wood in the future. (Courtesy of Noel Vietmeyer.)

cubic feet per year demands large-scale planning operations to ensure future supplies. Until 1950, virtually all wood was harvested by exploiting natural forest lands. Now, most American companies are trying to practice various methods of silviculture. Among these are selective harvesting of the mature trees in an area followed by reseeding or with little enough disturbance to ensure natural regeneration, a practice which is like thinning the stands of trees (Fig. 17-26). Even clear cutting (the complete cutting of all woody plants) can be effectively practiced if measures are taken to prevent erosion and to reseed the area.

One solution has been the planting of fast-growing pines ("supertrees") and species of *Eucalyptus* (Myrtaceae) in plantations. A previously unused genus, *Leucaena* (Fabaceae, Fig. 17-27) has been suggested as a potential new source of wood. In tropical areas, trees of species of this genus, which has wood as dense as that of oak, can grow 8 m (over 24 ft) in height per year. Once cut, the stumps regenerate, making reseeding unnecessary. Likewise, a woody herb from which we obtain the fiber kenaf (*Hibiscus cannabinus,* Malvaceae) has been proposed as a possible source for paper pulp. One acre of this herb can supply five times the amount of pulp supplied by an acre of pine trees. Kenaf pulp lacks the resins and lignins that must be removed from conifer pulp in the production of good paper.

Many scientists, economists, and conservationists are concerned that prudent exploitation is not extensively enough practiced to regenerate the forests as fast as they are being destroyed. Management techniques are not completely successful in preserving natural habitats, and invariably soil and nutrients are lost. Management often leads to an unnatural dominance of one or two species of trees. This reduction in natural tree diversity leads to losses of herbaceous plant and animal species. If pesticides and fertilizers are used, they can lead to further extinctions of species. No action will probably be taken until a severe shortage is created. Hopefully, in the face of these projected shortages, our national forests will not be released from their protection in order to supply what will obviously be only a short-term solution.

ADDITIONAL READING

Allaby, M., and J. Lovelock. 1980. Wood stoves: The trendy pollutant. *New Scientist* 88:420–422.

American Fabrics and Fashion Magazine (eds.). 1980. *Encyclopedia of Textiles,* 3rd ed. Englewood Cliffs, New Jersey, Prentice-Hall.

Barney, G. O. (Study Director). 1980. *The Global 2000 Report to the President of the U.S. Entering the 21st Century,* vol. 1. *The Summary Report Special Edition with the Environment Projections and the Government's Global Model.* Washington, D.C., U.S. Government Printing Office.

Britt, K. W. (ed.). 1970. *Handbook of Pulp and Paper Technology,* 2nd rev. ed. New York, Van Nostrand Reinhold.

Clark, T. F. 1965. Plant fibers in the paper industry. *Economic Botany:* 19:394–405.

Constantine, A., Jr. 1975. *Know Your Woods.* Revised by H. J. Hobbs. New York, Scribner's.

Crane, J. C. 1947. Kenaf: Fiber plant rival of jute. *Economic Botany* 1:334–350.

Department of Energy. 1979. Project Retrotech. *Home Weatherization Manual.* Department of Energy Conservation Paper No. 28C (DOE/CS-0106). Washington, D.C.

Harrar, E. S. 1947. Veneers and plywood—Their manufacture and use. *Economic Botany* 1:290–305.

Hunter, D. 1947. *Paper Making: The History and Technique of an Ancient Craft.* 2nd ed. New York, Alfred Knopf. Very good coverage of early developments in the making of paper and paperlike materials.

Jane, F. W. 1970. *The Structure of Wood,* 2nd ed. Revised by K. Wilson and D. J. B. White. London, Adam and Charles Black.

Marden, L. 1980. Bamboo, the giant grass. National Geographic 158(4):502–529.

Panshin, A. J., E. S. Harrar, J. S. Bethel, and W. J. Baker. 1962. *Forest Products: Their Sources, Production, and Utilization.* 2nd ed. New York, McGraw-Hill.

Spurr, S. H. 1979. Silviculture. *Scientific American* 240(2):76–91.

United States Pulp Producers Association. 1955. *Wood Pulp: A Basic Fiber.* New York, U.S. Pulp Producers Assoc.

Chapter 18
Ornamental Plants

*O*rnamental plants delight our eyes, titillate our noses, and in our increasingly urban environment, often constitute our only link with the natural, growing, plant world. While humans have planted gardens since the dawn of civilization, it is the growth of modern cities that has made horticulture the multibillion dollar business it is today. In this chapter we examine the philosophies of gardening that have shaped American horticulture and the characteristics that have been required of ornamental plants. We also look into the origins of our common horticultural species and the changes in them that have occurred under the influences of artificial selection. Finally, we look at some alarming trends and encouraging directions in the future use of ornamental plants.

QUALITIES OF ORNAMENTAL PLANTS

In contrast to plants that yield a tangible product, ornamental plants are those which are appreciated for their own aesthetic qualities or which are used to beautify the appearance of a primary object (Fig. 18-1). Beauty is, of course, a subjective thing. In many cases, aspects of practicality and utility (considered to be components of beauty by some but not by others) are also important factors in the choice of ornamental plants. Criteria of what is beautiful are in large part culturally determined and differ between countries, within parts of a country, between individuals, and from generation to generation. There are, however, a few basic elements of beauty common to art forms in general which can be applied to ornamental plants as well.

One of the primary elements of beauty is *color*. We perceive color by hue, value, and intensity. *Hue* is the actual spectral wavelength which is emitted by an object and received by our eyes as blue, yellow, red, etc. *Value* is the lightness or darkness of color, and *intensity* refers to the depth, or amount of light saturation, of a color. All of these factors combine in subtle ways to influence our choice of ornamental plants.

Texture is a component of beauty that is often subconsciously perceived. From birth, we learn to associate certain visual patterns (Fig. 18-2) with tactile sensations. Eventually, we develop visual impressions of textures that allow us to "feel" through our eyes. A good example is the visual contrast between a violet and a pansy. Both species belong to the genus *Viola* (Violaceae) and both can have flowers of the same color. However, without feeling a pansy, we know that the petal surface is velvety because of the subtle play of color across the petal surfaces. We have learned to associate this shaded appearance with soft coverings of soft upright hairs (or threads in the case of true velvet). A similar process is involved when we see shaggy or smooth-barked trees or leaves with waxy or fuzzy surfaces (Fig. 18-3). We often decide the texture without actually feeling how rough or slick they are.

Line is another aspect of beauty often not consciously appreciated except by designers or architects. Yet, our choice of plants is often determined by their natural, or inducible, sizes or outlines. Vertical branching patterns of trees or shrubs guide the eye upward, whereas horizontal branches lead it toward the ground. We trim hedges in order to produce precise lines to guide the eye or to frame specific areas (Fig. 18-4).

FIGURE 18-1
Plant motifs appear in many forms in art and architecture.

Form is the last component of beauty applicable here. It is a three dimensional quality that includes both shape and structure (Fig. 18-5). Differences in form can help determine a plant's use as an ornamental (Fig. 18-6). Take, for example, a weeping willow, a poplar, and an elm. Weeping willows are used as landscape trees where their unusual branching pattern can develop and be appreciated.

FIGURE 18-2
Without having to touch them, one can perceive that these agave and cactus stems have very different textures.

FIGURE 18-3
The size, shape, and coloration of leaves all contribute to our perception of foliage texture.

FIGURE 18-4
A tree's branching pattern, the curve of a walk, and the edges of hedges, buildings, and plantings can all work as elements of line in landscape design.

The poplar is a fast-growing, narrow tree, often planted to create hedges or screens. Elms have long been favorite street trees because of their vase-like shapes with high canopies that allow free movement of pedestrians and traffic below (Fig. 18-5).

MAJOR KINDS OF ORNAMENTAL PLANTS

Ornamentals can be divided into three major groups: nursery plants, florist crops, and house plants. Nursery plants are used for outdoor plantings, often to complement a structure or to beautify an area. Florist crops are those which yield flowers or foliage for cut arrangements. House plants are growing plants that are confined to pots and used for interior decoration. Different qualities are sought in plants belonging to each of the groups because they serve varied purposes. But one quality, aesthetic appeal, is common to all. We discuss each of these major groups of ornamentals separately, but before we begin, we digress to discuss the ways in which ornamental plants are named.

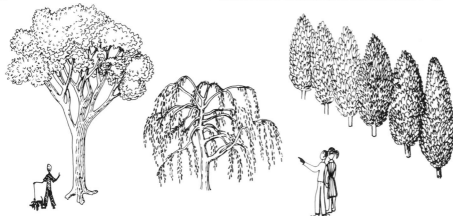

FIGURE 18-5
With age, most trees and shrubs assume a form characteristic of their species. By knowing what this shape will be, or by pruning and training plants into desired shapes, the landscaper can use plants as architectural elements to create outdoor spaces. At the bottom from left to right are trees of elm, willow, and poplar.

THE NAMING OF ORNAMENTAL PLANTS

Plants selected for use as ornamentals are often more difficult to assign to a biological species (see Chapter 2) than other crops because they are frequently entities created by humans. Some are produced by crossing (hybridizing) species. Such hybridizations can involve a series of elaborate steps if the plants to be

FIGURE 18-6
These sketches show how the
arrangement of plants with
different forms can create (top to
bottom) open, static, linear, or
intimate spaces.

FIGURE 18-6 These sketches show how the arrangement of plants with different forms can create (top to bottom) open, static, linear, or intimate spaces.

crossed are quite different genetically. Many of the horticultural hybrids produced are sterile and are propagated by cuttings or grafts. Horticulturalists also often seize upon bizarre mutants because of their novelty appeal and perpetuate them by asexual means. Since many ornamental plants cannot, or do not, reproduce naturally and were created via artificial hybridization, they are beyond the realm of natural species. Nevertheless, they are fitted as well as possible into the system of binomial nomenclature by using special provisions of the *International Code of Botanical Nomenclature* (see Chapter 2) that are stated in the *International Code of Nomenclature of Cultivated Plants.*

A horticultural crop can be given the Latin name of the species from which it was selected if a single parent was involved. If the crop is a hybrid, and one or both of the parents are known, a special designation called a *formula name* is used. This kind of name indicates the hybrid origin of a crop and specifies the two parents. For example, a hybrid produced by a cross between *Rosa alba* and *R. lutea* would be given the formula name *Rosa alba* × *R. lutea*. A hybrid between *Rosa alba* and an undetermined second species would be written *Rosa* × *alba*. Intergeneric hybrids are given a formula name preceded by an × and formed by combining parts of the two generic names (e.g., × *Heucherella tiarelloides* is a plant produced by hybridizing *Heuchera* × *brizoides* with *Tiarella cordifolia*).

Often, plants selected for trade purposes are given appealing names such as 'Dutchess of Windsor,' 'Crimson Star,' or 'Twinkle Toes.' The single quotation marks, which are usually omitted, indicate an artificial (and often patented) form. When buying an ornamental plant at a nursery, this cultivar name is often the only one given. We might, therefore, buy 'Andrew Jackson' azaleas or 'Purple Dawn' African violets with no real species name attached, although, of course, each plant is referable to one (or a hybrid) species and has a proper Latin designation. In some cases, cultivars of a species or genus are grouped into types that indicate a general kind of flower or growth form. Such groupings are commonly used for very popular plants such as roses or chrysanthemums.

NURSERY CROPS

The first major group of ornamentals is nursery crops, plants sold to be grown around houses, in parks and playgrounds, along highways and greenbelts, and as part of the landscaping for urban developments. While often not thought of as such, all of these places constitute kinds of gardens, public or private plots of land where plants are cultivated. It has been said that the gardens of each age express an image of paradise for the people who create them. Consequently, the historical development of gardening styles often parallels that of the philosophical thinking of civilizations. Similarly, we find the roots of our own landscaping ideas shaped by those of previous cultures. For this reason, we outline the development of landscaping as an art form before we discuss the major nursery crops in use today.

An appreciation for natural beauty seems common to all people, and flowers, leaves, and fruits have been used as ornaments for thousands of years. The development of gardens, however, had to wait until humans were settled. Only with sedentism could people lay out, plant, and tend the flowers and trees of a garden. The first true Western gardens were planted in ancient Egypt. Eastern gardening started in China. Early records show that even at the inception of gardening in these areas, two very different concepts of gardens existed. Our modern American gardens have elements of both these styles but have been

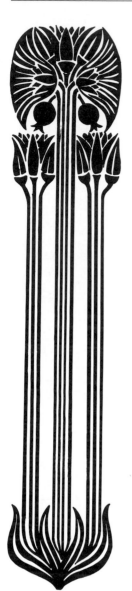

predominantly drawn from Western traditions. Most of our discussion of the developments of gardens thus centers around those of Western civilizations.

The Development of Western Gardens

Egyptian interest in botany and gardens is well documented in wall paintings and hieroglyphics drawn as early as 2200 B.C. (Fig. 18-7). The Egyptians developed a concept of a garden as an enclosed space, placing their houses within garden walls to keep out intruders and provide protection from the searing desert winds (Fig. 18-8). Within the walls, date palms, figs, pomegranates, and grape arbors provided shade and sustenance, while T-shaped, or rectangular, pools were planted with lotus and papyrus and stocked with fish. While it has been suggested that the pattern of irrigation ditches dictated the layout of these gardens, the geometric, stylized forms of the paths and planting beds are consistent with the formal, architectural style present in the rest of Egyptian art. In their search for plants to use in their gardens, the Egyptians organized the first plant collecting expeditions (see Fig. 11-23). An early catalog of Egyptian botanical knowledge can still be seen on the walls of the temple at Karnak.

  The formal Egyptian concept of gardens spread to Syria, Persia, India, and other parts of the western world. In Persia, autocratic rulers ordered their

FIGURE 18-7
A stylized ornament incorporating the forms of papyrus and lotus, two highly esteemed plants in ancient Egypt.

FIGURE 18-8
Sun and limited water availability were the major influences in the design of the first Western gardens. This artist's concept of a wealthy Egyptian's garden shows the formal, symmetric arrangement of pools, arbors, and partitions to create a cool, shady oasis within the desert. (Based on a drawing by D. Lancelot in *Histoire des Jardins*, Tours, 1883.)

FIGURE 18-9
An artistic reconstruction of the fabled Hanging Gardens of Babylon. Since the Babylonians used brick, rather than stone, for building, earthquakes and wars have left only legends of the most famous gardens of the ancient world. The fact that the Babylonians engineered lakes, canals, and reservoirs to try to control the erratic Euphrates River lends credence to the tales of a hydraulic device which drew water from the river and pumped it to the top of a pyramidal stack of hollow arches. These arches were constructed as huge planters filled with soil and planted with trees, shrubs, and flowers to create the illusion of a verdant hillside. (Based on a drawing by D. Lancelot in *Histoire des Jardins*, Tours, 1883.)

subjects to plant groves of trees which became pleasure gardens and hunting preserves. These gardens were the forerunners of our modern public parks. The word park arose when the Greeks mistranslated the Persian word *pardes,* meaning "paradise" or "garden."

The most famous gardens of ancient times are probably the Hanging Gardens of Babylon (Fig. 18-9). These gardens were the creation of King Nebuchadnezzar in 605 B.C., who commissioned the building of a terraced garden to recreate the hillside scenery that his favorite, but homesick, Persian wife craved. The Greeks considered the Babylonian gardens to be one of the seven wonders of the world, but they did little themselves to advance gardening art. Their democratic system of government did, however, encourage the development of public parks which became popular places to hold philosophical

FIGURE 18-10
Topiary, the creative pruning of plants, is shown at its most whimsical here at Green Animals, a garden maintained by the Newport Preservation Society, Rhode Island.

discussions. The Greeks also planted kitchen gardens in small, open courtyards, but the purpose of these gardens was utilitarian, not aesthetic.

Many artistic and cultural contributions of the Romans dwindle beside those of the Hellenistic Greeks, but they excelled at gardening and became the most accomplished gardeners of the ancient world. This expertise can be attributed to several factors. The Roman Empire lasted long enough for the development of a distinctive style of garden art. In addition, the Italian peninsula has a mild climate and good soils on which gardens thrive. Romans had ample inspiration for their gardens from the tales brought back by soldiers from all parts of Europe, western Asia, and northern Africa.

Because Rome was an urban civilization, most of its citizens lived in large city apartments. By planting window boxes and painting flowers on courtyard walls, people brought nature into this new urban world. Wealthy Romans implemented garden designs that fit the grounds of villas outside of the city, and, for the first time, urban planners incorporated greenbelts into city designs. As Rome prospered, the cultivation of ornamentals thrived. The Romans built the first greenhouses using mica or glass for windows. In these houses, they learned to grow tender, exotic species and to force plants to bloom out of season. The art of topiary (Fig. 18-10) was another Roman innovation for which they used native species such as juniper, rosemary, and lavender. When Rome fell, horticulture declined. For the next 1,000 years, gardening in Christian Europe was confined to monasteries where monks planted medicinal gardens and grew altar flowers (primarily lilies and roses).

In the Moslem world, gardening continued to flourish during the Dark Ages, and Muslims carried their visions of paradise and garden design to every land they conquered, including Spain which was invaded by the Moors in the first century A.D. Spanish hillsides were terraced to create verdant displays, and villa gardens were laid out in a cruciform pattern to reflect the Muslim belief of the founding of civilization at the confluence of four rivers. Vine-covered arbors ("glorietas"), colonnades, fountains, and colorful tiles, still popular motifs in Spanish gardens, are legacies from this period of Islamic domination (Fig. 18-11).

During the Crusades (1095 to 1291), interest in gardening was rekindled in other parts of Europe. The French revived the concept of pleasure gardens in the twelfth century. These small gardens, enclosed by walls and planted with beds of flowers, clipped hedges, or even mazes of trimmed shrubbery, were often designed to enhance their function as places to entertain ladies.

As the Renaissance slowly infiltrated all of Europe, the revival of Classical ideas that permeated the fine arts extended to gardening and landscaping as well. Italians began to look at plants as architectural or sculptural objects, and added perspective to garden design. Plants were considered building materials to create outdoor corridors, vistas, and plazas on Italy's sloping hillsides (Fig. 18-12). The French readily adapted the Italian designs to their flatter terrain, producing elaborate formal gardens and a kind of carpet bedding called "parterre" (Fig. 18-15). The most famous French gardens of this period are those of Versailles (Figs. 18-13, 18-14).

In 1660, French gardening ideas crossed the English Channel and were emulated by Charles II (ca. 1660) and other nobles in England. However, by the end of the seventeenth century, the stylized formal gardens of the Renaissance were being criticized as too costly and difficult to maintain (Fig. 18-14). Philosophers began to lean toward naturalism. This swing toward natural expression was reflected in gardens which were reshaped with winding paths, water channels, and copses of trees and shrubs (Fig.18-16).

Plants and ideas brought back to Europe by explorers had a great impact on seventeenth century gardens. The "exotics" piqued the curiosity of intellectuals, and people of the "upper classes" started considering horticulture an

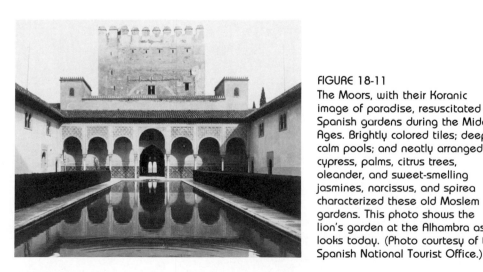

FIGURE 18-11
The Moors, with their Koranic image of paradise, resuscitated Spanish gardens during the Middle Ages. Brightly colored tiles; deep, calm pools; and neatly arranged cypress, palms, citrus trees, oleander, and sweet-smelling jasmines, narcissus, and spirea characterized these old Moslem gardens. This photo shows the lion's garden at the Alhambra as it looks today. (Photo courtesy of the Spanish National Tourist Office.)

FIGURE 18-12
The Italians built terraces so that they could use their naturally hilly terrain. The main axes of these gardens ran downslope so that shady evergreen corridors could be created for the display of sculpture and to orient the strollers toward particular objects. Artificial waterfalls, fancy waterworks, pergolas, and grottos were popular garden features of the day. (Photo taken at the Tivoli Gardens; courtesy of the Italian Government Travel Office.)

indispensable part of a well-rounded education. The wealthy began to maintain large private gardens in which to display new plants and animals. Public gardens such as the Royal Botanic Gardens at Kew (Fig. 18-17) are legacies of this era. Many species were first formally described from live specimens growing in these gardens, and, in a few cases (e.g., coffee, Chapter 14), seeds from European gardens were used to start plantations far from their native homes. For the most part, however, exotic plants in these gardens were like rare animals in zoos.

Not only the wealthy but also the common people began to engage in gardening as egalitarianism spread to all aspects of life. Everyone with a patch of land in England seemed to have a cottage garden. With the great variety of

FIGURE 18-13
Like other sixteenth and seventeenth century French gardens, Versailles is best looked at from above, where the forms of the ornate flower beds can best be seen. This *fleur de lis* "parterre" is executed in clipped boxwood and flowering annuals.

FIGURE 18-14
Gardens of the French romantic period reflect the influence of the Age of Reason. The classical symmetry of these "ordered universes" is obvious in this engraving of the Versailles gardens from about 1684.

blooms and colors available, species were chosen that would provide a succession of colors and flowers. Since land was at a premium, careful planning went into the choice and arrangement of the flowering plants. The rules governing the design of the English perennial border today are an outgrowth of practices developed for these small cottage gardens.

In the eighteenth century, exchanges between Europe and Asia resulted in an influx of oriental ideas into Western art. Pagodas and ponds began to appear in gardens, but such additions remained, in large part, only discordant elements in gardens originally laid out in a Western style.

American gardening did not come into its own until the late nineteenth century when the United States finally emerged as a major industrial power. Until this time, most gardening in America was pragmatic or copied from the native homelands of the many immigrants that flooded into the country. With industrialization, however, came a new lifestyle. Cities grew to sizes never

FIGURE 18-15
This floral clock, from the Jardin des Acclimatation in Paris, is the modern version of an eighteenth and nineteenth century garden feature. European gardeners tried to create accurate timepieces by planting flowers that opened and closed at specific times. As a guide for this pursuit, Linnaeus published the *Horologium Florae* ("Sundial of the Plants"). It was soon discovered that climatic differences affect flowering times, so botanists from areas other than Sweden began to compile their own lists of plants useful for floral clocks. In the nineteenth century, sundials were created from sculpted plants clipped to cast shadows on numerals on the face of the clock. Today, buried electrical or mechanical clocks are used to tell time, and they are bordered by colorful displays of flowers.

before imagined. Some cities met the demand for natural areas within urban environments by incorporating public parks into city plans. Many of our famous large city parks such as Central Park in New York (1858), Prospect Park in Brooklyn (1866), Fairmont Park in Philadelphia (1855), Forest Park in St. Louis (1904), and the continuous park system of Boston (1880s) date from this period. Attitudes toward botanical gardens began to change as they became not only public gardens, but also sources of new material for potential horticultural use. All of the major botanical gardens in the United States (e.g., the Arnold Arboretum, Boston, Massachusetts; the Morris Arboretum, Philadelphia, Pennsylvania; the Morton Arboretum, Lisle, Illinois; and the Missouri Botanic Garden, St. Louis, Missouri) are today actively concerned with the propagation of plants that have potential as beautiful and functional ornamentals.

Our twentieth century American landscaping is drawn from various components of traditional Western gardens, but the modern constraints of urban

FIGURE 18-16
During the English naturalistic movement, gardeners tried to create the kind of "picture perfect" landscapes popularized by painters of the time. For romantic effect, temples, ruins, hermitages, and even dead trees were placed within the gardens. This photo of the garden at Stourhead, Wiltshire, was provided by the British Tourist Authority.

FIGURE 18-17
The conservatory at Kew Gardens.
(Photo courtesy of the British
Tourist Authority.)

life make many features of conventional European gardens inappropriate. Consequently, we have developed a renewed interest in oriental landscape design. The reduced scale of Eastern gardens, their dependency upon a few, well-chosen elements rather than mass plantings, and their use of tolerant evergreens, make them particularly suited to urban environments. In many cases, however, we seem to have been able to incorporate elements of oriental design without really understanding the philosophies behind them.

Oriental Gardening Philosophies

Evidence indicates that the Chinese were cultivating food and medicinal plants in enclosed garden areas four centuries before their Egyptian counterparts. Although both kinds of gardens were precisely designed, the desired effect in Chinese gardens was naturalistic. Throughout the history of the Oriental garden, this naturalistic emphasis, reflecting a deep reverence for nature, is apparent. The Chinese were the first people to create true pleasure gardens, and by 190 B.C. they were constructing extensive parks and public gardens. The Chinese considered landscaping a fine art interrelated with poetry and landscape painting, and they conceived of the plants they used in them as symbolic, rather than as architectural objects.

When Buddhist missionaries carried their religion to Japan in the sixth century A.D., the Japanese adopted Chinese attitudes toward gardening, but they adapted it to their own climate, terrain, and indigenous flora. As a result, Japan gradually developed its own unique styles of gardening. In the twelfth century, the *Sakuteiki,* considered to be the oldest gardening manual in the world, was published. It provided advice for the orientation of a garden and how to move and place various elements within it.

While we tend to think of Japanese, or even Oriental, gardening as a single style, there are in fact, many different kinds of Chinese and Japanese gardens. Zen Buddhist philosophy espouses introspection and contemplation. Zen gardens were therefore designed to foster meditation, but in keeping with Zen asceticism, they reduced everything to its essential form. The ultimate expression of Zen reductionism is the *karesansui,* or dry garden. In these gardens, all trees, shrubs, and even water have been eliminated. Carefully arranged rocks, moss, and sand suggest natural landscapes.

During the fifteenth century, Zen monks also developed the tea ceremony, which was eventually moved into its own shelter within a specially designed

FIGURE 18-18
Hand washing is a ritual act of
purification that helps one to
prepare spiritually for the tea
ceremony. (Photo courtesy of the
Japan National Tourist
Organization.)

garden. The tea house and garden were built in the image of a rustic mountain retreat where one could contemplate quiet beauty. Within these gardens everything is precisely placed to guide strollers through a series of gates which promote mystery and symbolize internal change (Fig. 18-18). Garden features are chosen and placed to bring to mind the Buddhist virtues of humility, purification, and enlightenment. Except for the simple flower arrangements in the tea house, these gardens tend to be monochromatic with a few well-pruned evergreens and clumps of moss.

The rustic nature of the tea gardens and the visual treats of the earlier paradise gardens combined to create the stroll gardens that are today emulated in Japanese city and country gardens. The features of these gardens, streams which flow from the north, "picture windows" where shaped trees or structures frame a view, the use of borrowed scenery to increase the garden's perspective, and paths designed to produce a sense of privateness in the stroller show the synthesis of centuries of oriental garden art (Fig. 18-19). The "Japanese" gardens most frequently constructed as parts of American public gardens have been based on stroll gardens designed by oriental masters. Americans, ambling through one of these gardens, are usually oblivious to the care that has gone into the placement of each item in the garden or to the symbolism of each element. However, the feeling of quiet beauty and tranquility is impossible to ignore. It is this feeling of a natural harmony with plants that has infiltrated into American gardens and softened, to some extent, the linear, architectural styles inherited from European cultures.

Uses and Types of Nursery Plants

Our placement of nursery plants today is thus a conglomeration of aesthetic ideas and styles drawn from a wide array of cultures, but nursery plants often serve functional purposes as well. They enclose areas, provide privacy or form partitions, cover the ground, prevent erosion, and furnish shade. Recent studies have shown that the proper use of landscape plants can substantially modify

FIGURE 18-19
Gardening is a fine art in Japan. The plantings try to create a naturalistic landscape and, at the same time, represent spiritual ideas symbolically. In the Nitobe Garden at the University of British Columbia, the garden represents a journey through life. Here the scene represents the longer, more adventurous life path that can be taken after adolescence. If this path is selected, "you must retrace your steps and cross the yatsu-hashi bridge safely. Spiritual growth can begin only when this zigzagging crossing is accomplished, because, as mythology has it, the devil, who can travel only in a straight line, will then be left behind." (Photo courtesy of John Neff; quote by permission of the Botanic Gardens, University of British Columbia.)

the climate of a localized area and consequently reduce energy costs. Lawns have been shown to be 5 to 8°C (10 to 14°F) cooler than bare soil and 14 to 17°C (25 to 30°F) cooler than asphalt. Turf also absorbs sunlight and reduces glare (Fig. 18-20). Although the use of plants as windbreaks has long been appreciated, the effect of trees and shrubs as insulating factors for buildings has only recently been emphasized (Fig. 18-21). Proper use of shrubs or vines which shade in the summer and lose their leaves in cold winter months can save between 10 and 40 percent in heating and cooling costs (Fig. 18-22). Designs for solar heating of homes often include glassed areas for plants which aid in the collection and circulation of heat from the sun.

For home use, ornamentals are usually chosen for the area to be landscaped. Often, a yard is divided into a public access area, which is, for most homes, the frontyard (Fig. 18-23). Plantings in such an area tend to be rather formal and showy. A second part of the yard includes the driveway and work areas which

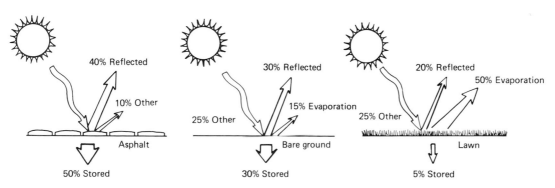

FIGURE 18-20
This diagram shows how different surfaces react to sunlight. The grass-covered surface cools itself by evaporation, while asphalt and bare ground store heat and reradiate it to the air above. Convection and conduction account for the percentage labeled "other" in these drawings.

FIGURE 18-21
Windbreaks can be arranged to deflect cold winter winds (*upper two figures*) while allowing summer breezes (*lower two figures*) to cool your house. Both effects can be achieved because prevailing wind directions often differ in the summer and winter. The height, density, and shape of the planting will affect how much wind reduction is achieved. In general, a moderately dense (60 percent coverage) mass of evergreens will perform better than an impenetrable barrier which tends to form vacuums on its leeward side. As a rule of thumb, the greatest wind reduction occurs immediately in front of the barrier and beyond it for a distance of five to seven times the height of the barrier (H). Windbreaks have been shown to affect wind speeds for a distance of up to 30 times their heights.

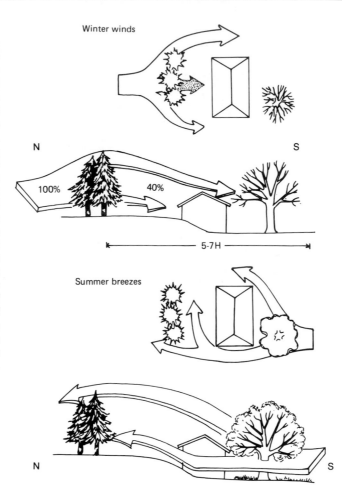

are usually minimally planted or planted to screen trash or storage areas. Finally, the family living area, usually the backyard, is designed to reflect a family's life style. In some cases, the presence of a patio and swimming pool leaves little room for gardens. In other cases, a family may decide to plant a series of flower beds that supply cut flowers throughout the growing season or to allow as much area as possible to serve as lawn for playing. All of these decisions about land use require different kinds of plants with different characteristics.

For urban use, primary emphasis is often on ease of maintenance and durability. An abundantly flowering or fruiting tree might be rejected for city planting even though it is beautiful if it causes a serious clean-up problem with fallen branches, flowers, or fruits. Examples of formerly popular street trees that are no longer used are mulberry trees, which drop masses of blue-black fruits on sidewalks, streets, and cars, and silver maples, which easily break in wind storms. In cities with high pollution levels from automobiles or industry, or which use salt on icy roads, plants have to be chosen which can tolerate high levels of acid rain, carbon monoxide, smog (hydrocarbons plus nitrous oxide), and/or salt-impregnated soil. Plantings in public areas must also withstand relatively high levels of physical abuse suffered when they are bumped, broken,

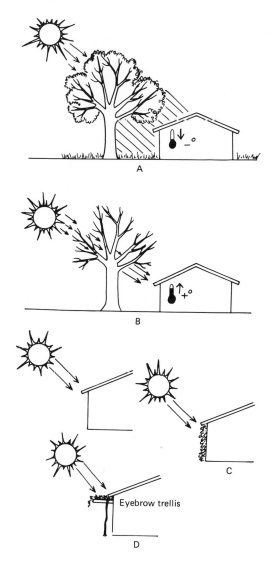

FIGURE 18-22
Judicious landscaping can modify the climate around your home. Deciduous trees on the south and west sides (A) will provide shade and reduce temperatures as much as 8 to 10°F inside. When these trees lose their leaves in winter (B), the sun's rays are able to reach the house and provide heat. Walls may be shaded by deciduous vines (C) growing against a wall or trellis. An eyebrow trellis (D) keeps out the sun when it is high in the sky (summer) but allows the winter sun in to heat the house.

carved upon, or visited by dogs. Highway plantings must require a minimum of maintenance (Fig. 18-24) and must also serve to deter erosion on embankments.

With these ideas about the concepts of gardening and the functional necessities of ornamental plants in mind, let us look at the major kinds of nursery plants used: turf, ground covers, bedding plants, and deciduous and evergreen trees and shrubs.

Turf

The turf, or nonforage grass, industry in the United States is based on a very limited number of species but constitutes a major part of the nursery trade. Our high esteem for a well manicured lawn is a vestige of the gardening heritage developed during the 17th and 18th centuries in England. What started in the

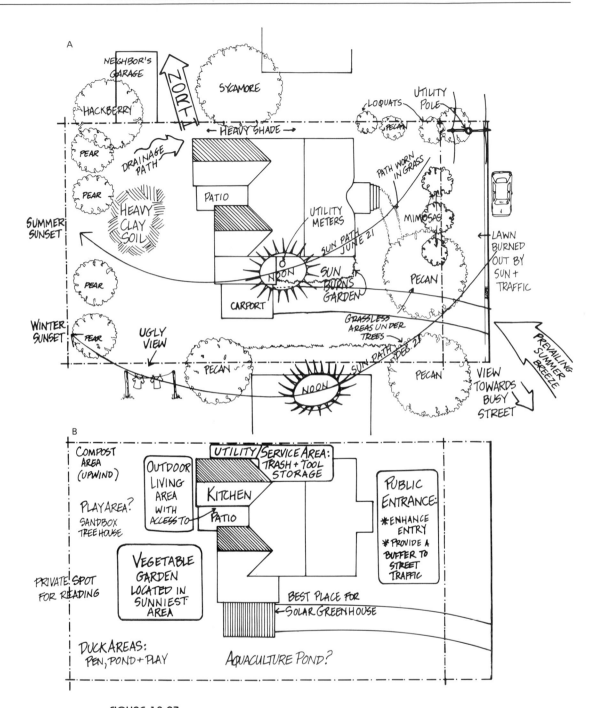

FIGURE 18-23
The first step in coming up with a landscaping design is to choose a scale and draw a base plan indicating the existing site conditions (A). Included on this plan should be all the relevant features: directional orientation, property lines and easements, existing vegetation and objects, utilities, views, slopes, and seasonal sunpaths. Second is a list with energy and space needs on a tissue overlay. (B) These will differ with each family and might include a formal entrance, driveway, parking space, areas for outdoor living, service, and allowances for privacy, shading, and windbreaks. From these, one can

Middle Ages as an informal patch of grass kept trim by household livestock, expanded during this era into lush manor lawns maintained with the extensive hand labor of crews who cut the fields with scythes. In 1831, an English textile maker adapted the design of a carpet cutting machine (a steel drum with a spiral cutting tool) to create the reel mower. With this tool, a neatly clipped carpet of bentgrass was possible for even the small homeowner.

By breeding adaptable native grasses we have been able to create a reasonable facsimile of the English sward in this country. Unfortunately, in the absence of England's cool, moist climate, maintenance of lawns on suburban yards and golf courses is a costly endeavor. In 1970, $4 billion per year were spent on urban and suburban turfs. Seventy percent of this was spent for residential lawns, 11 for highways, 8 for cemeteries, and 6 for golf courses (Fig. 18-25). Sixty-two million dollars were spent for seed and 100 million for sod. A hundred million dollars were spent for fertilizers, 120 million for pesticides,

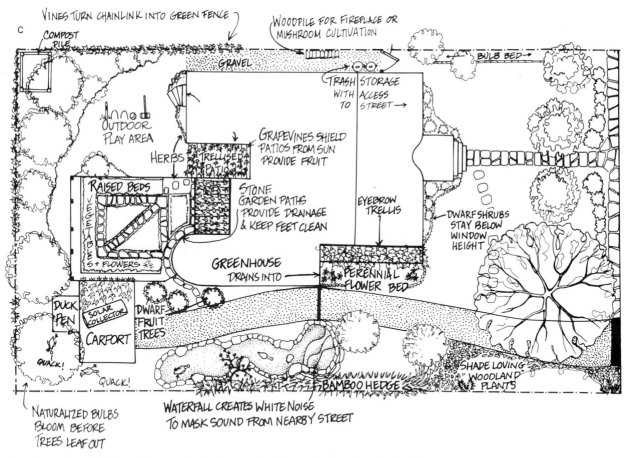

develop a plan (C) that will accomplish the goals while trying to relate the building to its site. finally, plants should be chosen to execute the design. In this example for an architect and botanist, the family was interested in taking advantage of the climate to grow the greatest possible array of plants. The couple wanted a large kitchen garden and a fenced area for ducks. Eyebrow trellises on the south and west facades support deciduous vines that keep the living spaces cool in the summer.

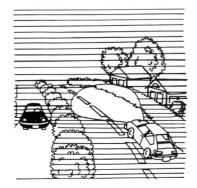

TABLE 18-1

Turf (Poaceae) Species Commonly Used in the United States

LOCATION, COMMON NAME	SCIENTIFIC NAME	NATIVE RANGE
Northern states:		
Bent grass	*Agrostis* spp.	Northern hemisphere
Fine or réd fescue	*Festuca rubra*	North America
Kentucky bluegrass	*Poa pratensis*	Eurasia
Southern states:		
Bahia grass	*Paspalpum notatum*	Mexico, South America
Bermuda grass	*Cynodon dactylon*	Old World
Carpet grass	*Axohopus* spp.	Americas
Centipede grass	*Eremochloa ophiuroides*	Southeast Asia
St. Augustine	*Stenotaphrum secundatum*	Tropical Americas
Zoysia	*Zoysia* spp.	Asia, New Zealand

and the rest for equipment, most of which (375 million) was spent for lawn mowers. Nevertheless the mania for lawns is understandable in terms of what they provide. Lawns are easy to walk on, cooling because of the water they transpire, suitable for large open spaces, and effective in preventing the formation of dust and mud. Table 18-1 lists the major grasses used for turfs in the United States today.

Ground Covers

Ground covers are often used in place of grass in areas where low maintenance is required, where there is too much shade for grass to grow well, or where slopes are too steep to mow. Ground covers are also used because they provide a texture different from that of grass and, in a few cases, a color contrast. All such covers, however, share the feature of low, creeping growth which provides a dense carpet over the surface of the ground. Unlike grass, ground covers are not suited to areas over which people walk because their stems are easily broken and leaves crushed. In addition, many of them provide an irregular cover about 19 cm above the ground, making walking difficult. They are, however, well suited for planting along highway embankments and on other types of slopes. Several of the most common ground covers used in the United States are given in Table 18-2.

FIGURE 18-24
Highway plantings can help to protect oncoming cars or nearby residents from automobile lights and traffic noises (top), keep drivers from being blinded by the sun's glare (middle) or lights from oncoming cars, and control erosion (bottom).

TABLE 18-2

Common Ground Covers Used in the United States

COMMON NAME	SCIENTIFIC NAME	FAMILY	NATIVE RANGE
Ice plant	*Mesembryanthemum crystallinum*	Aizoaceae	Southwest Africa
Ivy	*Hedera helix*	Araliaceae	Europe, Africa
Japanese Honeysuckle	*Lonicera japonica*	Caprifoliaceae	East Asia
Pachysandra	*Pachysandra* spp.	Buxaceae	Japan, North America
Periwinkle	*Vinca minor*	Apocynaceae	Old World
Winter creeper	*Euonymus fortunei*	Celastraceae	China

FIGURE 18-25
Turf. (Courtesy of the Lawn Institute.)

FIGURE 18-26
The geometric keystone pattern of this parterre is a modern version of an ancient Egyptian design.

Bedding Plants

Bedding plants are relatively small, flowering, usually annual, plants grown in gardens, window boxes, or hanging baskets to provide displays of color. They are also used for carpet bedding, a special form of planting that produces designs (Fig. 18-26). Despite their small size, they constitute an important item in the nursery trade, in part, because people tend to buy large quantities of them each year. As in the case of ornamental plants, relatively few species dominate the trade. One annual, *Petunia × hybrida* (Solanaceae), accounts for 50 percent of the plants sold as bedding plants (Fig. 18-27). Table 18-3 lists this and the other common species used.

Trees and Shrubs

Trees and shrubs constitute the important structural features of ornamental plantings. While both trees or shrubs can be pruned to desired shapes, trees are usually allowed to assume their normal growth form and shrubs are often trimmed. Trees and shrubs can be chosen for their attractive flowers, fruits, or foliage, but trees are primarily used to provide shade. Because of the slow growth of trees and the fact that they are less important as flowering plants

FIGURE 18-27
Despite the ever-increasing number and variety of plants available, petunias remain the most popular bedding plant sold.

TABLE 18-3

Some Common Bedding Plants Used in the United States

COMMON NAME	SCIENTIFIC NAME	FAMILY	NATIVE TO
ANNUALS			
Impatients	*Impatiens balsamina*	Balsaminaceae	Asia
Marigold	*Tagetes erecta*	Asteraceae	Central America
Pansy	*Viola × wittrockiana*	Violaceae	Europe
Petunia	*Petunia × hybrida*	Solanaceae	Europe
Phlox	*Phlox drummondii*	Polemoniaceae	North America
PERENNIALS			
Ageratum	*Ageratum* spp.	Asteraceae	New World tropics
Candytuft	*Iberis* spp.	Brassicaceae	Europe
Coleus	*Coleus × hybridus*	Lamiaceae	Old World tropics
Columbine	*Aquilegia* spp.	Ranunculaceae	North temperate region
Geranium	*Geranium* spp. or *Pelargonium* spp.	Geraniaceae	Cosmopolitan
Heliotrope	*Heliotropium* spp.	Boraginaceae	Tropics
Larkspur	*Delphinium* spp.	Ranunculaceae	North temperate
Lobelia	*Lobelia cardinalis*	Lobeliaceae	North America
Primrose	*Primula vulgaris*	Primulaceae	Europe
Sweet alyssum	*Alyssum* spp.	Brassicaceae	Eurasia
BULBS			
Crocus	*Crocus* spp.	Iridaceae	Mediterranean
Daffodil	*Narcissus* spp.	Amaryllidaceae	Europe, North Africa
Hyacinth	*Hyacinthus orientalis*	Liliaceae	Mediterranean
Iris	*Iris* spp.	Iridaceae	North temperate region
Tulip	*Tulipa* spp.	Liliaceae	Old World

than shrubs, native species of trees are more widely used (or allowed to remain) than native species of shrubs. Ornamental shrubs are usually exotics chosen from the variety of hardy types offered by local nurserymen. Consequently, as in the case of cut flowers, bedding plants, and food crops, the same ornamental species are used over and over again throughout the United States. The exact group of shrubs differs somewhat in different hardiness zones (Fig. 18-28) across the country, but compared to the number of shrub species available, only a handful are widely planted. Table 18-4 lists some of the ornamental shrubs commonly used in the United States.

In contrast to ornamental shrubs, the species of trees planted or retained as parts of public or private landscape designs are so numerous that it is impractical to assemble a short list. Several of the references listed at the end of the chapter provide descriptions of common tree species used in the various climatic regions of this country.

FLORIST CROPS

While production of ornamental plants on a commercial basis is a product of urbanization, the cutting of flowers and foliage for personal and ceremonial use dates to prehistoric times. Excavations of Paleolithic burial sites have shown

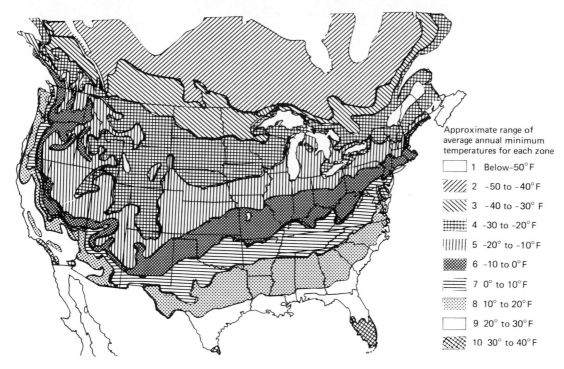

FIGURE 18-28
Plant hardiness zones (numbered 1-10) as circumscribed by the U.S. Department of Agriculture help gardeners and farmers determine which species and varieties to plant.

TABLE 18-4

Some Ornamental Shrubs Commonly Cultivated in the United States

COMMON NAME	SCIENTIFIC NAME	FAMILY	NATIVE TO
Azalea	*Rhododendron* spp.	Ericaceae	Primarily Asia
Barberry	*Berberis verruculosa*	Berberidaceae	Old World
Bayberry	*Myrica pensylvanica*	Myricaceae	Eastern North America
Boxwood	*Buxus sempervirens*	Buxaceae	Old World
Butterfly bush	*Buddleia davidii*	Loganiaceae	China
Deutzia	*Deutzia scabra*	Saxifragaceae	Japan
Fire thorn	*Pyracantha coccinea*	Rosaceae	Eurasia
Forsythia	*Forsythia × intermedia*	Oleaceae	Old World
Holly	*Ilex* spp.	Aquifoliaceae	Old and New World
Honeysuckle	*Lonicera* spp.	Caprifoliaceae	Eastern Asia
Hydrangea	*Hydrangea* spp.	Saxifragaceae	North and South America
Lilac	*Syringa* spp.	Oleaceae	Eurasia
Mock orange	*Philadelphus × virginalis*	Saxifragaceae	North temperate region
Ocotillo	*Fouquieria splendens*	Fouquieriaceae	New World
Oleander	*Nerium oleander*	Apocynaceae	Mediterranean
Oregon grape	*Mahonia aquifolium*	Berberidaceae	North America
Pieris	*Pieris japonica*	Ericaceae	Japan
Prickly pear	*Opuntia* spp.	Cactaceae	New World
Privet	*Ligustrum* spp.	Oleaceae	Old and New World
Rhododendron	*Rhododendron* spp.	Ericaceae	Primarily Asia
Rose of Sharon	*Hibiscus syriacus*	Malvaceae	Eastern Asia
Viburnum	*Viburnum* spp.	Caprifoliaceae	Usually North American species

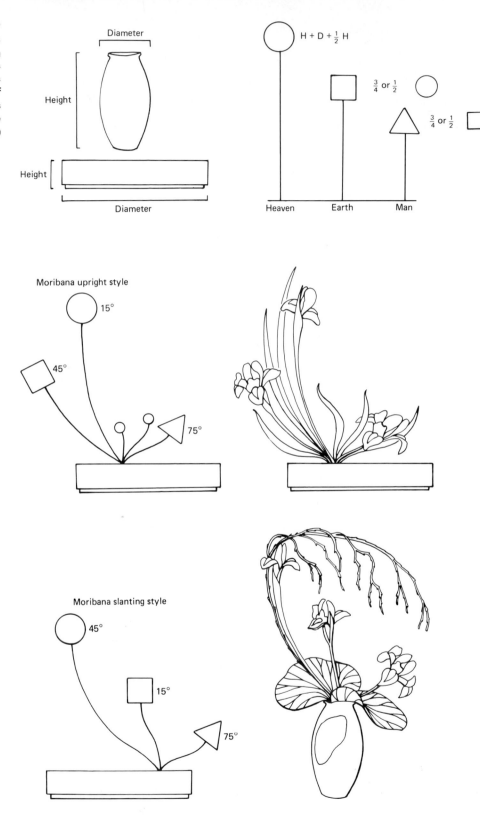

FIGURE 18-29
In Japanese flower arranging, the three main elements symbolizing heaven, earth, and man are arranged at precise angles to one another and are cut to lengths of specified proportions. Shown are two variations of the *Moribana* style. (H = height, D = diameter)

that sprigs of flowers, apparently important in the interment rites, were placed around the bodies. Since these very early times, differing cultures have adopted diverse species as favorite ornamental plants and developed personal styles in the use of cut flowers. In Japan, flower arranging developed into an art form that is an integral part of many aspects of the culture such as the tea ceremony (Fig. 18-29). In Victorian England, flowers were given symbolic connotations and were used to send messages (Table 18-5).

Traditionally, cut flowers were available only in the summer, or at very high cost from local greenhouses at other times of the year. Today, because of the efficiency of modern transportation, we are no longer limited to flowers produced locally. Cut flowers of temperate species grown in California and Florida are available all year at reasonable costs. Exotic species from Hawaii, the West Indies, or even Africa frequently appear in local shops (Fig. 18-30). Yet, despite the numerous possibilities now open to the cut flower industry, it remains dominated by six groups of species: chrysanthemums, roses, carnations, orchids, snapdragons, and gladiolus (Fig. 18-31).

Chrysanthemums (*Chrysanthemum* spp., Asteraceae) have been prized in the Orient as ornamentals and symbols of long life since at least 500 B.C. Over the period of their use, chrysanthemums have changed greatly from the original forms which had few ray flowers and a central button of numerous disk flowers. Most of the modifications have involved the selection of forms with an increased number of ray flowers (usually sterile) and fewer, or no, disk flowers (Fig. 18-32). The petaloid part of the ray flowers has also been greatly modified under the action of selection yielding exotic, showy heads. Within the modern horticultural cultivars, the major groups are singles (which retain the button, daisy-look), semidoubles, anemones, pompoms, commercial decoratives, spiders, and 'Shasta' daisies. The genus itself, with about 200 species, is native to the Old World.

To many people, roses constitute the pinnacle in cut flowers. Roses continue to be the ultimate extravagance and emphatic declaration of admiration that a gentleman can tender his beloved. The rose genus *Rosa* (Rosaceae) contains about 125 species occurring naturally throughout the north temperate regions. Different species and selected rose cultivars have been prized for centuries and, in various contexts, have been linked with human history for over 5,000 years. Few flowers other than the rose can boast of entire gardens devoted to their culture. Four states in the United States (North Dakota, Georgia, Iowa, and New York) have chosen a rose as their state flower. The American Rose Society is the largest specialty plant society in the United States. It publishes several journals and sponsors flower shows around the country. Each year, it awards the coveted designation of American Beauty Roses' 'Rose of the Year' to the best roses exhibited. The variety of flower colors and sizes, their delicate fragrances, and the ability of roses to last for several days once cut, have all contributed to their popularity. The wild ancestors of cultivated roses had only a few (or one) rows of petals, numerous stamens, and many separate pistils (Fig. 18-33). During the process of selection, mutants in which the stamens had become petaloid were chosen and perpetuated until, in many cases, the blossoms consisted of an almost infinite series of petals and no stamens.

Modern classes of roses include those with clusters of blossoms such as polyanthas and floribundas, the classic tea roses (forms of the Chinese *R.*

TABLE 18-5

Some Common Flowers and the Sentiments They Represent

COMMON NAME	SENTIMENT	SCIENTIFIC NAME	FAMILY
Almond	Indiscretion	*Prunus amygdalus*	Rosaceae
Aloe	Sorrow, bitterness	*Aloe barbadensis*	Liliaceae
Apple	Temptation	*Malus pumila*	Rosaceae
Azalea	Temperance	*Rhododendron* spp.	Ericaceae
Bachelor's button	Hope in love	*Centaurea cyanus*	Asteraceae
Basil	Hatred	*Ocimum basilicum*	Lamiaceae
Belladonna	Imagination	*Atropa belladonna*	Solanaceae
Blue violet	Modesty	*Viola odorata*	Violaceae
Cabbage	Profit	*Brassica oleracea*	Brassicaceae
Carnation	Pride and beauty	*Dianthus caryophylus*	Caryophyllaceae
Cashew	Perfume	*Anacardium occidentale*	Anacardiaceae
Chrysanthemum	Cheerfulness	*Chrysanthemum* spp.	Asteraceae
Cowslip	Pensiveness	*Caltha palustris*	Ranunculaceae
Daffodil	Deceitful hope	*Narcissus* spp.	Amaryllidaceae
Dandelion	Oracle	*Taraxacum officinale*	Asteraceae
Evening primrose	Inconstancy	*Oenothera biennis*	Onagraceae
Flax	Domestic industry	*Linum usitatissimum*	Linaceae
Hellebore	Female inconstancy	*Veratrum viride*	Liliaceae
Hollyhock	Fruitfulness	*Althea rosea*	Malvaceae
Honeysuckle	Bonds of love	*Lonicera japonica*	Caprifoliaceae
Hop	Injustice	*Humulus lupulus*	Cannabaceae
Hoya	Sculpture	*Hoya carnosa*	Asclepiadaceae
Hyacinth	Play	*Hyacinthus orientalis*	Liliaceae
Iris	Message	*Iris* spp.	Iridaceae
Jasmine	Availability	*Jasminum grandiflorum*	Oleaceae
Lavender	Distrust	*Lavandula angustifolia*	Lamiaceae
Lily of the valley	Return of happiness	*Convallaria majalis*	Liliaceae
Madder	Calumny	*Rubia tinctoria*	Rubiaceae
Marigold	Mental anguish	*Tagetes erecta*	Asteraceae
Mountain laurel	Ambition	*Kalmia latifolia*	Ericaceae
Nightshade	Falsehood	*Solanum dulcamara*	Solanaceae
Orange blossom	Purity equal to loveliness	*Citrus sinensis*	Rutaceae
Pansy	Think of me	*Viola* × *wittrockiana*	Violaceae
Periwinkle	Sweet remembrance	*Vinca minor*	Apocynaceae
Poppy	Consolation of sleep	*Papaver somniferum*	Papaveraceae
Rhododendron	Danger—beware	*Rhododendron* spp.	Ericaceae
Rose	Love	*Rosa odorata*	Rosaceae
Saffron	Mirth	*Crocus sativus*	Iridaceae
Snapdragon	Presumption	*Antirrhinum majus*	Scrophulariaceae
Sunflower	False riches	*Helianthus annuus*	Asteraceae
Sweet William	Gallantry	*Dianthus barbatus*	Caryophyllaceae
Tulip tree	Fame	*Liriodendron tulipifera*	Magnoliaceae

Source: Adapted from L. Gordon. 1977. *Green Magic.* New York, Viking Press. During the Victorian era, entire messages were written using flowers. The reader had to decipher the meaning by using the sentiments represented by the flowers pictured. Gordon gives many other flowers and their sentiments as well as examples of some messages in his book. We include here only flowers common in the United States, in particular, those mentioned in this text. The scientific names appear to be those that belong to the common names given by Gordon.

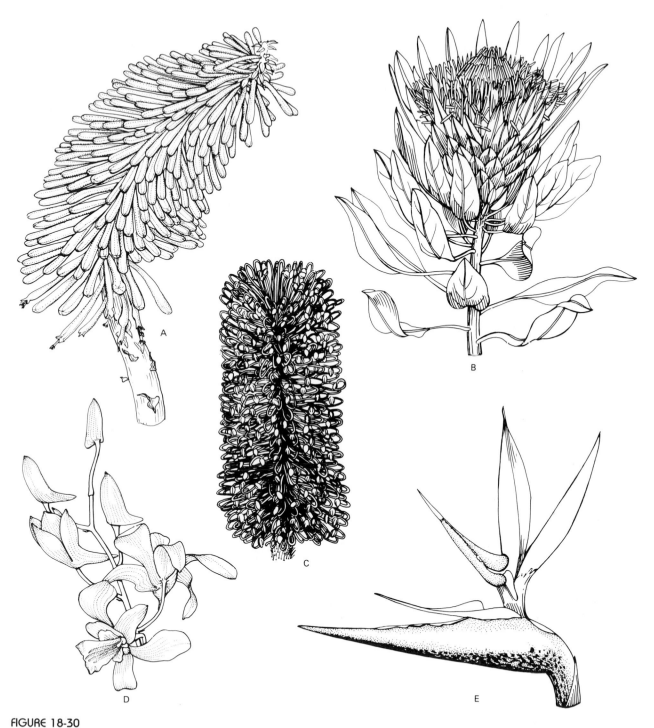

FIGURE 18-30
Some of the exotic blooms used in the cut flower industry include: (A) *Tritoma* (Liliaceae),
(B) *Protea cynaroides* (Proteaceae), (C) *Protea* spp. (Proteaceae), (D) *Dendrobium* spp.
(Orchidaceae), and (F) *Strelitzia regiia* (Musaceae). B and C after Baillon, A, D, and E
from nature.

FIGURE 18-31
Chrysanthemums (A), roses (B), carnations (C), orchids (D), snapdragons (E), and
gladiolus (F) are the most popular flowers in the floral industry today.

FIGURE 18-32
Chrysanthemums have been modified from the original form most similar to that on the center right to produce a whole array of showy flowers.

odorata), hybrid perpetuals, grandifloras, miniatures, tree (or standard), and climbing roses. While native American species have played some role in hybridizations, most of the cultivars are still based on Old World species.

Carnations, also called "pinks" because the edges of their petals look as if they had been cut by pinking shears, rival roses as the world's most loved

FIGURE 18-33
By selecting mutants with petaloid
stamens, the tea rose with its
many whorls of petals was
developed from the wild rose
similar to this one, which had one
whorl of petals and many separate
stamens and pistils.

flowers. They seem inextricably linked with suave gentlemen, tuxedos, and white sports coats. Included within the carnation genus *Dianthus* (Caryophyllaceae) are the grass-leaved, tall, clove-scented florist flowers and the humble Sweet William used as bedding plants (Fig. 18-34). Greenhouse cultivars are all derived from *D. caryophyllus,* a Eurasian species. Sweet William has been selected from *D. barbatus,* a native of western Europe. Because carnations produce flowers best under high light conditions, most of the U.S. supply is grown in sunny Denver, Colorado.

The orchid family, Orchidaceae, contains not only more species than any other family of angiosperms, but also many of the most bizarre and beautiful flowers known (Fig. 18-35). Because of their past costliness relative to other cut flowers, orchids came to be associated with very special occasions such as Easter Sunday, weddings, or the Senior Prom. Most florists' orchids are tropical species, but less showy ground orchids extend into Arctic areas. The most popular orchids sold in florist shops are hybrids of the genus *Cattleya* (Fig. 18-31), but in recent years, the popularity of the purple and white cattleyas has been challenged by smaller orchids such as species of *Cymbidium, Dendrobium, Odontoglossum,* and *Phalaenopsis* (Fig. 18-35).

In nature, orchid species are often reproductively isolated from one another because of specialized relationships with specific insect pollinators. It is this specificity that accounts for many of the elaborate floral forms exhibited by orchids (Fig. 18-36). In many cases, there is no genetic incompatibility separating species of orchids; they remain isolated because pollinators are relatively constant to one species. By making crosses, humans have succeeded in producing thousands of artificial hybrids between species and even genera. Superior hybrids are propagated vegetatively. As one can imagine, the naming of orchids has become quite complex.

Snapdragons (species of *Antirrhinum,* Scrophulariaceae) may at first seem like an odd flower to rank among the major florist crops, but they maintain a consistent position in the trade because of their extensive use in large arrangements (Fig. 18-37). There are species of snapdragons native to North America as well as Europe, but virtually all of the fine hybrids used in floral displays are derived from *A. majus,* a perennial species of the Mediterranean region.

Like so many of the major florist crops, gladiolus (*Gladiolus,* Iridaceae) are not native to North America. The genus occurs naturally in Africa and Europe

FIGURE 18-34
Sweet William is in the same genus as the spicy-scented florist's carnation.

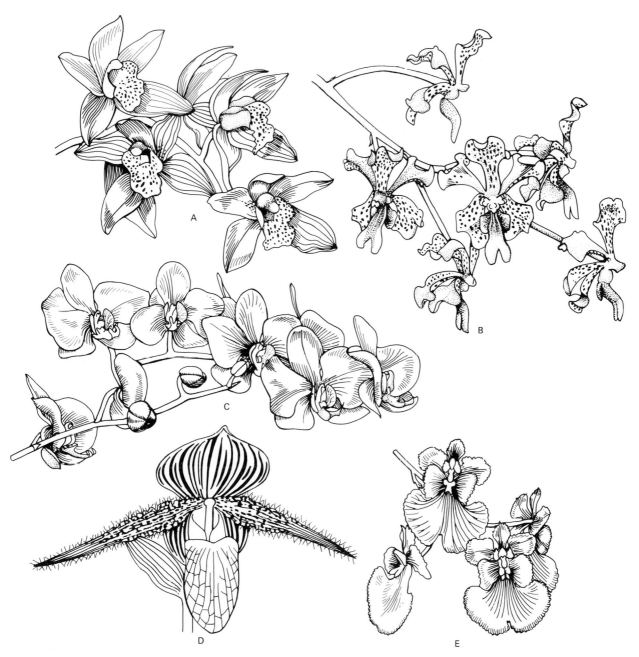

FIGURE 18-35
Orchids of many species are now found in florist shops. (A) *Cymbidium*, (B) *Vanda*, (C) *Phalaenopsis*, (D) *Paphiopedilum*, (E) *Oncidium*.

around the Mediterranean Sea. Hybridizations between species were first recorded from the early nineteenth century and have subsequently produced hundreds of varieties with every conceivable color (except true blue), variegation, petal ruffling, and size. Since gladiolus plants produce tall spikes of large flowers, they are generally used in large, rather formal arrangements.

FIGURE 18-36
The bizarre form of this *Coryanthes* orchid is a reflection of its unusual pollination biology. In this flower, the labellum has been modified into a lipped bucket which fills with secretions from glandular appendages above. Visiting male bees fall into the bucket while collecting volatile oils and can only escape by crawling through a hole at the lip past the anther and stigma thus depositing any pollen they may be carrying and picking up pollen to carry to another flower.

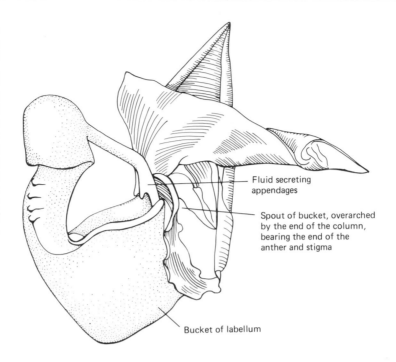

Fluid secreting appendages

Spout of bucket, overarched by the end of the column, bearing the end of the anther and stigma

Bucket of labellum

FIGURE 18-37
Snapdragon production. (Photo courtesy of the Ball Seed Company).

HOUSEPLANTS

Houseplants are often sold in florist shops, but they differ from traditional florist crops because they are growing. In addition they are chosen not for their flowers, but for their durability under the growing conditions imposed by human dwellings. Most of our foliage plants are native to tropical areas where they generally occur in understory habitats. Because they are adapted to growing in the shade of forest trees, they can tolerate the low light conditions found in

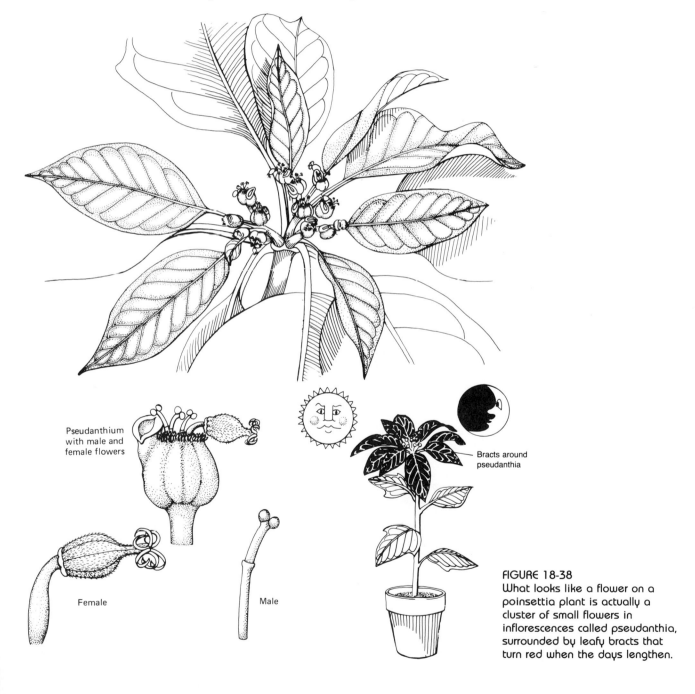

Pseudanthium with male and female flowers

Female

Male

Bracts around pseudanthia

FIGURE 18-38
What looks like a flower on a poinsettia plant is actually a cluster of small flowers in inflorescences called pseudanthia, surrounded by leafy bracts that turn red when the days lengthen.

TABLE 18-6

Principal Potted Plants Used in American Homes and Offices

COMMON NAME	SCIENTIFIC NAME	FAMILY	NATIVE TO
	FLOWERING PLANTS		
African violet	*Saintpaulia ionantha*	Gesneriaceae	Tanzania
Begonia	*Begonia* spp.	Begoniaceae	New World tropics
Coleus	*Coleus* spp.	Lamiaceae	Old World tropics
Crown of thorns	*Euphorbia milii*	Euphorbiaceae	Madagascar
Geranium	*Pelargonium* spp.	Geraniaceae	South Africa
Jade plant	*Crassula argentea,* *C. arborescens*	Crassulaceae	Africa
Kalanchoe	*Kalanchoe integra*	Crassulaceae	Africa
Lantana	*Lantana camara*	Verbenaceae	Tropical America
Wandering jew, spiderwort	*Tradescantia* spp. or *Zebrina pendula*	Commelinaceae	New World tropics
Wax plant	*Hoya carnosa*	Asclepiadaceae	East Africa
	FOLIAGE PLANTS		
Air plant	Bromeliads, e.g., *Cryptanthus* spp., *Quesnelia marmorata,* *Vriesea carinata*	Bromeliaceae	Brazil
Aralia	*Schefflera* spp.	Araliaceae	Old World tropics
Asparagus fern	*Asparagus densiflorus*	Liliaceae	South Africa
Corn plant	*Dracaena fragrans*	Agavaceae	Guinea
Dumb cane	*Dieffenbachia* hybrids	Araceae	Tropical America
Ivy	*Hedera helix*	Araliaceae	Africa
Jade plant	*Crassula argentea*	Crassulaceae	Africa
Philodendron	*Philodendron* spp.	Araceae	Tropical America
Rubber plant	*Ficus,* primarily *F. elastica*	Moraceae	Asia
Screw pine	*Pandanus veitchii*	Pandanaceae	Polynesia
Snake plant	*Sansevieria trifasciata*	Agavaceae	Asia, Africa
Split-leaved philodendron	*Monstera deliciosa*	Araceae	Central America
Spider plant	*Chlorophytum comosum*	Liliaceae	South Africa
Wandering Jew	*Tradescantia* spp.	Commelinaceae	Americas
Watermelon	*Peperomia argyreia*	Piperaceae	Tropical South America
Weeping fig	*Ficus benjamina*	Moraceae	Southeast Asia, Australia
	FORCED-FLOWERING PLANTS		
Azalea	*Rhododendron* spp.	Ericaceae	Primarily Azalea
Camellia	*Camellia japonica*	Camilliaceae	Japan, Korea
Christmas cactus	*Schlumbergera bridgesii*	Cactaceae	Brazil
Chrysanthemum	*Chrysanthemum frutescens* or *C.* × *morifolium*	Asteraceae	China, Canary Islands
Crocus	*Crocus* spp.	Iridaceae	Eurasia
Daffodil	*Narcissus* spp.	Amaryllidaceae	Eurasia
Easter lily	*Lilium longiflorum*	Liliaceae	Japan
Gardenia	*Gardenia jasminoides*	Rubiaceae	China
Hyacinth	*Hyacinthus orientalis*	Liliaceae	Eurasia, Africa
Poinsettia	*Euphorbia pulcherrima*	Euphorbiaceae	Mexico
Primrose	*Primula sinensis*	Primulaceae	China
Tulip	*Tulipa* spp.	Liliaceae	Eurasia

Note: The nomenclature used here follows that of *Hortus Third*. Liberty Hyde Bailey Hortorium. 1976. New York, Macmillan. Specialty plants such as orchids and the many kinds of cacti sold are not included.

most buildings. Such plants must also be able to withstand small soil volumes (due to the constraints of the pots), and the variable humidity caused by central heating and air conditioning. Since houseplants receive continuous muted light and are restricted in size by their containers, they rarely flower. There are exceptions such as African violets, but the majority of species of houseplants never flower inside a house, apartment, or office. Still, houseplants have become tremendously popular in the last 10 years. The fact that the wholesale value of foliage houseplants has increased by over 800 percent during this period reflects the importance they have in interior design where they soften both home and office environments. Table 18-6 lists a few of the most common houseplants and their places of origin.

In contrast to regular houseplants some potted plants are sold strictly for their flowers. These plants last only for a limited time because they are forced to bloom for a particular holiday, season, or market period. Examples of such potted plants are poinsettias, red tulips, and Easter lilies, all raised by nursery owners who time their bloomings to coincide with Christmas (Fig. 18-38), Valentine's Day, and Easter. While the flowers of these plants will last much longer than those of cut flowers, the plants themselves usually die when they finish blooming unless they are subsequently carefully tended.

FUTURE TRENDS

Several philosophical attitudes toward the use of ornamental landscape and garden plants have been emerging in the last few decades. While at first they appear to be diametrically opposed, they need not be. The recent increase in costs of fertilizer, pesticides, and even water has prompted man to reevaluate his use of high maintenance, exotic ornamental species. In response, man is experimenting with artificial plants, native species and even landscaping that you can eat.

The logical extreme of the tendency to choose low-maintenance or care-free plants is the employment of artificial plants such as astroturf and plastic plants. While such a trend is understandable in terms of resolving the conflict between the human desire for elements of nature and the practical constraints of space and maintenance, it somehow seems a contradiction to use a manu-factured object to help people forget their alienation from nature. In some cases, the use of artificial plants has actually created a new set of problems. Astroturf, for example, requires the use of special sports shoes and produces a much harder surface than grass, leading to comparatively more injuries. In addition, it becomes so hot that it is necessary to spray it with water to cool it. Grass naturally transpires (gives off water) and thus remains comparatively cool.

A second trend is toward the incorporation of more and more native plants into the repertoire of ornamental species. In many instances, the use of particular ornamental plants has been dictated mostly by custom and not by any intrinsic superiority of the species. Particularly if selections and hybrids are made with local, native species, numerous beautiful plants could be added to the list of bedding plants and shrubs that are now used. Since native trees, shrubs, and herbs are usually adapted to the local pests and climatic conditions of an area

FIGURE 18-39
One can apply design criteria to this edible landscape. The geometric form of the stone paths, the rows of onions bordering the paths, and upright corn stalks function as line in the design. Textural interest is provided by the contrast between the delicate foliage of the thyme, the fernlike parsley, the broad leaves of leeks and sorrel, and the coarse foliage of the artichokes. Color is contributed by red chard, okra, and beet leaves, the interplanting of marigolds, the edible flowers of squash, violets, and nasturtiums, and the varying shades of the ripening tomatoes, peppers, and eggplants. Grape vines growing on a trellis mask the carport and work area behind. Note that this garden is drawn from the plan described by Figure 18-23.

the use of fertilizers, pesticides and irrigation water is much less than that necessary to maintain introduced ornamentals. Encouraging the use of indige-nous species will not only increase the variety of plants from which to choose and conserve costs incurred by trying to grow exotic crops, but it will also help to provide a means of preserving many species which might otherwise become extinct.

A final direction in garden design is that of edible landscaping. While this may seem no different than an update of the earlier kitchen gardens, there is a distinction. In "foodscaping," the plants are chosen and arranged for their aesthetic value, with flavor and nutrition being ancillary qualities. Herbs and leafy vegetables and fruit bearing annuals, shrubs and trees are evaluated for their color, texture, line and form just as ornamentals would be (Fig. 18-39), but edible landscapers welcome the extension to the original definition of ornamentals as plants valued solely for their artistic merit. They emphasize that it's time to challenge the unwritten taboo against mixing flowers and vegetables and keeping food plants out of our yard areas as an unnecessary waste of resources. Rosalind Creasy dubs this taboo, our "edible complex" and argues that devoting resources that would otherwise be put into purely decorative landscapes into the production of food plants is one way to make our landscapes more environmentally sound.

ADDITIONAL READING

Bailey, L. H., and E. Z. Bailey. 1976. *Hortus Third. Revised and Expanded.* New York, Macmillan.

Bush-Brown, J., and L. Bush-Brown. 1980. *America's Garden Book.* Revised by the New York Botanical Garden. New York, Scribner's Sons.

Creasy, R. 1982. *The Complete Book of Edible Landscaping.* San Francisco, Sierra Club Books.

Foster R. S. 1978. *Homeowner's Guide to Landscaping that Saves Energy Dollars.* New York, David McKay.

Graf, A. B. 1978. *Exotic Plant Manual,* 5th ed. East Rutherford, New Jersey, Roehrs.

Halfacre, R. C., and J. Barden. 1979. *Horticulture.* New York, McGraw-Hill.

Janick, J. 1979. *Horticultural Science.* 3d ed. San Francisco, Freeman.

Johnson, H. 1973. *The International Book of Trees.* New York, Simon and Schuster.

Robinette, G. O. 1972. *Plants, People and Environmental Quality.* Washington, D.C., U.S. Department of the Interior.

Taylor, N. 1965. *The Guide to Garden Shrubs and Trees.* New York. Bonanaza Books.

Wyman, D. 1972. *Trees for American Gardens.* New York, Macmillan, Revised and enlarged.

Wyman, D. 1974. *Dwarf Shrubs: Maintenance-Free Woody Plants for Today's Gardens.* New York, Macmillan.

Wyman, D., and G. Koller (eds.). 1982. Handbook on Ground Covers and Vines. New York, Brooklyn Botanic Garden.

Chapter 19

The Uses of Plants in the Future

*I*n addition to the fact that plants are the primary producers of food for themselves and most of the other life on earth, they are commonly considered "renewable" resources. But because this renewability has limits we are forced to ask: Will our plant resources be able to support the world's burgeoning population in the future? Can we increase our food supplies in face of the dwindling reserves of fossil fuels? Will there be sufficient plant diversity left in the future world to allow us to choose new crops or improve the existing ones? By the year 2000, the population of the world is projected to increase more than 50 percent above the 4 billion estimated for the year 1975. Concomitant with this growth to 6.65 billion will come increased, and sometimes conflicting, needs for more space, more food, and more fuel (Fig. 19–1).

Scientists and national agencies are aware of the critical problem of feeding the people of the world and the potential dangers of allowing the diversity of life developed over millions of years to slip away forever. Awareness of these problems has engendered international cooperation. In many countries there are active research programs directed toward the investigation of previously unused or underused plants of potential use to humans. Other programs are aimed at making better use of plants that are already major crops by testing new methods of agriculture that will use less energy but still produce yields equivalent to, or greater than, those of modern highly mechanized farms. There is now an international cooperative effort to gather, catalog, and protect as wide a base of genetic variability as possible for the world's major crop plants. In this chapter we look at some of the future problems we face in terms of plant resources and examine some of the possible solutions.

FIGURE 19-1
Speculation about the future is a favorite human pastime.

OUR LIMITED FOOD SUPPLY AND THE PROBLEM OF INCREASING YIELDS

FIGURE 19-2
This child is a victim of the prolonged drought that brought famine to East Africa in the 1970s. He shows the stunted growth, swollen belly, and edema characteristic of kwashiorkor, a disease caused by protein deficiency (United Nations photo by Arild Vollan, issued by FAO.)

In the United States it is hard to realize that over half of the world's population goes to bed hungry each day (Fig. 19-2). While not all of our people are well fed, the average American consumes 5 lb of grain per day. Only about one-fourth of this is consumed directly; almost 80 percent is fed to animals. In underdeveloped countries, an average individual consumes a pound of grain per day, usually directly in the form of the whole or milled cereal.

This imbalance suggests that we could substantially increase our food supply by eating "lower" on the food chain. That is, by eating more plant products directly and eliminating the caloric waste encumbered when plant material is converted into meat (Fig. 19-3). Such a shift in dietary practices might increase the food supply in countries that now have the luxury of eating large quantities of meat, but it would do little to alleviate the hunger in countries where sustenance is already almost entirely derived from plant carbohydrates.

Problems of adequate nutrition in underdeveloped countries are exacerbated by the lack of effective birth control methods, and compounded by the decreased productivity of the land as overuse robs it of nutrients and even soil. The irony in the economies of underdeveloped countries is that they must maintain, or even increase, exports in order to support a trade balance vital to development. Yet this export is gained at the cost of putting land that could be used for the production of foodstuffs needed within the country into exportable cash crops.

Solutions must be sought that permit underdeveloped countries to raise their standards of living. An increase in the living standard has historically been accompanied by a lowering of the birth rate. While the first step toward improving conditions must be the provision of adequate nutrition, it does not now appear that the adoption of high-technology American farming methods (Fig. 19-3) is the answer to increasing food production in most of these countries because they lack, and cannot afford to buy, the machinery or fuel needed for this kind of agriculture. Instead, solutions involving improved or new crops and/or modified methods of agriculture must be found.

SELECTING FOR BETTER CROPS

Perhaps the oldest approach to increasing food supplies has been to increase yields of established cultigens by artificial selection. Crop improvements started with rudimentary agriculture. As soon as humans began preferentially to plant certain seeds, they were effectively exerting a form of selection. Transformation of crop plants by continually taking seeds from only a specific group of plants would have led to the morphological changes that we perceive as evidences of domestication in fossil remains (see Chapter 3). This type of selection is essentially all that was practiced for thousands of years, and it led to many local or land races (cultivars) as well as crops that were unlike any wild forms.

Modern methods of plant breeding were not employed until the end of the nineteenth century when population pressure began to create an urgent need

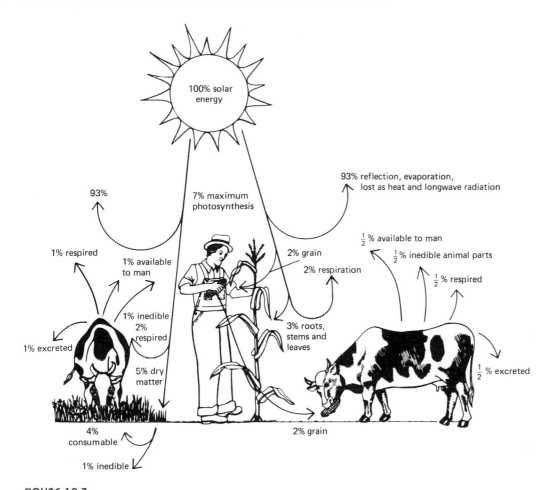

FIGURE 19-3

(A) A diagrammatic representation of energy flow shows why eating lower on the food chain is energy efficient. Of the solar energy that reaches the earth, 93 percent is reflected back toward space or converted to heat. Under ideal conditions, the remaining 7 percent is used by plants for photosynthesis. Two percent of this is respired by the plant, leaving about 5 percent of the original solar input being finally converted into plant biomass. Four-fifths of this biomass in pasture grass is available to grazing animals. In cereals two-thirds of the energy converted into biomass are in their grains. If the grain is fed to livestock, the percentage recovery by humans of the original 7 percent of the sun's energy is greatly reduced. (Modified by permission from R. Loomis. 1976. Agricultural systems. *Scientific American* 235:98–105 all rights reserved.)

for higher yields. By the end of that century, many scientists, Darwin among them, had put forth theories that provided explanations for the changes that occurred during selection. These ideas were based on the concept of differential survival. Darwin's ideas drew attention to the fact that individuals with desirable characters (from the human point of view) tended to produce offspring that also exhibited those characters. Consequently, instead of simply taking seed from a particularly desirable plant, two desirable plants could be crossed to produce offspring with a combination of good characters. By choosing not only

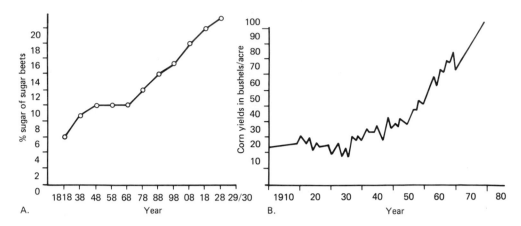

A.   B.

FIGURE 19-4
Increases in yields of different crops brought about by artificial selection. (A) Changes in sugar concentration in the sugar beet over time caused by selection. (B) Increases in corn yields over time due to human selection. (From D. Pimentel et al. 1973. Food production and the energy crisis. *Science* 182:443–449.)

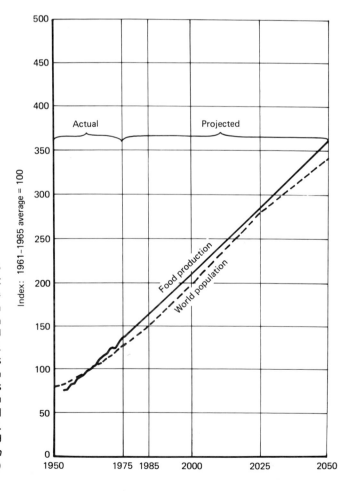

FIGURE 19-5
While this diagram indicates that increases in food production have kept pace with population increases, there is not an equitable distribution of food throughout the world. Consequently, there is mass starvation in some countries and a surplus of food in countries such as our own. Projections are that by the year 2050, this imbalance will be exaggerated. (Adapted from S. Wortman. 1976. Food and agriculture. *Scientific American* 235:31–39.)

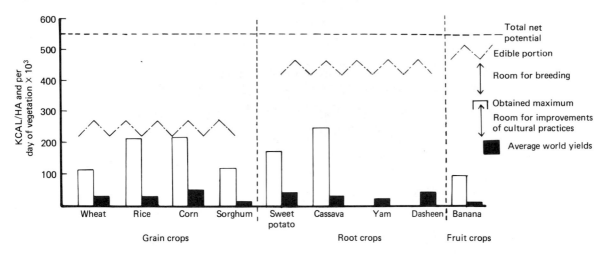

FIGURE 19-7
The columns in black indicate the average yields of crops, and the outlined columns, the maximum yields realized as of 1975. The difference between the two indicates the potential increases in world yields with improved cultivation techniques. Above the black columns is an estimate of the potential increases in yields that could be brought about in the future by selective breeding. (Based on T. T. Poleman. 1975. World food: A perspective. *Science* 188:510–518.)

THE NEED FOR GENETIC VARIABILITY

One problem with believing that breeding will continuously create ever superior crops is the assumption that enough variation exists within the breeding complex of crop species to allow selection for desired traits. While this is theoretically possible, an examination of the history of agriculture shows a continued decline in the number of crop species used, and in the amount of genetic variability present within them. It has been estimated that humans have used over 3,000 species of plants for food but that only 150 of these eventually came to be commercially exploited. Even within this limited number, about 12 dominate the world's agriculture, and the combined production of the top four (wheat, rice, corn, potatoes) exceeds that of all of the others listed in Fig. 19-8 added together.

The reduction of the number of crops on which people depend, and the loss of genetic variation within them, is a vicious cycle. As new, superior strains of crops have been developed, they have been distributed to farmers who naturally used them to replace their lower yielding or more disease prone varieties. Even in areas where a crop was first domesticated, local varieties have often been replaced by hybrids usually developed on another continent. The result has been the worldwide propagation of a very few selected genotypes. Wild forms are often actively eradicated to prevent crossing with the developed cultivars. Tragically, local varieties, once completely destroyed, can never be regained.

Almost all our hybrid corns, for example, are essentially derived from selections made by the Reid family in the 1840s and 1850s in the central United

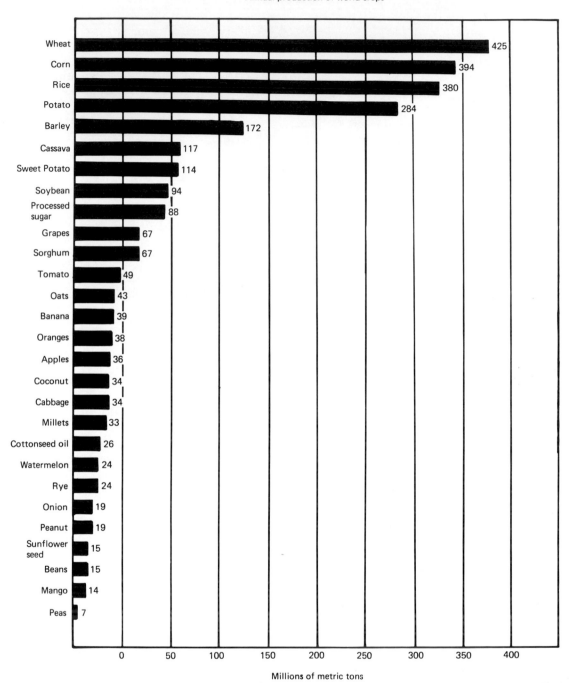

Annual production of world crops

Millions of metric tons

FIGURE 19-8

Annual production of the major world crops in terms of acreage devoted to cultivation. (Data from the *FAO Production Yearbook*, vol. 33. figures are for 1983.)

States. Almost all hard red wheat comes from a hybrid cross made in Canada around the turn of the century. Pima cotton was selected from a single plant of a cultivar developed in Egypt in 1910. Other crops have similar histories.

The unforeseen consequences of *monoculture*, the widespread planting of a single, often uniform cultivar, were dramatically demonstrated in the 1970s when much of the U.S. corn crop was destroyed. The excellent hybrid corn variety that was developed and planted by almost all of the American farmers had been produced using male sterile plants. These functionally unisexual plants eliminated the need for hand detasseling to ensure cross-pollination. Male sterility was conferred by a single gene which also had another effect, it made plants very susceptible to the southern corn leaf blight (caused by a fungus, *Helminthosporium maydis*). It was known before the seeds of the male sterile-derived cultivar were released that the gene reduced blight resistance, but agronomists believed that the fungus, a native of warm regions, would not be a pest in temperate areas. This assumption proved to be quite wrong. In 1970, the blight began to spread rapidly and devastatingly across North America. Since about 80 percent of the U.S. corn crop that year was planted with male sterile hybrids, the damage was great. Some southern states lost 50 percent of their corn crop, and the total U.S. loss was over 15 percent.

BANKING ON GENES

As hints of disasters like the loss of the corn crop appeared, agriculturalists began to appreciate the uniformity of our cultivated crops and how desperately we needed to preserve variability for breeding. Consequently, in the mid 1960s, scientists started to exert pressure for a worldwide cooperative effort to retain gene pools of agricultural crops for breeding purposes. The Food and Agricultural Organization (FAO) of the United Nations listened and began to take action. Initial meetings in 1961 led eventually to the establishment of an International Board for Plant Genetic Resources in 1971. Crops were ranked in terms of their importance to humans. For those with top priority, institutes were established or, if already present, were asked to take on increased responsibility. These institutes, shown in Table 19-1, were instructed to collect and store as much of the variation as possible left in the agricultural crops with which they were working.

Procedures for collecting and storing genetic variation vary somewhat from crop to crop, but all essentially involve sending researchers to areas in which crops were first domesticated and/or areas in which wild or local races still persist. The workers collect seeds from individuals or populations and carefully label each batch with data about the time and place of collection, the ecology of the area, and the characteristics of the plants from which the seeds were taken. Each batch must contain a large number of seeds, because once sent to the institute, a sample is tested for germination ability. If too few germinate, the batch is discarded because storage would be too expensive. If the seeds are good, they are placed in a seed bank, a storage area maintained at temperatures between 5 and $-16°C$ to keep seeds dormant but alive (Fig. 19-9). Since seeds will not remain viable (able to germinate) forever, samples are periodically

TABLE 19-1

Research and Germplasm Centers for the World's Major Crops

CENTER	LOCATION	CROPS AND PROCESSES STUDIED
CIAT (International Center for Tropical Agriculture)	Palmira, Colombia	Cassava, common beans, maize, and rice
CIMMYT (International Center for the Improvement of Maize and Wheat)	El Batan, Mexico	Wheat, triticale, barley, and maize
CIP (International Potato Center)	Lima, Peru	Potatoes
IBPGR (International Board for Plant Genetic Resources)	FAO, Rome, Italy	Conservation of plant genetic material, especially cereals
ICARDA (International Center for Agricultural Research in Dry Areas)	Lebanon	Study of mixed farming systems, especially sheep, barley, wheat, and lentils
ICRISAT (International Crops Research Institute for the Semi-arid Tropics)	Hyderabad, India	Sorghum, pearl millet, pigeon peas, chick-peas, and peanuts
IITA (International Institute of Tropical Agriculture)	Ibadan, Nigeria	Farming systems; relay station to IRRI and CIMMYT for rice and maize, cowpeas, soybeans, lima beans, pigeon peas, cassava, sweet potatoes, and yams
IRRI (International Rice Research Institute)	Los Banos, Philippines	Irrigated rice, upland rice; multiple cropping systems
WARDA (West African Development Association)	Monrovia, Liberia	Regional cooperative effort in rice research

Source: Adapted from N. Wade. 1975. International agricultural research. *Science* 188:585–589.

FIGURE 19-9
A geneticist at the USDA germplasm collection center at Urbana, Illinois, holds one of the 3,000 samples of soybean seeds stored at the facility. (Courtesy of the USDA.)

removed and tested for germinability. At intervals determined by these tests, entire lots are germinated, grown to maturity, and seeds recollected for storage. Care is taken to prevent crossing between the strains being stored. In some cases, strains can also be stored by preserving pollen or tissue cultures, small amorphous clumps of cells that will regenerate a new plant if given the proper hormones and environmental conditions (Fig. 19-10). Because of their slow growth, tree germplasm conservation presents a problem. If variation were stored as seed, it would take years to grow plants that could provide pollen or ovules for breeding. Consequently, variation present in trees in often maintained by planting "orchards" of the different strains.

Obviously the maintenance of these centers, generally referred to as *gene banks* or *germplasm centers*, requires enormous amounts of work, but it appears to be necessary if we are to have future variation available for breeding purposes.

BEYOND TRADITIONAL METHODS

Even if we were to preserve all of the variation currently present in our important crops, there would be limits to the changes that we could bring about by conventional selection practices. Selection acts by allowing differential reproduction of individuals with certain traits. Traits "chosen" via selection have to be among those present in the genotypes of individuals in a population. In some cases, a species simply does not have the traits we would like it to have. In other cases, the traits might have once been present but have inadvertently been lost. Some crops do not lend themselves to breeding programs. Crops such as bananas and potatoes are almost always vegetatively propagated since they have extremely high ovule or seed abortion rates. Finally, crosses can generally be made only within species or between closely related species. A gene that produces a particularly effective insect repellent in a milkweed cannot be transferred to corn with traditional methods. Consequently, new approaches are being sought that will allow the rapid detection and proliferation of useful mutant types, the transferral of particular genes from one species to a totally unrelated species, or the direct alteration of the genetic structure of individuals.

One technique that permits the study of potentially useful somatic mutations, and the use of the mutant cells to produce entire plants with the desired characters of these cells, is that of *protoplast manipulation* (Fig. 19-11). In this procedure, a piece of tissue is removed from a plant and the cells separated from one another. The cell walls are then dissolved, leaving simple protoplasts (membrane-bound cells). These protoplasts can then be assayed like bacterial cells for traits such as disease resistance. Once a resistant type has been identified, the cells are put in a solution that causes them to form amorphous clumps of callus tissue. Properly treated, pieces of the callus can give rise to entire plants with every cell genetically and cytoplasmically identical to the original protoplasm.

Attempts are also now being made to fuse the protoplasts of unrelated species to make new cells with novel combinations of characters. Protoplast culture and protoplasm fusion have shown promise for the rapid generation of disease-resistant potatoes.

FIGURE 19-10
This test tube baby will grow up to become a mimosa tree. The plant was grown from a single cell using tissue culture techniques. (Photo by E. N. Crizaldo. Courtesy of the USDA.)

FIGURE 19-11
In the cloning of potato plants: (A) small, newly produced leaves are removed and placed in an enzymatic solution which causes the cells to disassociate (B). The single cells (C) are transferred to another solution that separates the protoplasm from the cell wall (D). The protoplasts, which become spherical once freed from the cell wall can be screened for particular characters. Desirable cell types can then be placed in a nutrient medium where they begin to divide (E). After a series of divisions, an amorphous clump of cells called a callus (F) is formed. If placed in the proper conditions, the tissue will produce new shoots (G) and, ultimately, a new plant (H and I) genetically identical to the plant from which the leaf was taken. (Adapted by permission from J. F. Shepard. 1982. The regeneration of potato plants from leaf-cell protoplasts. *Scientific American* 246:154–166. All rights reserved.)

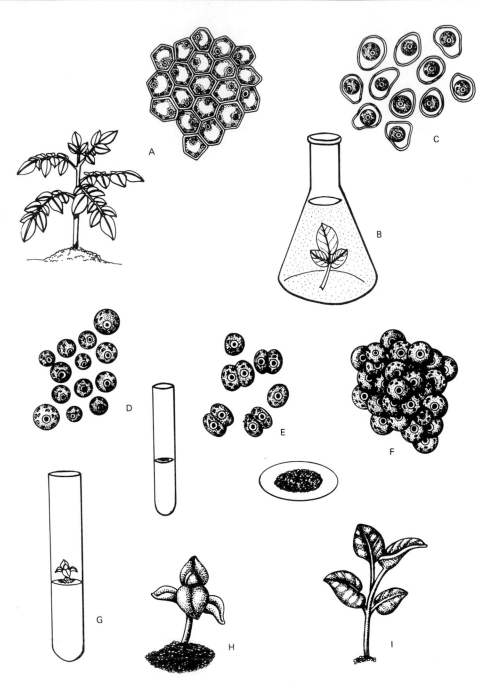

Other possibilities involve various aspects of genetic engineering, an enterprise in which humans directly alter the genetic makeup of individuals. One kind of genetic engineering employs plasmids, tiny circular pieces of bacterial DNA, as carriers of genes from one species to another. The process is quite involved (Fig. 19-12), but basically it involves cleaving the DNA of a donor chromosome on both sides of the desired gene and attaching it to the

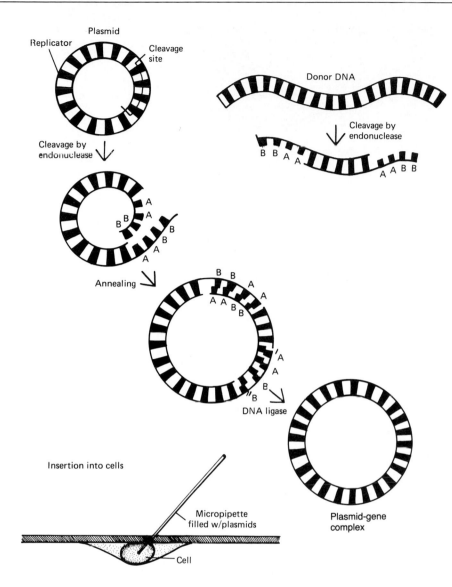

FIGURE 19-12
One of the techniques involved in genetic engineering, plasmid transfer, is illustrated in this diagram. A plasmid, a small circular piece of DNA from a bacterial cell, is placed with a donor chromosome in an enzymatic solution. The enzyme is chosen because it cleaves the chromosome on two sides of a desired gene and cleaves the plasmid at a precise and complementary location. The chromosome fragment then fuses with the unattached ends of the plasmid, producing a plasmid-gene complex. The complex is reinserted into a bacterial cell where it undergoes replication with each cell division. Since bacteria divide very rapidly, a large supply of the desired gene is soon produced. In some cases, the end product of the gene, for example a particular enzyme, is desired. If the gene is able to function in the bacterial cells, the transfer technique provides a method for rapid production of quantities of the enzyme. (Modified by permission from S. Cohen. 1975. The manipulation of genes. *Scientific American* and W. F. Anderson and E. Diacumakos. 1981. Genetic engineering in mammalian cells. *Scientific American* 245:106–122. All rights reserved.)

plasmid DNA, which has also been split open at a specific place. The plasmid-gene complex can then be introduced into the cells of another species. The hope is that the introduced DNA segment will continue to code for the desired products (Fig. 19-13). The technology of plasmid transfer is well advanced, and genes have been successfully transferred into higher plants. One such gene transferred into corn confers resistance to herbicides. This particular trait appears to be under relatively simple genetic control.

Major problems with using plasmid transfer to alter characters arise when dealing with traits that are controlled by more than one gene. The transfer of several genes, most of which may occur on different chromosomes, appears to be impossibly difficult. Moreover, as the number of genes per trait increases, the probability that all will function successfully in a foreign background decreases.

FIGURE 19-13
The kinds of experiments shown
here are designed to help
understand how soybeans
synthesize protein and fats. By
using such data, scientists may be
able to improve the nutritional
quality of other crop plants.
(Courtesy of the USDA.)

FIGURE 19-13
The kinds of experiments shown here are designed to help understand how soybeans synthesize protein and fats. By using such data, scientists may be able to improve the nutritional quality of other crop plants. (Courtesy of the USDA.)

EXPANDING TRADITIONAL AGRICULTURE

Efforts to increase food production have not been limited to manipulating major crops in highly advanced, temperate countries. Both wet, tropical lowland and arid areas are potential regions into which agriculture could expand (see Fig. 19-18). Research is being directed toward utilizing crops better adapted to these areas than the traditional crops developed in temperate latitudes. Likewise, new farming methods are being sought to help overcome the problems of poor soils, high leaching, high predation levels, and/or aridity and salt buildup that have thwarted the use of these regions in the past.

Over the last four decades, scientists have attempted to employ traditional methods of breeding and selection to adapt temperate crops to tropical environments. In the cases of wheat and rice, many of the original problems of growing these crops in humid tropical areas have been overcome, but it is still potentially more profitable to develop crops that are native to such habitats.

TABLE 19-2

Plants Discussed in Chapter 19

COMMON NAME	SCIENTIFIC NAME	FAMILY	CHROMOSOME NUMBER
Amaranth	*Amaranthus* spp.	Amaranthaceae	$2n = 32, 34$ diploid
Buffalo gourd	*Cucurbita foetidissima*	Cucurbitaceae	$2n = 40, 42$
Eelgrass	*Zostera marina*	Potamogetonaceae	$2n = 12$
Gopher weed	*Euphorbia lathyrus*	Euphorbiaceae	$2n = 20$
Opuntia	*Opuntia* spp.	Cactaceae	Variable
Prickly pear	*Opuntia* spp.	Cactaceae	Variable
Quinoa	*Chenopodium quinoa*	Chenopodiaceae	$2n = 36$ tetraploid
St. John's wort, klamath weed	*Hypericum perforatum*	Hypericaceae	$2n = 32, 36$
Strawberry	*Fragaria xananassa*	Rosaceae	$2n = 56$ octoploid
Winged bean	*Psophocarpus tetragonolobus*	Fabaceae	$2n = 18, 22, 26$
Zostera	*Zostera marina*	Potamogetonaceae	$2n = 12$

Some tropical crops such as sweet potatoes, cassava, and even white potatoes (if tropical montane areas are included) are already major crops. There are, however, many others that have been neglected.

Expanding Agriculture in Tropical Regions

In a booklet entitled "Underexploited Tropical Plants with Promising Economic Value," the National Academy of Sciences recently described numerous species that could potentially expand the tropical agricultural base. A few of the species mentioned in this report such as quinoa (*Chenopodium quinoa,* Chenopodiaceae) have already been the objects of intense agricultural research (Table 19-2). Quinoa produces fruits that strikingly resemble grains and even taste rather like rice. It is native to the South American Andes and was an important food item of the Inca Empire. In contrast to most edible crops, quinoa can be grown at extremely high elevations, over 4,000 m (13,120 ft). The seeds are rich in protein and have a balanced amino acid profile.

FIGURE 19-14
The winged bean is a tropical legume discovered in rural gardens in Papua New Guinea and southeast Asia. Some scientists believe that the bean could become the tropical equivalent of the soybean, with which it compares favorably in terms of protein and oil content. When young, the pods can be eaten as a green vegetable and the leaves can be used as potherbs. The tendrils have been compared to asparagus, and the steamed flowers are reported to taste like mushrooms. The vines emerge from a storage root that has a nutty flavor, and four times as much protein as potatoes. The mature seeds contain up to 40 percent protein and 18 percent oil. (Courtesy of N. Vietmeyer, National Academy of Science.)

FIGURE 19-15
Grain amaranths are not really grains at all but the fruits of *Amaranthus* species in the amaranth family. They resemble grains in their flavor and the methods by which they can be cooked. Amaranths flourish in arid and even saline habitats and represent a potentially important crop in the future that can be grown on otherwise unusable land. (Photo courtesy of Rodale Research Center.)

Other potential food crops that could profitably be used include the winged bean (*Psophocarpus tetragonolobus,* Fabaceae; Fig. 19-14); various grain amaranths (*Amaranthus* spp., Amaranthaceae; Fig. 19-15); eelgrass (*Zostera marina*, Potamogetonaceae, Fig.19-16); and buffalo gourd (*Cucurbita foetidissima*, Cucurbitaceae; Fig. 19-17).

Expanding into Arid Areas

Most of the uncultivated land across the world that is not covered by ice or tropical rain forest is unused because of aridity (Fig. 19-18). Conventional methods of agriculture and traditional crops often cannot successfully be employed in truly dry regions because of saline soils. Even if an area is not initially saline, salt can accumulate when a normally dry region is irrigated because of the unnatural, repeated applications of large amounts of water. Irrigation water flooded onto a field percolates rapidly through the soil, dissolving inorganic salts as it moves downward. However, because the air is so dry, there is intense evaporation at the soil surface, which soon begins to pull the water and dissolved salts upward again. As the water vaporizes into the air, the salts it contained are left behind, forming a deposit on the surface of the soil,

In several areas, notably in California and Israel, new methods of irrigation which combat salt accumulation by using a drip technique have been successfully employed. Although requiring large amounts of labor, drip irrigation greatly reduces the amount of water needed and drastically lowers salt accumulation. Drip irrigation is effected by applying water drop by drop through emitters spaced along hoses that are laid in the fields near the plant bases. The water is quickly absorbed by the soil and from there by the plants. Such small amounts of water are administered at a time that large-scale percolation and subsequent evaporation is prevented and as a consequence, salt deposition does not occur. An additional benefit of drip irrigation is that fertilizers can be added to the irrigation water when needed. In 1979 there were 500,000 hectares using drip irrigation on a worldwide basis, but by 1981 over half a million hectares in the United States alone, and over a million in the world, were irrigated by this method.

Another scheme for conserving water and increasing yields is *hydroponics* (Fig. 19-19), the growing of plants in nutrient solutions. Because there is no loss of soil and minerals from leaching and runoff, and evaporation from leaf surfaces is reduced, hydroponic growers boast of using only 10 percent of the water necessary for soil agriculture. In addition, they reap harvests that would normally be produced on 30 times as much land. In most hydroponic operations, the plants are "rooted" in an inert substance (e.g., marble, gravel, sand) to give them support while a solution of water, minerals, and air is pumped over them. Since hydroponic operations are enclosed, the environment can be controlled and pest problems are reduced. The drawback of hydroponic cultivation is that

FIGURE 19-16
Eelgrass, once used as a source of food by Indians of the American southwest, is now being reinvestigated as a possible crop of brackish areas. (Photo courtesy of R. R. Phillips.)

FIGURE 19-17
American Indians have used the seeds of the buffalo gourd as a source of oil for over 10,000 years. Highly tolerant of heat and drought, this squash relative has baseball-sized fruits that can contain 300 seeds each. It is estimated that under cultivation, 950 pounds of oil and 400 pounds of protein can be obtained from the fruits harvested from an acre. The large underground storage roots could provide 11,900 pounds per acre of starch suitable for industrial purposes. Here, experimentation with the gourd is conducted in Lebanon. (Photo by L. C. Curtis. Courtesy of the National Academy of Sciences.)

it is expensive, requiring large amounts of energy and hand labor. Hydroponics also does not lend itself to the production of crops such as wheat, corn, or rice, which are usually grown on huge acreages. Nevertheless, in regions with little available arable soil, adequate sunlight throughout the year, and a premium on fresh vegetables, such systems have proved to be profitable.

Problems of Expanding Agriculture into New Areas

Putting new areas, such as the lowland tropics and deserts, into cultivation carries a price. It will necessarily lead to a further reduction in the number of

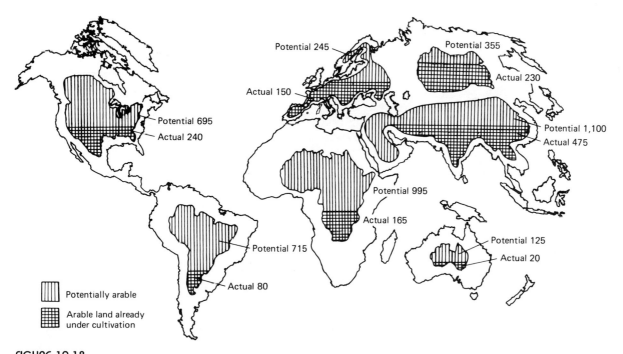

Potential 245
Actual 150

Potential 355
Actual 230

Potential 695
Actual 240

Potential 1,100
Actual 475

Potential 995
Actual 165

Potential 715
Actual 80

Potential 125
Actual 20

▦ Potentially arable

▦ Arable land already under cultivation

FIGURE 19-18
Potentially arable land compared to the area of each continent already under cultivation on each continent. The numbers indicate the hectares and are based on figures in D. Pimentel et al. 1973. Food production and the energy crisis. *Science* 182:443–449.

FIGURE 19-19
This 9 m by 41 m hydroponic greenhouse operation produces 1,000 heads of lettuce every 28 to 35 days. Because the water is recycled, 90 to 97 percent less water is used than in equivalent, open field operations. (Courtesy of Hydroculture, Inc.)

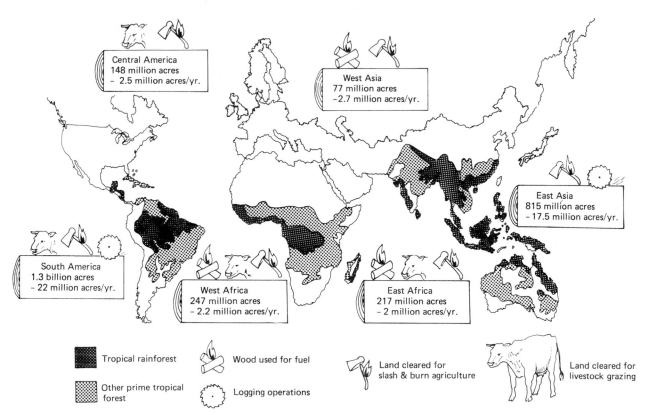

FIGURE 19-20
The world's forests are being cut down at an astonishing rate because of the demands for building materials, fuelwood, pulp, or simply for the land they cover. At the present rate of deforestation, we stand to lose 40 percent of our forests by the year 2000. If these predictions hold, there will be massive loss of topsoils and desertification in tropical regions. (Estimates from the Global 2000 Report. U.S. Interagency Task Force on Tropical Forests.)

species from which people might find new sources of food, medicines, and genetic material. As previously unused land is cleared for cultivation or timber, vast areas of natural vegetation, and with them many plant and animal species, will be eliminated. Predictions are that by the end of the century, about 40 percent of the present forest cover will be destroyed. The 1978 FAO estimate of forest cover was 4 billion hectares. A 40 percent reduction would leave only 160 million hectares by 2000 A.D. (Fig. 19-20). Most of the projected large-scale forest removal will be in the tropics. A recent National Academy of Sciencces report on priorities in tropical biology stressed that research efforts must be made now in these highly vulnerable regions.

For a variety of ecological and historical reasons, the lowland wet tropics contain the highest diversity of life (measured as the number of species per unit area) of any kind of habitat on earth (Fig. 19-21). Strange as it may seem,

FIGURE 19-21
This engraving of a tropical rain forest gives some indication of the denseness and diversity of the natural vegetation in lowland humid forests. (From K. von Marilaun. 1896. *The Natural History of Plants*. London, Blackie and Sons.)

much of this diversity still remains essentially unknown. Perhaps as many as 500,000 of the estimated 3 million species (plant and animal) in the tropics have never been adequately described (have no scientific name or proper documentation in a written text). While speculative, it is possible that many of these undescribed species could produce substances of importance to humans. The current rate of destruction of tropical areas will assure extinction of most of these species before they can be collected and studied.

While extinctions in tropical areas will undoubtedly reach unprecedented high levels, species in countries like our own are not immune. The Endangered Species Act passed by Congress in 1982 tries to protect both animals and plants from extinction. The Endangered Species Office of the Smithsonian Institution, charged by Congress with the task of enumerating the plants of the United States that were in danger of extinction, reported that 14 percent of the flora of the continental United States and 50 percent (one-half!) of the flora of the Hawaiian Islands is already extinct, in danger of extinction, or threateningly close to becoming endangered. It is hoped that the Endangered Species Act will protect some of these species, but the process of having each species examined, voted upon, and placed on the list of official endangered species has proved to be slow and arduous, and the responsibility for the enforcement of these laws has not been delegated.

HOW TO GROW MORE USING LESS ENERGY FROM FOSSIL FUELS

The task of producing greater supplies of food in temperate agricultural areas, or even in previously unused arid, or wet tropical, regions of the earth must be accomplished in the face of ever dwindling supplies of fossilized fuels and the products obtained from them. Almost all estimates of possible increases in food production assume a technology equivalent to that now used on highly productive U.S. farms. On these, large machines, fueled primarily by diesel oil, are essential for plowing, cultivating, and harvesting. In some areas, huge irrigation circles which are run by electricity (generally derived from fossil fuels) keep production levels high. Pesticides and fertilizers, synthetically produced in factories that use large amounts of energy, are routinely applied. By the time American food crops have been planted, harvested, shipped, processed, and put on the table, six times as much energy has been expended in their production than will be derived from their consumption (Figs. 19-22 and 19-23). With the increasing shortages of fossil fuels, the costs of this energy imbalance will continue to rise. Costs of food production have risen to the point where even affluent countries are investigating non-energy-intensive methods of agriculture.

In search of such methods, attention has been focused on some primitive agricultural techniques such as the *chinampa* agriculture of the Mayans (Fig. 19-24). This system was similar to what is today hailed as the "modern" French intensive, or raised bed, method of crop production. Likewise, increasing interest is being expressed in Chinese methods of agriculture such as that often employed in the growing of rice. Where possible, the Chinese introduce *Azolla*, an aquatic

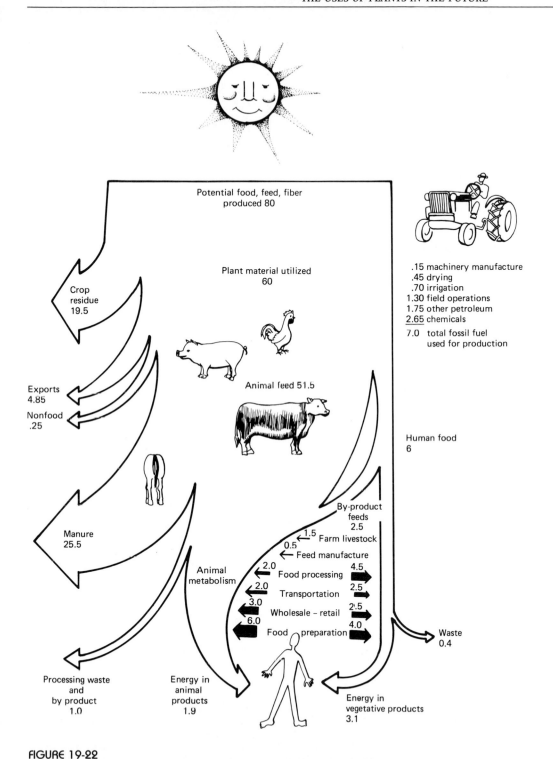

FIGURE 19-22
This diagram shows the energy flow in the U.S. food system expressed in billions of joules (a measure of energy). (Adapted from *The State of Food and Agriculture*. 1976. FAO. 1977; and F. C. Stickler et al. 1975. *Energy from Sun to Plant to Man.* Deere and Co.)

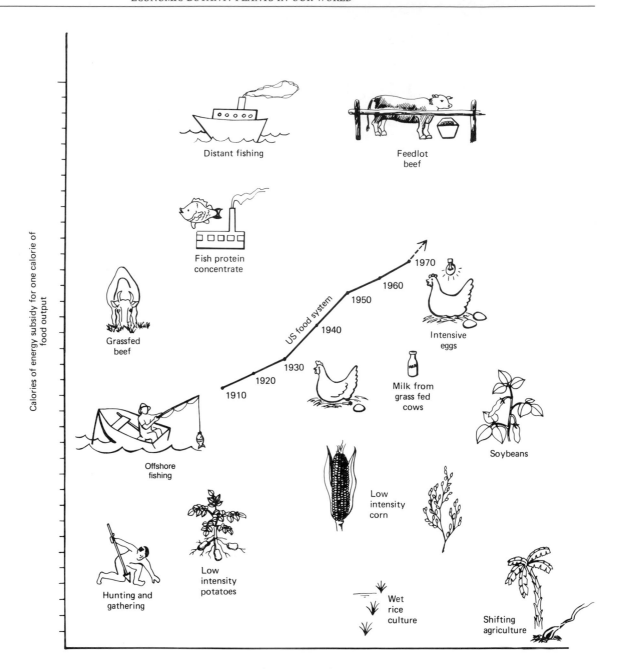

FIGURE 19-23
Superimposed on this chart of the energy subsidies for different food crops (expressed as input of calories per calories of food output) are graphs plotting the increased energy input into the U.S. food system as compared to the caloric content of the foods we consume (numbers are in kilocalories × 10^{12}). Placement of the cartoon figures is in terms of caloric input/output ratios. Placement relative to the horizontal axis is arbitrary.

floating fern, into rice paddies. These tiny plants do not interfere with the emergent rice in the flooded fields. The significant feature of *Azolla* is not that it can coexist with rice, but that it has a symbiotic association with a nitrogen-fixing (see Chapter 7) blue-green alga. When the rice is harvested, the fields are drained, and the ferns, with a rich supply of nitrogen in their nodules, sink to the ground and decompose, becoming, in essence, a green manure crop.

Agriculturalists are also interested in ways of reducing the consumption of fossil fuels on land that is now heavily used. One way to lower fuel use is to reduce the number of times tractors or cultivators cross the fields and to reduce the use of synthetically produced fertilizers and pesticides. Among the most successful methods now employed for reducing the use of heavy machinery is the practice of no-till, or low-till, agriculture. Tillage is the passing through the fields to remove weeds that are competing with crops. Tractors usually plow just below the soil surface between the crop rows in order to sever the roots of the weeds. In low- or no-tillage agriculture, all cultivating is eliminated. Plowing, seeding, and the application of herbicides are done with a single, initial operation.

Herbicides, chemicals that inhibit plant growth, are used in grain fields to prevent the growth of broad-leaved weeds (dicotyledons), and in fields of dicotyledonous crops, to prevent infestation by weedy grasses. The most commonly used broad-leaved plant killers are 2,4-D, 2,4,5-T, and triazine. A common herbicide for grasses is dalpon. In addition to lowering the fuel consumption of farm machinery, low-till agriculture reduces runoff, soil erosion, and consequently the leaching of fertilizers. Less fertilizer is ultimately needed and pollution of water systems is reduced. Herbicides are, of course, manufactured, and fuel is used during their synthesis and application. On the whole, total energy consumption of a farm is substantially reduced using low-tillage methods.

Organic gardeners advocate other methods for avoiding cultivating and/or plowing. These include the application of mulches and the planting of perennial crops. Mulches can be layers of organic matter (grass, leaves, or compost) or synthetic materials like black plastic. Mulches around plants prevent compaction of the soil by rain water, act as a sun shield, discourage weed growth, and prevent evaporation from the soil surface. Although the application of organic mulches is labor-intensive, they are practical where there is a supply of manure or crop residues. Incorporating these substances into the soil not only solves the problem of how to dispose of them, but, as they degrade, they contribute to the humus of the soil with a net effect of improving both the soil structure and its holding capacity. Plastic, a substitute for natural mulches, is manufactured from petroleum products and generally must be replaced each year. It is therefore impractical to use it for large acreages of crops, but it can be effective for row crops.

Many crop species have both annual and perennial varieties or are perennials grown as annuals. As pointed out in numerous previous chapters, humans have tended to select for annual crops because they invest so much of their energy in fruit production. Perennials must apportion their energy between fruit production, maintenance, and reserves for the next season. Yields per year of perennial varieties are thus less than those of annual varieties. Yet, as energy costs mount and the costs of planting new crops each year increase, the differences in yield between annuals and perennials diminish. A farmer might realize a greater total profit by growing a perennial variety that needs to be replanted only once every several years, than by replowing and resowing a high-yielding "annual" variety each year. There is now interest in experimenting with some crops such as corn and cotton as perennial crops.

Perennial plants have a more serious drawback, however, than low yields per year. In monoculture, they are more susceptible to pests than annuals. When annuals are plowed under at the end of the season, many pests are destroyed. The use of annual crops also allows for crop rotation on a yearly basis. By rotating crops in a field, problems with specific soil predators and pathogens are greatly reduced. Using perennial crops might reduce planting costs, but costs of pesticides often increase. Pesticides are usually synthetics produced in high energy consumptive factories and have themselves become one of the most expensive items in modern agriculture. Agricultural costs and, ultimately, the dependency on fossil fuels can thus be substantially reduced by limiting the use of pesticides. Unfortunately, along with droughts, pests (weeds, insects, and pathogens) constitute the greatest threat to high yields for the average farmer (Fig. 19-25). Losses to insects probably cause the greatest reduction in potential yields, followed by competition from weeds and then damage by bacterial and fungal pathogens. To combat these pests, we have developed an array of herbicides, insecticides, and fungicides.

The use of modern pesticides (the general term for all these chemicals) has been credited with increasing yields about 33 percent simply by reducing insect damage. In areas where huge acreages are devoted to a single crop, pesticides have become mandatory since pest populations can build up to large sizes and spread very rapidly. When DDT was first developed, it was hailed worldwide not only because it prevented crop destruction, but also for its role in eliminating the malaria mosquito. Millions of lives were saved, and there was optimism that humans would continue to develop chemicals that would solve all pest problems. When Rachael Carson published her classic *Silent Spring*, the world was shocked. She convincingly argued that pesticides were not the panacea they appeared to be and that there were insidious dangers involved in the haphazard use of agricultural chemicals (Fig. 19-26).

FIGURE 19-25
The timeless competition between man and plant pests is dramatically illustrated in this engraving of a locust plague in Europe.

We are now aware of the numerous drawbacks of extended pesticide use. Even with repeated applications, it is estimated that 20 percent of the possible U.S. crop production is still lost to various pests. Populations of all kinds of pests can become resistant to pesticides because natural selection favors mutations that confer resistance. DDT has been ineffective against the common housefly since 1946, and group after group of other insects has become resistant to one insecticide after another (Fig. 19-27).

Sometimes different kinds of pesticides act antagonistically toward one another. Likewise, the loss of natural weed predators through the use of a general insecticide can increase pressure from weed competition, or the use of herbicides can force more general plant feeding insects onto crop species. Beneficial insects such as pollinators or insect pest predators can also be as detrimentally affected by insecticides.

Humans are also not immune to the effects of pesticides. Some pesticides have been shown to cause cancer in laboratory animals and presumably could

FIGURE 19-26
Pesticide residues in the United
States. (Drawn from data presented
in P. E. Cornellussen. 1969.
Pesticide residues in total diet
sample (IV). *Pesticides Monitoring
Journal* 2:140–152.

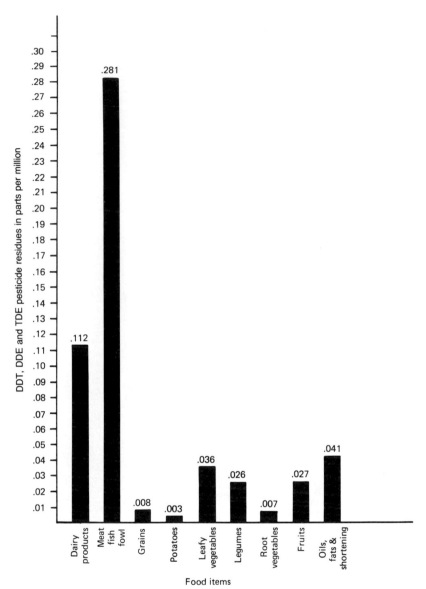

cause cancer in humans as well. Short-term data on the modern pesticides that are now GRAS (Generally Recommended As Safe in FDA language) indicate that most are broken down quickly and thus pose little threat to humans, but no one knows for sure what the long-term effects will be (Fig. 19-28). While economic benefits to agriculture through the use of pesticides seem to outweigh their disadvantages at the moment, the balance may turn as energy costs force the cost of pesticides (and synthetic fertilizers) higher and higher.

Active research is now being directed toward developing agricultural methods that will reduce the need for manufactured pesticides. The collective group of practices that reduce pest pressure without the need for heavy pesticide applications is known as Integrated Pest Management (IPM). Individual methods involved in IPM have their own particular goals and efficacy in given situations,

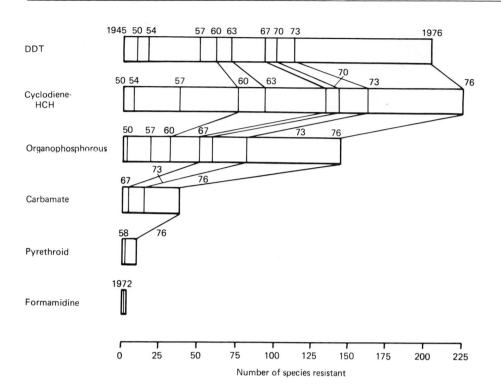

FIGURE 19-27
This time scale shows the development of resistance to insecticides with zero representing the introduction of each major group of chemicals. (From R. J. Wood. 1981. Insecticide resistance: genes and mechanisms. In J. A. Bishop and L. M. Cook. *Genetic Consequences of Man Made Change*. New York, Academic Press.)

but the strength of the management system lies in the simultaneous use of as many of the methods as possible. The major kinds of control that are currently integrated into this system are listed in Table 19 3.

Classic biological control, one of the methods of IPM, involves the introduction of a pest's predator to keep the populations of the pest at a relatively low level. This system has worked admirably in several situations. One example was the control of scale insect on citrus crops in California. Lady bugs (ladybird beetles), a natural predator of cottony scale, were introduced into California and managed to control the scale insects. Saint John's wort, or klamath weed, (*Hypericum perforatum,* Hypericaceae), an introduced weed of U.S. range land, was finally brought under control by the purposeful introduction of its predators, European beetles of the genus *Chrysolina*. The rampant, aggressive spread of *Opuntia* across Australia after its introduction from North America was likewise checked by the importation of its natural predator, a moth, *Cactoblastus cactorum*.

Unfortunately, such simple solutions are not possible in most cases. Biological control of this type is generally effective for pests that have been introduced into an area. Native pests have already reached an equilibrium with their native predators. For classic biological control to work effectively, the pest must have a specific predator that can successfully be introduced into the area where the pest is a problem. The predator should be specific, because it must concentrate its efforts on the pest organisms and it should not pose a threat to desirable organisms. Establishment of such a predator, even if one is found, is often impossible because the proposed area of introduction may lack the environmental conditions it needs for establishment and reproduction.

Monocultures of crops can lead to rapid population size increases of the native pests sometimes with, sometimes without, subsequent increases in

FIGURE 19-28
In a process known as biomagnification, pesticides tend to accumulate as they progress through the food chain. This accretion is represented by the increase in shading from the primary producer level at the bottom (plankton) to the top consumer level (duck).

TABLE 19-3

Kinds of Control Employed in Integrated Pest Management

KIND OF CONTROL	ACTION
Classic biological	Introduces a pest predator that will reproduce and survive in the area in which the pest is active. Usually this method is effective only if pest has also been introduced into the area.
Inundative biological	A periodic release of mass reared predators of the pest that will cause a massive reduction in pest populations. These predators usually cannot maintain high natural population densities in the areas where the pest is active.
Conservative biological	Practices the conservation of natural pest predators in the area where the pest is a problem. Plants that help maintain high parasitoid densities are encouraged. These aims are partly accomplished by multiple cropping and the retention of natural areas.
Competitive	Uses innocuous organisms that compete with pests for crop plants. Included in this category is the use of male sterile insects and the so-called trap-plants which divert pests from crops.
Biorational	Uses chemicals that modify the behavior of pests. These include pheromones, juvenile hormones, attractants, and antifeedants.
Chemical	Uses natural or synthetic compounds such as traditional pesticides as well as hormones, chemical sterilants, growth regulators, and microbial toxins.
Cultural	Manages agricultural systems by the physical techniques of proper tillage, irrigation, light traps, crop isolation, pruning, timing, and quarantine.

Source: Modified from S. Batra. 1982. Biological control in agroecosystems. *Science* 215:134–139.

predator population sizes. One of the methods employed in IPM is the maintenance of native predators through the preservation of natural areas and restrained use of broad-spectrum pesticides that eliminate them as well as the pests.

Other methods which can be employed as parts of an IPM scheme include the release of massive quantities of laboratory-reared predators. Large-scale releases can sometimes knock back outbreaks of pests, but such rearing programs are costly. Similar to this approach is the release of laboratory-produced male sterile individuals of an insect pest species. The theory behind this procedure is that the females of the pest species (hopefully a species in which the females mate only once) will mate with the sterile males and not produce any viable offspring. Male sterilization techniques have been successful in some situations as a control measure, but as in the case of pesticides, natural selection eventually leads to circumvention of the technique. Females that avoid mating with the male steriles are selected, and researchers eventually have to screen the strains that are released in particular localities very carefully.

Another measure that often meets with some success in keeping pest populations at a low level is *intercropping*, or the growing of unrelated species intermixed in fields. By avoiding large expanses of a single crop, this procedure tends to thwart large build-ups of insect pest populations. For pest species that feed preferentially on a given crop, intercropping makes the preferred species difficult to find. In other cases, intercropping consists of interplanting species repellent to pests with preferred host species.

While none of these methods, or the others listed in Table 19-3, can usually eliminate pests by itself, the integrated use of several techniques combined with prudent observations should be able to keep pest levels low with a minimum of pesticide use. As the current favorable economic balance on the side of sole reliance on pesticides diminishes, more and more of the practices included in IPM should begin to become incorporated into our agricultural systems.

NONFOOD USES OF PLANTS IN THE FUTURE

While the need to feed the increasing numbers of people on the earth will continue to stimulate the most active research on plants, there are other potential uses of plants that are also actively being investigated. In Chapter 10 we mentioned the use of jojoba as an important seed oil of the future; in Chapter 11, the potential guayule holds as a source of natural rubber; and, in Chapter 17, we examined plants that may be used in the future for their fibers. Other, and perhaps most important, future uses of plants will be as sources of energy.

Various species including *Euphorbia lathyrus* (Euphorbiaceae) and *Brickellia* spp. (Asteraceae) have been investigated as sources of hydrocarbons. Energy can be obtained from plants by various methods, including direct burning, extraction of burnable organic compounds such as hydrocarbons, or the production of ethanol via microbial fermentation (see Chapter 15). Table 19-4

TABLE 19-4

Potential Uses of Plants as Energy Sources

PLANT SOURCE	METHOD OF PREPARATION	PRODUCTS
Biomass		
Dry: wood (e.g., eucalyptus, poplar firs, pines, casaurina Leucaena	Combustion Gasification	Heat, electricity Gaseous fuels (e.g., methane, hydrogen)
Dry: residues (e.g., wood residue, cane tops, straw, bagasse)	Pyrolysis Hydrolysis, distillation	Oil, charcoal, gas Ethanol
Wet: sewage, aquatic plants (e.g., water hyacinth, reeds, rushes)	Anaerobic digestion	Methane
Sugars from juices and hydrolyzed starch (e.g., cane, beet, maize, sorghum, cassava)	Fermentation and distillation	Ethanol
Direct sources of organic compounds (e.g., *Euphorbia lathyrus*)	Extraction, often with fractionation	Terpenoids, alcohols, long-chain hydrocarbons

Source: Adapted from D. O. Hall. 1982. Solar energy through biology. Fuels from biomass. *Experientia* 38:3–10.

provides a more complete list of suggested ways that angiosperms and gymnosperms can potentially be used as energy sources.

The hydrocarbons produced by such plants can be used directly or split to produce hydrocarbons of chain lengths similar to those of petroleum. Plants can also be extracted to yield an array of compounds that can be "cracked" (split chemically) to yield straight-chain hydrocarbons. Yet, while seemingly feasible, the use of plants, even those with high hydrocarbon contents, is still too expensive to justify exploitation.

Production of ethanol from plants, particularly grains and crop residues (such as the remains of sugarcane after crushing), is already underway in countries such as Brazil which have no native sources of fossil fuels. As in the case of hydrocarbon extraction, the problem with alcohol production is the cost. Many cost-benefit studies have recently been made, but they draw differing conclusions. Production costs are incurred because energy is needed to distill the fermented liquid. If a cost-benefit analysis assumes that an external source of energy (electricity or fossil fuels) is required, then alcohol production is too expensive to be feasible. If energy for distillation is supplied by solar energy, burning "trash" from crops, or even the residues left after fermentation, there may be a small net gain. Recently, experimentation has suggested that various additives to the fermented liquor can absorb water and mimic distillation by leaving relatively pure alcohol. These methods are far from perfected, but they do show some promise. There seems to be little doubt that within the next 25 years, the production of alcohol, and possibly plant-derived petroleum substitutes, will become important uses of plants in many parts of the world.

ADDITIONAL READING

Bishop, J.A., and L.M. Cook (eds.). 1981 *Genetic Consequences of Man Made Change.* New York, Academic.

Calvin, M. 1977. The sunny side of the future. *Chemtect* 7(6):352–363.

Chrispeels, M.J., and D. Sadava. 1977. *Plants, Food and People.* (Chapter 9, Plant breeding and the green revolution). San Francisco, Freeman.

Clark, W. 1980. China's green manure revolution. *Science* 80,1(6):68–73.

Harlan, J.R. 1975. Our vanishing genetic resources. *Science* 188:618–621.

Harrison, P.D., and B.L. Turner II (eds.). 1978. *Pre-Hispanic Maya Agriculture.* Albuquerque, New Mexico, University of New Mexico Press.

Jennings, P.R. 1974. Rice breeding and world food production. *Science* 186:1085–1088.

National Academy of Sciences (U.S.A.) 1975. *Underexploited Tropical Plants with Promising Economic Value.* Washington, D.C., National Academy of Sciences.

National Academy of Sciences (U.S.A.). 1980. *Research Priorities in Tropical Biology.* Washington, D.C., National Academy of Sciences.

Poleman, T.T. 1975. World food: A perspective. *Science* 188:510–518.

Schwanitz, F. 1966. *The Origin of Cultivated Plants.* Cambridge, Massachusetts, Harvard University Press.

Vietmeyer, N. 1979. The greening of the future. *Quest* 3 (September): 25–32.

Wade, N. 1975. International agricultural research. *Science* 188:585–589.

Walton, S. 1980. Global 2000 projects grim future. *BioScience* 30:629–632.

Wittwer, S.H. 1979a. Future technological advances in agriculture and their impact on the regulatory environment. *BioScience* 29:603–610.

Wittwer, S.H. 1979b. The blue revolution. *Natural History* 88(9):8–18.

Index

Index